Halid Bikkin, Igor I. Lyapilin

Non-equilibrium Thermodynamics and Physical Kinetics

Also of Interest

Multiscale Thermo-Dynamics. Introduction to GENERIC
Michal Pavelka, Václav Klika, Miroslav Grmela, 2018
ISBN 978-3-11-035094-4, e-ISBN (PDF) 978-3-11-035095-1,
e-ISBN (EPUB) 978-3-11-038753-7

Statistical Physics
Michael V. Sadovskii, 2019
ISBN 978-3-11-064510-1, e-ISBN (PDF) 978-3-11-064848-5,
e-ISBN (EPUB) 978-3-11-064521-7

Thermoelectric Materials. Principles and Concepts for Enhanced Properties
Ken Kurosaki, Yoshiki Takagiwa, Xun Shi, 2020
ISBN 978-3-11-059648-9, e-ISBN (PDF) 978-3-11-059652-6,
e-ISBN (EPUB) 978-3-11-059453-9

Physics of Energy Conversion
Katharina Krischer, Konrad Schönleber, 2015
ISBN 978-1-5015-0763-2, e-ISBN (PDF) 978-1-5015-1063-2,
e-ISBN (EPUB) 978-1-5015-0268-2

Electrochemical Energy Storage. Physics and Chemistry of Batteries
Reinhart Job, 2020
ISBN 978-3-11-048437-3, e-ISBN (PDF) 978-3-11-048442-7,
e-ISBN (EPUB) 978-3-11-048454-0

Halid Bikkin, Igor I. Lyapilin

Non-equilibrium Thermodynamics and Physical Kinetics

DE GRUYTER

Physics and Astronomy Classification Scheme 2020
05.10.-a, 05.20.-y, 05.20.Dd, 05.30.-d, 05.30.Fk, 05.40.-a, 05.45.-a, 05.60.-k, 05.60.Cd, 05.60.Gg, 05.65.+b, 05.70.-a, 05.70.Ce, 51.10.+y, 52.25.Dg, 52.25.Fi, 52.35.-g, 72.10.-d, 72.10.Bg, 72.10.Di, 72.10.Fk, 72.15.-v, 72.15.Eb, 72.15.Gd, 72.15.Jf, 72.15.Lh, 72.15.Nj, 72.20.Dp, 72.20.Ht, 72.20.My, 72.20.Pa, 72.25.Ba, 72.25.Dc, 72.25.Rb, 72.30.+q, 74.40.-n

Authors
Prof. Dr. Halid Bikkin
Ural Federal University
named after the First President
of Russia B. N. Yeltsin
Ulitsa Mira, Dom 19
Yekaterinburg 620002
Russian Federation
kh.m.bikkin@urfu.ru

Prof. Dr. Igor I. Lyapilin
Russian Academy of Sciences
M. N. Mikheev Inst. of Metal Physics
Ulitsa Sofii Kovalevskoj, Dom 18
Yekaterinburg 620108
Russian Federation
lyapilin@imp.uran.ru

ISBN 978-3-11-072706-7
e-ISBN (PDF) 978-3-11-072719-7
e-ISBN (EPUB) 978-3-11-072738-8

Library of Congress Control Number: 2021937481

Bibliographic information published by the Deutsche Nationalbibliothek
The Deutsche Nationalbibliothek lists this publication in the Deutsche Nationalbibliografie; detailed bibliographic data are available on the Internet at http://dnb.dnb.de.

Cover image: shulz / iStock / Getty Images Plus
Typesetting: VTeX UAB, Lithuania
Printing and binding: CPI books GmbH, Leck

www.degruyter.com

Authors' introduction

This book is based, in large part, on lectures delivered over a number of years at the Physics Department of the Ural Federal University.

The main goal pursued by the authors in writing this book is to set forth systematic and consistent principles of non-equilibrium thermodynamics and physical kinetics in the form primarily available to both students and undergraduates who begin their study of theoretical physics and postgraduates and experienced research scientists working in a new field of study.

First of all, we point out principles that have guided us in selecting appropriate material. Physical kinetics or the theory of transport phenomena is a very broad and rapidly developing subject area of physics. On this subject matter, there are a sufficient number of educational papers as well as monographic studies, which discuss various aspects of the kinetic theory. However, most publications are expected to be understood by a reader who has a substantial scientific background rather than by third year students.

Therefore, there is an acute shortage of literature for "beginners" where a natural balance between general postulates of the theory and simple examples of the practical implementation would be observed. Another guiding approach in writing the present book consists in the authors' attempting to avoid as far as possible such turns of speech as "obviously" and "it is easy to show". It is no secret that cumbersome, time-consuming calculations are very often hidden behind these words.

The authors were trying to write the text in such a manner that those phrases should acquire their original meaning. Perhaps, the authors did not always succeed in doing so. Finally, also the authors have done their best to present different techniques to describe non-equilibrium systems and construct schemes of the theory of transport phenomena but they would not like to focus on the problem of a calculation of the kinetic coefficients for various model systems. This approach allows one to illustrate contemporary directions of non-equilibrium statistical mechanics which are now developed along with "classical" sections of kinetics.

The present book can be divided into four parts.

The first one is devoted to methods, describing both non-equilibrium systems and phenomenological non-equilibrium thermodynamics.

The second part covers the substantiation and application of the kinetic equation method in non-equilibrium statistical mechanics. Here, the kinetic equation for electrons and phonons in conducting crystals is considered as an example.

The third part discusses the linear response theory of a system to an external mechanical perturbation.

The fourth part, which can be called "modern methods of non-equilibrium statistical mechanics", contains the presentation of the method of the non-equilibrium statistical operator and the basic kinetic equation ("master equation").

It is worth dwelling on the contents in more detail.

https://doi.org/10.1515/9783110727197-201

Chapter 1 deals with the principles of constructing thermodynamically non-equilibrium systems of electrons in conducting crystals in a linear approximation of external forces. Also, the generalized kinetic coefficients and the Onsager symmetry relations were analyzed. This chapter acquaints the reader with a classification of the kinetic effects in the conducting crystals. Besides, here one can find not only methods of describing thermodynamics of highly non-equilibrium systems and their aufbau principle, but also the formation of dissipative structures in such systems. In addition, this chapter deals with extremely important issues of non-equilibrium statistical mechanics concerning orbital, structural, and asymptotic stability of a solution of equations, describing the dynamics of non-equilibrium parameters.

Chapter 2 considers the behavior of a Brownian particle under the influence of random forces. Here, we have derived the Fokker–Planck equation describing the dynamics of a single Brownian particle, and the reader can become familiar with a solution of this mathematical equation. This simple example of the above equation shows how one can introduce a detail-omitting (due to a time-averaging) description of the dynamics of non-equilibrium systems.

Chapter 3 is devoted to the method of kinetic equations in non-equilibrium statistical mechanics. Based on the chain of the Bogolubov equations for s-particle distribution functions, the substantiation of the quasi-classical kinetic equations is presented in this section. Examples of the kinetic equations are the Vlasov equation and the Boltzmann equation, which are here formulated. Also, this chapter reviews different methods for solving the Boltzmann equation.

Chapter 4 presents the kinetic equation for electrons and phonons in conducting crystals within the relaxation time approximation. This part of the book draws special attention to the procedure for calculating the kinetic coefficients, describing the thermoelectric, thermomagnetic, and galvanomagnetic phenomena in metals and semiconductors and the effect of electron dragging by phonons.

The description of the method of the linear response of a non-equilibrium system to an external mechanical disturbance constitutes the contents of Chapter 5. Here, the authors apply both the method of Green's functions and the mass operator method for calculating the kinetic coefficients. Chapter 5 deals with the computation of high-frequency magnetic susceptibility of an electron gas as well as electrical conductivity, including conductivity in a quantizing magnetic field on the basis of this procedure.

Chapter 6 examines the method of the non-equilibrium statistical operator (NSO). The equations of motion can be obtained with this method both for a non-equilibrium distribution and for equations of motion concerning effective parameters (similar to the Chapman–Enskog equations). Furthermore, the NSO-method can be regarded as a quantum-statistical method of constructing thermodynamics of non-equilibrium systems. The book discusses only principles of the construction of the non-equilibrium thermodynamics with the NSO and methods of deriving the balance equations for the parameters for the describe the non-equilibrium distribution. In this chapter of the scientific monograph, a diligent reader can trace the process of the derivation of linear relaxation equations that allow one to find the spectrum of collective excitations in the non-equilibrium system. The chapter sets forth in detail the Mori projection operator method to construct the equations of motion for hydrodynamic quasi-integrals. Also, it should be noted that the authors present the description of this technique to calculate transport coefficients.

The second edition of the book contains a new chapter devoted to the physical foundations of spintronics. This term originated in the early 1990s and currently stands for an interdisciplinary field that incorporates science and technology. The key of the chapter is the in-depth study and active use of the spin degrees of freedom of electrons in solid-state systems. Interest in research in this field largely concerns storing and transmitting information through the spin of an electron, apart from its charge, as an active element. Electron-spin-based units can, in a great measure, dislodge or supplement various conventional microelectronic devices, including ones for quantum computing and quantum information transfer. The main advantage of spin electronics against traditional techniques is to record and store a unit of information with less energy. This is because the change in the orientation of the electron spin is not associated with the electron motion and inevitable Joule losses during this process.

The transfer of mechanical and magnetic moments in magnetic metals and semiconductors being brought about not only by mobile electrons but also by magnons (spin waves), it is natural that spintronics also involves magnonics, a branch of spintronics that investigates the physical properties of magnetic micro- and nanostructures, the properties of propagating spin waves, as well as the possibility of utilizing spin waves to build the hardware components relying on new physical principles; this is required for processing, transmitting and storing information. In a broad sense, spintronics can also include solid-state optoelectronics and the creation and design of nano-electronic devices using spin degrees of freedom. Thus, over the past 20 years, spintronics has become one of the most important trends in the development of solid-state physics and microelectronics. Microelectronic spin-transport-based devices such as spin valves, magnetoresistive memory cells, and spin transistors have been developed and successfully applied. These instruments have made it possible to enhance the density of information recording on magnetic storage media, the speed of reading and writing into magnetic memory cells with random access, and to reduce heat release.

It is quite obvious that there is no possibility to cover all aspects of this rapidly burgeoning field of physical electronics in such a small-scope chapter. Therefore, it would be reasonable to dwell upon only such main points of the spin transport kinetics as the nature of a spin current, spin accumulation, giant magnetoresistance, and the spin Hall effect. The issues discussed in the chapter deals with only a small part of the problems of spintronics. However, they are fundamental for this science and allow one to understand the principles of operation of most of the known devices for spin microelectronics.

It should also be underscored that the chapter can also be regarded as a useful application of the theoretical approaches developed in Chapters 1–6 to solve applied problems of physical kinetics.

Chapters 8 and 9 focus on the response of a highly non-equilibrium system to a weak mechanical perturbation and master equation approach, respectively. As an example, in both cases, the authors suggested a procedure for calculating the static electrical conductivity of highly non-equilibrium system. The material presented in these chapters goes beyond the classic courses of physical kinetics, but the book demonstrates potential "growth points" of the theory of kinetic phenomena.

In each chapter, one can find examples represented in the form of problems to illustrate the theoretical statements under consideration. Most of them are offered to students either for a solitary effort or for assignments to be submitted during examination sessions. In spite of understanding that the examples for solving specific problems of physical kinetics are extremely useful for practical learning, the authors did not think it necessary to include into the book most of those problems to avoid increasing the book volume.

We should note that many of the issues discussed on pages of this book are extremely complex and not always taken into account with necessary degree of rigor. Nevertheless, the authors tried as far as possible to discuss applicability conditions for certain approximations completely enough and consistently, and note those "pitfalls" that can be encountered in practical implementation of the above methods.

It is worth saying a few words about references to literature resources. Owing to the nature of the material the authors did not seek to provide a link with original scientific papers or give an exhaustive bibliography on the issues at hand. Therefore, only those papers which, on the one hand, are readily available, on the other hand, contain enough information about the themes in question, are presented in the references.

All compilers of the book hope that this book will allow both students and graduate students to adequately get acquainted with "kitchen" methods of the contemporary theory of transport phenomena and prepare them for independent scientific work in the field of the quantum-statistical theory of transport phenomena.

In preparing the second edition of this book, the authors have corrected the typos noticed.

The authors are grateful to Yu. G. Gorelykh for the translation of the first and second editions of the book into English.

Contents

1 Phenomenological thermodynamics of irreversible processes

1.1 Main postulates of non-equilibrium thermodynamics

1.1.1 Thermodynamic description of equilibrium and non-equilibrium systems

The thermodynamic description of a many-particle system in equilibrium thermodynamics is based on an assumption that there are few macroscopic parameters, characterizing the system as a whole. The number of those parameters should be sufficient to precisely determine a state of the system. Averaged physical quantities such as the average energy or average momentum of the particles forming the system, the components of the electric polarization and magnetic induction per unit volume are usually chosen as those parameters to characterize the system.

There are two interrelated approaches to describing the thermodynamic system: the state equation method and the method of thermodynamic potentials. In the case, for example, of an ideal gas, a set of thermodynamical equations can be written as

$$
\begin{aligned}
TdS &= dE + pdV, \\
E &= C_V T, \\
pV &= \nu RT.
\end{aligned}
\tag{1.1}
$$

The first equation (1.1) is the fundamental equation of thermodynamics. Here S is for the entropy, T is for the temperature, E is for the internal energy, p is for the pressure, V is for the volume of an ideal gas.

The second equation is known as a calorific equation of state of an ideal gas and determines a relationship between internal energy, volume and temperature. The third equation of the set (1.1) is so-called the thermal equation of state and allows for finding the gas pressure p as a function of volume, and temperature; ν being the number of moles of an ideal gas.

The set of equations (1.1) contains three equations and five unknowns (S, E, V, p, T). As is known [1, 2], the state of a system in equilibrium thermodynamics is defined by external parameters and temperature. For an ideal gas, it suffices to specify only one external parameter—the volume. Thus, in this case, the simplest case, having two parameters (temperature and volume, for example), with the help of (1.1) one can find all quantities, characterizing an ideal gas.

To describe a more complex one-phase thermodynamic system whose state is defined by n external parameters $x_1, x_2, \ldots, x_n$ and temperature, equations (1.1) should be modified. This modification reduces simply to the fact that it is necessary to take into account the work of the system in the course of change of all the generalized thermodynamic coordinates $x_1, x_2, \ldots, x_n$, but not only volume. Also, to close the set of

https://doi.org/10.1515/9783110727197-001

equations it is necessary to write down n thermal equations of state, expressing the relationship between the generalized thermodynamic forces F_i with the generalized thermodynamic coordinates and temperature:

$$TdS = dE + \sum_{i=1}^{n} F_i\, dx_i,$$
$$E = E(x_1, x_2, \dots, x_n, T),$$
$$F_i = F_i(x_1, x_2, \dots, x_n, T), \quad i = 1, 2, \dots, n. \tag{1.2}$$

The thermal equations of state and caloric one must be found experimentally or obtained within the limits of statistical mechanics through a model-based analysis.

Another alternative method for an analytical description of a many-particle system in equilibrium thermodynamics is associated with the method of thermodynamic potentials (functions of state) [3]. In this case, one of the possible functions of state should be found by using the methods of equilibrium statistical mechanics.

It can be claimed that, if at least one of the thermodynamic potentials is defined as a function of its natural variables, the thermodynamic properties of the systems are determined completely. This is accounted for by the fact that all thermodynamic quantities characterizing this system can be obtained as partial derivatives of the thermodynamic potential. It is free energy Φ that is usually used as a function of temperature T, volume V and the number of particles N_i of i-th sort for the analysis of multicomponent systems with a constant number of particles:

$$d\Phi = -SdT - pdV + \sum_{i=1}^{k} \zeta_i dN_i. \tag{1.3}$$

The Gibbs thermodynamic potential G is also suitable as a function of temperature, pressure and the number of particles:

$$dG = -SdT + Vdp + \sum_{i=1}^{k} \zeta_i dN_i. \tag{1.4}$$

In equations (1.3), (1.4), the quantity ζ_i is the chemical potential of particles of i-th sort.

Thus, in equilibrium thermodynamics, there exist quite simple universal methods for an analytical description of the many-particle systems. Special attention should be paid to the fact that equations (1.1)–(1.4) allow one to analyze, in turn, the thermodynamics of quasi-steady state processes.

To explain the above, let us consider a simple system consisting of a gas in a cylindrical vessel with the characteristic size l the upper base of which is a movable piston. The pressure inside the vessel is changed as the piston moves. The time of sound wave propagation $\tau_p \simeq l/v_s$, where v_s is the velocity of sound in a gas may be taken as the characteristic time of establishing the equilibrium pressure. Therefore, if

$$\frac{dp}{dt}\tau_p \ll \Delta p,$$

where Δp is an acceptable accuracy of the pressure measurement, one may assume that the pressure will be the same (equilibrium) throughout the system while the piston is moving.

Also, processes of establishing the equilibrium values of the other macro parameters should be analyzed because it is not clear beforehand which of them is the slowest (limiting).

In any case, one can always find conditions of thermodynamic description applicability for the phenomena in a many-particle system. If these conditions are not met, processes proceeding in such a system when the equilibrium is established should be analyzed in more detail.

1.1.2 Local equilibrium principle

The natural step in this direction is to generalize results of equilibrium thermodynamics in respect of non-equilibrium case by introducing the concept of local-equilibrium. As was noted above, thermodynamic parameters are, in substance, physical quantities which characterize a many-particle system. If we average the physical quantities over the portion of the system which consists of rather small, but at the same time, macroscopic areas, then we also get macroscopic parameters. However, values of these parameters depend both on time and on a position of the selected volume (coordinates). As to the time dependence, it is two-fold. On the one hand, it is caused by natural fluctuation of the physical quantities in sufficiently small volumes. The time scale of these fluctuations is comparable to an atomic time scale.

On the other hand, the time dependence of the local averages has a distinctly different scale and is associated with more slow relaxation processes in a macroscopic system. The characteristic time scale of these changes is close in order of magnitude to $\tau_p \simeq l/v_s \simeq 10^{-5}$ s. Performing the additional time-averaging of the local macro parameters, one can eliminate the fluctuating component, leaving only the part that describes a slow change in these parameters due to relaxation processes.

Thus, for a non-equilibrium system, one can introduce local-equilibrium thermodynamic parameters, which will characterize some small enough volume of a macroscopic system. These parameters depend on coordinates and time. The local-equilibrium parameters can be considered as a continuous function of coordinates and time subject to changing the parameters slightly as the transition from one physical small volume to another happens.

At first glance, the condition of quasi-equilibrium does not impose significant constraints on applicability of the thermodynamic approach to describing non-equilibrium phenomena, since volume over which the averaging is produced can be made arbitrarily small. In fact, that is not the case.

To see this, consider a sample of the semiconducting material or metal, placed in the field of a temperature gradient. If a subsystem of conduction electrons is selected as the thermodynamic system, in this case, the electron mean free path l defines the characteristic spatial scale to determine physically small volume over which the averaging is produced. The passage to the local-equilibrium description is possible, if the following inequality is valid:

$$\left|\frac{l}{T}\frac{dT}{dx}\right| \ll 1.$$

This inequality ensures the condition of a small temperature change along the mean free path of electrons.

In this chapter basic postulates of thermodynamics of irreversible processes to be applied to the system of conduction electrons in a metal or semiconductor explain the deviation of the system from its equilibrium state caused by action of an external electrical field and a temperature gradient.

1.1.3 Entropy balance equation and conservation laws

Assuming that the use of the local-equilibrium approach is fair, we can write down the fundamental equation of the thermodynamics for a physically small volume of a system:

$$dE(\vec{r},t) = T(\vec{r})\,dS(\vec{r},t) + \left[\zeta(\vec{r}) + e\varphi(\vec{r})\right] dn(\vec{r},t), \tag{1.5}$$

where $E(\vec{r},t)$, $S(\vec{r},t)$, $n(\vec{r},t)$ is for the internal energy density, entropy, and the number of particles of the system at the point with coordinate $\vec{r}$ at the moment of time t, respectively. $T(\vec{r})$, $\zeta(\vec{r})$ is for the local temperature and chemical potential of electron system; $\varphi(\vec{r})$ is for the potential of electrostatic field; e is for the electron charge.

We find an expression for the entropy production $dS(\vec{r},t)/dt$, supposing that the local-equilibrium state of the system under study is stationary:

$$\frac{dS(\vec{r},t)}{dt} = \frac{1}{T(\vec{r})}\frac{dE(\vec{r},t)}{dt} - \frac{\zeta(\vec{r}) + e\varphi(\vec{r})}{T(\vec{r})}\frac{dn(\vec{r},t)}{dt}. \tag{1.6}$$

The particle number density $n(\vec{r},t)$ and density of the internal energy $E(\vec{r},t)$ meet the conservation laws, which have the form of the equations of continuity:

$$\begin{aligned} &\frac{dn(\vec{r},t)}{dt} + \operatorname{div}\vec{J}_n = 0,\\ &\frac{dE(\vec{r},t)}{dt} + \operatorname{div}\vec{J}_E = 0, \end{aligned} \tag{1.7}$$

where $\vec{J}_n$, $\vec{J}_E$ is for flux density of the particles and energy flux, respectively.

Problem 1.1. Obtain the first equation of continuity (1.7), using the definition of the particle number density:

$$n(\vec{r},t) = \sum_{i=1}^{n} \delta(\vec{r} - \vec{r}_i(t)),$$

where $\vec{r}_i(t)$ is the coordinate of the i-th particle.

Solution. Consider a change in the number of particles in the small volume v, inside which the conditions of local equilibrium are fulfilled. Then

$$\begin{aligned}\int dv \frac{dn(\vec{r},t)}{dt} &= \int dv \sum_{i=1}^{n} \frac{d}{d\vec{r}_i} \delta(\vec{r} - \vec{r}_i(t)) \frac{d\vec{r}_i}{dt} \\ &= -\int dv \sum_{i=1}^{n} \frac{d}{d\vec{r}} \delta(\vec{r} - \vec{r}_i(t))\vec{v}_i = -\int dv \operatorname{div} \vec{J}_n. \end{aligned} \tag{1.8}$$

In writing the last expression we have used the definition of the flux density of the particles:

$$\vec{J}_n = \int dv \sum_{i=1}^{n} \vec{v}_i \delta(\vec{r} - \vec{r}_i(t)). \tag{1.9}$$

Thus, the particle number density really satisfies the continuity equation

$$\frac{dn(\vec{r},t)}{dt} + \operatorname{div} \vec{J}_n = 0.$$

The second equation of continuity (1.7) is derived by analogy.

Let us return to the transformation of equation (1.6) for the entropy production. Substituting the continuity equation (1.7) into the equation for the entropy production (1.6) and performing simple transformations we can obtain two equivalent representations:

$$\frac{dQ}{dt} = -\operatorname{div} \vec{J}_Q + \vec{J}\vec{\varepsilon}, \tag{1.10}$$

$$\frac{dS}{dt} = -\operatorname{div} \vec{J}_S - \vec{J}_Q \frac{\vec{\nabla} T}{T^2} + \vec{J}\frac{\vec{\varepsilon}}{T}, \tag{1.11}$$

where the heat flux $\vec{J}_Q$ and the entropy flow $\vec{J}_S$ are determined by the relations

$$\vec{J}_Q = \vec{J}_E - \left(\frac{1}{e}\zeta + \varphi\right)\vec{J}, \tag{1.12}$$

$$\vec{J}_S = \frac{\vec{J}_Q}{T}, \quad \vec{J} = e\vec{J}_n. \tag{1.13}$$

For reduction, in formulas (1.10), (1.11) we have introduced the variable $\vec{\varepsilon}$, which has the meaning of the electrochemical gradient:

$$\vec{\varepsilon} = -\vec{\nabla}\left(\varphi + \frac{1}{e}\zeta\right). \tag{1.14}$$

Throughout the book, we will omit the dependence of thermodynamic functions on coordinates and time where it does not cause confusion. Equation (1.10) is the law of conservation of heat. This fact becomes especially obvious in case of integrating both sides of equation (1.10) over small closed volume. Then, using the Ostrogradski–Gauss theorem, one can reveal that the change in an amount of heat per unit time inside a closed volume equals the volumetric heat generation minus the heat transferred per unit time across a surface bounding the volume. Similarly, it makes sense to speak of equation (1.11) which is a key for non-equilibrium thermodynamics (similar to master equation of thermodynamics for thermodynamics systems) as a local entropy balance equation.

A distinctive feature of the thermodynamics of irreversible processes is the appearance of the thermodynamic fluxes $\vec{J}_S, \vec{J}, \vec{J}_Q$, which, in our case, are caused by the external thermal thermodynamic forces $(\vec{\nabla}T, \vec{\varepsilon})$. For this reason, equations (1.10), (1.11) are not closed. Indeed, even if we assume that the thermodynamic forces $\vec{\varepsilon}/T$ and $\vec{\nabla}T/T^2$ are external parameters and defined, all the same, equation (1.11) still contains three the unknown thermodynamic quantities $(S, \vec{J}_Q, \vec{J})$. We remind that a similar situation exists in equilibrium thermodynamics where the fundamental equation of thermodynamics requires adding the thermal equations of state together with a caloric one to close the set of equations. As was already mentioned, these equations cannot be obtained in the context of the thermodynamics, but must be determined either within the statistical molecular-kinetic theory or empirically.

In non-equilibrium case, the equations of the type (1.10), (1.11) should also be supplemented by equations relating the thermodynamic fluxes and thermodynamic forces. Moreover, by in complete analogy with the case of equilibrium, these equations can be found by generalizing the experimental data or obtained by means of methods of non-equilibrium statistical mechanics.

To close equation (1.11), it is necessary to expand the thermodynamic fluxes into a series in the thermodynamic forces. Assuming that the non-equilibrium state is weak, we should restrict ourselves to linear terms of the expansion. As a result of this, one obtains two vector equations:

$$\vec{J} = \hat{\sigma}\vec{\varepsilon} - \hat{\beta}\vec{\nabla}T,$$
$$\vec{J}_Q = \hat{\chi}\vec{\varepsilon} - \hat{\kappa}\vec{\nabla}T, \tag{1.15}$$

which define a linear relationship between the thermodynamic fluxes and thermodynamic forces. The zeroth terms in the expansion with respect to the thermodynamic forces in equations (1.15) are absent because if there are no thermodynamic forces, then there are no fluxes. $\hat{\sigma}$, $\hat{\beta}$, $\hat{\chi}$, $\hat{\kappa}$, which are the proportionality coefficients, are referred to as transport coefficients or kinetic coefficients. In our case of the linear relationship, they are tensors of a second rank. Kinetic coefficients as parameters of the phenomenological theory of transport phenomena should be determined within the microscopic theory of transport phenomena, therefore the large part of the present book is devoted to setting forth the methods of this theory.

1.1.4 Generalized flows and generalized thermodynamic forces

In previous paragraphs we have formulated basic ideas of thermodynamics of irreversible processes and introduced the basic equation (1.11) of irreversible process thermodynamics which is based on the local equilibrium principle and equation (1.15), relating the thermodynamic fluxes and thermodynamic forces.

Having analyzed the above expressions we can easily note that there is certain arbitrariness in choosing the thermodynamic fluxes and thermodynamic forces. As far as there is no possibility of getting rid of this arbitrariness completely within the framework of phenomenological non-equilibrium thermodynamics, it is necessary to do a number of significant clarifications in the case of a linear relationship between fluxes and thermodynamic forces (Onsager, 1931).

Consider a system whose equilibrium state is defined by a set of macroparameters $a_1, a_2, \ldots, a_n, b_1, b_2, \ldots, b_m$. Assume that the parameters a_i are even ones with respect to time reversion operation ($\vec{r} \to \vec{r}$, $\vec{p} \to -\vec{p}$, $\vec{s} \to -\vec{s}$, where $\vec{p}$, $\vec{s}$ are the vectors of momentum and spin, respectively), variables b_i being odd parameters relative to this operation. Let us suppose that, in the equilibrium state, the system is characterized by equilibrium values of the a_i^0 and b_i^0 parameters. Now we introduce small deviations of the non-equilibrium parameters from the equilibrium ones: $\alpha_i = a_i - a_i^0$, and $\beta_i = b_i - b_i^0$, and expand the expression for the system entropy into a series in small deviations of the macroparameters α_i and β_i. Taking into account the entropy invariance in respect of time reversion operation and leaving only the first terms in expansion for entropy of the non-equilibrium system S, we can obtain the following equation:

$$S = S^0 - \frac{1}{2}\sum_{i,k=1}^{n} A_{ik}\alpha_i\alpha_k - \frac{1}{2}\sum_{i,k=1}^{m} B_{ik}\beta_i\beta_k, \tag{1.16}$$

where S^0 is the entropy of the equilibrium system, A_{ik} and B_{ik} is for some symmetric matrices of the positive-definite coefficients:

$$A_{ik} = -\left.\frac{d^2S}{d\alpha_i d\alpha_k}\right|_{\substack{\alpha_i=0,\\ \beta_j=0}} \qquad B_{jl} = -\left.\frac{d^2S}{d\beta_j d\beta_l}\right|_{\substack{\alpha_i=0,\\ \beta_j=0}} \qquad \begin{array}{l} i = 1,2\ldots,n, \\ j = 1,2,\ldots,m. \end{array}$$

Positive definiteness of the matrices A_{ik}, B_{jl} follows from the extremality of entropy in an equilibrium state (entropy is a concave upward function of its parameters α and β, and has a maximum at the point $\alpha = \beta = 0$).

Now, one can determine the generalized thermodynamic forces X_i^α and X_i^β and the generalized thermodynamic fluxes I_i^α and I_i^β by the following relations:

$$X_i^\alpha = \frac{dS}{d\alpha_i} = -\sum_{k=1}^{n} A_{ik}\alpha_k, \quad X_i^\beta = \frac{dS}{d\beta_i} = -\sum_{k=1}^{m} B_{ik}\beta_k, \tag{1.17}$$

$$I_i^\alpha = \frac{d\alpha_i}{dt}, \quad I_i^\beta = \frac{d\beta_i}{dt}. \tag{1.18}$$

In virtue of the definition of the expressions (1.17), (1.18), provided that the time reversion operation takes place, we have $X_i^\alpha \to X_i^\alpha$, $X_i^\beta \to -X_i^\beta$, $I_i^\alpha \to -I_i^\alpha$, $I_i^\beta \to I_i^\beta$.

Consider the generalized fluxes I_i^α and I_i^β. Having assumed that the parameters $\alpha_i(t)$ and $\beta_i(t)$ introduced characterize the non-equilibrium state of the system completely, it becomes obvious that the generalized fluxes I_i^α and I_i^β also are functions of these parameters:

$$I_i^\lambda = I_i^\lambda(\{\alpha_k\}, \{\beta_l\}), \quad \lambda = \alpha, \beta,$$

where $\{\alpha_k\}$ and $\{\beta_l\}$ is for complete sets of α_k and β_l parameters.

Then, having expanded the fluxes in the powers of α_i and β_j up to linear terms, we can express them via deviations of the thermodynamic parameters from the state of equilibrium:

$$\begin{aligned} I_i^\alpha &= \sum_{k=1}^{n} \lambda_{ik}^{(\alpha\alpha)}\alpha_k + \sum_{k=1}^{m} \lambda_{ik}^{(\alpha\beta)}\beta_k, \\ I_i^\beta &= \sum_{k=1}^{n} \lambda_{ik}^{(\beta\alpha)}\alpha_k + \sum_{k=1}^{m} \lambda_{ik}^{(\beta\beta)}\beta_k. \end{aligned} \tag{1.19}$$

Using the definition of the thermodynamic forces (1.17), we can express the parameters α_k and β_k in formula (1.19) via the thermodynamic forces X_i^α and X_i^β, since this operation can be reduced to solving a system of linear algebraic equations. Thus, under assumptions made about the weak non-equilibrium it is always possible to write down linear relations between the generalized thermodynamic fluxes I_i^γ, and generalized thermodynamic forces X_k^δ, introducing the generalized kinetic coefficients $L_{ik}^{(\gamma\delta)}$:

$$\begin{aligned} I_i^\alpha &= \sum_{k=1}^{n} L_{ik}^{(\alpha\alpha)} X_k^\alpha + \sum_{k=1}^{m} L_{ik}^{(\alpha\beta)} X_k^\beta, \\ I_i^\beta &= \sum_{k=1}^{n} L_{ik}^{(\beta\alpha)} X_k^\alpha + \sum_{k=1}^{m} L_{ik}^{(\beta\beta)} X_k^\beta. \end{aligned} \tag{1.20}$$

Definitions (1.17), (1.18) allow one to obtain a useful expression for entropy production. Indeed, if there is no entropy flow through the boundary of a volume, the differentiation over time of the expression (1.16) for the system's entropy gives a simple relation provided that we use the definitions of the thermodynamic fluxes (1.18) and thermodynamic forces (1.17)

$$\frac{dS}{dt} = \sum_{i,\lambda} I_i^{\lambda} X_i^{\lambda}. \tag{1.21}$$

Thermodynamic fluxes and thermodynamic forces, satisfying the relation (1.21), are often called conjugate fluxes and forces.

1.1.5 Generalized transport coefficients and the Onsager symmetry relations

The $L_{ik}^{(\gamma\delta)}$ coefficients of a linear relationship between generalized thermodynamic fluxes and generalized thermodynamic forces are often referred to as the Onsager coefficients. In phenomenological non-equilibrium thermodynamics an explicit form of these coefficients is not disclosed. Their physical meaning and an explicit expression for different systems can only be found in the context of the molecular-kinetic theory. We draw the reader's attention to the fact that in the framework of linear non-equilibrium thermodynamics, the Onsager coefficients are calculated by averaging over an equilibrium state of the system. Therefore, the dissipation process of large-scale equilibrium fluctuations should be described by using the same factors. This facilitates the analysis of properties of transport coefficients because during the analysis of fluctuations in the equilibrium system one can use properties arising from its symmetry.

The symmetry properties of the $L_{ik}^{(\gamma\delta)}$ coefficients were first determined by Onsager [4]. We give these symmetry relations in the presence of an external magnetic field $\vec{H}$ without proof it will be represented later (see Chapter 5):

$$\begin{aligned} &L_{ik}^{(\lambda\gamma)}(\vec{H}) = \varepsilon_\lambda \varepsilon_\gamma L_{ki}^{(\gamma\lambda)}(-\vec{H}), \\ &\lambda = \alpha, \beta, \quad \gamma = \alpha, \beta, \quad \varepsilon_\alpha = 1, \quad \varepsilon_\beta = -1. \end{aligned} \tag{1.22}$$

Furthermore, we can make sure that the Onsager symmetry relations (1.22) are really valid if the kinetic coefficients are well-defined.

As an example of an application of equations (1.22), we ascertain the symmetry relations for the kinetic coefficients involved in equation (1.15). In the given case, the thermodynamic forces $\vec{\varepsilon}$ and $\vec{\nabla} T$ are even in respect of time reversion operation, $\vec{J}$ and $\vec{J}_Q$ fluxes are odd ones, and therefore, one should take $\lambda = \gamma = \alpha$ in equation (1.22). Then, after the denotation $L_{ik}^{\alpha\alpha} = L_{ik}$ instead of (1.22), the new equation reads

$$L_{ik}(\vec{H}) = L_{ki}(-\vec{H}). \tag{1.23}$$

Formula (1.15) contains two vector flows, so the set of equations (1.20) for this case can be written down as follows:

$$\vec{I}_1 = L_{11}\vec{X}_1 + L_{12}\vec{X}_2,$$

$$\vec{I}_2 = L_{21}\vec{X}_1 + L_{22}\vec{X}_2. \tag{1.24}$$

In the expression (1.24) the L_{ij} coefficients are tensors of a second rank.

Let us determine generalized fluxes and generalized thermodynamic forces in such a way that the ratio (1.21) for entropy production should coincide with the expression (1.11) provided that there is no entropy flow through the surface, that bounds the volume. Comparing (1.11) and (1.21), one should take the following system of definitions:

$$\vec{I}_1 = \vec{J}, \quad \vec{I}_2 = \vec{J}_Q, \quad \vec{X}_1 = \frac{\vec{\varepsilon}}{T}, \quad \vec{X}_2 = -\frac{\vec{\nabla} T}{T^2}. \tag{1.25}$$

Obviously, the generalized fluxes and generalized thermodynamic forces can be determined in another way, since the expression for entropy production (1.21) contains only binary combinations of the generalized fluxes and generalized forces. Once the generalized fluxes and generalized thermodynamic forces are found, using the relations (1.21), (1.24), we can calculate the Onsager coefficients:

$$L_{11} = \hat{\sigma} T, \quad L_{12} = \hat{\beta} T^2, \quad L_{21} = \hat{\chi} T, \quad L_{22} = \hat{\kappa} T^2. \tag{1.26}$$

In the isotropic case, when the kinetic coefficients are scalar quantities, the undoubtedly important relation $\chi = \beta T$ follows from the relations (1.26). For anisotropic systems, by introducing additional tensor indices for the kinetic coefficients:

$$\begin{aligned} &L_{11}^{ik} = \sigma_{ik} T, \quad L_{12}^{ik} = \beta_{ik} T^2, \\ &L_{21}^{ik} = \chi_{ik} T, \quad L_{22}^{ik} = \kappa_{ik} T^2, \quad i, k = x, y, z, \end{aligned} \tag{1.27}$$

basing on the formula (1.22) we can obtain the following corollaries of the symmetry principle of the Onsager kinetic coefficients:

$$\begin{aligned} &\sigma_{ik}(\vec{H}) = \sigma_{ki}(-\vec{H}), \quad \beta_{ik}(\vec{H}) = \beta_{ki}(-\vec{H}), \quad \chi_{ik}(\vec{H}) = \chi_{ki}(-\vec{H}), \\ &\kappa_{ik}(H) = \kappa_{ki}(-\vec{H}), \quad \chi_{ik}(\vec{H}) = T\beta_{ki}(-\vec{H}). \end{aligned} \tag{1.28}$$

Other applications of the symmetry principle of the Onsager kinetic coefficients can be found in the book written by Gurov. This book has small volume, but is of splendid style [5].

1.1.6 Variational principles in linear non-equilibrium thermodynamics

Basic laws of irreversible process thermodynamics were established by generalizing the results of equilibrium thermodynamics and phenomenological transport laws such as Fourier's law, which relates the heat flux and the temperature gradient. These phenomenological laws allow one to determine the kinetic coefficients and equations of interrelationship of thermodynamic fluxes and thermodynamic forces. Along with the inductive method, there is another, path-deductive one, when equations of non-equilibrium thermodynamics are derived from some variational principle, just as it is done in mechanics or electrodynamics. For the sake of better understanding the essence of the variational principles, we list again the basic postulates of linear Onsager's thermodynamics.

1. A linear relationship between generalized thermodynamic forces and generalized fluxes is

$$I_i = \sum_k L_{ik} X_k.$$

2. The Onsager symmetry relations (reciprocity) have the form

$$L_{ik} = L_{ki},$$

   if only odd fluxes take place in respect of the time reversion operation, a magnetic field being absent.

3. If there is no entropy flux through a surface, the system's entropy production is determined by the positive-definite symmetric quadratic form of generalized forces:

$$\dot{S} = \sum_i X_i I_i = \sum_{i,k} L_{ik} X_i X_k \geq 0.$$

Onsager was the first to show that the relations listed in points 1–3 can be derived from some variational principle. We determine the dissipative function:

$$\varphi(X, X) = \frac{1}{2} \sum_{i,k} L_{ik} X_i X_k, \tag{1.29}$$

which as well as the entropy production is a measure of the intensity of irreversible processes occurring in a system. Next we write down the expression for entropy production at fixed external fluxes in the form of the binary combination of the generalized fluxes and generalized thermodynamic forces $\dot{S}(I, X) = \sum_i X_i I_i$. And we determine the functional $\mathfrak{L}(I, X)$ by the relation

$$\mathfrak{L}(I, X) = \dot{S}(I, X) - \varphi(X, X). \tag{1.30}$$

According to Onsager, the functional (1.30) has maximum for a process really occurring in a system as compared with other processes that have the same fluxes I, but different conjugate forces X:

$$\delta(\dot{S}(I, X) - \varphi(X, X)) = \delta\Big(\sum_i I_i X_i - \frac{1}{2} \sum_{i,k} L_{ik} X_i X_k\Big) = \sum_i \Big(I_i - \sum_k L_{ik} X_k\Big) \delta X_i. \tag{1.31}$$

In deriving the second part of (1.31), the symmetry property of kinetic coefficients has been used. If the generalized forces X_i acting actually provide an extremum of the functional $\mathfrak{L}(I, X)$ for the specified generalized fluxes I_i, the variation (1.31) should be equal to nought. Because the variations δX_i are arbitrary, the linear relationship between generalized fluxes and generalized thermodynamic forces immediately follows from the second part of formula (1.31):

$$I_i = \sum_k L_{ik} X_k.$$

Thus, the symmetry principle (reciprocity) of the kinetic coefficients allows formulating the variational principle to derive linear equations of interrelationship between generalized fluxes and generalized thermodynamics forces.

1.1.7 Minimum entropy production principle for weakly non-equilibrium steady states

There are other formulations of the variational principle. A more detailed discussion of this issue can be found in monographs [6, 7]. Let us consider the formulation of the variational principle for steady-state systems, when thermodynamic fluxes are constant. In practical terms, this important special case is realized in open non-equilibrium systems. Which physical quantity has the extreme properties under these conditions? An answer to this question is provided by Prigogine's variational principle: a stationary weakly non-equilibrium state of an open system where an irreversible process proceeds is characterized by minimum entropy production under specified external conditions that impede to an attainment of the equilibrium.

As an instance of the application of the Prigogine's variational principle, we consider the process of transferring both heat and matter between two phases when there is a difference in temperatures between them. Let I_1 be the heat flux and I_2 be the flow of matter, X_1 and X_2 are the corresponding thermodynamic forces conjugate to these fluxes. Entropy production for this system can be mathematically presented as a positive definite quadratic form. Taking into account the Onsager reciprocity relations, one obtains

$$\dot{S} = L_{11} X_1^2 + 2L_{12} X_1 X_2 + L_{22} X_2^2. \tag{1.32}$$

Formally, considering the extremality conditions and varying the entropy production (1.32) in the thermodynamic forces X_1 and X_2, we can write down two equations:

$$\left.\frac{d\dot{S}}{dX_1}\right|_{X_2} \delta X_1 = I_1 \delta X_1 = 2(L_{11} X_1 + L_{12} X_2)\delta X_1 = 0, \tag{1.33}$$

$$\left.\frac{d\dot{S}}{dX_2}\right|_{X_1} \delta X_2 = I_2 \delta X_2 = 2(L_{22} X_2 + L_{12} X_1)\delta X_2 = 0. \tag{1.34}$$

Equality to zero in the expression (1.33) is satisfied if the system is under conditions where the force X_2 is controllable. Then, by virtue of arbitrariness of the variation δX_1, the formula (1.33) implies that the flux of $I_1 = d\dot{S}/dX_1 = 0$. Similarly, if it is possible to realize the condition when the force X_1 is controllable, then the flux I_2 will be equal to zero, which follows from equation (1.34). The thermodynamic force conjugated to the heat flux I_1, is $X_1 \sim \nabla T$. Obviously, the condition of the temperature gradient constancy is quite easy to implement. The thermodynamical force X_2 is proportional to the gradient of the chemical potential $X_2 \sim \nabla\zeta$ and the condition of the chemical potential constancy under the variation of entropy production in force X_1, most likely, is unrealistic. In this situation, from Prigogine's principle of minimum entropy production we can see that the heat flux $I_1 \neq 0$ and the flow of matter $I_2 = 0$. The state found corresponds exactly to a minimum of entropy production (1.32), since this extreme point is a minimum for the function of two variables:

$$\dot{S} = L_{11}X_1^2 + 2L_{12}X_1X_2 + L_{22}X_2^2$$

if the following condition is valid:

$$L_{11}L_{22} - L_{12}^2 > 0.$$

This condition coincides with the quadratic form positivity condition (1.32) and, therefore, it is fulfilled automatically.

Prigogine's principle of minimum entropy production can be generalized for the case of N independent forces, when k of them remain constant due to any external factors. Moreover, the minimum entropy production principle provides equality to zero of $N - k$ fluxes and constancy of k fluxes (fluxes corresponding to non-fixed forces disappear). If none of the forces is fixed, then all fluxes will be equal to zero and the system is still in the equilibrium state.

The principle of minimum entropy production in steady states allows for making a conclusion about stability of weakly non-equilibrium stationary states. After a while, the system's steady state with minimum entropy production will be set under the influence of time-independent external forces. When a sufficiently small change in the system state is the result of fluctuations of some parameter characterizing its non-equilibrium state, in the system there arise processes leading to restoration of the stationary non-equilibrium state. In other words, fluctuations occurring in a system are dissipated not disturbing the stationary non-equilibrium state of this system. It is thought that the mechanisms of the system's response to fluctuations in macroparameters as well as the action of external forces are identical. Then, if the system at the non-equilibrium steady state experiences the action of an external force, there take place processes that tend to weaken (or even to compensate) this imposed change (the principle of Le Chatelier–Braun). It is natural to assume, for this reason, that the steady state of the weakly non-equilibrium system is stable.

1.2 On the application of the Onsager theory

1.2.1 Thermoelectric phenomena. The Peltier, Seebeck, Thomson effects and their relationship

There are a large number of different effects that occur in the presence of a magnetic field, electric current, and temperature gradient. These effects can be ascertained under various measurement conditions, at different combinations of the thermal and the electric current. It suffices to say that only in a transverse magnetic field there exist theoretically about 560 different effects [8]. We discuss only some fundamental kinetic phenomena for analyzing properties of solids.

Let us consider the kinetic effects in an isotropic conductor, when there is only an external electric field and a temperature gradient. The set of equations (1.15) to be rewritten in such a way that the electric current density $\vec{J}$ and the temperature gradient $\vec{\nabla}T$, which are under control in the experiment, may appear in the right-hand side of the phenomenological equations

$$\vec{\varepsilon} = \hat{\rho}\vec{J} + \hat{\alpha}\vec{\nabla}T, \quad \vec{J}_Q = \hat{\Pi}\vec{J} - \hat{\tilde{\kappa}}\vec{\nabla}T, \tag{1.35}$$

$$\hat{\rho} = \hat{\sigma}^{-1}, \quad \hat{\alpha} = \hat{\sigma}^{-1}\hat{\beta}, \quad \hat{\Pi} = \hat{\chi}\hat{\sigma}^{-1}, \quad \hat{\tilde{\kappa}} = \hat{\kappa} - \hat{\chi}\hat{\alpha}. \tag{1.36}$$

In writing the phenomenological transport equations (1.35) the new transport coefficients have been introduced: the electrical resistivity $\hat{\rho}$, the Seebeck coefficient (the coefficient of differential thermopower) $\hat{\alpha}$, the Peltier coefficient $\hat{\Pi}$, and the coefficient of thermal conductivity $\hat{\tilde{\kappa}}$. The physical interpretation of these coefficients and also the conditions under which they can be experimentally determined are discussed below.

The Peltier effect

Let $\vec{J}$ be the electric current flowing through a sample and a temperature gradient be equal to zero. As it follows from the second equation of (1.35), this causes the heat flux $\vec{J}_Q = \hat{\Pi}\vec{J}$ in the sample. In any homogeneous material, this heat flux is impossible to detect, but when the current is made to flow through the conductive circuit of two materials with different coefficients $\hat{\Pi}_1$ and $\hat{\Pi}_2$, then in addition to the Joule heat, some amount of the Peltier heat Q is released or absorbed (depending on the current direction) at the points of contact of two dissimilar conductors with different values of the Peltier coefficients:

$$Q = (\hat{\Pi}_1 - \hat{\Pi}_2)tS_k. \tag{1.37}$$

The release of the Peltier heat at the junction of the two materials with the Peltier coefficients $\hat{\Pi}_1$ and $\hat{\Pi}_2$ being different in value bears the name of the Peltier effect. As follows from formula (1.37) the amount of the Peltier heat is directly proportional to both the contact area S_k and the time t, during which the electric current is passed through the circuit.

The Peltier effect can be qualitatively explained in terms of a scheme of the band model of the conductor near the contact (Figure 1.1). Consider the case when there is a contact between a metal and an electronic semiconductor in which free electrons obey the Maxwell–Boltzmann statistics.

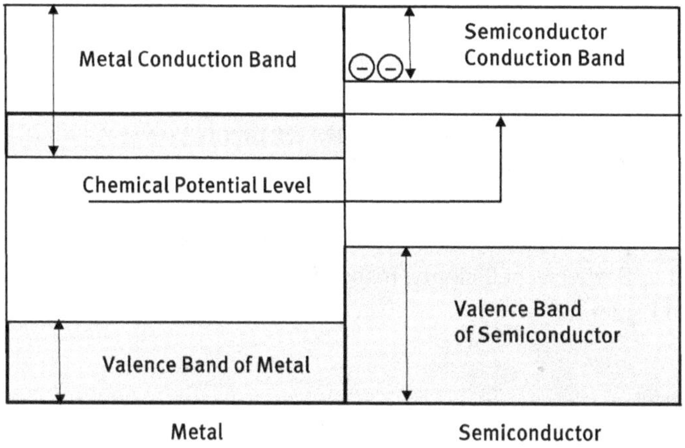

Figure 1.1: Scheme of the band structure in semiconductor–metal contact.

A condition for the equilibrium of the electron gas would be the equality of the chemical potentials in both materials. Since the chemical potential lies below the bottom of the conduction band in the non-degenerate semiconductor, the conductivity electrons in the semiconductor have a higher energy level than the Fermi energy, as in the metal their energy is equal to the Fermi energy. Therefore, the passage of each electron from the semiconductor into the metal in the contact region is characterized by the release of additional energy. This electron transfer from the metal to the semiconductor is accompanied by overcoming the potential barrier and only electrons possessing sufficient kinetic energy are able to surmount it.

This process leads to a decrease in the number of high-speed electrons in the contact region. Thermal equilibrium, in this case, becomes broken, and its recovery requires a heat delivery, consequently, the contact region will be cooled. The Peltier effect is the underlying technology that permits one to construct various types of refrigerators which are widely used for cooling electronic devices, including PC processors.

The Seebeck effect

Now, we assume that $\vec{J} = 0$, and $\vec{\nabla} T \neq 0$. In this case, from the first equation (1.35), after recalling the definition of the electrochemical potential gradient

$$\vec{\varepsilon} = \vec{E} - \frac{1}{e}\vec{\nabla}\zeta,$$

the thermopower can be written in the form

$$\vec{E}(\vec{r}) = \frac{1}{e}\vec{\nabla}\zeta(\vec{r}) + \hat{\alpha}\vec{\nabla}T. \tag{1.38}$$

The essence of the phenomenon of the thermopower or the Seebeck effect consists in the fact that in an electric circuit that is characterized by a series connection of conductors there arises an electromotive power (thermopower), provided that the junctions of the circuit conductors are sustained at different temperatures. In the simplest case, if such a circuit consists of two conductors, it is called a thermoelement or thermocouple. The $\hat{\alpha}$ coefficient, which virtually determines the thermopower value with temperature difference in 1 K between the junctions, bears the name of the differential thermopower coefficient or the Seebeck coefficient.

Consider a thermocouple composed of isotropic samples of metal and semiconductor with the α_m and α_s Seebeck coefficients, respectively. Such a thermocouple is schematically shown in Figure 1.2.

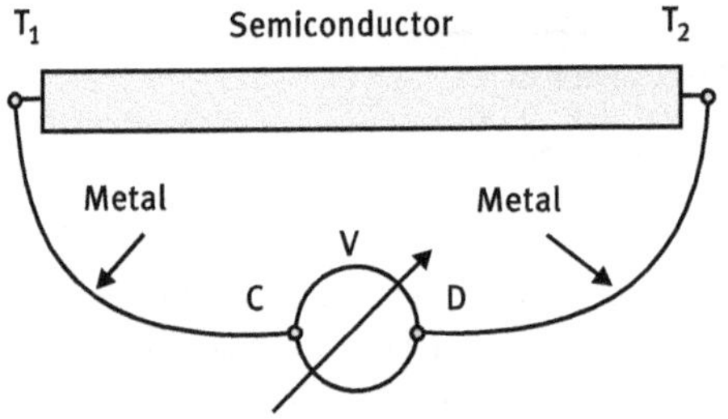

Figure 1.2: The Seebeck effect measurement scheme.

We find the potential difference between points C and D at the scheme in Figure 1.2, via using formula (1.38); as a result we obtain

$$V_{CD} = \int_C^D \vec{E}(\vec{r})\, d\vec{r} = \int_C^D \alpha(\vec{r})\vec{\nabla}T\, d\vec{r} + \frac{1}{e}\int_C^D \vec{\nabla}\zeta\, d\vec{r}. \tag{1.39}$$

The second integral on the right-hand side of formula (1.39) does not contribute, since the points C and D are assumed to be in the isothermal cross-section with the chemical potential ζ being the same.

Passing from integration over coordinates to integration over temperature in the second term of (1.39), after the following obvious transformations V_{CD} acquires the form

$$V_{CD} = \int_{T_1}^{T_2} (\alpha_s(T) - \alpha_m(T))\, dT \simeq (\alpha_s - \alpha_m)(T_2 - T_1). \tag{1.40}$$

The second equality in the formula (1.40) is written for the case, when a temperature dependence of both α_s and α_m in the temperature range from T_1 to T_2 can be neglected.

As follows from formula (1.40), it is impossible to experimentally find the coefficient of the differential thermopower for one of the materials, since the potential difference measured is determined by the difference of simultaneously both of the Seebeck coefficients for materials forming the thermocouple. However, one can select a thermocouple with sufficient accuracy for practical purposes so that one of the coefficients (α_s) be much larger than the other (α_m). Indeed, as will be shown in the next chapters if temperatures are rather low, then $\alpha_m/\alpha_s \simeq k_B T/\zeta_m \approx 10^{-2}$ (k_B is for the Boltzmann constant, ζ_m is for the Fermi energy of electrons in the metal). In this case, the value of the Seebeck coefficient can be found for any semiconducting material with good accuracy. Greater accuracy can be achieved only within a low temperature range by using a thermocouple when the material of one of the arms of the thermocouple is in a superconducting state (differential thermopower is zero in the superconducting state).

The Thomson effect

The Thomson phenomenon is that, if an electrical current is passed through a homogeneous conductor, the current-carrying conductor with a temperature gradient applied along it, releases heat called the Thomson heat Q in addition to the Joule heat. The Thomson heat is proportional to both the electric current density and the temperature gradient:

$$Q = -\sigma_T \vec{J}\vec{\nabla} T t, \tag{1.41}$$

where σ_T is the Thomson coefficient.

We express the coefficient σ_T via the kinetic coefficients that enter the phenomenological transport equations (1.35). Consider equation (1.10), expressing the balance of heat, and insert into it the expressions for the electrochemical potential gradient $\vec{\varepsilon}$ and heat flux density $\vec{J}_Q$. After simple transformations we obtain the following formula:

$$\frac{dQ}{dt} = \operatorname{div}(\hat{\kappa}\vec{\nabla} T) + \vec{J}\hat{\rho}\vec{J} + \vec{J}\left(\hat{\alpha} - \frac{d\hat{\Pi}}{dT}\right)\vec{\nabla} T. \tag{1.42}$$

In deriving this equation we have used an assumption of the constancy of the charge flux density $\vec{J}$ along the sample and taken into account that the Peltier coefficient depends on the coordinates only in terms of temperature: $\hat{\Pi}(\vec{r}) = \hat{\Pi}(T(\vec{r}))$. Then,

$$\operatorname{div}(\hat{\Pi}\vec{J}) = \vec{J}\frac{d\hat{\Pi}}{dT}\vec{\nabla} T.$$

The first summand on the right-hand side of equation (1.42) determines the heat flux through a surface of the volume of the conductor due to the phenomenon of heat conduction. The coefficient $\hat{\kappa}$ has the meaning of the heat conduction of an electron system. The second summand describes the volumetric generation of Joule heat as the coefficient $\hat{\rho}$ has meaning of the electrical resistivity. The third summand describes the release of the Thomson heat, therefore

$$\hat{\sigma}_T = -\left(\hat{\alpha} - \frac{d\hat{\Pi}}{dT}\right). \tag{1.43}$$

The expressions (1.28), (1.36) imply a simple relationship between the $\hat{\Pi}$ and $\hat{\alpha}$ transport coefficients: $\hat{\Pi} = \hat{\alpha}T$. Substituting this result into the expression (1.43), we have another definition of the Thomson coefficient:

$$\hat{\sigma}_T = T\frac{d\hat{\alpha}}{dT}. \tag{1.44}$$

It is easy to understand the physical cause of the Thomson effect in terms of a pictorial presentation about motion of conduction electrons in the sample with a temperature gradient. Figure 1.3 shows two isothermal cross-sections of the sample marked with numbers 1 and 2. The gradient-depicting filling of the rectangle schematically represents the temperature gradient. Let us suppose that in the cross-section 1 of the sample the temperature T_1 is higher than the temperature T_2 in the cross-section 2, and motion of electrons coincides with the direction of the arrow.

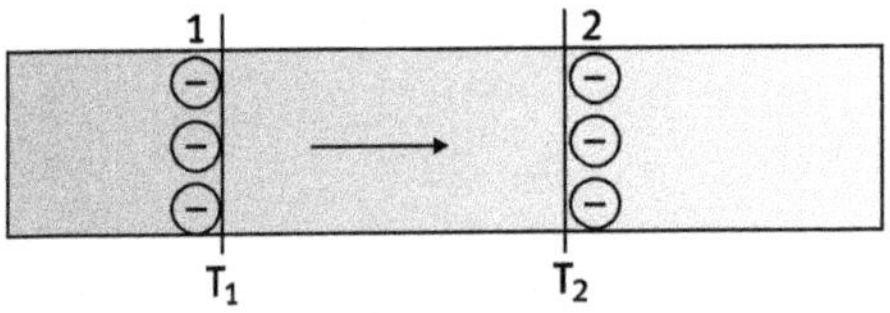

Figure 1.3: Scheme illustrating the essence of the Thomson effect.

We assume that the distance between the cross-sections 1 and 2 is comparable with the free path of electrons in the sample and when moving between these cross-sections, electrons do not experience collisions with the lattice. Electrons in the cross-section 1 were in a thermal equilibrium state with the lattice and by going over to the cross-section 2 they are turned into carriers of the excess kinetic energy. Thermalizing they give away the excess energy to their surroundings in the cross-section 2. It is this energy that is released in the form of Thomson heat. To heat the middle of the current-carrying sample is the easiest way to observe the Thomson effect. In this case, the temperature difference at the edges of the conductor can be easily discerned.

The Peltier effect, the Thomson effect, and the Seebeck effect are closely related between themselves. Therefore, the dependence between differential thermoelectric power and temperature allows one to find both the Peltier coefficient and the Thomson coefficient. At the same time, the expression (1.44) allows one to determine the dependence $\alpha(T)$, if the values of the Thomson coefficient are known over the wide temperature range. Indeed, upon integrating (1.44) one obtains

$$\alpha(T) = \int_0^T \frac{\sigma_T(T)}{T} dT. \tag{1.45}$$

The above case is difficult enough to implement in practice due to complex problems in high-precision measurement of the Thomson effect.

Concluding this brief review of thermoelectric effects, consider an example of application of the Onsager reciprocal relations as far as an analysis of thermoelectric phenomena is concerned.

Problem 1.2. Generalized thermodynamic fluxes and generalized thermodynamic forces can be determined in different ways. If there is only an electric field and a temperature gradient, entropy production (without regard to the flux through a surface), as is known, is determined by formula (1.11):

$$\frac{dS}{dt} = -\vec{J}_Q \frac{\vec{\nabla} T}{T^2} + \vec{J} \frac{\vec{\varepsilon}}{T}.$$

Therefore, two generalized fluxes can be written as

$$I_1 = \vec{\varepsilon}, \quad I_2 = \vec{J}_Q.$$

And their conjugate generalized thermodynamic forces are calculated with formulas:

$$X_1 = \frac{\vec{J}}{T}, \quad X_2 = -\frac{\vec{\nabla} T}{T^2}.$$

In spite of determining in such a way, the generalized forces and generalized fluxes can still be calculated by the relations (1.21).

Expressing the generalized kinetic coefficients L_{ij} via tensor quantities involved in equation (1.35), using the above determination of the generalized fluxes and generalized thermodynamic forces, we establish, also, an interrelationship of the tensors $\hat{\Pi}$ and $\hat{\alpha}$, via employment of the Onsager relation of reciprocity.

Solution. We write down the transport equations in two ways either using the generalized fluxes and forces being defined in the statement of problem or using formulas (1.35). As a result, we obtain

$$\vec{\varepsilon} = L_{11} \frac{\vec{J}}{T} + L_{12} \left(-\frac{\vec{\nabla} T}{T^2} \right),$$

$$\vec{J}_Q = L_{21} \frac{\vec{J}}{T} + L_{22} \left(-\frac{\vec{\nabla} T}{T^2} \right), \tag{1.46}$$

$$\vec{\varepsilon} = \hat{\rho} \vec{J} + \hat{\alpha} \vec{\nabla} T, \quad \vec{J}_Q = \hat{\Pi} \vec{J} - \hat{\kappa} \vec{\nabla} T. \tag{1.47}$$

Upon comparing the expressions (1.46) and (1.47), it is easy to find the values of the L_{ij} coefficients:

$$L_{11} = \hat{\rho}T, \quad L_{12} = -\hat{\alpha}T^2,$$
$$L_{21} = \hat{\Pi}T, \quad L_{22} = \hat{\kappa}T^2. \tag{1.48}$$

To establish the relationship between the Peltier tensors and differential thermoelectric power, we use the Onsager reciprocity relations. As far as the thermodynamic flux I_1, in given case, is even term with respect to time-reversal operation and the flux I_2 is odd, the symmetry relations in accordance with formula (1.22) give the equality $L_{12} = -L_{21}$, whence it follows that the desired interrelation takes the form: $\hat{\Pi} = \hat{\alpha}T$.

1.2.2 Effects in an external magnetic field

The generation of an external magnetic field leads to the appearance of additional anisotropy of crystal properties. Indeed, let a medium before switching the external field be isotropic. However, when the external field is present, the system becomes anisotropic and a direction of the anisotropy coincides with that of the external magnetic field (axis Z). The motion of a charged particle along the magnetic field is a quasi-free motion, whereas charged particles with non-zero velocity experience an action of the Lorentz force in a plane perpendicular to the axis Z.

All directions remain equivalent in the plane perpendicular to $\vec{H}$. Frenkel suggested calling such an anisotropy: gyrotropy. Physical properties in such a gyrotropic medium are unchanged under the rotation of the coordinate system around the axis Z to an arbitrary angle. Hence it follows that all the kinetic coefficients should be invariant with respect to this transformation. From these considerations, one finds, for example, the following equality for the electrical resistivity tensor:

$$\rho'_{ik} = \alpha_{il}\alpha_{km}\rho_{lm} = \rho_{ik}, \tag{1.49}$$

where ρ'_{ik}– value of the electrical resistivity tensor after transformation of the coordinate system rotation around the axis Z to an arbitrary angle; α_{il} is for matrix that defines this transformation.

The invariance requirement for the electrical resistivity tensor with respect to the transformation of rotations around the axis Z to an arbitrary angle (1.49) is reduced to the fact that the structure of this tensor must have the form:

$$\rho_{ik} = \begin{pmatrix} \rho_{xx} & \rho_{xy} & 0 \\ -\rho_{xy} & \rho_{xx} & 0 \\ 0 & 0 & \rho_{zz} \end{pmatrix}. \tag{1.50}$$

Obviously, the tensor structure for other kinetic coefficients in the magnetic field will be exactly the same.

We rewrite the set of equations (1.35) for the case when $\vec{H} \neq 0$ by considering the tensor structure of the kinetic coefficients (1.50):

$$\varepsilon_x = \rho_{xx}J_x + \rho_{xy}J_y + \alpha_{xx}\nabla_x T + \alpha_{xy}\nabla_y T, \tag{1.51}$$

$$\varepsilon_y = -\rho_{xy}J_x + \rho_{xx}J_y - \alpha_{xy}\nabla_x T + \alpha_{xx}\nabla_y T, \tag{1.52}$$

$$\varepsilon_z = \rho_{zz}J_z + \alpha_{zz}\nabla_z T. \tag{1.53}$$

$$J_{Qx} = \Pi_{xx}J_x + \Pi_{xy}J_y - \tilde{\kappa}_{xx}\nabla_x T - \tilde{\kappa}_{xy}\nabla_y T, \tag{1.54}$$

$$J_{Qy} = -\Pi_{xy}J_x + \Pi_{xx}J_y + \tilde{\kappa}_{xy}\nabla_x T - \tilde{\kappa}_{xx}\nabla_y T, \tag{1.55}$$

$$J_{Qz} = \Pi_{zz}J_z - \tilde{\kappa}_{zz}\nabla_z T. \tag{1.56}$$

Using the set of equations of phenomenological transport phenomena (1.51)–(1.56), we may proceed to discuss the principal phenomena in a magnetic field.

Effects in a longitudinal magnetic field

As follows from formulas (1.53), (1.56), the magnetic field does not lead to additional effects, if a temperature gradient and an electric field are directed along the axis Z. In reality, the magnetic field can alter longitudinal components of the electrical resistivity, differential thermopower, and electronic thermoconductivity. In semiconductors, nature of these effects is usually associated with the influence of the magnetic field on the state of scatterers, which determine the relaxation both momentum and energy of current carriers. Longitudinal effects may occur also in metals with a complex structure of the Fermi surface, where they are used for studying its structure. In any case, interpretation of these effects is beyond the frame of the introductory course in transport phenomena, and further the effects in a longitudinal field will not be considered.

Galvanomagnetic phenomena. The Hall effect

One can distinguish isothermal and adiabatic effects depending on experimental conditions. If an investigated sample is placed in a heat bath, the effect is called the isothermal, but if this sample is placed in an insulated environment, the effect is said to be adiabatic. It would be expedient to start the consideration of isothermal phenomena with the Hall effect.

The Hall effect consists in the appearance of the electric field E_y if the following conditions are met: the passage of the electric current J_x through an electrical conductor and equality to zero of temperature gradients in it (assuming that the magnetic field H is applied along the axis Z). The typical observation geometry of the Hall effect is shown in Figure 1.4.

The Hall effect is usually characterized by the Hall constant R. On the basis of equation (1.52), provided that $J_y = 0$, $\nabla_x T = 0$, $\nabla_y T = 0$, we have

$$R = \frac{\varepsilon_y}{J_x H} = \frac{E_y}{J_x H} = \frac{-\rho_{xy}}{H}. \tag{1.57}$$

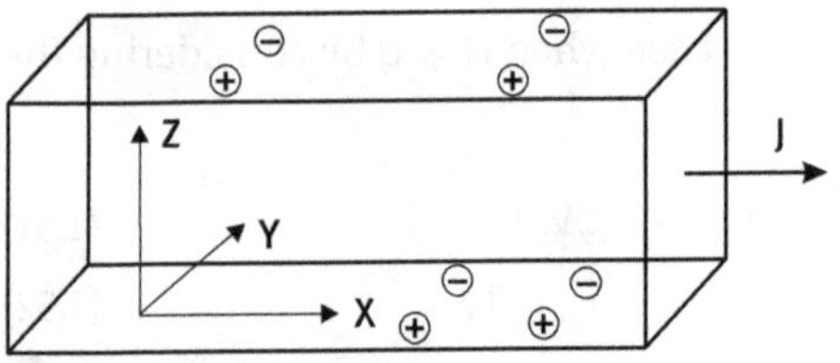

Figure 1.4: Scheme to observe the Hall effect. Potential difference occurs between the front and rear walls of the sample.

It would seem that the nature of the Hall potential difference is quite obvious: the passage of an electric current through a conductor along the X-axis induces the Lorentz force acting on the electrons along the Y-axis. Therefore, an excess negative charge is generated on the rear wall of the sample, with an excess positive charge arising on the front wall of that (see Figure 1.4). Such an elementary reasoning does not stand up to scrutiny because the velocity components v_x of all electrons directed along the X-axis are assumed to be the same. If the speeds of all electrons are identical, a charge concentration gradient leads to the production of the electric field, which entirely compensates an action of the magnetic component of the Lorentz force.

In fact, electrons are distributed in velocities, therefore, the total compensation of the Lorentz force by the Hall field is not observed, i. e. high-speed electrons move to the rear wall of the sample, whereas slow-speed ones head on towards the front wall (see Figure 1.4). This is easily seen by considering the distribution of the current carries over their velocities in the interpretation of galvanomagnetic phenomena. For example, the change of the electrical resistance in a magnetic field or the adiabatic Ettingshausen effect just would be equal to zero, unless one does not consider the velocity electron distribution. We recall that the Ettingshausen effect is the creation of a temperature gradient $\vec{\nabla}_y T$ between two points of a wire when an electric current, J_x, flows through this conductor.

Change of electrical resistance in a magnetic field

As was mentioned above, the electrical resistance change in a magnetic field can be explained only by considering the velocity distribution of electrons. In this case, the magnetic component of the Lorentz force is compensated by an action of the Hall potential difference only for electrons, having a certain average speed. High-speed and slow-speed electrons move in circular segments of the Larmor orbits between two scattering events. This invariably leads to a decrease in their mean free path along the direction of the electric field. For this reason, one should expect increasing electrical resistance in the magnetic field.

The change of the transversal resistance is usually characterized by the relationship

$$\frac{\Delta\rho_{xx}}{\rho_{xx}} = \frac{\rho_{xx}(H) - \rho_{xx}(0)}{\rho_{xx}(0)}. \tag{1.58}$$

Adiabatic galvanomagnetic phenomena. The Ettingshausen effect

Let us proceed to the consideration of adiabatic galvanomagnetic phenomena. If $J_x \neq 0$, $J_y = 0$, $J_{Qy} = 0$, $\nabla_x T = 0$, equation (1.55) implies creation of a temperature gradient in the Y-direction when passing an electric current along axis X. This phenomenon is referred to as the Ettingshausen effect and characterized by the coefficient P:

$$P = -\frac{\nabla_y T}{HJ_x} = \frac{\Pi_{xy}}{H\tilde{\kappa}_{xx}}. \tag{1.59}$$

The temperature difference between the front and rear walls (see Figure 1.4) of the sample is caused by the fact that high-speed electrons ("hot" electrons) are deflected to the rear wall of the conductor, whereas slow-speed electrons, which have a velocity of motion lesser than some average velocity, are drawn to the front wall of this conductor. It should be emphasized that the magnetic component of the Lorentz force for the average velocity of the electrons is compensated by the Hall field. Thermalizing, high-speed electrons give away their excess energy to the lattice thereby increasing temperature of the sample face. On the contrary, slow-speed electrons, in virtue of the energy relaxation processes, absorb part of the energy from the lattice, which results in that the temperature of the front wall of the sample decreases. Thus, the temperature difference between the two opposite faces of the sample is generated.

The Hall effect measured under adiabatic conditions

If conditions of the adiabatic insulation ($J_{Qy} = 0$) are met along axis Y and only the component of the current J_x, is non-zero, then putting also that $\nabla_x T = 0$, we can obtain the following relation from equations (1.55) and (1.52):

$$\nabla_y T = -\frac{\Pi_{xy} J_x}{\tilde{\kappa}_{xx}}; \tag{1.60}$$

$$\varepsilon_y = -\left(\rho_{xy} + \frac{\alpha_{xx}\Pi_{xy}}{\tilde{\kappa}_{xx}}\right)J_x. \tag{1.61}$$

Having determined the Hall coefficient measured under the adiabatic conditions by simple ratio $R_{\text{ad}} = \varepsilon_y / J_x H$, one can proceed to:

$$R_{\text{ad}} = -\frac{1}{H}\left(\rho_{xy} + \frac{\alpha_{xx}\Pi_{xy}}{\tilde{\kappa}_{xx}}\right). \tag{1.62}$$

The difference between the isothermal and adiabatic Hall effects comes about due to a temperature gradient arising in the Y-direction under adiabatic conditions. Due to the presence of the Seebeck effect this temperature gradient produces an additional thermoelectric field in this direction.

The Nernst effect

The Nernst effect is the production of a temperature gradient along X-axis without heat flux along it. This effect is measured under the following conditions: $J_x \neq 0$, $J_y = 0$, $J_{Qx} = 0$, $\nabla_y T = 0$. In this case, from equation (1.54) the Nernst coefficient B is

$$B = \frac{\nabla_x T}{J_x} = \frac{\Pi_{xx}}{\tilde{\kappa}_{xx}}. \tag{1.63}$$

It should be noted that the Nernst effect is also possible in the absence of an external magnetic field. A magnetic field only changes the coefficient B. The physical nature of the Nernst effect is quite simple: the passage of electric current through a sample is accompanied by a heat flux (see formula (1.35)). This leads to the heating of one and cooling of the other end of the sample in conditions of the adiabatic insulation in the direction of the axis X. The temperature difference at the ends of the sample grows up as long as the heat flux, which arises due to the presence of the temperature gradient, is compensated by heat flux associated with the Peltier phenomenon.

When measuring the Nernst effect another situation is possible, namely, when instead of the condition $\nabla_y T = 0$ the condition $J_{Qy} = 0$ is satisfied (the adiabatic Nernst effect). In this case, equations (1.54), (1.55) enable obtaining the relation

$$B_{\mathrm{ad}} = \frac{\nabla_x T}{J_x} = \frac{\Pi_{xx}\tilde{\kappa}_{xx} + \Pi_{xy}\tilde{\kappa}_{xy}}{\tilde{\kappa}_{xx}^2 + \tilde{\kappa}_{xy}^2}. \tag{1.64}$$

Transverse resistivity under adiabatic conditions

Let there be fulfilled the conditions as $J_x \neq 0$, $J_y = 0$, $J_{Qy} = 0$, $\nabla_x T = 0$. Define the component of the resistivity tensor in adiabatic conditions of measurement ρ_{xx}, by the condition $\rho_{xx\mathrm{ad}} = \varepsilon_x / J_x$. Then, using equations (1.51), (1.60), we have

$$\rho_{xx\mathrm{ad}} = \rho_{xx} - \frac{\alpha_{xy}\Pi_{xy}}{\tilde{\kappa}_{xx}}. \tag{1.65}$$

The second term on the right-hand side of formula (1.65) is caused by the thermomagnetic transversal Nernst–Ettingshausen effect, leading to an additional electric field in the X-direction in the presence of a temperature gradient in the Y-direction.

Thermomagnetic phenomena. The transverse Nernst–Ettingshausen effect

Thermomagnetic phenomena occur in the presence of a temperature gradient along one of the axes of a sample and can be detected under both isothermal and adiabatic conditions. It is worth emphasizing that under isothermal conditions the rest faces of the sample are in contact with a thermostat, whereas under adiabatic conditions they are in conditions of adiabatic insulation. In this case, heat flux along different directions is zero. We start the consideration with the isothermal transverse Nernst–Ettingshausen effect.

The transverse Nernst–Ettingshausen effect consists in the appearance of a transverse potential difference in the Y-direction in the presence of the temperature gradient $\nabla_x T$ in the X-direction. Let the conditions be $J_x = 0, J_y = 0, \nabla_y T = 0$. Then, using equation (1.52), we can express the quantity ε_y as

$$\varepsilon_y = -\alpha_{xy}\nabla_x T. \tag{1.66}$$

Typically, the transverse Nernst–Ettingshausen effect is characterized by the coefficient $Q_{\mathrm{NE}} = \alpha_{xy}/H$.

It is of interest to discuss in detail the causes of the Nernst–Ettingshausen effect, as well as factors, defining a sign of the effect. Let us consider the n-type semiconductor sample along which a constant temperature gradient $\nabla_x T$ is sustained and a magnetic field is applied along the axis Z.

The free time of electrons between collisions, as it will be shown in Chapter 4, depends on the velocity (energy) of the electrons. This time may either increase or decrease with increasing energy of the electrons depending on what kind of scattering mechanism determines the electron momentum relaxation time. If we consider some cross-section of the sample in the direction perpendicular to the X-axis, then the projection of the thermal velocity on the X-axis for electrons moving towards the cold end of the sample will be higher than the projection of the thermal velocity for electrons traveling in opposite direction. For this reason, these electrons will be differently deflected by an external magnetic field. Consequently, in the Y-direction there arises a non-zero electric current. This results in an excess charge that creates the electric field E_y. The sign of the effect depends on some factors that can be illustrated by the following simple model [8].

We suppose that the sample (conductor) contains two groups of electrons: electrons in amount of n_1 moving from the cold towards the hot end of the sample, have a velocity v_{1x} and electrons in amount of n_2 moving in opposite direction have a velocity v_{2x} ($v_{2x} > v_{1x}$). All the electrons move only along X-axis, in the absence of the external magnetic field, with the following ratio to be obligatorily fulfilled in the stationary state:

$$n_1 v_{1x} = n_2 v_{2x}. \tag{1.67}$$

When placing the conductor into a magnetic field there arises the charge flux in the Y-direction:

$$J_y = e(n_1 v_{1y} - n_2 v_{2y}) = (n_1 v_{1x} \mathrm{tg}\varphi_1 - n_2 v_{2x} \mathrm{tg}\varphi_2), \tag{1.68}$$

where $\varphi_1 = \omega_0\tau_1$, $\varphi_2 = \omega_0\tau_2$ – angles characterizing the velocity vector change of electrons between two successive collisions, or the Hall angles for slow-speed and high-speed groups of electrons; ω_0 is for frequency of the Larmor precession in a magnetic field, τ_1 and τ_2 – the times of covering the free path for electrons with the velocities v_1 and v_2.

Each of the parameters $\varphi_1 = \omega_0\tau_1$, $\varphi_2 = \omega_0\tau_2$ in a non-quantizing magnetic field is much less than unity and, therefore, $\mathrm{tg}\varphi_1 \simeq \varphi_1$, $\mathrm{tg}\varphi_2 \simeq \varphi_2$. Leaving in equation (1.68) only terms of first order in the small parameter $\omega_0\tau$, instead of (1.68) one finds the simple equation

$$J_y = en_1 v_{1x}\omega_0(\tau_1 - \tau_2), \tag{1.69}$$

which implies that the sign of the transverse Nernst–Ettingshausen effect depends on an increase or a decrease in the relaxation time τ with increasing the electron energy. Thus, when varying temperature, the change in sign of the Nernst–Ettingshausen effect indicates that the mechanism of the electron momentum relaxation has been modified. The qualitative conclusions based on the formula (1.69) are valid and confirmed completely in the calculation of the quantity Q_{NE} on grounds of the solution of the kinetic equation.

The longitudinal Nernst–Ettingshausen effect

The longitudinal Nernst–Ettingshausen effect is a change of the thermoelectric power under an external magnetic field. Let the following conditions be met: $\nabla_x T \neq 0$, $\nabla_y T = 0$, $J_x = J_y = 0$. In this case from equation (1.51) we can obtain

$$\varepsilon_x = \alpha_{xx}(H)\nabla_x T. \tag{1.70}$$

It is of interest to elucidate on what kind of factors the sign of the change of the differential thermopower in a magnetic field depends at all. We examine the same conductor model, used for the analysis of the transverse Nernst–Ettingshausen effect. When generating an external magnetic field the components of velocities v_{1x} and v_{2x} are changed and instead of balance equation (1.67) in stationary conditions, one can write down the following formulas:

$$n_1' v_{1x}(H) = n_2' v_{2x}(H),$$
$$v_{1x}(H) = v_{1x}\cos\omega_0\tau_1, \quad v_{2x}(H) = v_{2x}\cos\omega_0\tau_2, \tag{1.71}$$

where n_1' and n_2' is for the number of electrons with velocities $v_{1x}(H)$ and $v_{2x}(H)$, respectively.

In expanding $\cos\omega_0\tau$ in a series in formula (1.71) and retaining the first non-vanishing terms in small parameter $\omega_0\tau$, with taking into account the ratio (1.67), we have

$$\frac{n_1'}{n_2'} = \frac{n_1}{n_2}\left(1 + \frac{\omega_0^2}{2}(\tau_1^2 - \tau_2^2)\right). \tag{1.72}$$

When generating the magnetic field, the number of electrons n_1' at the cold end of the conductor will increase and the change of thermoelectric power will be positive, if the relaxation time turns out to be high for slow-speed electrons. And vice versa the effect will be negative, if the electron relaxation time increases with growth of the velocity $\vec{v}$.

The Maggi–Righi–Leduc effect

The Maggi–Righi–Leduc effect consists in changes in thermal conductivity while placing a conductor in a magnetic field. The effect is determined by the conditions $\nabla_x T \neq 0$, $\nabla_y T = 0, J_x = J_y = 0$. Taking into account the above conditions and equation (1.54) we obtain the following equality:

$$\tilde{\kappa}_{xx}(H) = -\frac{J_{Qx}}{\nabla_x T}. \tag{1.73}$$

A physical reason for the electronic component change of the thermal conductivity is actually the same as in the case of the change of the transverse electrical resistivity, a decrease in the length projection of the electron free path in a magnetic field on the direction of a temperature gradient.

Adiabatic thermomagnetic effects

All of the above-mentioned isothermal effects have their adiabatic analogs measured under conditions: $J_x = J_y = 0, J_{Qy} = 0, \nabla_x T \neq 0$. The expressions for finding coefficients that generally characterize the adiabatic transverse and longitudinal Nernst–Ettingshausen effects and the adiabatic Maggi–Righi–Leduc effect are

$$Q_{\mathrm{NE\,ad}} = \frac{1}{H}\left(\alpha_{xy} - \frac{\alpha_{xx}\tilde{\kappa}_{xy}}{\tilde{\kappa}_{xx}}\right), \tag{1.74}$$

$$\alpha_{xx\,\mathrm{ad}} = \alpha_{xx} + \frac{\alpha_{xy}\tilde{\kappa}_{xy}}{\tilde{\kappa}_{xx}}, \tag{1.75}$$

$$\tilde{\kappa}_{xx\,\mathrm{ad}} = \tilde{\kappa}_{xx} + \frac{\tilde{\kappa}_{xy}^2}{\tilde{\kappa}_{xx}}. \tag{1.76}$$

The difference between the isothermal and the adiabatic effects is that in adiabatic insulation conditions there arises an additional temperature gradient, directed along the Y-axis and acting as a new thermodynamic force. The temperature gradient leads to the potential difference in the Y-direction in the transverse Nernst–Ettingshausen effect and determines the appearance of the second terms on the right-hand sides of formulas (1.75), (1.76). The appearance of the temperature gradient in the Y-direction is due to the Righi–Leduc effect.

Indeed, taking into consideration the conditions in which the above adiabatic thermomagnetic phenomena are detected, equation (1.55) logically implies the equality below

$$\nabla_y T = \frac{\tilde{\kappa}_{xy}}{\tilde{\kappa}_{xx}} \nabla_x T. \tag{1.77}$$

Typically, this effect is characterized by the Righi–Leduc coefficient

$$S_{\mathrm{RL}} = \frac{\nabla_y T}{H \nabla_x T}.$$

Based on this definition, it follows from (1.77) that the Righi–Leduc coefficient can be expressed via components of the tensor of heat conduction

$$S_{\mathrm{RL}} = \frac{\tilde{\kappa}_{xy}}{H \tilde{\kappa}_{xx}}. \tag{1.78}$$

After substituting the formulas (1.77) into (1.51), (1.52), (1.54), the previously given expressions (1.74)–(1.76) can easily be obtained for the coefficients, characterizing the thermomagnetic phenomena in the adiabatic conditions of measurement.

In conclusion of this paragraph we establish some relationships between the kinetic coefficients determined in the theory of the above effects. The most obvious is the fact that all adiabatic effects find expression via kinetic coefficients that are determined upon isothermal conditions. Indeed,

$$R_{\mathrm{ad}} = R - \alpha_{xx} P, \tag{1.79}$$

$$\rho_{xx\,\mathrm{ad}} = \rho_{xx} - H^2 Q_{\mathrm{NE}} P, \tag{1.80}$$

$$Q_{\mathrm{NE\,ad}} = Q_{\mathrm{NE}} - \alpha_{xx} S_{\mathrm{RL}}, \tag{1.81}$$

$$\alpha_{xx\,\mathrm{ad}} = \alpha_{xx} + H^2 Q_{\mathrm{NE}} S_{\mathrm{RL}}, \tag{1.82}$$

$$\tilde{\kappa}_{xx\,\mathrm{ad}} = \tilde{\kappa}_{xx}(1 + H^2 S_{\mathrm{RL}}^2). \tag{1.83}$$

To derive additional relations between the kinetic coefficients, it is necessary to make use of the ratio $\hat{\Pi} = \hat{\alpha} T$ which follows from the Onsager reciprocity principle, a number of additional equations of interrelationship:

$$B \tilde{\kappa}_{xx} = T \alpha_{xx}, \tag{1.84}$$

$$P \tilde{\kappa}_{xx} = T Q_{\mathrm{NE}}, \tag{1.85}$$

$$B_{\mathrm{ad}} \tilde{\kappa}_{xx\,\mathrm{ad}} = T \alpha_{xx\,\mathrm{ad}}. \tag{1.86}$$

There are also other relations that do not follow from the general principles of linear non-equilibrium thermodynamics but are a corollary of the model used.

1.3 Self-organization in highly non-equilibrium systems

1.3.1 Non-equilibrium dissipative structures

The development of physics, chemistry, and biology during the last century allowed accumulating a sufficiently large number of examples of highly non-equilibrium systems, in which the non-equilibrium state is a source for the order establishment. As far back as 1901 a phenomenon, which came later to be known as the formation of Benard cells, was experimentally discovered. This phenomenon may serve as a classic example of the peculiar structure of convective motion in fluid in the presence of a temperature gradient, directed along the field of gravitational forces. Another example of a non-equilibrium system through generation of electromagnetic oscillations when passing a direct current in the Gunn diodes. The main distinguishing feature of the systems demonstrating the self-organization is that there appear ordered structures that were absent in the equilibrium state. The formation of the above structures takes place due to energy transferred from outside. Such structures can be sustained only at the expense of the inflow of energy or matter and, therefore, it would be natural enough to call them dissipative structures. For instance, a large city or even terrestrial civilization as a whole can serve as an example of dissipative structures.

Self-organization is characterized by creating spatial, temporal or spatial-temporal structures. Obviously, the self-organization is possible if there is a cooperative behavior in such systems whose different parts interact. All this served as a basis to separate out self-organization phenomena occurring in highly non-equilibrium systems into a special science, which the German physicist Haken called *synergetics* (from the Greek word synergeia means joint action or cooperation).

The special literature [9, 10, 11, 12] enables one to find both multiple examples of other systems with dissipative structures, as well as existing methods to describe them.

One should make a distinction between the mechanism of formation of dissipative structures and the mechanism of formation of equilibrium structures. Namely, the principle of an entropy increase works for isolated equilibrium systems, and maximum entropy corresponds to a steady equilibrium state. The principle of minimal free energy is observed for equilibrium systems that are in contact with thermostat (electric or magnetic domains may serve as examples). Therefore, in the given case there is a possibility of creating the spatial structures whose appearance does not contradict to the principles of equilibrium thermodynamics.

Approaches based on the principles of equilibrium thermodynamics in no case are applicable to dissipative structures. For example, the appearance of convective dissipative structures of Benard should be considered as the demonstration of convective instability of a liquid. From this point of view, natural convection in the liquid exists as sufficiently weak fluctuations, which are not coordinated at low values of a temperature gradient but they damp out over periods less than the time required for the coordination of these fluctuations. When the temperature gradient exceeds a certain critical threshold there takes place a bifurcation, which in result leads to convective motion in the system. The term bifurcation is derived from Latin "bifurcus". Bifurcus means "split in two parts" and refers to qualitative changes in behavior of a system upon changing of some governing parameter.

1.3.2 The Glansdorff–Prigogine universal evolution criterion

As was noted above, different variational principles can be formulated for linear non-equilibrium processes. For example, it can be mentioned the Onsager's principle of least dissipation of energy. This postulate claims that the functional (1.30) is maximum while varying in the generalized forces under constant flow conditions. As for the systems being under stationary conditions, Prigogine's variational principle can be formulated. According to the Prigogine principle, a weakly non-equilibrium state of an open system in which an irreversible process takes place, is characterized by minimum entropy production under given external conditions that impede the achievement of equilibrium. These principles show a rather heuristic character and do not permit a researcher to describe any system, but they make it possible to elucidate whether a constructed theory contradicts some general tenets or principles.

When considering nonlinear effects, it is assumed that the entropy production can still be written as a sum of products of fluxes and their conjugate thermodynamic forces:

$$\dot{S} = \int \sum_i I_i X_i \, dv. \tag{1.87}$$

Moreover, it is usually assumed that generalized kinetic coefficients can be determined by relations (1.24), but for nonlinear systems they should be calculated when the system is in a non-equilibrium state and, therefore, these coefficients are to be a function of generalized thermodynamic forces. As far as the kinetic coefficients prove to be a function of the generalized forces in the nonlinear case, a direct application of the variational principle of Onsager or Prigogine for such systems is inappropriate because neither the functional (1.30), nor the entropy production (1.32) possess extreme properties.

The time derivative of the entropy production (1.87) can be divided into a the part that is due to changes in flow and the part that is due to the change of thermodynamic forces in time:

$$\frac{d\dot{S}}{dt} = \int \sum_i \frac{dI_i}{dt} X_i \, dv + \int \sum_i I_i \frac{dX_i}{dt} \, dv. \tag{1.88}$$

The behavior of the first summand in formulas (1.88) is non-single-valued, whereas the second summand satisfies the inequality of a general nature known in the literature as the Glansdorff–Prigogine evolution principle. According to this principle, processes with fixed boundary conditions in any non-equilibrium system occur in such a manner that the entropy production change rate caused by changes in thermodynamic forces decreases with time:

$$\frac{d_X \dot{S}}{dt} = \int \sum_i I_i \frac{dX_i}{dt}\, dv \leq 0. \tag{1.89}$$

The Glansdorff–Prigogine evolution criterion (1.89) is referred to as the Universal Criterion of Evolution, because situations for which the inequality (1.89) is violated have not yet been revealed.

Problem 1.3. Check the validity of the Glansdorff–Prigogine evolution criterion (1.89) for the thermal conductivity with a fixed value of the temperature gradient at the boundary of a sample.

Solution. In the case of heat conduction there is a single generalized flow

$$I_1 = \vec{J}_Q$$

and its conjugate thermodynamic force

$$X_1 = \vec{\nabla}\frac{1}{T}.$$

We write down the expression (1.89) in more detail with respect to the above case:

$$\frac{d_X \dot{S}}{dt} = \int \vec{J}_Q \frac{d}{dt} \vec{\nabla}\frac{1}{T}\, dv. \tag{1.90}$$

Let us transform the term under the integral sign in the right side of the expression (1.90) to obtain

$$\vec{J}_Q \frac{d}{dt}\vec{\nabla}\frac{1}{T} = \operatorname{div}\left(J_Q \frac{d}{dt}\frac{1}{T}\right) - \left(\frac{d}{dt}\frac{1}{T}\right) \operatorname{div}\vec{J}_Q.$$

Insert this result into expression (1.90). Converting a volume integral into a surface integral, we have

$$\frac{d_X \dot{S}}{dt} = \oint_S \vec{J}_Q \frac{d}{dt}\frac{1}{T}\, d\vec{S} - \int_V \left(\frac{d}{dt}\frac{1}{T}\right) \operatorname{div}\vec{J}_Q\, dv. \tag{1.91}$$

As far as the temperature at the boundary of the sample is assumed to be fixed, then the surface integral on the right-hand side of formula (1.91) vanishes. This condition may be fulfilled in open systems, when the non-equilibrium system in question is surrounded by external bodies. To transform the volume integral on the right-hand side of formula (1.91), we use the heat balance equation (1.10), writing it in relation to the previously mentioned conditions as

$$\frac{d\rho_0 C_v T}{dt} = -\operatorname{div}\vec{J}_Q,$$

where ρ_0 is for the sample density, C_v is for the sample specific heat.

After simple transformations one obtains the following expression:

$$\frac{d_X\dot{S}}{dt} = -\int_V \frac{\rho_0 C_v}{T^2}\left(\frac{dT}{dt}\right)^2 dv \leq 0. \tag{1.92}$$

Here, the equality sign corresponds to the steady state. After calculating we have shown that the entropy production change rate due to changes in the external forces in the strongly non-equilibrium system in which there is heat transfer, is negative.

1.3.3 Ways of describing strongly non-equilibrium systems

Similar to how equations of equilibrium thermodynamics are constructed within the framework of equilibrium statistical theory it would be desirable to construct equations that describe a behavior of highly non-equilibrium systems directly from the first principles. When solving this problem it is then natural to pose a question of whether the behavior of such systems is irreversible. How can equations that describe a non-equilibrium system be obtained, basing on, for example, the dynamic equations of Newton that are reversible in time? Until recently, it seemed impossible within the current paradigm to overcome a deep disparity between the completely determinate mechanical and the statistical description. However, as far back Poincaré in his works expressed ideas of deterministic chaos that have allowed one to bridge the gap. In the second half of the last century, Kolmogorov, Arnold, Sinai, Zaslavsky, and other scientists have developed an elegant theory that formulates conditions under which a dynamic description of the system becomes meaningless, and therefore, to describe such systems, the statistical approach is needed.

The approach based on the concept of dynamic chaos is very useful in understanding the principles of non-equilibrium statistical mechanics, and we will discuss these ideas at the end of the section devoted to self-organization in highly non-equilibrium systems. It has to be admitted, however, that the approach is not appropriate in solving particular problems of dynamics of highly non-equilibrium systems.

At the end of the last century, the rapid growth of the number of papers concerning the self-organization theory was mainly associated with the emergence of a new direction in mathematics, which has originated at the junction of two disciplines such as topology and theory of differential equations (mathematical analysis). Both these disciplines have merged into a single profound theory owing to works of the French mathematician Thom. Thanks to him, the efforts of predecessors Whitney (topology) and Poincaré, Lyapunov, Andronov (qualitative theory of differential equations) became integrated. The English mathematician Zeeman was responsible for coining the term "catastrophe theory" for this new branch of mathematics.

The sonorous name of this theory generated a huge number of speculations of mystic nature which have nothing to do with mathematics or physics. Actually, it is believed that the catastrophe is a sudden shift in behavior caused by small changes in external conditions. In most cases of interest for applications in physics, there takes place a qualitative restructuring (bifurcation) of the character of the solutions of differential equations provided that one of the control parameters is changed gradually.

The essence of the new approach, which determines its practical significance lies in the fact that, as was noted by Poincaré, there is very often no necessity to obtain a complete solution of complex nonlinear differential equations. Therefore, it is sufficient to know information about only the qualitative behavior of solutions. The complete solution, even be it obtained with great effort, would only complicate the behavior analysis of such systems.

After the above remarks one may raise the question of how we should describe highly non-equilibrium systems capable of self-organization. It is clear that means of mechanics is not appropriate to describe these systems, since the mechanical description with coordinates and velocities of particles making up the systems is too fine structured. Therefore, such a description of the systems typical of cooperative behavior proves to be too complex. However, the thermodynamic approach, as was previously mentioned, is also unacceptable to describe these systems.

For this reason, in most cases it is accepted that one describe highly non-equilibrium systems with self-organization via establishing evolution of a chosen set of macroscopic variables provided that some dynamic motion equations for them have been already previously found. Strictly speaking, this procedure omits the most complex step of derivation of the equations describing the evolution of the highly non-equilibrium systems from first principles of non-equilibrium statistical classical mechanics, replacing the step mentioned by semi-phenomenological derivation of appropriate dynamical equations.

If a system is spatially uniform, it should be described by first-order differential equations with respect to time. Equations of higher order can be always reduced to a set of first-order equations. After finding the set of equations the approach based on the catastrophe theory is used for a qualitative of the character of its solutions.

Consider for example, the derivation of equations of the Lotka–Volterra predator–prey model describing the size of predator population (tuna) and that of a victim (sardines) which are constituted into a single food chain. This model was proposed by Volterra in 1920. Mathematical equations found by him coincided with Lotka's equations for describing a hypothetical reaction scheme with the formation of an intermediate unstable state. This system is referred to as the Lotka–Volterra "predator–prey" model. At present, all papers where self-organization theory is discussed include this model.

Let n_1 be the number of "herbivores" in the population; n_2 be the number of "predators". Then the population dynamics of "predators" and "herbivores" is determined by the equations

$$\begin{aligned} \dot{n}_1 &= \gamma_1 n_1 - \beta n_1 n_2, \\ \dot{n}_2 &= \beta n_1 n_2 - \gamma_2 n_2. \end{aligned} \tag{1.93}$$

According to equations (1.93) the reproduction rate of "herbivores" is proportional to the number n_1 and depends on the constant γ_1, regulating the reproduction rate. On the other hand, the population reduction rate is proportional to the number of the "predators" and the number of "victims" where β is a certain constant. The "predator" population increase rate depends on the product of n_2 n_1, defined both by the number of "predator" zooids and by the presence of food. The "predator" extinction rate depends on their number and is determined by the constant γ_2.

It is typical of (1.93) that, the system (1.93) is nonlinear. The time dependence $n_1(t)$ is shown in Figure 1.5. The population n_2 has a similar periodic time dependence with some shift relative that of n_1 along the time scale.

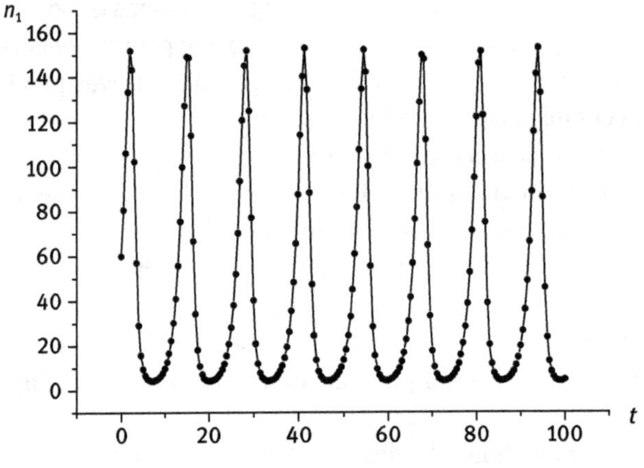

Figure 1.5: Periodic fluctuations in population n_1 in the "predator–prey" problem: $n_1(0) = 60$, $n_2(0) = 20$; parameters $\gamma_1 = 0.3712$, $\beta = 0.0097$, $\gamma_2 = 0.3952$.

Instead of studying the time dependence we can plot a phase pattern of the system. In the case of the "predator–prey" model the phase space is the coordinate plane with the axes of n_1 and n_2. Any point in phase space corresponds to each state of the system and a multiplicity of points, reflecting the state of the system at different moments of time, represents the phase pattern. Figure 1.6 depicts the phase pattern of the "predator–prey" problem.

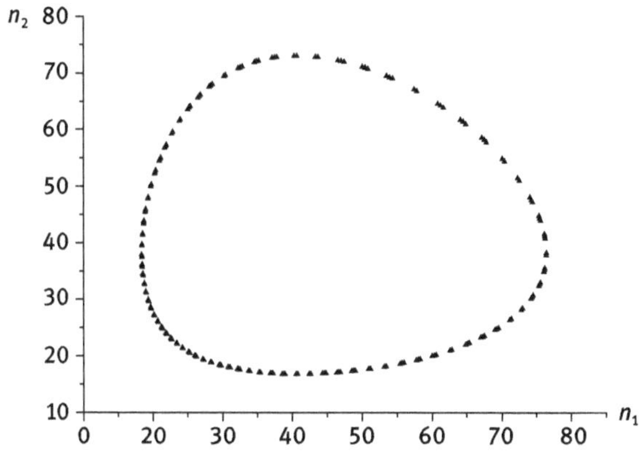

Figure 1.6: Phase pattern of the "predator–prey" problem (model parameters are the same as in Figure 1.5).

The phase pattern is a closed curve that resembles a circumference. Such a form of the curve indicates the existence of almost periodic oscillations in a "predator–prey" model. The phase pattern analysis is a very common technique for studying systems with self-organization.

Summarizing the above results, we will describe a highly non-equilibrium state with a set of variables $q_1(\vec{r}, t)$, $q_2(\vec{r}, t) \ldots$, $q_n(\vec{r}, t)$, which are dependent on the coordinates and time. Quantities $q_1, q_2, \ldots, q_n$ in aggregate determine the system state vector $\vec{q}$ or the point in phase space which univalently characterizes the system's state in the phase space.

The time dependence of the quantities $q_1(t)$, $q_2(t), \ldots, q_n(t)$ defines evolution of the system in time. However, it is sufficiently difficult to plot a phase pattern of the system for three or more dynamical variables. In this case, the behavior of the system can be studied by cutting the phase space by a plane and plotting the points where the trajectory intercepts this plane.

Methods for obtaining information about qualitative behavior of solutions of nonlinear sets of equations will be considered in more detail in the next paragraphs.

1.3.4 Stability of states of highly non-equilibrium systems

Let us assume that a highly non-equilibrium system is described by a set of macroparameters $q_1(t)$, $q_2(t), \ldots, q_n(t)$, for which one can write the set of differential equations

$$\frac{dq_i}{dt} = f_i(q_1(t), q_2(t), \ldots, q_n(t), B); \quad i = 1, 2, \ldots, n, \tag{1.94}$$

where B is for some parameters that define the external and internal conditions.

The functions f_i are assumed to be differentiable. As can be seen from the expression (1.94), the right-hand side of equations do not contain an explicit time dependence. Such differential equations are commonly referred to as *autonomous*.

In virtue of the theorem on the existence and uniqueness of solution, through each point in the phase space there passes a unique trajectory. This means that phase trajectories never intersect.

Qualitative analysis of solutions usually begins with searching for the stationary points that satisfy the equations which are below:

$$\frac{dq_i}{dt} = 0; \quad f_i(q_1, q_2, \dots, q_n, B) = 0; \quad i = 1, 2, \dots, n. \tag{1.95}$$

Fixed points of phase space correspond to stationary states. If the functions $f_i(q_1, q_2, \dots, q_n, B)$ are nonlinear, there may exist a lot of solutions satisfying to the equations

$$f_i(q_1, q_2, \dots, q_n, B) = 0, \quad i = 1, 2, \dots, n.$$

Then the question arises: which of possible states will the system be in? This problem largely is physical rather than mathematical. Each real physical system has fluctuations of the parameters. Let a set of parameters q_i^s, $i = 1, 2, \dots, n$ defines some stationary special point, and a set of the equations $q_i(t) = q_i^s + \delta q_i$ defines a state that arises as a result of fluctuations near the steady state. If the stationary point is stable, the system located in such a state is not sensitive to small fluctuations. Conversely, if the stationary point is unstable, the fluctuations will grow and the system finally leaves the stationary point.

The issue of stability of stationary states admits many interpretations. Consider several different notions of stability.

Asymptotic stability stands for that the state is stable. Moreover, we can always find $\varepsilon > 0$ such that, if the inequality

$$|\vec{q}^{\,s} - \vec{q}_0'| < \varepsilon$$

is fulfilled, then

$$\lim_{t\to\infty} |\vec{q}^{\,s} - \vec{q}(\vec{q}_0')| = 0. \tag{1.96}$$

In the above formula the quantity $\vec{q}_0'$ is some point in the vicinity of the stationary state in which the system was at an initial time moment. If the stationary point is asymptotically stable, it means that all systems, whose phase points are located in a neighborhood of the stationary point, will be in this stationary point after a lapse of time. That is why asymptotically stable states are referred to as *attractors*, but stationary points satisfying the condition (1.96) are *attractive*. The multiplicity of the points attracted to $\vec{q}^{\,s}$ is called the domain of attraction for a given solution.

All thermodynamic equilibrium states which are not critical points are asymptotically stable.

1.3.5 The Lyapunov global stability criterion

A stationary singular point $\vec{q}^S$ is stable, if in the vicinity D of this point it is feaseble to construct some positive (negative) definite function $V(q_1, q_2, \ldots, q_n)$ such that its derivative dV/dt is a non-positive (non-negative) definite one in the entire domain D. Stability is asymptotic if the signs of V and dV/dt are opposite to each other.

In essence, the Lyapunov theorem generalizes the method of potentials for systems which have no potential. The importance of this theorem is that, if one succeeds in constructing such a function, it is not necessary to solve equations of motion when considering the stability problem but only the following functions need to be investigated:

$$V(q_1, q_2, \ldots, q_n) \quad \text{and} \quad \frac{dV}{dt} = \sum_{i=1}^{n} \frac{dV}{dq_i} f_i(q_1, q_2, \ldots, q_n), \tag{1.97}$$

because $\dot{q}_i = f_i(q_1, q_2, \ldots, q_n)$ follows from equations (1.94).

The practical significance of this theorem is not great, since the theorem is not constructive and does not lead to an understanding of how to plot up such a function. There are, however, several simple examples. At the beginning, we consider a function which describes an entropy behavior depending on generalized coordinates when the system is deflected from an equilibrium state. It follows from the maximum entropy condition that the deviation of entropy from its equilibrium value is $\delta S \le 0$. On the other hand, the entropy production is $\dot{S} \ge 0$ in an isolated system. Thus, the entropy S is a Lyapunov function for the isolated system near thermodynamic equilibrium, and an equilibrium state is asymptotically stable (attractor).

Another example of the Lyapunov function can be found in linear non-equilibrium thermodynamics. Here, entropy production plays the role of the Lyapunov function:

$$\dot{S} = \sum_i I_i X_i \ge 0, \quad \frac{d\dot{S}}{dt} \le 0. \tag{1.98}$$

The Lyapunov function can be also introduced for systems that are far from equilibrium. This is particularly evident for the systems in which the flows are stationary. In this case, the entropy production can play again the role of the Lyapunov function:

$$\begin{aligned} &\dot{S} = \sum_i I_i X_i \ge 0, \\ &\frac{d_X \dot{S}}{dt} = \frac{d\dot{S}}{dt} = \int \sum_i I_i \frac{dX_i}{dt}\, dv \le 0. \end{aligned} \tag{1.99}$$

The paper by Kaiser [13] presents the proof of Lyapunov's theorem for some special cases.

There are other stability criteria which can be formulated for solving differential equations.

The solution $\vec{q}(t)$ of the set of dynamical equations (1.94) is called stable (according to Lyapunov) if there exists the quantity $\eta = \eta(t_0, \varepsilon)$ for any t_0 and $\varepsilon > 0$ such that, if the inequality

$$|\vec{q}(t_0) - \vec{q}\,'(t_0)| < \eta,$$

is fulfilled, then we have the inequality

$$|\vec{q}(t) - \vec{q}\,'(t)| < \varepsilon \quad \text{at } t \geq t_0. \tag{1.100}$$

If the condition

$$|\vec{q}(t) - \vec{q}\,'(t)| \to 0$$

holds at $t \to \infty$, the solution is asymptotically stable. We can give a simple interpretation of the condition (1.100). A solution (or motion) is a *Lyapunov stable* one if all solutions (motions) that were in the immediate vicinity of this solution in the initial time remain close enough to it. A solution (motion) is asymptotically stable if all contiguous solutions asymptotically converge to it. The Lyapunov stability imposes rigid constraints on the nature of the solution, because the closeness of trajectories is needed for all t.

A notion of the orbital stability is less rigorous and more useful when considering limit cycles and chaotic trajectories. In the case of *orbital stability* trajectories initially very close to one another can be separated at all other moments in time. Here, a more benign condition replaces the condition (1.100): a minimum distance between the trajectories must be less than some value beforehand specified:

$$|\vec{q}(t) - \vec{q}\,'(t')| < \varepsilon \quad \text{if } t \geq t_0,\ t' \geq t_0. \tag{1.101}$$

A concept of the orbital stability is based on the claim that, if there are two closely adjacent circular trajectories, then the phase points of systems initially close to each other can mutually strongly diverge after a sufficiently long interval of time, for example, due to different periods of rotation. In this case, the solution is not stable in the sense of Lyapunov, but is said to be orbitally stable (Figure 1.7).

In many physically interesting cases, it can be seen that the right-hand side of the dynamical equations (1.94) depends on some set of the parameters B. Let b_k be one of these parameters. If the solution changes by the value $|\delta\vec{q}| \simeq \delta b_k$, while changing the parameter b_k by the value δb_k, then such a solution is referred to as *structurally stable*. When values of the parameter b_k are changed weakly and the phase pattern exhibits some quantitative differences, these values are called ordinary. If small changes in parameter b_k result in qualitative changes of the trajectories, then the values of b_k are called critical values or bifurcation points. The bifurcation points play an important role in structure formation in irreversible processes.

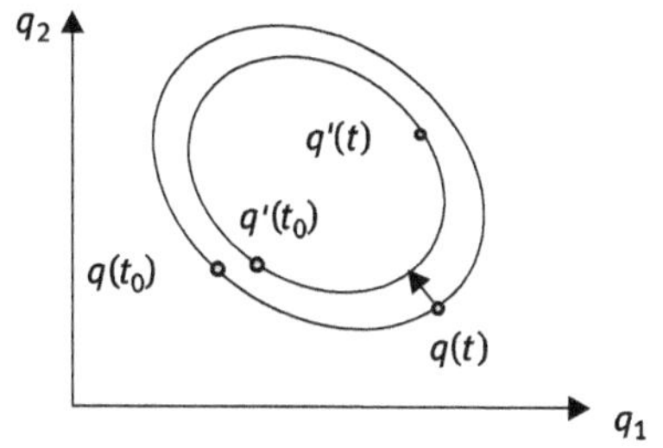

Figure 1.7: To the concept of orbital stability: two phase points $q(t_0)$ and $q'(t_0)$, close to each other at an initial moment of time, strongly diverge by time t, but a minimum distance between orbits (shown by an arrow) has remained small.

A complete structural stability analysis of dynamical systems can be performed only in cases of one or two degrees of freedom. To date, only gradient systems with an arbitrary number of degrees of freedom have been investigated, for which the dynamic equations of motion are written as follows:

$$\frac{dq_i}{dt} = -\frac{d}{dq_i}V(q_1(t), q_2(t), \ldots, q_n(t), B), \quad i = 1, 2, \ldots, n. \tag{1.102}$$

Studies by the famous French mathematician Thom who made many efforts to popularize the catastrophe theory were just devoted to structural stability of dynamical systems described by equations (1.102).

1.3.6 Dynamical systems with one degree of freedom

Consider a system whose dynamics is described by a single variable $q(t)$, which obeys the equation of motion:

$$\frac{dq}{dt} = f(q), \tag{1.103}$$

where $f(q)$ is for some function of the dynamical variable q (system is autonomous, so the right-hand side of equation (1.103) is explicitly independent of time).

In this case, the phase space is a line, and stationary points are determined from solving the equation $f(q) = 0$.

Having started off its motion out of a non-stationary state, a system cannot achieve a steady state over a finite time interval by the uniqueness theorem. Otherwise in violation regardless of the above theorem, equation (1.103) would have two solutions: $q(t)$ and a steady-state solution q^s. Therefore, the system can only asymptotically approach the stationary state if this state is stable. In order to investigate stability of the system near the stationary state points, we expand the function $f(q)$ into a series in the vicinity of these stationary points by restricting ourselves to the first non-vanishing term.

There are only three possible situations shown in Figure 1.8 for a one-dimensional system and therefore, for the stationary points q_1^s and q_2^s one may restrict oneself to the linear terms on the right-hand side of (1.103) while expanding $f(q)$. Then, for deviations of $x(t) = q(t) - q^s$ the following linearized equations can be obtained:

$$\frac{dx(t)}{dt} = f(q^s) + f'(q^s)x(t) + \cdots. \tag{1.104}$$

Taking into account $f(q^s) = 0$ and introducing the notation $f'(q^s) = p$, one obtains

$$\frac{dx(t)}{dt} = px(t); \quad x(t) = x(0)e^{pt}, \tag{1.105}$$

where $x(0)$ is the deviation of the system from a stationary state at the time moment $t = 0$.

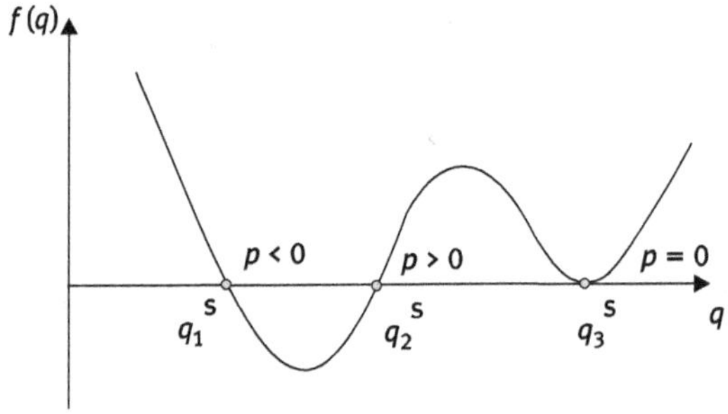

Figure 1.8: Possible types of stationary points for a dynamical system with one degree of freedom.

It follows from the expression (1.105) that, if the condition $df/dq = p < 0$ is met for a stationary point, then such a point is asymptotically stable (point q_1^s in Figure 1.8). But if $df/dq = p > 0$, then any small deviation of the quantity q from the stationary value q^s will grow with increasing time and the system will leave the neighborhood of the stationary point (point q_2^s).

Another case shown in Figure 1.8 (stationary point q_3^s) also corresponds to an unstable node. It is not hard to verify that the last case is valid, if we make an expansion of $f(q)$ to a second term in power of the deviation $x(t) = q(t) - q^s$ in the neighborhood of this point. As a result, we have the equation

$$\frac{dx(t)}{dt} = ax^2(t), \quad a = \frac{1}{2}f''(q^s). \tag{1.106}$$

The solution of equation (1.106) can be written in the form

$$x(t) = \frac{1}{1/x(0) - at}. \tag{1.107}$$

Therefore, if $x(0) < 0$, it follows from formula (1.107) that the point q_3^s is stable, and if $x(0) > 0$, the stationary point q_3^s is unstable. This stationary point should be classified as unstable if we determine stability in the sense of Lyapunov, namely, according to (1.96).

1.3.7 Dynamical systems with two degrees freedom

We now pass on to a qualitative analysis of the autonomous system with two degrees of freedom near the stationary points. We suppose that the system dynamics is described by two variables q_1 and q_2 whose dependence is determined by the equations

$$\frac{dq_1}{dt} = f_1(q_1, q_2),$$
$$\frac{dq_2}{dt} = f_2(q_1, q_2). \tag{1.108}$$

In the special case, equations (1.108) can coincide with a set of the Hamilton equations describing, for example, dynamics of a one-dimensional nonlinear oscillator. Then, it makes sense to consider the variable q_1 as a generalized coordinate, and q_2, as a generalized momentum.

Stationary points of the system (1.108) are determined from the equations

$$f_1(q_1, q_2) = 0, \quad f_2(q_1, q_2) = 0,$$

and the behavior of the phase trajectory can be described by the equation

$$\frac{dq_2}{dq_1} = \frac{f_2(q_1, q_2)}{f_1(q_1, q_2)}. \tag{1.109}$$

Equation (1.109) allows one to find the slope of the tangent to the trajectory in each given point of the phase space and plot the phase pattern based on these points. The direction of the phase point's motion can be found from the set of equations (1.108).

The detailed examination of stability is performed in the same way as in the one-dimensional case, i. e. via linearization of the equations of motion (1.108) in small deviations of the dynamic variables from their stationary values. Let us introduce new dynamic variables $x_1(t) = q_1(t) - q_1^s$ and $x_2(t) = q_2(t) - q_2^s$. After linearizing equation (1.108) with respect to x_1 and x_2, we obtain the following expressions:

$$\frac{dx_1(t)}{dt} = a_{11}x_1(t) + a_{12}x_2(t),$$
$$\frac{dx_2(t)}{dt} = a_{21}x_1(t) + a_{22}x_2(t). \tag{1.110}$$

Elements of the matrix

$$a_{ij} = \left.\frac{df_i(q_1, q_2)}{dq_j}\right|_{q=q^s}$$

are calculated for a stationary point and, therefore, are constant values.

To solve the set of the equations we use the substitution of Euler,

$$x_1(t) = Ae^{pt}, \quad x_2(t) = Be^{pt}.$$

The result is a system of homogeneous linear equations for determining the constants A and B. The condition of consistency of this system implies the equality of its determinant to zero,

$$\begin{vmatrix} a_{11} - p & a_{12} \\ a_{21} & a_{22} - p \end{vmatrix} = 0.$$

Expanding the determinant, one obtains the characteristic equation of a second power in p:

$$p^2 - (a_{11} + a_{22})p + a_{11}a_{22} - a_{12}a_{21} = 0. \tag{1.111}$$

In the general case, equation (1.111) has two complex conjugate roots:

$$p_{12} = \frac{T}{2} \pm \frac{1}{2}\sqrt{T^2 - 4\Delta},$$
$$T = a_{11} + a_{22}, \quad \Delta = a_{11}a_{22} - a_{12}a_{21}. \tag{1.112}$$

Let $\sqrt{T^2 - 4\Delta} \neq 0$ and the roots p_1 and p_2 of the characteristic equation (1.111) be different. Then the general solution of the system of equations (1.110) is a superposition of possible partial solutions and it can be written as

$$x_1(t) = A_1e^{p_1t} + A_2e^{p_2t},$$
$$x_2(t) = A_1K_1e^{p_1t} + A_2K_2e^{p_2t}. \tag{1.113}$$

The A_1 and A_2 constants are determined by initial conditions, and the constants K_1 and K_2 are roots of the equation

$$a_{12}K^2 + (a_{11} - a_{22})K - a_{21} = 0.$$

The last equation can be easily obtained if one assumes that $B = AK$.

The type of a stationary point depends on the roots (1.112) of the characteristic equation (1.111). All in all, there exist six types of stationary points corresponding to six variants. For two-dimensional systems, a schematic representation of the phase patterns is shown in Figure 1.9.

Let us consider these six variants.

(a) $T^2 - 4\Delta > 0$, $\Delta > 0$, $T < 0$. In this case, the roots p_1 and p_2 are real negative numbers. The system performs aperiodic damped motion, approaching the equilibrium position. Such a stationary point is called *an asymptotic stable node*.

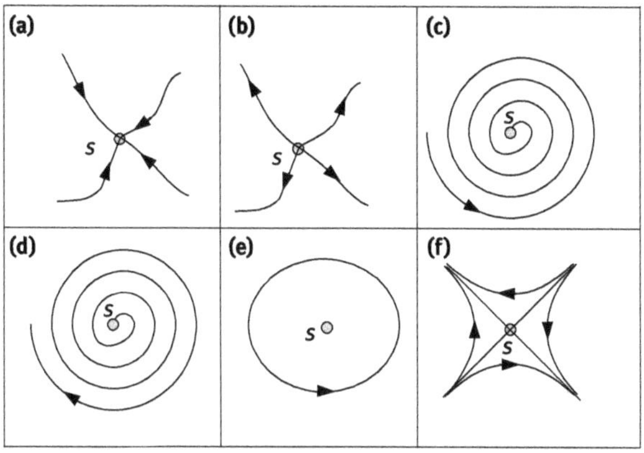

Figure 1.9: The main types of stationary points for a dynamical system with two degrees of freedom: (a) is for the asymptotically stable node; (b) is for the asymptotically unstable node; (c) is for the asymptotically stable focus; (d) is for the asymptotically unstable focus; (e) is for the a stationary point of center type; (f) is for the saddle stationary point.

(b) $T^2 - 4\Delta > 0$, $\Delta > 0$, $T > 0$. In this case, the roots p_1 and p_2 are real positive numbers. The stationary point is unstable. For any fluctuations, leading to the phase point shift from the stationary state, perturbation grows and the system leaves the neighborhood of this point (aperiodic self-excitation). Such a stationary point is called *asymptotic unstable node.*

(c) $T^2 - 4\Delta < 0$, $T < 0$. In this case, the roots p_1 and p_2 are complex numbers with negative real part. The system performs damped oscillations asymptotically approaching the stationary point. The phase pattern of such a system looks like a converging spiral. Such a stationary point is called *a stable focus.*

(d) $T^2 - 4\Delta < 0$, $T > 0$. In this case, the roots p_1 and p_2 are complex numbers with positive real part. The system demonstrates oscillations that grow in amplitude (self-excitation). The phase pattern of such a system resembles a diverging spiral. Such a stationary point is called an *unstable focus.*

(e) $\Delta > 0$, $T = 0$. In this case, the roots p_1 and p_2 are purely imaginary values. The system performs undamped oscillations near a stationary point. The phase pattern represents a closed curve. Such a stationary point is called *a center*. A singular point center is stable in the sense of Lyapunov but is not asymptotically stable.

(f) $T^2 - 4\Delta > 0$, $\Delta < 0$. In this case, the roots p_1 and p_2 are real numbers of opposite sense. Trajectories of the phase point are hyperbolic curves divided by separatrices (straight lines in Figure 1.9(f)). Such a stationary point is called *a saddle point*. Since at $t \to \infty$ the phase trajectories go to infinity, the saddle point is an unstable stationary point. Such systems are characterized by the presence of two states (a trigger type system). The above classification is based on the assumption that there are two different solutions of the characteristic equation (1.111). Such points are referred to as *points of the general position.*

There may well be situations when $\Delta = 0$. Such special points are called *multiple points*. Behavioral analysis of the phase trajectories in the vicinity of multiple special points may turn out to be sufficiently complicated, but fortunately, this case need not be analyzed, because "a small perturbation" (the system parameter change) is the cause of decomposition of the multiple singular points into two or more singular points of general position.

Problem 1.4. For the above Lotka–Volterra "predator–prey" model we have

$$\begin{aligned} \dot{n}_1 &= \gamma_1 n_1 - \beta n_1 n_2, \\ \dot{n}_2 &= \beta n_1 n_2 - \gamma_2 n_2, \end{aligned} \tag{1.114}$$

with the numerical values of parameters $\gamma_1 = 0.3712$, $\beta = 0.0097$, $\gamma_2 = 0.3952$. It is necessary to: determine the stationary values of populations n_1^s, n_2^s; find a solution of the characteristic equation for the linearized model; determine types of stationary points in this model; find a solution of the set of the linearized equations motion for small initial deviations of the population numbers from the stationary values; determine the phase point's motion in the neighborhood of the stationary points; ascertain if the types of the stationary points in this model are dependent on the numerical values of the parameters.

Solution. Stationary values of the populations can be found from the equations

$$\begin{aligned} \gamma_1 n_1 - \beta n_1 n_2 &= 0, \\ \beta n_1 n_2 - \gamma_2 n_2 &= 0. \end{aligned}$$

This set of equations has two solutions. The first solution is obviously: $n_1^s = 0$, $n_2^s = 0$. The second stationary point corresponds to values: $n_1^s = \gamma_2/\beta = 40.7423$; $n_2^s = \gamma_1/\beta = 38.2680$. Consider first the behavior of the system near the second stationary point.

We introduce new dynamical variables such as $x_1(t) = n_1(t) - n_1^s$ and $x_2(t) = n_2(t) - n_2^s$. By linearizing equation (1.114) with respect to x_1 and x_2, we get

$$\begin{aligned} \dot{x}_1 &= (\gamma_1 - \beta n_2^s)x_1 - \beta n_1^s x_2, \\ \dot{x}_2 &= \beta n_2^s x_1 - (\gamma_2 - \beta n_1^s)x_2. \end{aligned} \tag{1.115}$$

Comparing the expressions (1.115) and (1.110), it is easy to see that in given case we have $a_{11} = a_{22} = 0$, $a_{12} = -\beta n_1^s = -\gamma_2$, $a_{21} = \beta n_2^s = \gamma_1$, $T = 0$, $\Delta = a_{11}a_{22} - a_{12}a_{21} = \gamma_1\gamma_2 > 0$.

So, both roots of the characteristic equation are purely imaginary and the stationary point is a stable center. The phase trajectory of the linearized set of equations (1.115) presents a circular curve whose center is a stationary point. The phase trajectory of the original system (1.114) will also resemble a circumference in the case of small deviations (see Figure 1.6).

Using the general solution (1.113), we write down a parametric equation of trajectories in the vicinity of this stationary point. In the case under consideration the roots are $p_{1,2} = \pm i\omega$, $\omega = \sqrt{\gamma_1\gamma_2}$, $K_{1,2} = \pm i\sqrt{a_{21}/a_{12}} = \pm i\sqrt{\gamma_1/\gamma_2}$. In general case the constants A_1 and A_2 are complex quantities, and therefore, they can be presented in the form $A_1 = a_1 + ib_1$, $A_2 = a_2 + ib_2$.

Substituting the results obtained into the formula for the general solution of the system (1.113) and separating out the real part, we have the parametric equation trajectory:

$$x_1(t) = (a_1 + a_2)\cos\omega t + (b_2 - b_1)\sin\omega t,$$
$$\sqrt{\gamma_2/\gamma_1}x_2(t) = -(a_1 + a_2)\sin\omega t + (b_2 - b_1)\cos\omega t. \quad (1.116)$$

By changing the scale along the axis x_2, we will introduce a new variable $x_2^*(t) = \sqrt{\gamma_2/\gamma_1}x_2(t)$. The constants $(a_1 + a_2)$, and $b_2 - b_1$ are determined by the initial conditions: $(a_1 + a_2) = x_1(0)$, $b_2 - b_1 = x_2^*(0)$. Now it is easy to verify that, the condition is met:

$$(x_1(t))^2 + (x_2^*(t))^2 = (x_1(0))^2 + (x_2^*(0))^2,$$

if we raise both sides of each equation of (1.116) to the second power and sum up. Thus, we see that the equation of trajectory of the phase point is a circumference.

Now we can pass on to the behavior analysis of the system near the stationary point $n_1^s = 0$, $n_2^s = 0$. In this case, a set of the linearized equations has the form

$$\dot{x}_1(t) = \gamma_1 x_1(t),$$
$$\dot{x}_2(t) = -\gamma_2 x_2(t). \quad (1.117)$$

Therefore, $a_{11} = \gamma_1$, $a_{22} = -\gamma_2$, $a_{12} = a_{21} = 0$, $\Delta = -\gamma_1\gamma_2 < 0$, and in accordance with the above classification, this fixed point is an unstable saddle point. The phase pattern of the system in the vicinity of this stationary point for different initial conditions is shown in Figure 1.10 and in the particular case the coordinate axes x_1 and x_2 are separatrices.

From the above analysis, it follows that the types of stationary points for the Lotka–Volterra model do not depend on the particular numerical model parameters. Thus, either an unstable saddle or a stable center will be observed in the system depending on selection of initial conditions.

To analyze the behavior of solutions of dynamic equations in the vicinity of the stationary point in the systems with an arbitrary number of degrees of freedom the same method of the motion equation linearization is used. We suppose that a stationary point has the coordinates $q_i^s, i = 1, \dots, n$. Then, introducing the deviation of the dynamical coordinates from the stationary values of $x_i(t) = q_i(t) - q_i^s$, instead of the initial dynamical equations (1.94) we obtain a set of equations for the deviations of coordinates from stationary values in which we retain only terms up to second order

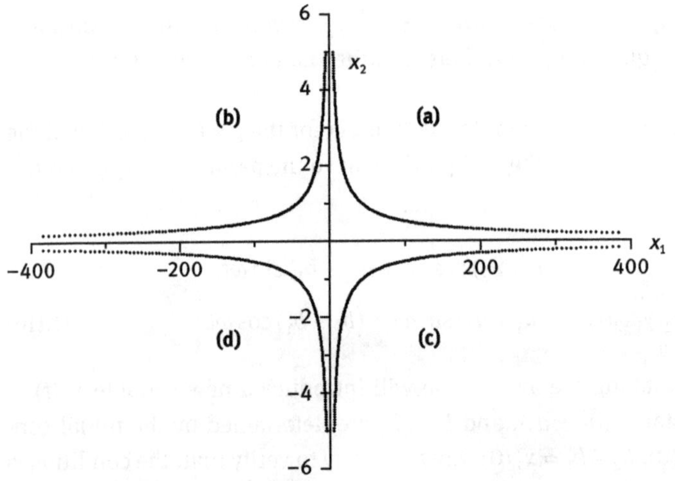

Figure 1.10: Phase pattern of the linearized system (1.114) in the vicinity of the stationary point $n_1^s = 0, n_2^s = 0$: (a) $x_1(0) = 5, x_2(0) = 5$; (b) $x_1(0) = 5, x_2(0) = -5$; (c) $x_1(0) = -5, x_2(0) = 5$; (d) $x_1(0) = -5, x_2(0) = -5$.

with respect to $x_i(t) = q_i(t) - q_i^s$:

$$\frac{dx_i(t)}{dt} = \sum_{j=1}^{n} a_{ij}x_j(t) + f_i^{(2)}(x_1(t), x_2(t), \dots, x_n(t)), \tag{1.118}$$

where $f_i^{(2)}$ is the function that contains the deviations $x_i(t)$ at least in the second power.

The terms of the second order and over in the expression (1.118) being neglected, the dynamics of the deviations $x_i(t) = q_i(t) - q_i^s$ is determined by solving a set of linear equations. The technique for solving such equations in two variables was discussed earlier. The roots of the characteristic equation are still determined by the equation

$$\det|a_{ij} - p\delta_{ij}| = 0.$$

In this case, the following assertions are valid:
1. If all roots of the characteristic equation have negative real parts, the stationary point $x_i = 0$ is stable regardless of the type of the function $f_i^{(2)}$.
2. If at least one of the roots of the characteristic equation has a positive real part, the stationary point is unstable regardless of the type of the function $f_i^{(2)}$.
3. If there are no roots with positive real parts, but there are only purely imaginary roots, then stability of the stationary point depends on the function $f_i^{(2)}$.

More detailed information about modern methods and problems of the dynamic description of nonlinear systems is presented in the lectures of Kuznetsov [14] that have been read for student-physicists at the Saratov State University. A large number of books of Russian and foreign authors on the issues covered here can be found in the electronic library, located on the site http://www.scintific.narod.ru/nlib/.

1.3.8 Dynamic chaos

The main purpose of excursion into the field of nonlinear dynamics is to explain how time-reversible dynamic equations, in particularly the Hamilton equations, can describe the irreversible behavior of real systems. Is there a probability of the irreversible behavior in the dynamic equations? Or should this idea be brought in from outside?

The development of the dynamical theory in the second half of the last century resulted in the remarkable fact that consists in the discovery of dynamic chaos. At first glance, chaos appears to be incompatible with the definition of a dynamical system whose state is unmistakably determined at any moment of time in accordance with the initial state.

In fact, there is no contradiction here because a hypersensitivity to initial conditions of the systems demonstrating chaos takes place. An arbitrarily small change in the initial conditions leads to a finite change in the system's state after a sufficiently long period of time. For this reason, in spite of the fact that the system still remains dynamic, it is impossible to forecast dynamics of its development accurately.

Edward Lorentz, an American meteorologist became the first to discover chaotic mode in systems of a few degrees of freedom when studying convective fluid motion in the experiments of Benard. He has succeeded in converting a set of hydrodynamic equations for density, velocity, and temperature of volume of a fluid to a set of three relatively simple equations for variables x, y, and z. Properties of the fluid under the experimental conditions are given in the Lorenz model with three parameters σ, r, and b:

$$\begin{aligned} \dot{x} &= -\sigma(x - y), \\ \dot{y} &= -xz + rx - y, \\ \dot{z} &= xy - bz. \end{aligned} \tag{1.119}$$

Consider the qualitative behavior of solutions of this set of equations without discussing the physical interpretation of the dynamical variables and input parameters. A more detailed derivation of the Lorenz equations can be found in the book mentioned earlier, written by Kuznetsov [14].

The qualitative behavior exhibited by solutions turns out to be dependent on the parameter r. If the condition $0 < r < 1$ is met, then there is a stable node at the origin. If the condition $r > 1$ is fulfilled, the attractor loses its stability and there appear two stationary points:

$$x_{1,2} = \pm\sqrt{b(r-1)}, \quad y_{1,2} = \pm\sqrt{b(r-1)}, \quad z_{1,2} = r - 1.$$

They characterize stationary convection of shafts of a fluid with opposite directions of rotation. The phase pattern of the system near one of these points is shown in Figure 1.11(a). It is worth emphasizing that the phase trajectory behaves in a strange manner at $r > r_{cr}$. This phase trajectory approaches one of the stationary points, makes a few loops and goes to another fixed point. The phase portrait of such a system for the values of the parameters $\sigma = 10$, $b = 2.666$, $r = 26.7$ is depicted in Figure 1.11(b). The number of loops around each node in each series is different, unpredictable, and depends on exact data of initial conditions.

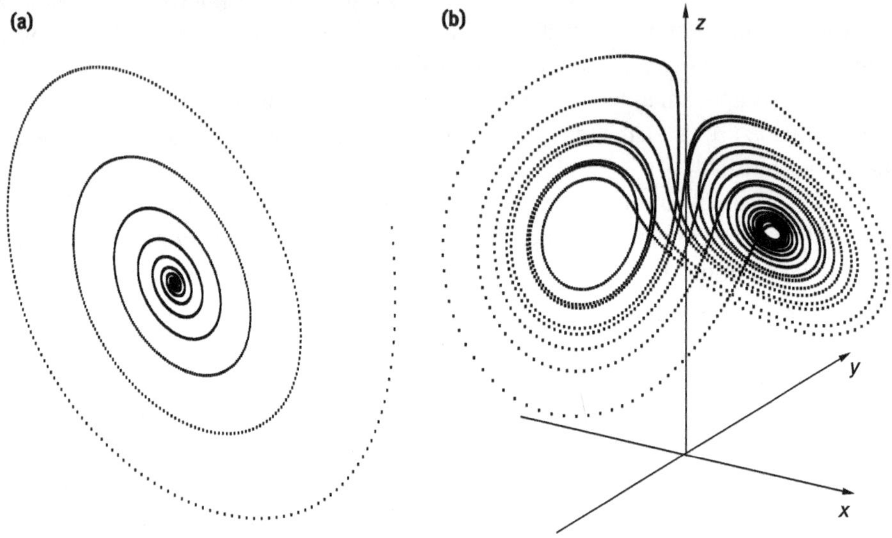

Figure 1.11: Phase pattern of the system (1.119) for the parameters: $\sigma = 10$, $b = 2.666$; (a) is for a stable focus at $r = 10$; (b) is for a strange attractor at $r = 26.7$.

Another remarkable feature of this system proved to be a contraction of volume in phase space over time and formation of a strange attractor. The phenomenon is accounted for by the fact that any classical system which obeys the Hamilton equations is conservative. This means that, if we take a small volume element $d\Omega_0$ of the phase space of the system containing some number of phase points at an initial time, the phase points in the process of evolution will have been in some volume $d\Omega_t = d\Omega_0$ by time t. This statement is well known in classical mechanics as Liouville's theorem.

Conservative systems present a sufficiently narrow class of dynamical systems. Most dynamical systems, describing real processes, are non-conservative and do not preserve phase volume. The system of Lorenz equations (1.119) has a direct bearing to such systems.

Let us consider a small element of phase volume of the system (1.119). The element is equal to the product of small changes of coordinates $\triangle\Omega = \triangle x \triangle y \triangle z$. We find the relative rate of change of the volume

$$\frac{1}{\Delta x\Delta y\Delta z}\frac{d}{dt}(\Delta x\Delta y\Delta z) = \frac{1}{\Delta x\Delta y\Delta z}(\Delta\dot{x}\Delta y\Delta z + \Delta x\Delta\dot{y}\Delta z + \Delta x\Delta y\Delta\dot{z})$$
$$= \frac{\Delta\dot{x}}{\Delta x} + \frac{\Delta\dot{y}}{\Delta y} + \frac{\Delta\dot{z}}{\Delta z}. \tag{1.120}$$

The above formula shows that the relative rate of change of the phase volume is determined by divergence of a velocity field of phase points.

In general, summarizing the result (1.120) we may write down a simple formula to determine the rate of change of the relative phase-space volume over time:

$$\frac{1}{\Omega}\frac{d\Omega}{dt} = \sum_i \frac{d\dot{x}_i}{dx_i} = \operatorname{div}\vec{v}, \tag{1.121}$$

where x_i, $i = 1, 2, \ldots, n$ is the set of the dynamic variables describing the system, $\vec{v}$ is the velocity vector of phase points in phase space.

If the system is conservative, the equality $\operatorname{div}\vec{v} = 0$ is realized. A dynamical system is referred to as *dissipative* if the condition $\operatorname{div}\vec{v} < 0$ is met.

For the system of the Lorenz equations a velocity vector of phase points is determined by right-hand sides of each equation (1.119):

$$v_x = -\sigma(x-y); \quad v_y = -xz + rx - y; \quad v_z = xy - bz,$$

but this vector has the divergence $\operatorname{div}\vec{v} = -\sigma - 1 - b$. Since σ and b are positive quantities, then, solving the equation

$$\frac{1}{\Omega}\frac{d\Omega}{dt} = \sum_i \frac{d\dot{x}_i}{dx_i} = -\sigma - 1 - b,$$

one obtains

$$\Omega_t = \Omega_0 e^{-(\sigma+1+b)t}. \tag{1.122}$$

It follows that all the phase points are concentrated at some volume which contracts to zero over time. Actually, this means that the phase flow in the three-dimensional Lorenz model generates a set of points whose dimension is less than three (the Hausdorff dimension of this attractor proves to be fractional and equals to 2.06). One of the signs which makes the attractor strange is the fractional dimension of the set of the points to which trajectories are attracted. We will discuss later how the Hausdorff dimension of the attractor can be determined empirically.

No matter how close the phase points have been spaced initially, they become separated from each other by a finite distance for a certain time interval. This is another feature of the strange attractor. In other words, there is a hypersensitive dependence to the initial conditions. It implies that the dynamic description of this system is impossible. In essence, the emergence of dynamic chaos is one of the prerequisites to go over to a statistical description. It is important to note that dynamic chaos is an intrinsic property of systems themselves. This property is not associated with an external factors effect. The emergence of dynamic chaos in the problem of convective motion in a fluid is not something exceptional. Firstly, there are enough many other problems of nonlinear dynamics which are reduced to the Lorenz model, in particular a problem of the transition of a single-mode laser into a lasing regime. Secondly, dynamic chaos arises also in simple Hamiltonian systems, for example in a system, modeled by two interacting nonlinear oscillators of Eno–Eyeless [14].

Thus, the concept of entropy S may be introduced for systems demonstrating dynamic chaos. In fact, entropy is a measure of the incompleteness of knowledge about the system's state:

$$S \sim -\sum_{i=1}^{n} P_i \ln P_i,$$

where P_i is the realization probability of the i-th state. If there is complete determinateness and the probability of finding the system in a state is equal to 1, the entropy is zero and maximum under conditions of a complete indeterminateness when all states are equiprobable.

As far back as 1954 Kolmogorov and Sinai have introduced the concept of entropy for dynamical systems. Let the dynamics be described by a set of differential equations. Let us define a distance $d(t)$ between two phase points in the phase space by the relation

$$d(t) = |\vec{x}_1(t) - \vec{x}_2(t)|.$$

The Kolmogorov–Sinai entropy S_{KS} for dynamic systems can be defined by the relation:

$$S_{KS} = \lim_{\substack{d(0)\to 0 \\ t\to\infty}} \frac{1}{t} \ln\left[\frac{d(t)}{d(0)}\right]. \tag{1.123}$$

It follows from this definition that, if phase points have been spaced close to each other initially and stand next close each other at the following moments of time or if the distance between these points increases, but not exponentially, then $S_{KS} = 0$. If dynamic chaos is realized and the condition

$$d(t) \simeq d(0)e^{\lambda t} \tag{1.124}$$

is met, where $\lambda > 0$, then the Kolmogorov–Sinai entropy takes a positive value. It is important to note that the Kolmogorov–Sinai entropy is a dimensional quantity that is proportional to the rate of information loss about the system. In principle, the inverse magnitude $1/S_{KS}$ determines time of chaotization when the dynamic description of the system becomes meaningless.

How can the probability of the appearance of a strange attractor be determined in accordance with the form of the dynamic equations? It is easy enough to answer this question, if we can manage to establish the connection of the indicator λ in equation (1.124) with eigenvalues of the characteristic equation of the linearized system (1.118). In the general case, one can always construct a normal coordinate system for which the matrix a_{ij} in equation (1.118) is diagonal. The real parts of the characteristic equation $\det|a_{ij} - p\delta_{ij}| = 0$ and $\lambda_i = \operatorname{Re} p_i$ are referred to as *Lyapunov exponents*. Obviously, the number of different roots of this equation coincides with the dimension of the matrix. Thus, the range of eigenvalues of a_{ij} determines also the range of characteristic values of the Lyapunov exponents.

The geometric meaning of the Lyapunov exponents can be easily understood. Let us imagine some small spherical region of characteristic radius ε_0 and filled with phase points in the normal coordinate space. In the course of time each phase point will move along its trajectory and the spherical region will be deformed. Then, if the values of λ_1, λ_2, and λ_3 of the Lyapunov exponents are known for this system, the full sphere will be turned into an ellipsoid with semi-axes l_1, l_2, and l_3 after a time t from the beginning of evolution. The semi-axes are

$$l_1 = \varepsilon_0 e^{\lambda_1 t}, \quad l_2 = \varepsilon_0 e^{\lambda_2 t}, \quad l_3 = \varepsilon_0 e^{\lambda_3 t}.$$

In the given case concerning attractors, the Lyapunov exponents possess the following important properties.

Firstly, the sum of Lyapunov exponents is equal to the divergence of flow of the velocities of the phase points:

$$\sum_i^k \lambda_i = \sum_i^k \frac{d\dot{x}_i}{dx_i}.$$

Therefore, the sum of the Lyapunov exponents for a dissipative system is always negative, and for a conservative system is zero.

Secondly, an attractor, different from a fixed point or node, must have at least one Lyapunov exponent equal to zero. This Lyapunov exponent characterizes the motion along a direction of no contraction of the phase points.

Consider a two-dimensional case. Here, if both Lyapunov exponents are negative, the contraction of the phase points in a node occurs. But if there is a limit cycle, this means that the phase points are concentrated in a confined region of the phase space. Such a situation takes place only when the average distance between them does not suffer a change, which in its turn implies the vanishing of one of the Lyapunov exponents. The book [14] reviews a more rigorous evidence of this statement.

For a one-dimensional system only singular points can be as attractors, if the condition $\lambda < 0$ is valid. Consequently, strange attractors are not possible in one-dimensional systems because here the phase points converge into a cluster rather than diverge. Two-dimensional systems have two types of attractors: stable fixed points and limit cycles. If both Lyapunov exponents λ_1 and λ_2 are negative, the phase points are contracted into a node. If one of the Lyapunov exponents is negative and another is equal to zero, the appearance of another type of attractors take place, namely, a limit cycle. No other attractors are in general possible when we address two-dimensional systems.

In three-dimensional systems, different combinations of signs for the Lyapunov exponents are possible:
1. $\{-,-,-\}$ – an attractive node;
2. $\{0,-,-\}$ – a limit cycle;
3. $\{0,0,-\}$ – a two-dimensional torus;
4. $\{+,0,-\}$ – a strange attractor.

Note that the sequence order of the signs does not matter, as combinations of the characters that differ only by order are identical. Let us draw attention to the fact that in the three-dimensional and more than three-dimensional systems strange attractors can also occur. In this case, the initial volume filled by phase points at an initial moment of time is stretched in one direction, compressed in the other direction, and remains unchanged along the third one.

The appearance of strange attractors is one of the possible mechanisms of generating chaotic dynamics. In the next sections of this chapter we will get acquainted with other scenarios of chaotic behavior of dynamic systems.

1.3.9 Dynamic chaos in one-dimensional mappings

Up until now we have had to deal with dynamical systems whose evolution is determined by differential equations of motion. There is another possibility to describe dynamics of a system by using finite difference equations. In this case a time step is assumed to be some finite value. It is easy to arrive at a finite-difference equation by analyzing the relation between coordinates of a phase point at successive passage of it through a Poincaré section. Consider only one particular and rather simple case of one-dimensional phase space when its mapping is given by the recurrence relation

$$x_{n+1} = f(x_n), \tag{1.125}$$

where $f(x) = rx(1-x)$ is a function dependent on of a single parameter r.

A mapping defined by the recurrence relation (1.125) is referred to as the *logistic mapping*. Even this simple case is very useful for understanding the challenges faced when studying dynamic chaos. The mapping defined by (1.125) transforms points of the segment $[0, 1]$ into points of the segment $[0, r/4]$. Therefore, if the condition $r \leq 4$ is met, then all points of the mapping lie on the segment $[0, 1]$.

The function $f(x)$, obviously has a maximum equal to $r/4$ in point $x = 1/2$. Stationary points of the mapping can be found from the following condition $x_c = f(x_c)$. Substituting the explicit expression of the function, one obtains the equation for determining the stationary values of x_c:

$$x_c^2 - x_c + \frac{1}{r} x_c = 0.$$

This implies that there are two stationary points $x_c = 0$ and $x_c = 1 - 1/r$. Since $0 \leq x \leq 1$, there is a stationary point $x_c = 0$ for $r < 1$. At the point $r = 1$ there occurs bifurcation, and there appear two stationary solutions:

$$x_c^{(1)} = 0, \quad x_c^{(2)} = 1 - \frac{1}{r}.$$

Let us determine which of the stationary points is stable for $r > 1$. For this, it is necessary to define a small deviation of the dynamic variable Δx_n from a stationary value and linearize the recurrence relation (1.125) in the vicinity of the stationary point. The result is the recurrent relation for small deviations from steady-state values:

$$\Delta x_{n+1} = r(1 - 2x_c)\Delta x_n. \tag{1.126}$$

If the value of $|r(1 - 2x_c)| < 1$, then the sequence (1.126) converges to a stationary point, and if it is greater than 1, then it leaves the vicinity of x_c. It follows that, if $r > 1$, the stationary point $x_c^{(1)} = 0$ is unstable and the stationary point $x_c^{(2)}$ is stable. It is worth noticing that the stability testing of the stationary point reduces to the calculation of the derivative function f at the stationary point:

$$\left|f'(x)\right|_{x=x_c} = \left|r(1 - 2x_c)\right| < 1.$$

If the absolute value of the derivative of the function at the stationary point is less than 1, then such a stationary point is stable.

Figure 1.12 shows a so-called bifurcation diagram in which the numerical values of the stationary points x_c are plotted along the Y axis depending on the parameter r. The first bifurcation, as it has been already indicated, occurs at the point $r = 1$. The second bifurcation occurs at the point $r = 3$ (see Figure 1.12). There are two stable stationary solutions for $3 < r < 1 + \sqrt{6}$ that satisfy the equations

$$\begin{aligned} x_c^{(1)} &= r x_c^{(2)}(1 - x_c^{(2)}), \\ x_c^{(2)} &= r x_c^{(1)}(1 - x_c^{(1)}). \end{aligned} \tag{1.127}$$

A solution of this set of equations can be easily obtained numerically by using, for example, a package of symbolic and numerical computations in Maple. The stationary

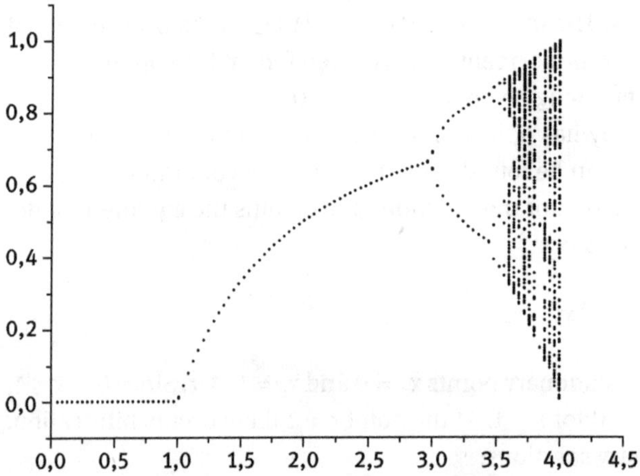

Figure 1.12: Bifurcation diagram of the logistic mapping: coordinates of the stationary points are plotted along the Y axis; the parameter r is displayed on the horizontal axis.

points $x = 0$ and $x = 1 - 1/r$ for $r > 3$ are unstable and therefore they are not depicted in Figure 1.12.

The next doubling bifurcation occurs at the point $r = 1 + \sqrt{6} \simeq 3.45$. The double-stable cycle at this point is replaced by a fourfold stable cycle:

$$\begin{aligned} x_c^{(1)} &= rx_c^{(2)}(1 - x_c^{(2)}), \\ x_c^{(2)} &= rx_c^{(3)}(1 - x_c^{(3)}), \\ x_c^{(3)} &= rx_c^{(4)}(1 - x_c^{(4)}), \\ x_c^{(4)} &= rx_c^{(1)}(1 - x_c^{(1)}). \end{aligned} \tag{1.128}$$

The dynamics of the logistic mapping for $r = 3.46$ is represented in Figure 1.13 by the *Lamerey diagram*. The straight line $y = x$ and the function describing the right-hand side of the logistic mapping $y = rx(1 - x)$ for $r = 3.46$ are shown. Suppose that the value $y \simeq 0.4$ has been obtained at some iteration step. This value is marked by the character 1 in Figure 1.13.

Let us find graphically the value of $x = y$, which should be inserted into the function $y = rx(1 - x)$ at the next iteration step. To do this, we draw a horizontal line to intersect with the line $y = x$. If we draw a vertical line from this point to intersect the curve $y = rx(1 - x)$, we will get the value $y \simeq 0.8$ at the next iteration step. Continuing this construction further, one obtains four stationary solutions marked with the numbers 1, 2, 3, 4 in Figure 1.13. These solutions will be repeated in sequence.

As r is further increased, the period doubling bifurcations will be repeated up to the value $r = r_\infty \simeq 3.5699$, at which there arises an attractive (stable) cycle to be an infinitely large period as all cycles with periods of $2^m, m = 1, 2, \ldots,$ become unstable. The dynamics becomes irregular at values $r_\infty < r < 4$ and aperiodic trajectories take

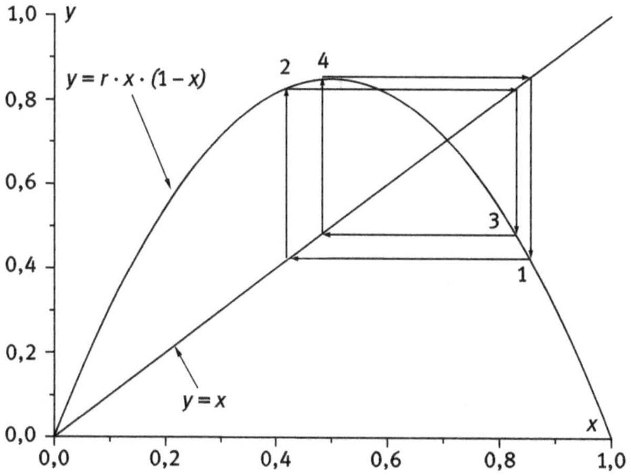

Figure 1.13: Formation of the fourfold cycle of the logistics mapping for $r = 3.46$.

place which cannot be reduced to cycles and in the same time at $r = 4$ dynamic chaos appears in the system.

At $r = 4$ the mapping $x_{n+1} = 4x_n(1-x_n)$ has an exact solution that can be presented as follows:

$$x_n = \sin^2(\pi\Theta_n) = \frac{1}{2}(1 - \cos(2\pi\Theta_n)), \quad \Theta_n = 2^n\Theta_0. \tag{1.129}$$

Thus, an initial angle is multiplied by two for sequential mappings. Measuring angles in radians, it would be advisable to restrict oneself to considering the initial angles in the interval $0 \le \Theta_0 \le 1$. Then the initial angle can be represented via a binary system of calculations:

$$\Theta_0 = 0 + \frac{a_1}{2} + \frac{a_2}{4} + \frac{a_3}{8} + \cdots = \sum_{\nu=0}^{\infty} a_\nu 2^{-\nu}, \tag{1.130}$$

where the coefficients a_ν are equal to zero or unity. Such a representation of the initial angle allows one to see that the sequential mappings can be obtained from the initial mapping by shifting the decimal just one position to the right. For example, having defined an arbitrary angle $\Theta_0 = 0.10100110\ldots$, one obtains the sequence of iterations $\Theta_1 = 1.0100110\ldots$, $\Theta_2 = 10.100110\ldots$, $\Theta_3 = 101.00110\ldots$. It is obvious that this sequence can be continued infinitely. Each new value of x_n is determined by valid digit which stands in the next place of the initial value Θ_0. By virtue of the periodicity of solutions of (1.129) the integer part of the value Θ_n does not contribute and therefore it can be dropped.

If an origin point is set up arbitrarily and values of significant digits are random, then a phase point will visit the neighborhood of any point of the interval $[0, 1]$ countless times. Essentially, this statement is equivalent to the statement of system ergodicity. Chapter 3 focuses at an ergodicity condition of systems in more detail.

Such systems whose behavior is completely determined by an initial value (code) may provide clues to understanding how the genetic code works. The number of specified significant digits is responsible for the number of time periods, for which behavior of a system can be predefined. If an error in the nth place of the binary number was equal to $\varepsilon = 1/2^n$, the system completely "forgets" its initial state after n temporal cycles and demonstrates a random behavior.

This phenomenon is observed by in numerical experiment. It is clear that the solution of (1.129) is cyclic at some initial angles, e. g. $\Theta_0 = 1/3$, $\Theta_0 = 1/5$, $\Theta_0 = 1/9$. In particular, there is a cycle with a period of 2 as the value of x_n periodically takes either the value of $x_n \simeq 0.345$, or the value $x_n \simeq 0.905$ when $\Theta_0 = 1/5$. But, when setting the angle, the error quite rapidly accumulates and information about the initial angle is completely lost after some number of iterations. The number of iterations depends on the accuracy of initial conditions (Figure 1.14).

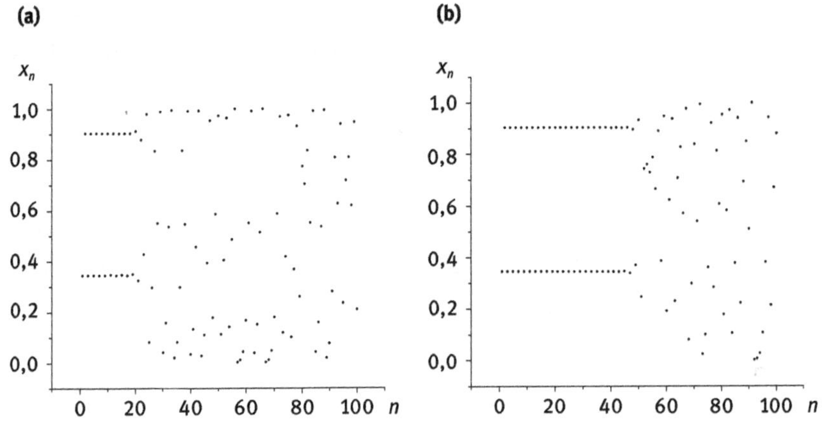

Figure 1.14: Emergence of chaos in the mapping (1.129) for the initial angle $\Theta_0 = 1/5$: (a) is for initial state is given with an accuracy of 8–9 decimal places; (b) is for initial state is given with an accuracy of 19–20 decimal places.

Concluding the brief acquaintance with the peculiarities of the one-dimensional logistic mapping, we should mention how a dimension of the point set of this mapping can be determined. All the more so, as it has been already mentioned above, exactly the same difficulty arises while analyzing the dimension of strange attractors in other problems. Here, we confine ourselves to a qualitative discussion of the problem. More detailed information is presented in the book by Schuster [16].

To determine the dimension of a set of imaging points that result from successive bifurcations of doubling the period 2^m, $m \to \infty$ is the most simple. It turns out that the set is self-similar, possessing a fractal structure. Besides, the dimension of this set is not equal to 1 and represents by itself a fractional value, of 0.543 [16].

Self-similar fractal multitudes are well known in mathematics. Among them the simplest multitude is *a set of Cantor*. The procedure of constructing the Cantor set begins by taking a line segment of unit length and repeatedly deleting the middle third of each line segment. The first three steps are depicted in Figure 1.15. The Sierpinski carpet is a two-dimensional generalization of the Cantor set. The Sierpinski sponge can serve as a case in point typical of three-dimensional space [14, 16].

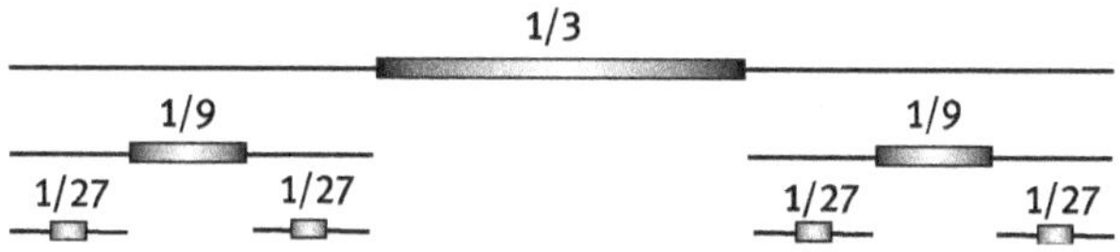

Figure 1.15: Construction of the Cantor set: crosshatched segments of the straight lines are discarded (fraction above implies the length of the segments deleted from the straight line).

Let us find the length l of the removed part of the unit segment while constructing the Cantor set. Using the formula for the sum of a geometric progression, one obtains

$$l = \frac{1}{3} + \frac{2}{9} + \frac{4}{27} + \cdots = \frac{1}{2}\sum_{k=1}^{\infty}\left(\frac{2}{3}\right)^k = \frac{1}{2}\frac{a_1}{1-q} = \frac{1/3}{1-2/3} = 1,$$

where $a_1 = 2/3$ is for the first summand of the geometric progression, $q = 2/3$ the denominator of the progression.

As far as the length of the removed part is equal to 1, the dimension of the Cantor set must not be an integer. The following procedure allows one to determine the dimension of a fractal set in the case of a phase space. Let A be a set of states in an n-dimensional phase space. Let $N(\varepsilon)$ be the number of the n-dimensional cubes of side ε needed to completely cover the set so that these cubes should contain all points of this set. We count these cubes. Then, the dimension $d(A)$ of the set A of phase points can be determined by the formula

$$d(A) = \lim_{\varepsilon\to 0}\left(\frac{\ln N(\varepsilon)}{\ln(1/\varepsilon)}\right). \tag{1.131}$$

It is easy to show that formula (1.131) gives correct results for the regular sets with dimension 1, 2 or 3. Consider a one-dimensional set of points that is a straight line of unit length. Then, $N(\varepsilon) = 1/\varepsilon$ segments of the length ε are needed to cover all points of this segment. Applying the formula (1.131), one obtains $d = 1$.

Similarly, one can easily find out that this formula gives the correct results in two-and three-dimensional cases. To calculate the dimension of the Cantor set, the formula (1.131) should be applied. In this case, the set is covered by intervals whose lengths are equal to $(1/3)^m$, $m = 1, 2, 3, \ldots$ and the number of the intervals for covering the set is respectively equal to $2, 4, 8, \ldots$. Therefore, in the case of the Cantor set the following holds: $\varepsilon = (1/3)^m$, $N(\varepsilon) = N(m) = 2^m$. The application of the formula (1.131) to the Cantor set yields

$$d = \lim_{m\to\infty}\left(\frac{m\ln 2}{m\ln 3}\right) \simeq 0,631.$$

Proceeding similarly, one can determine also a dimension of other fractal sets.

To determine on the basis of formula (1.131) the dimension of a set such as a strange attractor, numerical experiments may be used. To do this, we count the number $N(\varepsilon)$ hypercubes with side length ε that contain the phase points and of the are required to cover the phase space. Then, the procedure for counting the cubes is repeated for different cube sizes. Each side of the cubes can be diminished, for example, by factor of two, four, eight, etc. The results obtained should be presented graphically by plotting the values of $\ln(1/\varepsilon)$ and $\ln N(\varepsilon)$ along the X axis and Y axis, respectively. If the graph formed by plotting the points admit an approximation via a straight line, the tangent of the angle made by the straight line and the X-axis will be approximately equal to the dimension of the set of the points of this attractor. Naturally, all the calculations must be automated.

Concluding this chapter, it should be emphasized once more that irreversible behavior and self-organization by no means are an alternative to the dynamic description. These phenomena are inherent in full to those dynamical systems, in which dynamic chaos is realized. Dynamic chaos in dissipative systems is associated with strong instability of nonlinear dynamical systems. Besides, dynamic chaos is probable in systems with few degrees of freedom. As to systems with dynamic chaos, a set of points in phase space has a fractal structure. Until now it is not clear how this fact should be considered in a statistical description of properties of non-equilibrium systems.

The beginning of Chapter 3 is devoted to emergence of dynamic chaos in Hamiltonian systems and to noninvertible behavior of quantum systems. Recent results in research on chaotic behavior of quantum systems can be found in the monograph by Stockman [15].

1.4 Problems to Chapter 1

1.1. The plasma discharge is generated in an electron-ion system with a characteristic length of ≈ 1 cm. Temperature estimates of the plasma give values of 10^4 K. The electron energy relaxation is determined by electron-ion collisions. The order of magnitude of the relaxation time is $\tau = 10^{-11}$ s. Can we regard the electronic system as being locally equilibrium under such conditions? To obtain the correct result, one should assume that the velocity of electrons is defined by their average kinetic energy.

1.2. Determine both the temperature gradient and the cooling rate causing the violation of the local equilibrium principle for electrons in creating a metal film from the melt. One should assume that the mean free path of the electrons at the Fermi surface is 10^{-5} cm, the velocity being 10^8 cm/s, and the melting point is 10^3 K.

1.3. Derive a continuity equation for energy density using the definition of the latter for a system consisting of n particles

$$E(\vec{r},t) = \sum_{i=1}^{n} \frac{mv_i^2}{2} \delta(\vec{r} - \vec{r}_i(t)).$$

Give a definition for the energy flow density (an analogue of the expression (1.9)).

1.4. The set of n independent macroscopic parameters $\alpha_1, \alpha_2, \dots, \alpha_n$ defines the weakly non-equilibrium state of an adiabatic isolated system. Construct a general expression for entropy production in such a system to express it through generalized fluxes and generalized thermodynamic forces.

1.5. There are two free-external-environment subsystems having the energy-related and the particle interchange. The energy of the first subsystem is $E_1^0 + \alpha$, the second one is $E_2^0 - \alpha$; E_1^0, E_2^0 are energies of the first and second subsystems in steady state, respectively, α is the parameter characterizing the deviation from equilibrium. Similarly, the number of particles of the first subsystem is equal to $n_1^0 + \beta$, and the second one – $n_2^0 - \beta$; β being a non-equilibrium parameter.
Prove that entropy production of the whole system can be written in terms of small deviations from equilibrium as

$$\frac{dS}{dt} = \left(\frac{1}{T_1} - \frac{1}{T_2}\right)\frac{d\alpha}{dt} - \left(\frac{\zeta_1}{T_1} - \frac{\zeta_2}{T_2}\right)\frac{d\beta}{dt},$$

where T_1, T_2; ζ_1, ζ_2 is for temperatures and chemical potentials of the first and second subsystems, respectively.

1.6. Using the relation (1.20) prove that the Onsager reciprocal relations are a special case of the more general relations

$$\frac{\partial I_i}{\partial X_k} = \frac{\partial I_k}{\partial X_i}.$$

1.7. In a thermodynamic system the generalized forces $X_1, X_2, \dots, X_m$ are kept constant due to external influences, and the rest of them $X_{m+1}, X_{m+2}, \dots, X_n$ are not controlled.
Prove using the Prigogine minimum entropy production principle that in this case, the generalized fluxes $I_{m+1}, I_{m+2}, \dots, I_n$ thermodynamically conjugate to the uncontrolled generalized forces are equal to zero.

1.8. Describe an experiment that allows heat generated by the Joule effect and heat generated by the Thomson effect to be measured.

1.9. Electrical oscillations in a circuit containing active elements are described by a set of the Van der Pol equations

$$\frac{dU}{dt} = U'; \quad \frac{dU'}{dt} = -\gamma(1 - U^2)U' - U,$$

where γ is a parameter that takes both positive and negative values.
Applying the linearization method for equations of motion in respect to small parameter deviations from a steady state (see the expressions (1.110)–(1.113)), analyze stability of the Van der Pol system and find all possible types of stationary points.

1.10. It is well known that the Sierpinski carpet is a generalization of the Cantor set in two dimensions. This fractal set is the result of the following procedure: a square is divided into nine smaller equal squares and the central piece is removed for the first iteration; then the remaining eight squares are each divided into nine equal squares again, and the middle square is again ejected for the second iteration. This procedure is repeated.
Depict the first two steps of the construction of the above set on a sheet of paper. Prove using the definition of (1.131) that the fractal dimension of the Serpinski carpet is equal to ≈ 1.893.

2 Brownian motion

2.1 The Langevin equation for a Brownian particle

2.1.1 Nature of motion of a Brownian particle. Random forces

Chaotic motion of small solid particles with characteristic size R of the order of the wavelength of visible light suspended in a fluid is referred to as *Brownian motion*. This phenomenon was discovered by Robert Brown in 1827, when he observed chaotic motion of particles of pollen in a drop of water through a microscope.

In 1905 Einstein developed a quantitative theory of Brownian motion. Shortly after Einstein's paper on Brownian motion, in 1908 Langevin derived a simple enough phenomenological equation, using the concept of random forces acting on a Brownian particle. This equation allows one to reproduce the results found by Einstein. Since the concept of random forces has gained widespread acceptance in non-equilibrium statistical mechanics, it would be expedient to start consideration of a Brownian motion model with a derivation of Langevin's equations.

Let us assume that a Brownian particle has mass m and is a spherically symmetric particle with a characteristic size R. In this case, according to Stokes' law a friction force $\vec{F}_{fr} = -\gamma \cdot \vec{v}$ acts on such a particle moving with velocity $\vec{v}$ in a liquid. Here $\gamma = 6 \cdot \pi \cdot R \cdot \eta$, η is the shear viscosity coefficient. However, a force should not be overlooked caused by elastic collisions between molecules of the fluid and the particle. As far as the fluid is assumed to be homogeneous and isotropic, the resultant force of these elastic collisions can be related only with random fluctuations in their number. In other words, both magnitude and direction of this force $f(t)$ are random time-dependent variables.

As far as the medium is isotropic but the particle is spherically symmetric, it suffices to consider one-dimensional motion along X-axis. If we remain within the framework of classical mechanics, we can write down the equation of motion:

$$m \cdot \ddot{x} + \gamma \cdot \dot{x} = f(t). \tag{2.1}$$

In writing equation (2.1) it should be taken into account the fact that the resistance force is oriented in a direction opposite to the velocity. This equation is called the Langevin equation that involves a source of random forces on the right-hand side.

It is easy to obtain a formal solution of Langevin's equation. From the standpoint of the theory of differential equations, equation (2.1) is an inhomogeneous first-order linear equation relating to the velocity $v_x = \dot{x}$. The solution of this equation is the superposition of a general solution of the homogeneous equation and a particular solution of the inhomogeneous equation:

$$v_x(t) = v_x(0)e^{-\gamma/mt} + \frac{1}{m}\int_0^t e^{-\gamma/m(t-t_1)} f(t_1)\,dt_1. \tag{2.2}$$

https://doi.org/10.1515/9783110727197-002

Naturally, the formal solution of the expression (2.2) does not give any new results as long as the function $f(t)$ is unknown. To take a step further it is necessary to explore the properties of the functions $f(t)$.

The Langevin source is a random function of time. Therefore, if we choose a sufficiently long time interval T, the average value of this force will be zero:

$$\langle f(t)\rangle = \frac{1}{T}\int_0^T f(t_1)\,dt_1 = 0.$$

There are at least two time scales in this task. One of them is related to an interaction time of a single molecule with a Brownian particle. This characteristic time τ_0 can be estimated as the ratio of the radius of action of intermolecular forces $r_0 \sim 10^{-8}$ cm to the thermal velocity of molecules $v_t \sim 10^5$ cm/s:

$$\tau_0 \simeq \frac{r_0}{v_t} \simeq \frac{10^{-8}\,\text{cm}}{10^5\,\text{cm/s}} \simeq 10^{-13}\,\text{s}.$$

Another characteristic time is associated with the velocity relaxation time of a Brownian particle in a liquid. From formula (2.2) it follows that, if there are no random forces, the particle velocity

$$v_x(t) = v_x(0)e^{-\gamma/mt}$$

relaxes with the relaxation frequency $1/\tau \simeq \gamma/m$; $\tau \simeq m/\gamma$.

After plugging the typical values of quantities which, for example, Jean Perrin obtained in his experiments into the above expression, but namely: $R \simeq 10^{-7}$ m, $m \simeq 10^{-17}$ kg, viscosity of water $\eta \simeq 10^{-3}$ kg/m s, $\gamma = 6\pi R\eta \simeq 2\cdot 10^{-9}$ kg/s, one gets magnitude that is significantly greater than τ_0: $\tau \simeq 10^{-8}$ s. Therefore, the equations of motion (2.1) need to be averaged over a time interval τ when considering the motion of a Brownian particle over periods greater than τ, without taking into account individual collision events.

Let us consider a behavior of a random force when such averaging is required to apply. Obviously, the average value of this force $\langle f(t)\rangle$ at time interval τ is zero. However, the vanishing average does not give a complete description of the random variable. No less important characteristic is a correlation of values of this quantity at different times. To describe an interrelation the values of the random force, taken at different moments of time, we use the pair correlation function $K_f(t_1, t_2)$, which can be defined as follows:

$$\begin{aligned} K_f(t_1,t_2) &= \langle f(t_1)f(t_2)\rangle - \langle f(t_1)\rangle\langle f(t_2)\rangle \\ &\equiv \langle (f(t_1) - \langle f(t_1)\rangle)(f(t_2) - \langle f(t_2)\rangle)\rangle. \end{aligned} \tag{2.3}$$

It is obvious that in virtue of the homogeneity of time, the pair correlation function (2.3) depends only on the difference of time arguments $t_1 - t_2$: $K_f(t_1, t_2) = K_f(t_1 - t_2)$.

In the given case the averages of the random force are equal to zero: $\langle f(t_1)\rangle = \langle f(t_2)\rangle = 0$. Therefore we can believe that

$$K_f(t_1, t_2) = \langle f(t)f(0)\rangle,$$

where $t = t_1 - t_2$.

Based on the fact that there are two very different time scales in this problem, we may attempt simulating the behavior of the correlation function $K_f(t_1, t_2) = \langle f(t_1 - t_2)f(0)\rangle$. As far as the duration of each collision event has the order of τ_0, the random forces $f(t_1)$ and $f(t_2)$ are correlated only in the case when $t = t_1 - t_2 \le \tau_0$. When approximating the temporal behavior in the roughest manner, the correlation function is thought to be constant and equal to some variable C, if $|t| \le \tau_0$, and equal to zero if $|t| \ge \tau_0$:

$$K_f(t, 0) = \begin{cases} C & |t| \le \tau_0, \\ 0 & |t| \ge \tau_0. \end{cases} \tag{2.4}$$

The temporal behavior of the random force $f(t)$ is schematically shown in Figure 2.1(a), and the graph for the time-dependent correlation function $K_f(t, 0)$ given by equation (2.4) is depicted in Figure 2.1(b).

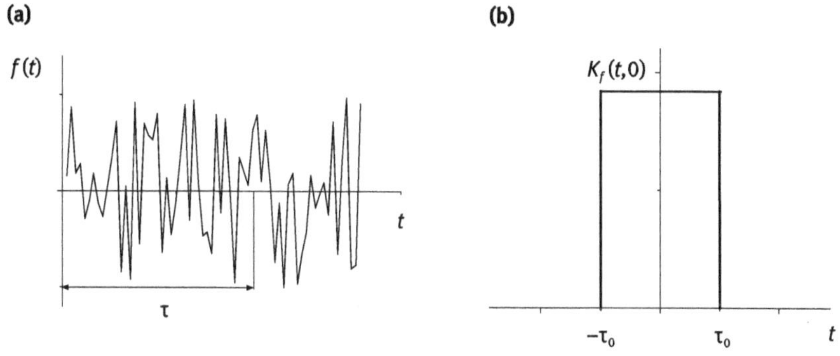

Figure 2.1: Temporal behavior of the random function $f(t)$ (a) and the correlation function $K_f(t, 0)$ (b).

Since the length of time interval τ_0 can be considered as very small in the rough time scale τ, then, simplifying the formula (2.4), we can accept that the random forces correlate only when their arguments are the same:

$$K_f(t_1 - t_2) = C\delta(t_1 - t_2). \tag{2.5}$$

Now we can pass on to the Fourier representation of $K_f(\omega)$ for the correlation function of random forces:

$$K_f(\omega) = \int_{-\infty}^{\infty} K_f(t)e^{i\omega t}\,dt = C\int_{-\infty}^{\infty} \delta(t)e^{i\omega t}\,dt = C. \tag{2.6}$$

The quantity $K_f(\omega)$ is often referred to as *spectral density* of the correlation function of random forces. From formula (2.6) it follows that $K_f(\omega) = C$ and it does not depend on the frequency. A random process whose a spectral density of the pair correlation function does not depend on frequency, is called *white noise*. White noise gets its name from white light in which the spectral density is the same for all frequencies.

From formula (2.6) it follows that the constant C determines the spectral intensity of the random force. The intensity can be expressed via the average square of the velocity fluctuations. It has already been indicated that the average value of the random force is zero at $t > \tau$. Therefore, by averaging equation (2.2) over the time interval $t \sim \tau$, one obtains

$$\langle v_x(t)\rangle = v_x(0)e^{-t/\tau}, \quad \frac{1}{\tau} = \frac{\gamma}{m}.$$

This implies that the velocity fluctuation is completely determined by random force:

$$v_x(t) - \langle v_x(t)\rangle = \frac{1}{m}\int_0^t e^{-\gamma/m(t-t_1)} f(t_1)\,dt_1. \tag{2.7}$$

Let us define the quantity that would be of the mean square velocity fluctuations:

$$D_v(t) = \langle (v_x(t) - \overline{v_x(t)})^2\rangle. \tag{2.8}$$

To simplify this formula we have used the notation $\langle v_x(t)\rangle \equiv \overline{v_x(t)}$. Substituting the expression (2.7) for the velocity fluctuations into formula (2.8), one obtains

$$D_v(t) = \frac{1}{m^2}\int_0^t\int_0^t e^{-(t-t_1)/\tau}e^{-(t-t_2)/\tau}K_f(t_1 - t_2)\,dt_1\,dt_2. \tag{2.9}$$

Given that $K_f(t_1 - t_2) = C\delta(t_1 - t_2)$, we perform the temporal integration over the argument t_2 in the expression (2.9):

$$D_v(t) = \frac{C}{m^2}\int_0^t e^{-2(t-t_1)/\tau}\,dt_1 = \frac{C}{m^2}e^{-2t/\tau}\int_0^t e^{2t_1/\tau}\,dt_1 = \frac{C\tau}{2m^2}(1 - e^{-2t/\tau}). \tag{2.10}$$

The expression (2.10) defines the square of velocity of a chaotically moving Brownian particle:

$$D_v(t) = K_v(t,t) = \langle (v_x(t) - \overline{v_x(t)})^2\rangle = \frac{C\tau}{2m^2}(1 - e^{-2t/\tau}).$$

This fact can be used to determine the constant C. The estimation of the magnitude of the spectral intensity of the random force C is obtained when applying the theorem on the uniform distribution of energy over degrees of freedom of the chaotic motion. The energy equal to $k_B T/2$ corresponds to one-degree of freedom, where k_B is the Boltzmann constant, T is the absolute temperature. Therefore, for times $t \gg \tau$, we have

$$\frac{m}{2}\left\langle \left(v_x(t) - \overline{v_x(t)}\right)^2 \right\rangle = \frac{C\tau}{4m} = \frac{k_B T}{2}.$$

This yields a simple estimate for the quantity C:

$$C = \frac{2k_B T m}{\tau} \equiv 2k_B T\gamma.$$

Now the final form of an expression for the pair correlation function of random forces appears as

$$K_f(t_1 - t_2) = 2k_B T\gamma\delta(t_1 - t_2), \quad K_f(\omega) = 2k_B T\gamma. \tag{2.11}$$

Equation (2.11) is well known in the literature as one of existing formulations of the fluctuation-dissipation theorem that relates the fluctuations of random forces in an equilibrium state with parameters, characterizing the irreversible processes. For example, the parameter γ determines a momentum relaxation frequency of a Brownian particle in a fluid.

The quantity (2.8) found above stands for the pair correlation function of velocity fluctuations of a Brownian particle which are taken at the same time: $D_v(t) = K_v(t, t)$. One can generalize this result and determine the correlation function of fluctuations of velocity component, taken in different moments of time:

$$K_v(t_1, t_2) = \left\langle (v(t_1) - \overline{v(t_1)})(v(t_2) - \overline{v(t_2)}) \right\rangle. \tag{2.12}$$

Problem 2.1. Using the expression (2.12), it is necessary to determine the temporal behavior of the pair correlation function of the velocity components of a Brownian particle.

Solution. We use the expression (2.7) for velocity fluctuations of a Brownian particle and plug it into (2.12). As a result of this operation, we have succeeded in expressing the correlation function of velocity components via the correlator of random forces $K_f(t_1, t_2)$:

$$K_v(t_1, t_2) = \frac{1}{m^2}\int_0^{t_1} dt \int_0^{t_2} dt' e^{-(t_1 - t)/\tau} e^{-(t_2 - t')/\tau} \langle f(t) f(t') \rangle. \tag{2.13}$$

Now we apply the fluctuation–dissipation theorem (2.11) according to which $\langle f(t)f(t') = K_f(t - t') = 2k_B T\gamma\delta(t - t')$. Substituting this result into the expression (2.13) and integrating over t', one gets

$$K_v(t_1, t_2) = \frac{2k_B T\gamma}{m^2} e^{-(t_1 + t_2)/\tau} \int_0^{t_1} dt e^{2t/\tau} = \frac{k_B T\tau\gamma}{m^2} e^{-(t_2 - t_1)/\tau}(1 - e^{-2t_1/\tau}). \tag{2.14}$$

Given that the mean square of the thermal velocity of a Brownian particle is equal to $\overline{v^2} = k_B T/m$, and the inverse relaxation time of the speed is $1/\tau = \gamma/m$, the expression (2.14) can be significantly simplified:

$$K_v(t_1, t_2) = \overline{v^2} e^{-(t_2-t_1)/\tau}(1 - e^{-2t_1/\tau}). \tag{2.15}$$

It is easy to see that, if $t_1 = t_2 = t$, then one can obtain the desired result (2.10).

2.1.2 Displacement of a Brownian particle

The displacement of a Brownian particle is easily determined by integrating the expression (2.2) for velocity:

$$x(t) - x(0) = v_x(0) \int_0^t dt_1 e^{-\gamma/m t_1} + \frac{1}{m} \int_0^t dt_1 \int_0^{t_1} dt_2 e^{-\gamma/m(t_1-t_2)} f(t_2). \tag{2.16}$$

Let us find the mean displacement of a Brownian particle located initially at the origin after the moment of time t. Carrying out the averaging of the both sides of the expression (2.16) and taking into account the fact that the average value of the random force is zero $\overline{f(t_2)} = 0$, we find

$$\overline{x(t)} = x(0) + v_x(0)\tau(1 - e^{-t/\tau}); \quad \tau = \frac{m}{\gamma}. \tag{2.17}$$

The formula (2.17) implies that if $t \ll \tau$ then $\overline{x(t)} = x(0) + v_x(0)t$. This means that the displacement of a Brownian particle at $t \ll \tau$ also obey the laws of classical dynamics.

Now let us calculate the variance of the displacement of a Brownian particle $D_x(t) = \langle (x(t) - \overline{x(t)})^2 \rangle$. Having changed the order of integration over the variables t_1 and t_2, we simplify the double integral on the right-hand side of the formula (2.16). Bearing in mind the expression (2.17), as a result, we get

$$x(t) - x(0) = v_x(0)\tau(1 - e^{-t/\tau}) + \frac{1}{m} \int_0^t dt_2 f(t_2) \int_{t_2}^t dt_1 e^{-(t_1-t_2)/\tau}. \tag{2.18}$$

The integral over the variable t_1 on the right-hand side of the expression (2.18) is estimated easily enough. The result is a simple formula for the displacement fluctuation:

$$x(t) - \overline{x(t)} = \frac{\tau}{m} \int_0^t dt_2 f(t_2)(1 - e^{-(t-t_2)/\tau}). \tag{2.19}$$

Substituting the last result in the displacement variance formula for a Brownian particle, one obtains

$$D_x(t) = \langle (x(t) - \overline{x(t)})^2 \rangle = \frac{\tau^2}{m^2} \int_0^t dt_1 \int_0^t dt_2 K_f(t_1 - t_2)(1 - e^{-(t-t_1)/\tau})(1 - e^{-(t-t_2)/\tau}). \tag{2.20}$$

Taking into consideration the δ-shaped nature of the source of random forces

$$K_f(t_1 - t_2) \sim \delta(t_1 - t_2),$$

we can now perform the integration over the variables t_1 and t_2. This yields an easily interpretable formula for the displacement variance of a Brownian particle:

$$D_x(t) = \left\langle \left(x(t) - \overline{x(t)}\right)^2 \right\rangle = \frac{2k_B T\tau}{m}\left(t - 2\tau(1 - e^{-t/\tau}) + \frac{\tau}{2}(1 - e^{-2t/\tau})\right). \tag{2.21}$$

From this expression it follows that $D_x = 0$ within the limit of small times $t/\tau \ll 1$ up to quadratic terms over a small parameter. Moreover, this formula implies a linear growth in variance as a function of time when $t \gg \tau$:

$$D_x(t) = \frac{2k_B T\tau}{m}\left(t - \frac{3}{2}\tau\right).$$

It is largely simple to verify experimentally the formulae for variance displacement of a Brownian particle calculated with respect to the initial coordinate x_0 rather than the average displacement $\overline{x(t)}$. Therefore, it is necessary to convert the variance formula (2.21) to a variance formula where the deviation is computed from the initial coordinate x_0. The formula of this conversion can be easily obtained singlehandedly:

$$\left\langle \left(x(t) - x_0\right)^2 \right\rangle = \left\langle \left(x(t) - \overline{x(t)}\right)^2 \right\rangle + \left(\overline{x(t)} - x_0\right)^2.$$

Given that $\overline{x(t)} - x_0$ can be found from the expression (2.17); one obtains the Langevin formula for the variance of the displacement of a Brownian particle:

$$\left\langle \left(x(t) - x_0\right)^2 \right\rangle = \left\langle \left(x(t) - \overline{x(t)}\right)^2 \right\rangle + (v_0\tau)^2(1 - e^{-t/\tau})^2. \tag{2.22}$$

Consider a behavior of the displacement variance over periods much less and much longer than the characteristic time of the momentum relaxation τ of a Brownian particle. Within limits of short periods $t \ll \tau$, the variance is $D_x(t) = 0$ up to quadratic terms in the parameter t/τ. Therefore, expanding the second term on the right-hand side of the expression (2.22) over the small parameter t/τ, we have

$$\left\langle \left(x(t) - x_0\right)^2 \right\rangle \simeq v_0^2 t^2, \quad \frac{t}{\tau} \ll 1. \tag{2.23}$$

At the limit $t/\tau \gg 1$ the second term on the right-hand side of the expression (2.22) can be neglected and we obtain Einstein's formula for the variance of the displacement of a Brownian particle with respect to the initial position:

$$\left\langle \left(x(t) - x_0\right)^2 \right\rangle \simeq \frac{2k_B T\tau}{m}t, \quad \frac{t}{\tau} \gg 1. \tag{2.24}$$

The above results concerning the temporal behavior of both the variance of the velocity and of the variance of the displacement of a Brownian particle with respect to x_0 are represented in Figure 2.2.

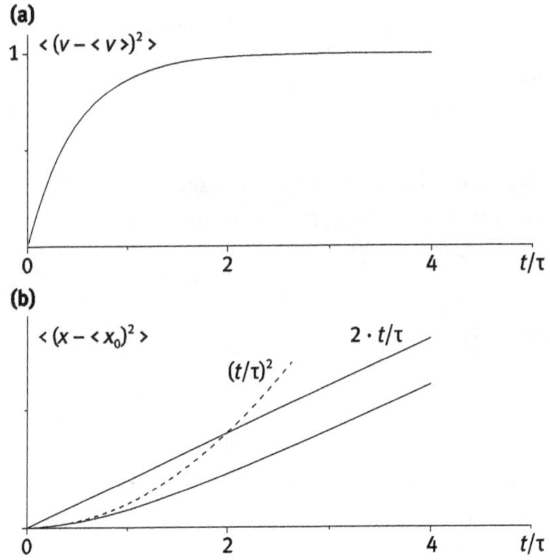

Figure 2.2: Temporal behavior of the velocity variance (a) and the displacement variance $\langle (x(t) - x_0)^2 \rangle$ (b) of a Brownian particle.

The velocity variance is measured in units of $k_B T/m$, and the displacement variance is measured in units of $k_B T/m\tau^2$. Figure 2.2(a) shows the velocity relaxation of Brownian particles to the Maxwellian distribution. It is seen that the Maxwell velocity distribution occurs when transferring from the mechanical description to a description in the rough time scale for the time $t \sim \tau/2$ and the particle "forgets" its the initial velocity v_0. At the same time the displacement of a Brownian particle continues to keep the features of the mechanical behavior, as far as the displacement variance of a Brownian particle is proportional to t^2 when $t \ll \tau$. See (2.23).

Figure 2.2(b) shows with a dotted line the behavior of the displacement variance at $t \ll \tau$. The straight line corresponds to the behavior of the displacement variance when $t \gg \tau$. The lower curve corresponds to the displacement variance, calculated in accordance with the Langevin formula (2.22) under the assumption that $v_0^2 = k_B T/m$ is valid.

Concluding the brief discussion of Brownian motion theory, formulated by Langevin, one should be once again focused on principal issues.

Firstly, the concise description has become possible due to the presence of two time scales in this problem. Instead of calculating the coordinates and velocities of a Brownian particle exactly, we have restricted ourselves to computing the averaged characteristics or moments of the distribution for two limiting cases when $\tau_0 \ll t \ll \tau$ and $t \gg \tau$. In the first case traces of mechanical motion are retained partially and the particle travels as if by inertia at a speed of v_0. Expanding over the parameter t/τ in formula (2.10) and substituting the value of constant C, we obtain for the first limiting case:

$$\left\langle \left(v_x(t) - \overline{v_x(t)}\right)^2 \right\rangle \simeq \frac{2k_B T}{m}\frac{t}{\tau},$$
$$\left\langle (x(t) - x_0)^2 \right\rangle \simeq v_0^2 t^2, \quad \tau_0 \ll t \ll \tau . \tag{2.25}$$

But when $t \gg \tau$, traces of the dynamic description are lost completely and Brownian motion can be treated as a diffusion process:

$$\left\langle \left(v_x(t) - \overline{v_x(t)}\right)^2 \right\rangle \simeq \frac{k_B T}{m},$$
$$\left\langle (x(t) - x_0)^2 \right\rangle \simeq \frac{2k_B T}{\gamma} t, \quad t \gg \tau . \tag{2.26}$$

We have restricted ourselves to calculating the moments of the second order (variance), but at least there is a possibility in principle of calculating the moments of higher order [6, 17]. Herewith, it is worthwhile noticing that only even moments are non-zero: the fourth, sixth, etc.

Secondly, in the problem of Brownian motion the temporal averaging along the phase trajectory was used. But to determine amplitude of spectral intensity of the random force (constant C) the ergodic hypothesis on equality of temporal and phase averages is required to apply. The next chapter discusses the ergodic hypothesis in more detail. As a result, the spectral intensity of the random force proved to be dependent on equilibrium temperature.

Finally, a passage to the description in the rough timeline proved to be possible only because a Brownian particle sufficiently fast "forgets" its the initial speed for the time $t \sim \tau/2$. When considered over such a period, the dynamic description is not only impossible, but meaningless because motion becomes chaotic. Over periods $t \gg \tau$, evolution of the motion of Brownian particles does not obey equations of mechanics and the process is referred to as a *Markov process*, that is, the system state in a given moment does not depend on the prehistory of this system.

2.2 The Fokker–Planck equation for a Brownian particle

2.2.1 Derivation of the Fokker–Planck equation

Consider the evolution of an ideal gas of Brownian particles, using an approach based on the application of the statistical distribution function. The analysis will be carried out in rough temporal scale, assuming the condition $t \gg \tau$ is met. As shown above, the momentum of a Brownian particle is thermalized during this time and the mean value of impulse coincides with the average heat impulse for the time interval $\sim \tau$. For this reason, there is no point maintain dependence of the distribution function on the momentum. Therefore, we assume that the distribution density $\rho(\vec{r}, t)$ depends only on the coordinates $\vec{r}$ and time t. Naturally the density must be normalized to 1:

$$\int \rho(\vec{r}, t)\, d\vec{r} = 1. \tag{2.27}$$

As far as Brownian particles obey the particle number conservation law, and then the distribution function must satisfy the continuity equation:

$$\frac{d\rho}{dt} + \operatorname{div}(\rho\vec{v}) = 0, \tag{2.28}$$

where $\vec{v}$ is the velocity of the Brownian particles.

Staying within the framework of the semi-phenomenological description, we represent the particle flux, which is consists of two parts:

$$\vec{v} = \vec{u}_0 + \vec{u}_{\text{ran}}.$$

The first part of the flux $\vec{u}_0$ is associated with the presence of external forces acting on the particles, and it can be called a regular part of this flux. In writing the Langevin equation (2.1), we assumed that a Brownian particle experiences a resistance force $\vec{F}_{fr} = -\gamma\vec{v}$.

Now, by arguing in a similar way, we assume that, if a Brownian particle is in the field of external forces $\vec{F}_{fr} = -\vec{\nabla}U$ with a potential U, these external forces will cause motion of the particle with the velocity:

$$\vec{u}_0 = \vec{F}_{fr}/\gamma = -\vec{\nabla}U/\gamma.$$

This result is not a corollary of mechanical but of hydrodynamic laws of motion.

The second part of the flux, associated with the random walk, has nature of the diffusion process. In the phenomenological theory diffusion is described by Fick's law. This law holds that the particle flux density $\vec{J}_{\text{ran}} = \rho\vec{v}_{\text{ran}}$ is proportional to the gradient of the particle number density or to the concentration gradient. Then, applying the distribution density function, the Fick law can be written as

$$\rho\vec{u}_{\text{ran}} = -D\vec{\nabla}\rho,$$

where D is the phenomenological diffusion coefficient.

Taking into consideration these two results, we find the expression for the total flux of Brownian particles:

$$\rho\vec{v} = -\left(\frac{\rho}{\gamma}\vec{\nabla}U + D\vec{\nabla}\rho\right). \tag{2.29}$$

Indeed, the quantities D and γ in the expression (2.29) for the particle flux are not independent phenomenological coefficients. There is a simple relationship between them and it can be easy ascertained. In equilibrium, the total flux (2.29) is zero. Therefore, equation (2.29) for the equilibrium state of the system can be regarded as an equation to determine the equilibrium distribution ρ. Equation components for ascertaining the equilibrium distribution ρ can be written as

$$\frac{d\ln\rho}{dx_\alpha} = -\frac{1}{\gamma D}\frac{dU}{dx_\alpha}; \quad \alpha = 1, 2, 3. \tag{2.30}$$

The variables in this equation are separated, so the solution may be written immediately:

$$\rho(\vec{r}) = \text{const}\exp\left\{-\frac{U(\vec{r})}{\gamma D}\right\}. \tag{2.31}$$

However, if the particles are in the field with the potential $U(\vec{r})$, the equilibrium distribution of these particles will have the form

$$\rho(\vec{r}) = \text{const}\exp\left\{-\frac{U(\vec{r})}{k_B T}\right\}. \tag{2.32}$$

Comparing the expressions (2.31) and (2.32), we find an expression for the diffusion coefficient D:

$$D = \frac{k_B T}{\gamma}, \quad \gamma = 6\pi R\eta,$$

where η is the shear viscosity coefficient; R is the radius of a Brownian particle.

Now we can return to the continuity equation (2.28). Substitution of the flux density of particles in the form of (2.29) in the above equation yields the Fokker–Planck equation for the density distribution of Brownian particles:

$$\frac{d\rho}{dt} - \frac{1}{\gamma}\operatorname{div}(\rho \operatorname{grad} U) - \frac{k_B T}{\gamma}\triangle\rho = 0, \tag{2.33}$$

where $\triangle$ is for the Laplace operator. Equation (2.33) allows one to find the distribution function of Brownian particles $\rho(\vec{r}, t)$, if initial and boundary conditions are set for this equation. Interpreting it, this equation describes the relaxation of the non-equilibrium distribution $\rho(\vec{r}, t)$ with respect to the equilibrium Boltzmann distribution, defined by (2.31).

2.2.2 The solution of the Fokker–Planck equation

Consider a simple case, allowing, on the one hand simply to solve the Fokker–Planck equation, on the other hand to obtain a Brownian particle motion picture, corresponding to the limit $t \gg \tau$ in the Langevin equation.

Let the potential of external forces be $U = 0$ and a system is assumed to be infinite and spatially homogeneous. In this case, it is sufficient to analyze the one-dimensional distribution $\rho(x,t)$. We suppose that a Brownian particle at the initial time was at the point with coordinate $x = 0$, and the distribution function density was described by the delta-function $\rho(x,0) = \delta(x)$. Then the further dynamics of the distribution will obey the Fokker–Planck equation:

$$\frac{d\rho}{dt} = \frac{k_B T}{\gamma}\frac{d^2\rho}{dx^2}. \tag{2.34}$$

In addition to the initial condition $\rho(x,0) = \delta(x)$, the solution of equation (2.34) must also satisfy both the normalization condition (2.27) and the distribution density trend condition to zero at an infinite distance from the starting point:

$$\lim_{x\to\pm\infty}\rho(x,t) = 0.$$

To solve equation (2.34) we define the Fourier transform of the distribution density $\rho_p(t)$ by the ratio

$$\rho(x,t) = \frac{1}{2\pi}\int_{-\infty}^{\infty}\rho_p(t)e^{ipx}\,dp \tag{2.35}$$

and we write down equation (2.34) for the Fourier transform $\rho_p(t)$:

$$\frac{d\rho_p(t)}{dt} + \frac{k_B T}{\gamma}p^2\rho_p(t) = 0, \quad \rho_p(0) = 1. \tag{2.36}$$

In equation (2.34) all coefficients are constants and the variables are separated. Therefore, taking into account the initial condition $\rho_p(0) = 1$, we can write down the solution:

$$\rho_p(t) = \exp\left(-\frac{k_B T}{\gamma}p^2 t\right). \tag{2.37}$$

To find the distribution function in coordinate presentation, we should substitute the result found in the determination (2.35):

$$\rho(x,t) = \frac{1}{2\pi}\int_{-\infty}^{\infty} e^{-k_B T/\gamma p^2 t}e^{ipx}\,dp. \tag{2.38}$$

If we perform the integration over p in this formula, one obtains a Gaussian distribution:

$$\rho(x,t) = \frac{1}{\sqrt{4\pi k_B T/\gamma t}}\exp\left(-\frac{x^2}{4k_B T/\gamma t}\right). \tag{2.39}$$

The procedure for estimating such a sort integrals will be discussed later (see Problem 2.2).

If we now take into account the fact that the Gaussian distribution (normal distribution) is determined by two parameters such as the average value of $\overline{x}$ and variance D_x and has the form

$$f(x) = \frac{1}{\sqrt{2\pi D_x}} \exp\left(-\frac{(x-\overline{x})^2}{2D_x}\right),$$

then in comparison to formula (2.39) the average value of $\overline{x}$ for the distribution (2.39) is equal to zero and the variance takes the following form:

$$D_x(t) = \frac{2k_B T}{\gamma} t.$$

The above quantity of D_x coincides with the result (2.26), found from the Langevin equation. The same result may be obtained by calculating the second moment of distribution:

$$D_x(t) = \langle (x-\overline{x})^2 \rangle = \int_{-\infty}^{\infty} (x-\overline{x})^2 \rho(x,t)\, dx.$$

It is interesting to analyze how the distribution (2.39) evolves with increase time t. Graphs of the distribution density function (2.39) for four values of the parameter t/τ are illustrated in Figure 2.3.

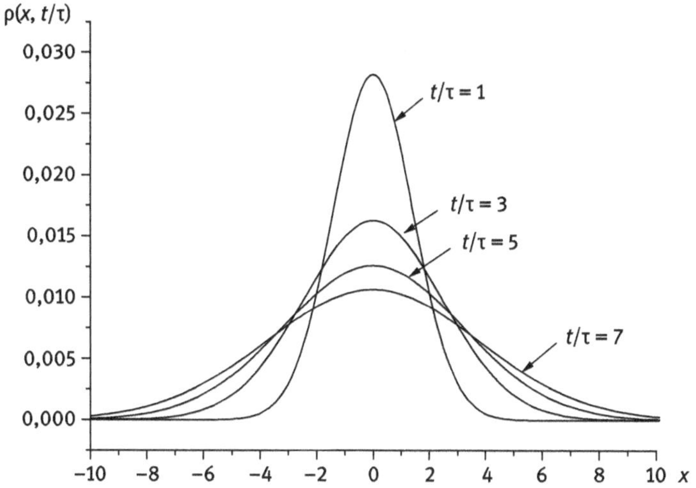

Figure 2.3: Distribution density (2.39) for different values of the parameter t/τ; the quantity x is measured in units of $v_0\tau$.

It is seen that with increase time t the evolution of the distribution is reduced to the broadening of the distribution. It becomes less concentrated, and the probability of finding a Brownian particle far enough away from the starting point increases.

Problem 2.2. Consider the calculation of the integral appearing in Fourier transform of a normal distribution (2.38). We set up the problem as follows: find the characteristic function of the standard normal distribution:

$$f(x) = \frac{1}{\sqrt{2\pi}} \exp\left(-\frac{x^2}{2}\right).$$

It is worth noticing that the Fourier transform $f(p)$ of this distribution is referred to as *characteristic function*:

$$f(p) = \frac{1}{\sqrt{2\pi}} \int_{-\infty}^{\infty} e^{ipx} \exp\left(-\frac{x^2}{2}\right) dx. \tag{2.40}$$

Solution. To find the integral in the definition of the Fourier transform of a normal distribution, we preliminarily estimate the Poisson integral:

$$I_P = \int_0^{\infty} e^{-x^2}\, dx.$$

The simplest and most elegant way of finding the Poisson integral is a reduction of it to the computation of a certain integral in polar coordinates over quadrant area:

$$I_P^2 = \int_0^{\infty} e^{-x^2}\, dx \int_0^{\infty} e^{-y^2}\, dy = \lim_{R\to\infty} \int_0^R \int_0^R e^{-x^2-y^2}\, dx\, dy.$$

The last integral can be regarded as an integral over the circle area with radius R, located in the first quadrant of a coordinate plane. While transferring to the polar coordinate system $x = r\cos\varphi$, $y = r\sin\varphi$, this integral can easily be taken:

$$I_P^2 = \lim_{R\to\infty} \int_0^{\pi/2} d\varphi \int_0^R re^{-r^2}\, dr = \frac{\pi}{4}.$$

It follows that

$$I_P = \int_0^{\infty} e^{-x^2}\, dx = \frac{\sqrt{\pi}}{2}.$$

But a similar integral has the following form:

$$\int_{-\infty}^{\infty} \exp(-(Ax)^2)\, dx = \frac{\sqrt{\pi}}{A}. \tag{2.41}$$

To calculate the integral (2.40) we will try to lead it to the form of (2.41) with the aid of the replacement of variables, having written the power of exponent $-x^2/2 + ipx$ in the subintegral function of the expression (2.40) as

$$-(Ax - B)^2 + C.$$

Comparing these two expressions, we find that $A = 1/\sqrt{2}$, $B = ip/\sqrt{2}$, $C = -p^2/2$. Using the result (2.41), we have

$$f(p) = \frac{1}{\sqrt{2\pi}} \int_{-\infty}^{\infty} \exp(-x^2 + ipx)\, dx = \exp\left(-\frac{p^2}{2}\right). \tag{2.42}$$

Thus, the characteristic function of the standard normal distribution has been found. The above-described method has been also used to evaluate the integral (2.38).

This concludes our brief introduction to the methods of a description of a Brownian particle motion. A more detailed discussion and examples for solving many problems concerning motion of a Brownian particle can be found in a book written by I. A. Kvasnikov [17].

In the next chapter we will return to a justification and an application of the Fokker–Planck equation as far as kinetic equations are concerned.

Concluding the chapter, let us summarize. The problem of motion of a Brownian particle is one of simple tasks of physical kinetics. It allows pictorially seeing how the roughing of a description of the dynamic system can occur. An accurate description of a Brownian particle motion in terms of the equations of classical mechanics is not only impossible but also meaningless, because the system forgets its initial momentum after a sufficiently small interval time. The further movement of this system resembles a diffusion process rather than mechanical movement. We discuss the reason for this phenomenon in detail in Chapter 1 in relation to the dynamics of dissipative systems. The natural question arises of how the roughing of the description of the dynamic systems can occur if they obey dynamic Hamilton equations. Therefore, at the beginning of the next chapter conditions under which a system cannot be described in terms of the dynamic equations of motion will be analyzed. Consequently, for this reason it is a statistical description of such a system that is required.

2.3 Problems to Chapter 2

2.1. Write down the Langevin equation for a harmonic oscillator in the field of a random force $f(t)$. Calculate the average value of the displacement and average momentum of the harmonic oscillator in the field of the random forces.

2.2. Applying the rule of "decoupling" for the correlation of random forces that satisfy the condition $\langle f(t_i)\rangle = 0$:

$$cK_f(t_1, t_2, t_3, t_4) = \langle f(t_1)f(t_2)\rangle \cdot \langle f(t_3)f(t_4)\rangle + \langle f(t_1)f(t_3)\rangle \cdot \langle f(t_2)f(t_4)\rangle + \langle f(t_1)f(t_4)\rangle \cdot \langle f(t_2)f(t_3)\rangle.$$

Prove that the fourth moment of the velocity fluctuation of a Brownian particle relates the particle velocity dispersion $D_v(t)$ by the simple relation

$$\langle v_x(t) - \overline{v_x(t)}\rangle^4 = 3D_v(t),$$

which holds for all variables with a Gaussian distribution.

2.3. Using the Langevin equation, derive equations of motion for the kinetic and the average kinetic energies of a Brownian particle.

2.4. Making use of the solution of the previous problem, define the temporal dependence of the average kinetic energy of a Brownian particle to hold for any arbitrary time point.

2.5. Solve the Fokker–Planck equation for a Brownian particle in a gravitational potential field $U(x) = mgx$, where x is a coordinate, m is mass of the particle, g being the acceleration of gravity provided that the Brownian particle distribution function has originally the form $\rho(0, x) = \delta(x)$. To clarify how the distribution changes over time, plot the sequence of the distributions for different time points (see Figure 2.3). The problems can be reduced to the case in the absence of a potential field by introducing the new variable ξ:

$$x = \xi - \frac{mg}{\gamma}t.$$

3 Kinetic equations in non-equilibrium statistical mechanics

3.1 Description of non-equilibrium systems in statistical mechanics

3.1.1 Integrable and nonintegrable dynamical systems

Unfortunately, insufficient attention is given to non-integrable systems within a university course of lectures on classical mechanics. Such a situation requires a short excursion into mechanics of classical systems.

As is known, a set of Hamilton dynamic equations,

$$\frac{\partial H}{\partial p_i} = \dot{q}_i, \quad \frac{\partial H}{\partial q_i} = -\dot{p}_i, \quad i = 1, 2 \ldots N, \tag{3.1}$$

is referred to as completely integrable if there exists a canonical transformation of the variables q_i, p_i that results in transferring from the generalized coordinates q_i and generalized momenta p_i to the variables J_i, α_i (action-angle) in terms of which the set of equations (3.1) may be written [18]

$$\frac{\partial H}{\partial J_i} = \dot{\alpha}_i, \quad \frac{\partial H}{\partial \alpha_i} = 0, \quad i = 1, 2 \ldots N. \tag{3.2}$$

In equations (3.1) and (3.2) H is Hamilton's function. The special role of such variables as action-angle is that the Hamiltonian in these variables depends only on integrals of motion J_i and does not depend on the angles α_i. Obviously, if these variables are found, the set of equations (3.2) is easily integrated:

$$J_i = J_i(0), \quad \alpha_i = \alpha_i(0) + \omega_i t, \quad \omega_i = \frac{\partial H}{\partial J_i}, \quad i = 1, 2 \ldots N. \tag{3.3}$$

For this reason, the main problem of mechanics is to find an appropriate canonical transformation, leading a system to the form of (3.2). Moreover, individual intuitive notions concerning behavior mechanical systems also have to do with exceptionally integrable systems.

Meanwhile, the number of systems which are integrable is not great. Integrable systems certainly include a system with one degree of freedom. It is true, there are a few particular cases when systems with two- or three degrees of freedom can be reduced to a system with one degree of freedom, consequently to be also integrable. For example, systems consisting of non-interacting particles or a set of harmonic oscillators, reacting on each other according to a harmonic law, may be regarded as systems with two- or three degrees of freedom. The above list of integrable systems has finished. All other systems are non-integrable, and their behavior may be very different

https://doi.org/10.1515/9783110727197-003

from the familiar integrable systems. Strictly speaking, such terms as determinateness, possibility of a dynamic description and time reversibility are referred to only integrable systems.

Combination of two harmonic oscillators interacting among each other according to a law that is not harmonical is the simplest system to address the difference in the behavior of integrable and nonintegrable systems. Such a system is described by the Hamiltonian of Eno–Eyless [12]:

$$H = \frac{1}{2m}(p_1^2 + p_2^2) + \frac{m\omega_1^2 q_1^2}{2} + \frac{m\omega_2^2 q_2^2}{2} + V\left(q_1^2 q_2 - \frac{1}{3} q_2^3\right), \tag{3.4}$$

where p_1, p_2 are the momenta and q_1, q_2 the coordinates, and ω_1, ω_2 the eigen frequencies of the first and second oscillators, respectively. Moreover, all particles are assumed to have the same mass. If the parameter V in equation (3.4) is zero, we get an integrable set of equations of motion. But if the same parameter $V \neq 0$, the set of equations is non-integrable and its solution can be obtained only by making use of numerical techniques for integrating the system of differential equations.

Before passing to the direct analysis of dynamics of the system with Hamiltonian (3.4), we recall the reader some important results of classical mechanics, relating to Hamiltonian systems (see [19]).

We now define the mechanical system state at given moment of time by the position of the phase point in the phase space $6N$ of variables q_i, p_i, $i = 1, 2, \dots, 3N$. In this case, the system evolution can be represented graphically with the aid of trajectory of the phase point in the phase space.

Consider a small region A of the phase space. The dynamics equations of the Hamiltonian (3.1) give a one-parameter group of transformations of the phase space G^t. These transformations shift the phase point $(\vec{q}(0), \vec{p}(0))$ in the new location $(\vec{q}(t), \vec{p}(t))$. Such a transformation is commonly referred to as phase flow and it results in moving the phase points belonging to region A to some region A^t at moment of time t, where $G^t A = A^t$.

As for conservative systems, due to Liouville's theorem [19] a phase flow preserves phase volume. In other words, volume of the region A is equal to the volume of A^t. Based on this theorem, Poincaré formulated a statement which at first glance is paradoxical. If the phase point of a system is in arbitrary small region of the phase space U, it may be in this area U as much as desired times in the process of evolution. This statement known as the Poincaré recurrence theorem, in essence holds that any system in the course of evolution must return to the original state after some time. The proof of the Poincaré theorem can be easily obtained.

Consider images of the phase region U at regular intervals of time τ, i. e., at moments of time $t, t+\tau, t+2\tau, \dots, t+n\tau$. The phase flow will transform the region U at time $t+n\tau$ in the region $G^{t+n\tau}U = U^n$. As far as the volumes of the regions $U^t, U^1, U^2, \dots, U^n$, according to Liouville's theorem, are equal, sooner or later the volumes of U^n and U^m

will overlap, provided that the phase system volume is not equal to infinity. Figure 3.1 schematically shows the evolution of the phase region U and the overlapping of these areas at some moment of time. The images of U arising during the evolution may have a different form, but preserve their volume.

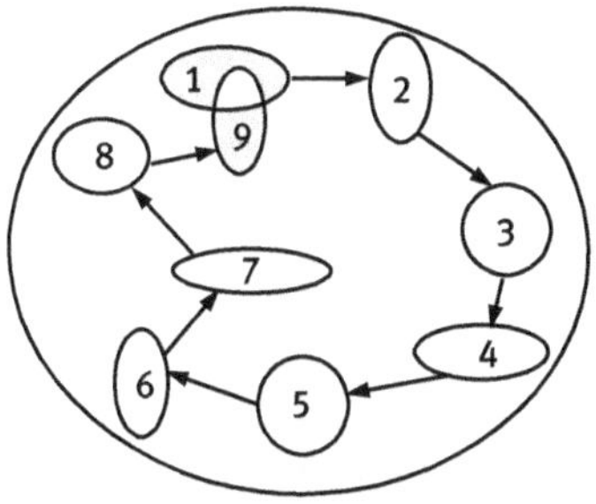

Figure 3.1: The proof of Poincaré's theorem: evolution of a small region of phase space; the partial overlapping of regions 1 and 9 is shown.

Despite the apparent clash of the Poincaré theorem and common sense, there can be several different explanations concerning the paradox about return of a mechanical system in initial state. One of possible explanations is reduced simply to estimate the time of return. Given that a huge number of possible states are equal to the order of $6N!$ and also the change velocity of the phase variables is finite; it is easy to obtain an estimate of the return time for a macroscopic system which exceeds greatly the lifetime of the Galaxy. This explanation was historically the first, but as we will see later, there are other reasons when the Poincaré recurrence theorem is not valid for the systems observed.

3.1.2 The evolution of dynamical systems in phase space

Let us now return again to the matter about behavior of integrable and nonintegrable systems and consider it from the standpoint of the evolution of a small region in the system's phase space. It should be distinguished three typical scenarios of the evolution of the small vicinity of the phase point. It should be noted that sometimes we are going to talk about a small neighborhood of the phase space as a phase point in the phase space.

In the first case, the phase trajectory is a closed line and the system undergoes a periodic motion. A combination of two noninteracting harmonic oscillators with a multiple ratio of eigenfrequencies ω_1 and ω_2 may serve as a case of point of such a system.

Hamiltonian of the system can be obtained if it is assumed that $V = 0$ in the expression (3.4), and frequencies ratio is equal to some rational number. Figure 3.2(a)

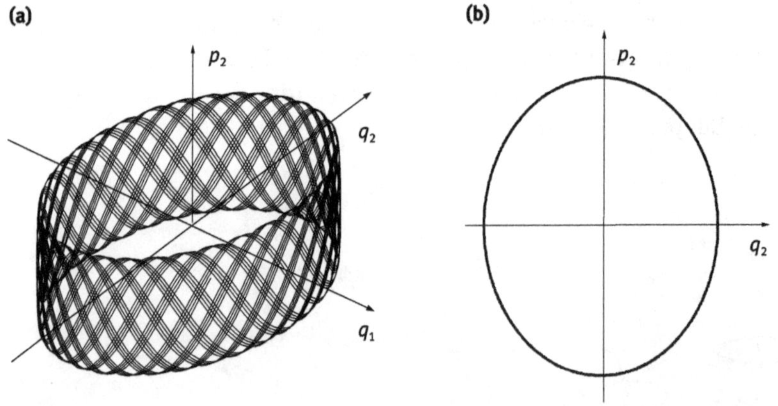

Figure 3.2: The phase portrait of the system (3.4): (a) the case of non-ergodic system, the phase trajectory does not cover the torus; (b) the phase portrait of an ergodic system in the Poincaré cross-section $\omega_1^2 = 65$, $\omega_2^2 = 4$.

shows the surface of constant energy of this system, which is a torus in the space of the variables p_2, q_1, q_2. In addition, the phase trajectory is a closed line, wound on the torus.

Dynamical description is possible for such a system, but statistical description is inappropriate. The system behavior information can be obtained by observing the appearance of phase points in the *Poincaré cross-section*. One of planes of the phase space, for example $q_1 = 0$ may serve as a case in point of such a cross-section (Figure 3.2(b)). If the frequencies are commensurable, then one obtains a discrete set of points in the Poincaré cross-section. But if the frequencies are incommensurable, the set of points where the phase curve "pierces" the plane $q_1 = 0$, will be represented as an ellipse.

Now we consider the case when the ratio of the frequencies of oscillators cannot be reduced to a rational number because the variables (ω_1 and ω_2 are incommensurable). In this case, the phase trajectory is not a closed line, which covers the torus completely. It is this circumstance that allows one to introduce a statistical description of the system.

Let us define the function $\rho(p,q)$ setting density of probability to find a phase point of system in an infinitesimal element of volume $dpdq$ in the neighborhood point, whose position in the phase space is defined by a set of values p, q. To do this, in the phase space of the system we separate out the volume element dp, dq and will record the time-fraction τ, in the course of that the phase point is inside the volume $dpdq$. To simplify the notations, the quantity ensemble p_i, q_i, $i = 1, 2, \ldots, N$ is needed to replace by the letters p and q, respectively. It is obvious that the ratio limit takes the following form:

$$\lim_{t\to\infty} \frac{\tau}{t} = \rho(p,q)dpdq, \tag{3.5}$$

where t is the time of observation of the system. This time defines a probability for detecting the phase point within the volume $dpdq$. From the probability density definition it follows that the probability is normalized to unity:

$$\int_{\Omega} \rho(p,q)\,dp\,dq = 1. \tag{3.6}$$

In the expression (3.6), the integration is performed over the isoenergetic surface Ω $H(p,q) = \text{const}$. In the future, such a quantity $\rho(p,q)$ may be called *the statistical operator* of the system.

If the statistical operator $\rho(p,q)$ is already known, the average value of any physical variable $f(p,q)$ can be found as the mathematical expectation of the quantity $f(p,q)$:

$$\langle f(p,q)\rangle = \int_{\Omega} f(p,q)\rho(p,q)\,dp\,dq, \tag{3.7}$$

where the integration is performed over an available region of the phase space of the system, i. e. over the constant-energy surface. If the system has K integrals of motion besides the energy, the dimension of the hypersurface over which integration is performed will be equal to $6N - K - 1$.

The average value of the quantity $f(p(t),q(t))$ can be also obtained by averaging of this quantity over time:

$$\overline{f} = \lim_{T\to\infty} \frac{1}{T}\int_0^T f(p(t),q(t))\,dt. \tag{3.8}$$

The quantity found by averaging in accordance with formula (3.7), $\langle f(p,q)\rangle$, can be called the statistical average, the quantity $\overline{f}$, calculated by formula (3.8) being the dynamic average.

It is necessary to emphasize that the statistical mechanics of equilibrium systems is based on a very rough simplification of formula (3.7) for the average over a phase space. The cornerstone of statistical mechanics of Gibbs is the hypothesis that holds for the quantity $\rho(p,q) = \text{const}$, if p and q belong to constant-energy surface Ω. Reasons underlying such a rough approximation are hidden in the peculiarity of dynamics of Hamiltonian systems. Therefore, this matter will not be treated until later.

It is usually assumed that the statistical and dynamic averages are equal. Given that the quantity $\rho(p,q)$ is constant on isoenergetic surface and also the normalization condition (3.6) is valid, we have

$$\langle f(p,q)\rangle = \overline{f} = \frac{\int_{\Omega} f(p,q)\,dp\,dq}{\int_{\Omega} dp\,dq}. \tag{3.9}$$

This statement is referred to as *ergodic hypothesis*. In truth, its validity defies strict proof; however, a corollary of the ergodic hypothesis is a possibility to arrange a thermodynamic description of equilibrium systems, which is in agreement well with experiment.

So, if a dynamic system is designed in such a way that a phase trajectory covers a whole constant-energy surface in the course of evolution for a sufficiently long time, then it is possible to statistically describe the system using the statistical operator $\rho(p, q)$. A significant simplification of description arises when it is assumed that $\rho(p, q) = \text{const}$ throughout the hypersurface of constant energy.

Let us pass on to another possible situation when $\omega_1 = \omega_2$ and $V = 1$ in the Hamiltonian (3.4). Then the Hamiltonian equations of motion are not to be integrable and the phase trajectory behavior changes completely. Now an isoenergetic surface of the system is not a torus in the phase space. The phase trajectory of the system obtained as a result of numerical integration of the motion equations is illustrated in Figure 3.3(a).

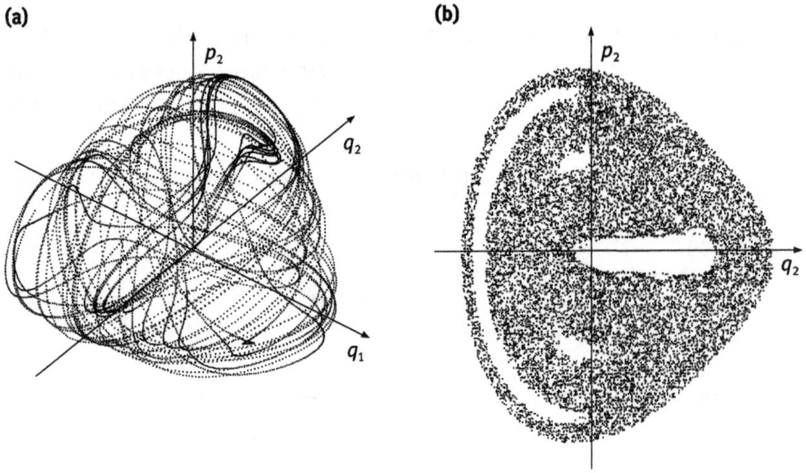

Figure 3.3: The phase portrait of the system (3.4): (a) the parameter $V = 1$, $\omega_1 = \omega_2 = 1$; (b) the phase portrait of the same system in the Poincaré cross-section $q_1 = 0$.

The trajectory resembles tangled threads and is not similar to regular motion of the phase point on the torus surface in the previous case.

More information about behavior of a system can be obtained by observing the appearance of phase points in the Poincaré cross-section, i. e. in cross-section of the phase space with plane $q_1 = 0$. The result of such a numerical experiment is shown in Figure 3.3(b). Each point in this figure corresponds to "a puncture" of the plane $q_1 = 0$ by the phase trajectory as the phase point moves along the positive direction of the axis q_1.

There are as yet virtually no rigorous analytical calculations even for such a simple model, while the results of numerical experiments of various hands point clearly that the model demonstrates a stochastic behavior. This is evidenced by the fact that if we enumerate the points appearing on the display, the sequence of points with adjacent numbers turns out to be randomly scattered over the isoenergetic surface. In integrating equations of motion, reducing the time-step does not change the situation. Simply put, such a system exhibits the stochastic behavior (or so-called dynamic chaos).

Let us now try to understand how dynamic chaos appears in a system described by the Newton equations. Consider a small region of the phase space A. In the case of integrable systems, a phase flow virtually shifts the region A to a new position on the isoenergetic surface and covers it completely with the lapse of time. In contrast, for non-integrable systems, the region A, preserving its volume, is stratified into thin threads and gradually is distributed throughout the isoenergetic surface for some characteristic time, which should be called the mixing time. The concept of mixing can be quantitatively described through a concept of measure.

Let us call the measure of the area A the ratio of volume of the area A to volume of the phase space, available for the system and denote it as $\mu(A)$. During the evolution, the volume of A is replaced by the volume A^t. However, the volume of the area A is the same size of A^t, therefore, it is obvious that $\mu(A) = \mu(A^t)$. We separate out some other arbitrary region B and we will think that it is nonmovable. It stands to reason that small parts of the area A will fall into the area B because of the mixing event. The mixing process will be complete if volume of the overlapping parts of the areas A^t and B, divided by the volume of B, will be equal to the relative volume of A. In the context of the concept of a measure, this condition of the complete mixing can be written as follows:

$$\mu(A) = \lim_{t\to\infty} \frac{\mu(A^t \cap B)}{\mu(B)}. \tag{3.10}$$

The mixing event occurs in such systems, where there is a strong divergence of two phase points located arbitrarily close to each other at some initial moment. Such systems are called unstable. The system instability, in turn, leads to unpredictability in its behavior over time. Indeed, if a position of the phase point is known with some accuracy in the initial moment, i. e. we know that it belongs to some region with the characteristic size ε, it is impossible to say where the phase point will be after some temporal interval t. There is a finite probability for this phase point being anywhere on the constant-energy surface.

When it comes to a divergence of phase points in systems with mixing, this is easy enough to imagine. In this case, one can analyze the behavior of the system's replicas whose initial conditions differ. The divergence of trajectories means that the systems exhibit hypersensitivity to initial conditions. But a question arises: why do we say of chaos in solving a set of differential equations for several particles and analyzing the motion of a single phase point? It would seem that the uniqueness theorem for solutions of differential equations must provide the deterministic behavior to calculate exact coordinates and momenta of all particles making up the system at any time.

Stochasticity here also arises due to extreme sensitivity of the system's dynamics to the assignment of the initial conditions. Unfortunately, we have no chance of analyzing this problem in detail, so we just give only a pictorial example demonstrating the essence of the issue. It is said that a good example is the best sermon.

The Sinai billiard may serve as the simplest model of the stochastic system. This is a flat table bounded by walls; a disk of radius R is located in the center of the table. Another mobile disk with smaller radius r is launched with some initial velocity $\vec{v}$ from an arbitrary place of the table. It is assumed that all strikes are absolutely elastic. Since (see Figure 3.4) the result of the scattering depends strongly on a direction of the initial velocity and initial position of the mobile disk, then any small change in the initial conditions leads in the long run to another pattern of motion.

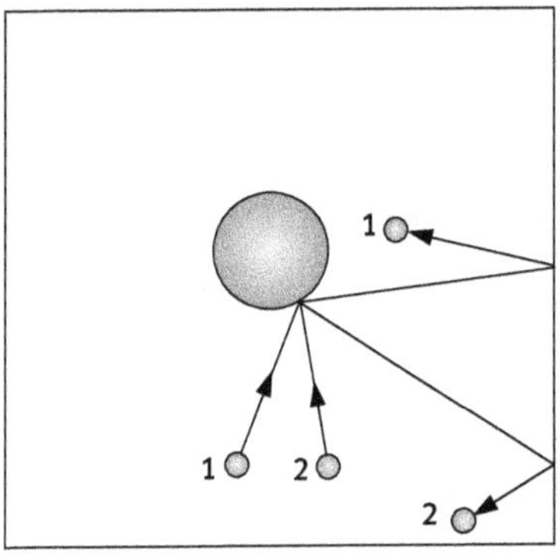

Figure 3.4: Billiards Sinai – the simplest mechanical system demonstrating chaotic behavior.

Thus, it is the extreme sensitivity to conditions of the scattering that leads to the stochastic behavior of the system. Then, when colliding with the central disk, the particle's behavior will not already depend on its initial position and velocity after some scattering events at any finite accuracy of calculations. Put it otherwise, the system "forgets" its the initial state and a dynamic description becomes impossible.

Only having calculated the probability of detecting the disk at any place on the table, description of the motion of such a puck can be possible. It is obvious that the probability will no longer depend on t after a time equal to the mixing time, and will be determined only by peculiarities of the system structure, in particular, by geometrical dimensions. Moreover, the motion of the particle of billiards can be assumed to be irreversible. Indeed, information loss about the initial conditions means that information entropy grows in an isolated system what is characteristic for irreversible behavior. The criterion, permitting to distinguish systems with the mixing event from integrable systems consists in difference of the Kolmogorov–Sinai entropy from zero (1.123).

Concluding this subject, one would like to draw the reader's attention to the following main aspects. The possibility, but more exactly, the necessity to introduce a statistical description is associated with weak stability of the dynamic systems. The statistical description can be assumed to become possible because a phase point visits a huge number of places scattered randomly throughout the phase surface for any macroscopic time of the measurement of the dynamic quantity. It is the mixing that allows one to apply an idea of that the distribution density of phase points on the isoenergetic surface is a constant (microcanonical distribution). In essence, this is a cornerstone of the statistical mechanics of Gibbs. The system ergodicity (3.9) is a necessary condition but not sufficient that of the statistical description applicability, because only in systems with the mixing, the distribution density of phase points proves to be the same throughout the isoenergetic surface of the system.

Statistical description for integrable systems is impossible, since a phase point moves along a trajectory and if one were to introduce the averaged description, the averaging would be performed along the trajectory, rather than the entire phase space.

Chaotic behavior arises both in the dynamical system described by the Hamilton equations and in the dissipative dynamical systems, with a mechanism of the emergence of dynamic chaos being the same, i. e. there is an ultra-high dependence of the motion pattern on the initial conditions.

It should be dwelt on one else matter. One should not think that a complexity of the system automatically guarantees the emergence of the mixing event in it. As far back as the dawn of the computer experiment, Ulam, Pasta and Fermi decided to test by numerical study, whether one of main hypotheses of statistical mechanics is fulfilled, but namely, the hypothesis of proportional distribution of energy over degrees of freedom. For these purposes there was taken the system of oscillators interacting against each other not harmonically. As was shown by that numerical study, on being exited of one of the vibrational modes at the beginning there occurred an intensive exchange of energy with other modes. Moreover, it would seem energy was distributed among all the vibrational modes, but after a while fluctuations of the original mode intensified again. The phenomenon observed is similar to return of the system to its original state, predicted by Poincaré. The Fermi–Pasta–Ulam problem solution was obtained by Kruskal and Zabuski in the early 1960s. They established that the Fermi–Pasta–Ulam system is a difference analog of the Korteweg–De Vries equation and solitonic nature of wave propagation in this system inhibits the uniform distribution of energy. The term soliton was proposed by Zabuski.

Finally, there is one more remark. Statistical mechanics of Gibbs comes from a fairly simple assumption about constancy density of phase points on the constant-energy surface. At the same time, as noted in Chapter 1, a phase space is fractal under conditions of dynamic chaos and has a fractional dimension. Unfortunately, until now it is not clear, whether this circumstance affects somehow the statistical properties of a system or not.

3.2 Substantiation of quasiclassical kinetic equations

3.2.1 The Liouville equation for the distribution function

Consider a gas of classical particles, consisting of N identical monatomic molecules enclosed in some volume V. Suppose for simplicity, the dynamic state each molecule is determined by the coordinate q and momentum p. We denote Cartesian projections of the vectors p and q, as p^α and q^α ($\alpha = 1, 2, 3$), respectively. Since we regard the gas of classical particles, their coordinates and momenta obey the Hamilton equations (3.1)

$$\frac{dp_i^\alpha}{dt} = -\frac{\partial H}{\partial q_i^\alpha}, \quad \frac{dq_i^\alpha}{dt} = \frac{\partial H}{\partial p_i^\alpha}, \quad i = 1, 2 \ldots N, \tag{3.11}$$

where H is the full Hamiltonian of the system, i is the index of the molecule.

As shown above, a state of the mechanical system at some point t is given by a set of values of coordinates and momenta of all particles making up the system. Thus, every time the system is represented as a point in $6N$-dimensional phase space. The system evolution may be described by investigating the motion of the point in the phase space.

After the manner of Gibbs, we proceed to describe the system dynamics in terms of a distribution function. To do this, instead of considering the evolution of particular system, we refer to the collection of quite identical dynamical systems which are distinct from each other only by the initial locations in the phase space. This set of systems is referred to as the Gibbs ensemble. It should be noted that the density of points in the phase space, normalized to unity, being denoted via by $\rho(p, q, t)$, the quantity $\rho(p, q, t)\, dpdq$ is the probability of finding the phase point in the volume element $dpdq$ of the phase space.

System description in the framework of the Gibbs method is a purely dynamic. It is easy to verify that the distribution function $\rho(p, q, t)$ would meet the certain sort of equations. As mentioned above in connection with the discussion of the Poincaré theorem recurrence, the motion of phase points in classical mechanics is *a phase flow*, which is defined by given a one-parameter group of transformations of the phase space:

$$\begin{aligned} &G^t(p_1(0), p_2(0), \ldots, p_N(0); q_1(0), q_2(0), \ldots, q_N(0)) \\ &\quad \to G^t(p_1(t), p_2(t), \ldots, p_N(t); q_1(t), q_2(t), \ldots, q_N(t)), \end{aligned}$$

where $p(t)$ and $q(t)$ are found from the solution of the Hamilton equations (3.11).

Consider the phase points, trapped at the time t by some volume element $dpdq$ of the phase space. These points are replaced into the volume element $dp'dq'$ of the phase space under action of the phase flow at the moment of time t'. As far as the phase points do not vanish and not reappear, one can write down the obvious equality:

$$\rho(p, q, t)dpdq = \rho(p', q', t')dp'dq'.$$

This expression holds for the conservation law of the phase points. Since, according to Liouville's theorem, the phase flow preserves the phase volume and $dpdq = dp'dq'$, then hence follows the distribution function constancy condition in the course of evolution of particles along the phase trajectory

$$\rho(p, q, t) = \rho(p', q', t').$$

Assuming that the temporal increment $dt = t' - t$ is infinitesimally small, we carry out the expansion of the function $\rho(p', q', t')$ with an accuracy to first order:

$$\rho(p', q', t') = \rho(p, q, t) + \left[\frac{\partial \rho}{\partial t} + \sum_{i=1}^{N} \frac{\partial \rho}{\partial q_i}\dot{q}_i + \sum_{i=1}^{N} \frac{\partial \rho}{\partial p_i}\dot{p}_i\right] dt. \tag{3.12}$$

It implies the vanishing of the expression in square brackets on the right side of (3.12). Taking into consideration the Hamilton equations, satisfied by the coordinates and momenta of the system particles, one obtains the Liouville equation for the classical function distribution:

$$\frac{\partial \rho}{\partial t} + [\rho, H] = 0, \quad \text{where } [\rho, H] = \sum_{i=1}^{N}\left(\frac{\partial \rho}{\partial q_i}\frac{\partial H}{\partial p_i} - \frac{\partial \rho}{\partial p_i}\frac{\partial H}{\partial q_i}\right). \tag{3.13}$$

Equation (3.13) yields a value of the function $\rho(p, q, t)$, if a value of the distribution function $\rho(p, q, 0)$ was given at the initial time. Moreover, no reduction of the description of the system in terms of the N-particle distribution function is possible to achieve. This description is so much detail as a dynamic description by the Hamilton equations. To proceed, it is necessary to pass on to a less detailed description of the system using a one-particle distribution function. Such a description may be made, it follows from the material presented in Section 3.1.1. Actually, the system has forgotten its initial state in the course of its evolution for the temporal period of the order of a characteristic mixing time. At the same time, correlation coefficients of higher orders vanish. So, when considered over periods greater than the time of chaotization, it doesn't make sense to describe the system's behavior in terms of the N-particle distribution function $\rho(p, q, t)$. It suffices to use a simplified description in the language of the one- or two-particle distribution functions. Bogoliubov was the first to show such an approach for deriving kinetic equations which were investigated in his work "Problems of dynamic theory in statistical physics" [20].

3.2.2 The chain of the Bogoliubov equations

Since the N-particle distribution functions contain redundant and useless information about particle correlations of higher orders, it would be advisable to introduce more simple the s-particle distribution functions $F_s(t, x_1, x_2, \ldots, x_s)$, $s = 1, 2\ldots, s \ll N$ having defined them so that the quantity

$$\frac{1}{V^s} F_s(t, x_1, x_2, \ldots, x_s) dx_1 dx_2 \ldots dx_s \tag{3.14}$$

would give a probability of the dynamical states of s molecules situated into the infinitely small volume $dx_1 dx_2 \ldots dx_s$ near the point $x_1, x_2, \ldots, x_s$ at time t. To determine the s-particle distribution function, one needs to integrate $\rho(x_1, x_2, \ldots, x_N)$ over all "extra" variables:

$$F_s(t, x_1, x_2, \ldots, x_s) = V^s \int \rho(t, x_1, x_2, \ldots, x_N)\, dx_{s+1}\, dx_{s+2} \ldots dx_N. \tag{3.15}$$

Here and below, the quantity x_i means an aggregate of a coordinate and momentum of the i-th particle, V being volume of the system.

The goal of the authors is to derive the equation for the one-particle distribution function $F_1(t, x)$. Nevertheless, it would be proper to start with a derivation of the motion equation for s-particle distribution function by simplifying it at the final stage.

For this purpose, let us apply the Liouville equation (3.13). Suppose that the system is a dilute gas of freely moving molecules, with the interaction between them being determined by short-range potential $\Phi(|q_i - q_j|)$, depending only on the distance modulus between particles. In this case the Hamiltonian system in a potential field $U(q)$ can be written as follows:

$$H = \sum_{1\le i\le N} H_1(x_i) + \sum_{1\le i<j\le N} \Phi(|q_i - q_j|), \quad H_1(x_i) = \frac{p_i^2}{2m} + U(q_i). \tag{3.16}$$

Using the Hamiltonian (3.16), we write down the Liouville equation (3.13) for the full distribution function:

$$\frac{\partial \rho}{\partial t} = \sum_{1\le i\le N} [H_1(x_i), \rho] + \sum_{1\le i<j\le N} [\Phi(|q_i - q_j|), \rho]. \tag{3.17}$$

Let us multiply both sides of equation (3.17) by V^s and integrate them over the variables $x_{s+1}, x_{s+2}, \ldots, x_N$, except but the integration for each variable of x_i is performed over all possible values of coordinate and momentum of the i-th particle. This results in obtaining the following expression:

$$\begin{aligned}\frac{\partial F_s}{\partial t} &= \sum_{1\le i\le s} V^s \int [H_1(x_i), \rho]\, dx_{s+1}\, dx_{s+2} \ldots dx_N \\ &+ \sum_{s+1\le i\le N} V^s \int [H_1(x_i), \rho]\, dx_{s+1}\, dx_{s+2} \ldots dx_N \\ &+ \sum_{1\le i<j\le s} V^s \int [\Phi(|q_i - q_j|), \rho]\, dx_{s+1}\, dx_{s+2} \ldots dx_N\end{aligned}$$

$$+ \sum_{\substack{1\le i\le s\\ s+1\le j\le N}} V^s \int [\Phi(|q_i - q_j|), \rho]\, dx_{s+1}\, dx_{s+2} \dots dx_N$$

$$+ \sum_{s+1\le i<j\le N} V^s \int [\Phi(|q_i - q_j|), \rho] dx_{s+1} dx_{s+2} \dots dx_N. \tag{3.18}$$

While writing (3.18) we have made allowance for the representation of the double sum holding for any symmetric function Φ_{ij} with respect to permutations of the indices:

$$\sum_{1\le i<j\le N} \Phi_{ij} = \frac{1}{2}\left(\sum_{i=1}^{s} + \sum_{i=s+1}^{N}\right)\left(\sum_{j=1}^{s} \Phi_{ij} + \sum_{j=s+1}^{N} \Phi_{ij}\right)$$
$$= \sum_{1\le i<j\le s} \Phi_{ij} + \sum_{\substack{1\le i\le s\\ s+1<j\le N}} \Phi_{ij} + \sum_{s+1\le i<j\le N} \Phi_{ij}.$$

For further transformation of equation (3.18) we take into account the following identities (see the suggestions in Problem 3.1):

$$\int [H_1(x_l), \rho]\, dx_l = 0; \tag{3.19}$$

$$\int [\Phi(|q_i - q_j|), \rho]\, dx_i\, dx_j = 0. \tag{3.20}$$

These identical equations are valid, if the distribution density ρ tends to zero at the boundaries of the phase region, with $|q| \to \infty$ and $|p| \to \infty$. Consider one after another each summand on right side of equation (3.18).

Bearing in mind the definition of (3.15), operation of integrating the first term yields the new form

$$\sum_{1\le i\le s} V^s \int [H_1(x_i), \rho]\, dx_{s+1}\, dx_{s+2} \dots dx_N = \sum_{1\le i\le s} [H_1(x_i), F_s].$$

The second term in line with the identity (3.19) vanishes and does not contribute. The third term on the right side of (3.18) in consideration of the definition of (3.15) is easily transformed into

$$\sum_{1\le i<j\le s} V^s \int [\Phi(|q_i - q_j|), \rho]\, dx_{s+1} \dots dx_N = \sum_{1\le i<j\le s} [\Phi(|q_i - q_j|), F_s].$$

In the fourth term, one may observe that all summands by summing over the index j by virtue of the particles identity, which leads to the distribution function invariance relating to permutation of the particle coordinates x_j and x_s:

$$\rho(x_1, x_2, \dots, x_s, \dots, x_j, \dots, x_N) = \rho(x_1, x_2, \dots, x_j, \dots, x_s, \dots, x_N)$$

can have the same form, if the change of variables in the course of integration takes place. The number of such summands, obviously, is equal to $N - s$. Therefore, we get

$$\begin{aligned}
&\sum_{\substack{1\le i\le s\\ s+1\le j\le N}} V^s \int [\Phi(|q_i - q_j|), \rho]\, dx_{s+1}\, dx_{s+2} \ldots dx_N \\
&= (N-s) \sum_{1\le i\le s} V^s \int [\Phi(|q_i - q_{s+1}|), \rho]\, dx_{s+1}\, dx_{s+2} \ldots dx_N \\
&= \frac{(N-s)}{V} \sum_{1\le i\le s} \int [\Phi(|q_i - q_{s+1}|), F_{s+1}]\, dx_{s+1}.
\end{aligned}$$

Finally, the fifth term on the right side of (3.18) taking into account the identity (3.20) is zero and does not make any contribution.

Thus, the equation for the s-particle distribution function can be written in the following form:

$$\begin{aligned}
\frac{\partial F_s}{\partial t} &= \Big[\sum_{1\le i\le s} H_1(x_i) + \sum_{1\le i<j\le s} \Phi(|q_i - q_j|), F_s\Big] \\
&+ \frac{N-s}{V} \int \Big[\sum_{1\le i\le s} \Phi(|q_i - q_{s+1}|), F_{s+1}\Big]\, dx_{s+1}.
\end{aligned} \tag{3.21}$$

We define the Hamiltonian set of molecules s by the relation

$$H_s = \sum_{1\le i\le s} H_1(x_i) + \sum_{1\le i<j\le s} \Phi(|q_i - q_j|) \tag{3.22}$$

and in equation (3.21) we pass to the thermodynamic limit $N \to \infty$, $V \to \infty$, $N/V = n =$ const, where n is for the particle number density. Then the equation for the s-particle distribution function can be written in more compact form:

$$\frac{\partial F_s}{\partial t} = [H_s, F_s] + n \int \sum_{1\le i\le s} [\Phi(|q_i - q_{s+1}|), F_{s+1}]\, dx_{s+1}. \tag{3.23}$$

Having written down this equation, we have not yet progressed toward reducing the description. In fact, we have got a chain of coupled equations for the distribution functions. In the sense of information completeness this chain is equivalent to the initial Liouville equation. The idea of applying the aggregate of coupled equations of motion for a sequence of the distribution functions or correlation functions of the form (2.3) is often used in non-equilibrium statistical mechanics to build the reduced description. Similar ideas were expressed in work of Born, Green, Kirkwood, Yvon. Therefore in the corresponding literature, the motion equations (3.23) are very often called the Bogoliubov–Born–Green–Kirkwood–Yvon hierarchy (BBGKY).

To obtain a closed equation, it is required to express the distribution function, such as F_{s+1}, via the distribution functions of lower orders. Then the set of equations is closed up and we obtain the reduction in description. In the following paragraphs we regard various variants of equations for the one-particle distribution function based on a chain of equations (3.23).

Problem 3.1. Using the definition of the classical Poisson brackets (3.13), one needs to prove the validity of the identities (3.19), (3.20), provided that ρ tends to zero at the boundaries of the phase region.

Solution. Consider the identity (3.19). Using the definition of Poisson brackets, we have

$$\int [H_1(p,q),\rho(p,q)]\,dp\,dq = -\int\left(\frac{\partial H_1(p,q)}{\partial p}\frac{\partial \rho(p,q)}{\partial q} - \frac{\partial H_1(p,q)}{\partial q}\frac{\partial \rho(p,q)}{\partial p}\right)dp\,dq. \quad (3.24)$$

We integrate the first and second terms on the right side of the last expression over parts;

$$\int \frac{\partial H_1(p,q)}{\partial p}\frac{\partial \rho(p,q)}{\partial q}\,dp\,dq = \int dp\frac{\partial H_1(p,q)}{\partial p}\rho(p,q)|_{q\to\infty} - \int dp\,dq\frac{\partial^2 H_1(p,q)}{\partial p\partial q}\rho(p,q); \quad (3.25)$$

$$\int \frac{\partial H_1(p,q)}{\partial q}\frac{\partial \rho(p,q)}{\partial p}\,dp\,dq = \int dq\frac{\partial H_1(p,q)}{\partial q}\rho(p,q)|_{p\to\infty} - \int dp\,dq\frac{\partial^2 H_1(p,q)}{\partial p\partial q}\rho(p,q). \quad (3.26)$$

The first summands into the right-hand side of the expressions (3.25) and (3.26) yield no contribution, and the second summands proved to be the same. Therefore, there is a minus sign between the two identical terms on the right-hand side of (3.24).

To prove the identity (3.20), we use the definition of the Poisson bracket (3.13). Since the potential of the pair interaction between the particles depends only on the coordinates, we obtain the expression

$$\begin{aligned}&\int [\Phi(|q_i - q_j|),\rho]\,dx_i\,dx_j\\ &= \int\left(\frac{\partial \Phi(|q_i - q_j|)}{\partial q_i}\frac{\partial \rho}{\partial p_i} + \frac{\partial \Phi(|q_i - q_j|)}{\partial q_j}\frac{\partial \rho}{\partial p_j}\right)dp_i\,dq_i\,dp_j\,dq_j.\end{aligned} \quad (3.27)$$

Upon integrating by parts of each term on the right side of (3.27), it is easy to observe that the right side of (3.27) is zero and identity (3.20) is indeed satisfied.

3.2.3 Equation for the one-particle distribution. The relaxation time approximation

Let us obtain the equation of motion for the one-particle distribution function $F_1(x,t)$. At the beginning, consider the Poisson bracket $[H_1, F_1]$. Going over to the vector notations, one obtains the following expression:

$$[H_1, F_1] = \frac{\partial H(\vec{r},\vec{p})}{\partial \vec{r}} \frac{\partial F_1(t,\vec{p},\vec{r})}{\partial \vec{p}} - \frac{\partial H(\vec{r},\vec{p})}{\partial \vec{p}} \frac{\partial F_1(t,\vec{p},\vec{r})}{\partial \vec{r}}$$
$$= -\vec{F}(\vec{r})\vec{\nabla}_p F_1(t,\vec{p},\vec{r}) - \frac{\vec{p}}{m}\vec{\nabla}_r F_1(t,\vec{p},\vec{r}); \; \vec{F}(\vec{r}) = -\frac{\partial H(\vec{r},\vec{p})}{\partial \vec{r}}. \tag{3.28}$$

Given this result, based on equation (3.23), the one-particle distribution function $F_1(t,\vec{p},\vec{r})$ is given by the equation

$$\left(\frac{\partial}{\partial t} + \vec{F}(\vec{r})\vec{\nabla}_p + \frac{\vec{p}}{m}\vec{\nabla}_r\right) F_1(t,\vec{p},\vec{r})$$
$$= n \int dr'\, dp' [\Phi(|\vec{r} - \vec{r}'|), F_2(t,\vec{p},\vec{r},\vec{p}',\vec{r}')]. \tag{3.29}$$

Equation (3.29) is still an accurate dynamic equation; its left side is a change rate of the one-particle distribution function due to its explicit dependence on the time and motion of particles in the coordinate and momentum spaces. In other words, the total derivative over time of the function F_1 is recorded on the left side (3.29). This derivative, in contrast to the N-particle distribution function, is not equal to zero, but it is equal to a change of the distribution function with sacrifice in binary collisions with other particles. For this reason, the right side of equation (3.29) is often referred to as *a collision integral*. From all has been said it follows that the equation for the one-particle distribution function can be written as the below form, replacing the right side of (3.29) by the collision integral:

$$\frac{\partial F_1}{\partial t} + \vec{F}(\vec{r})\vec{\nabla}_p F_1 + \frac{\vec{p}}{m}\vec{\nabla}_r F_1 = -\left(\frac{\partial F_1}{\partial t}\right)_{\text{col}}. \tag{3.30}$$

Various methods of constructing closed kinetic equations differ essentially only in a way how to build the collision integral. We suggest considering some of these ways later, but now we start with a relaxation time approximation as the simplest approach.

The relaxation time approximation comes from a simple assumption of that a spatially homogeneous system in the absence of external forces will relax towards equilibrium with a certain characteristic time τ. Otherwise speaking, the equation

$$\frac{\partial F_1}{\partial t} = -\left(\frac{\partial F_1}{\partial t}\right)_{\text{col}} \tag{3.31}$$

must describe the relaxation of the non-equilibrium distribution of $F_1(t)$ towards the equilibrium distribution function f_0 of the system. It is easy to see that the integral collision written in the form

$$\left(\frac{\partial F_1}{\partial t}\right)_{\text{col}} = \frac{F_1(t) - f_0}{\tau}$$

meets all these conditions.

The solution of equation (3.31) in this case takes the form

$$F_1(t) - f_0 = C(0)e^{-t/\tau},$$

where the constant $C(0)$ is determined from the initial conditions for the function F_1.

As a result, the kinetic equation in the relaxation time approximation appears as

$$\frac{\partial F_1}{\partial t} + \vec{F}(\vec{r})\vec{\nabla}_p F_1 + \frac{\vec{p}}{m}\vec{\nabla}_r F_1 = -\frac{F_1 - f_0}{\tau}. \tag{3.32}$$

It would be advisable to note that there is no good probative evidence for considering the relaxation to be exponential. Nevertheless, this approach in virtue of its simplicity is widely used, particularly in the qualitative interpretation of the experiment results. As to analysis of transport phenomena in metals and semiconductors, an application of concept of the relaxation time often gives noticeable results. In this case the quantity τ acts as an adjustable parameter. Issuing from the first principles, in some cases one can construct closed expressions for the relaxation time τ and thus justify the use of the relaxation time approximation. Details will be discussed in Chapter 4.

3.2.4 The Vlasov kinetic equation for a collisionless plasma

To obtain a closed equation for the one-particle distribution function from the Bogoliubov hierarchy (3.29) it is necessary to represent the two-particle distribution function in the form of a functional depending only on the one-particle distribution function. It stands to reason that further progress is impossible without invoking any additional physical ideas relating to properties of the interaction potential or behavior of F_2. Therefore, it is worth regarding two opposite cases: $R_0^3 n \ll 1$ and $R_0^3 n \gg 1$, where R_0 is characteristic radius of the interaction of microparticles, n being the number of particles in volume unit. The first case corresponds to a low-density gas, when the characteristic radius of particle interaction forces is much less than the average distance between particles. But we leave this question aside so far.

The second case occurs in ionized plasma, where the quantity R_0 has meaning of the Debye screening radius (the Debye length) of charged particles.

Consider a system of particles with the Coulomb interaction potential

$$\Phi(|\vec{r} - \vec{r}'|) = \pm\frac{e^2}{|\vec{r} - \vec{r}'|}.$$

The system as a whole is thought to be electrically neutral. The special feature of the Coulomb interaction is that the interaction potential decays too slowly with distance between particles. So, one is forced to take into account the interaction effect of the test particle with all other particles of the system. Moreover, the effect of pair interaction of the test particle with any other particle of the system proves to be much less than the interaction effect of this particle with an effective field, created by a set of the remaining $N-2$ particles. Thus, in the case of the Coulomb interaction potential, the interaction effect of the test particle with the average field of other particles is more important than pairwise interactions. This causes of a significant simplification.

Obviously, the two-particle distribution function can be always written in the form

$$F_2(t,\vec{p},\vec{r},\vec{p}\,',\vec{r}\,') = F_1(t,\vec{p},\vec{r})F_1(t,\vec{p}\,',\vec{r}\,') + G_2(t,\vec{p},\vec{r},\vec{p}\,',\vec{r}\,'), \tag{3.33}$$

where the function $G_2(t,\vec{p},\vec{r},\vec{p}\,',\vec{r}\,')$ takes into account paired correlations. As noted above, the pair correlations turned out to be less important than the influence of the effective field, so the pair correlation function G_2 in (3.33) can be neglected. This simplification allows immediately interrupting the Bogoliubov hierarchy and obtaining a closed equation for the one-particle distribution function.

In real systems, such as electron plasma, the Coulomb potential is screened by mobile electrons, consequently, the line of reasoning proposed above is valid only for distances $r \ll r_d$, where the inverse Debye length q_0 is given by

$$q_0 = \frac{1}{r_d} = \sqrt{\frac{4\pi n e^2}{k_B T}}.$$

On the other hand, in order to concept of an average field would have a right-to-life, it is necessary that a lot of particles would be inside the Debye sphere: $n r_d^3 \gg 1$. Substituting the estimate of the Debye length, one obtains the condition $k_B T \gg 4\pi e^2 n^{1/3}$. As far as $n^{1/3} \sim 1/a_0$, where a_0 is a quantity of the order of the average distance between particles, the condition written above is easily interpreted: kinetic energy of the motion of the particles must be much larger than the Coulomb interaction between neighboring particles:

$$k_B T \gg \frac{4\pi e^2}{a_0}.$$

Thus, the pair correlations being neglected, one finds the two-particle distribution function as the product of the one-partial functions

$$F_2(t,\vec{p},\vec{r},\vec{p}\,',\vec{r}\,') = F_1(t,\vec{p},\vec{r})F_1(t,\vec{p}\,',\vec{r}\,'). \tag{3.34}$$

Then one can substitute this expression into the right side of formula (3.29), which yields

$$\begin{aligned} n \int d\vec{r}\,' \, d\vec{p}\,' [\Phi(|\vec{r}-\vec{r}\,'|), F_2(t,\vec{p},\vec{r},\vec{p}\,',\vec{r}\,')] \\ = n \int d\vec{r}\,' \, d\vec{p}\,' \frac{\partial \Phi(|\vec{r}-\vec{r}\,'|)}{\partial \vec{r}} \frac{\partial F_1(t,\vec{p},\vec{r})}{\partial \vec{p}} F_1(t,\vec{p}\,',\vec{r}\,'). \end{aligned} \tag{3.35}$$

Another summand that arises when opening the Poisson brackets in which the derivatives are computed by $\vec{r}'$ and $\vec{p}'$ is zero because of (3.20). Now the right side of expression (3.35) can be written as

$$\frac{\partial \widetilde{U}(\vec{r})}{\partial \vec{r}} \frac{\partial F_1(t,\vec{p},\vec{r})}{\partial \vec{p}},$$

where the effective potential $\widetilde{U}$ is determined by the expression

$$\widetilde{U}(t,\vec{r}) = n \int d\vec{p}'\, d\vec{r}'\Phi(|\vec{r}-\vec{r}'|)F_1(t,\vec{p}',\vec{r}').$$

Substituting this result into the equation for the one-particle distribution function (3.29), one obtains a closed equation for the function F_1 with a self-consistent field:

$$\frac{\partial F_1}{\partial t} + \frac{\vec{p}}{m}\frac{\partial F_1}{\partial \vec{r}} - \frac{\partial (U(t,\vec{r}) + \widetilde{U}(t,\vec{r}))}{\partial \vec{r}} \frac{\partial F_1}{\partial \vec{p}} = 0. \tag{3.36}$$

In writing this equation we have assumed that the external force $\vec{F}(t,\vec{r}) = -\vec{\nabla} U(t,\vec{r})$, where $U(t,\vec{r})$ is potential of the field of external forces. Equation (3.36) is the Vlasov equation, obtained by him in 1938. Note several key features of this equation.

First, the integro-differential Vlasov equation is time reversible. Time reversal is a natural consequence of failure to account for the interaction between particles.

Second, a one-component plasma cannot really exist. Therefore, the equation for the electron distribution function should be added to the equation for the ion distribution function except for a model case when the density of the ion distribution is homogeneous and constant.

Third, the motion of charged particles leads to the appearance of an alternating electromagnetic field. So, the Vlasov equation must be supplemented by the Maxwell equations for the components of electric and magnetic fields, respectively. Thus, equations (3.36) in fact should be regarded as a kind of program, which requires serious efforts. Let us consider how to solve a practically important problem using the linearized Vlasov equation.

Problem 3.2. Determine spectrum of longitudinal oscillations of electron plasma using the linearized Vlasov kinetic equation provided that the positively charged ions are immobile and uniformly distributed.

Solution. In the conditions of this problem we can confine ourselves only considering the electron motion. We represent the one-particle distribution function F_1 as the sum of the equilibrium distribution function $f_0(v)$ and the non-equilibrium increment $f(t,\vec{v},\vec{r})$:

$$F_1(t,\vec{v},\vec{r}) = f_0(v) + f(t,\vec{v},\vec{r}).$$

The electron gas will be assumed to be nondegenerate. In this case, equilibrium distribution f_0 is the Maxwell–Boltzmann distribution:

$$f_0(v) = \left(\frac{m}{2\pi k_B T}\right)^{3/2} \exp\left(-\frac{mv^2}{2k_B T}\right), \tag{3.37}$$

where m is the electron mass, and v is the electron speed. The distribution (3.37) is normalized to unity:

$$\int dv_x\, dv_y\, dv_z \left(\frac{m}{2\pi k_B T}\right)^{3/2} \exp\left(-\frac{mv^2}{2k_B T}\right) = 1.$$

Upon choosing such a normalization of the distribution function, integration in the formula for self-consistent potential $\widetilde{U}$ should perform over $\vec{v}'$, rather than over $\vec{p}'$.

If non-equilibrium is weak and $f(t,\vec{v},\vec{r})/f_0(v) \ll 1$, equation (3.36) can be linearized. Now we analyze forces acting on an electron. According to the closed equation (3.36) one can observe that both the external force of interaction with the positively charged background $-\vec{\nabla}U(\vec{r})$ and the force being determined by the gradient of the self-consistent field $-\vec{\nabla}\widetilde{U}(t,\vec{r})$ act on the electron. In the equilibrium state, these forces must compensate each other. Therefore, the net force acting on the electron will be determined only by non-equilibrium increment $f(t,\vec{v},\vec{r})$:

$$-\vec{\nabla}(U + \widetilde{U}) = e\vec{E}(\vec{r},t) = -n\vec{\nabla}_r \int d\vec{v}'\, d\vec{r}'\, \frac{e^2}{|\vec{r}-\vec{r}'|} f(t,\vec{v}',\vec{r}'). \tag{3.38}$$

In this case, magnetic field, which occurs when there is the motion of charged particles, does not contribute because the summand

$$[\vec{v}\vec{H}]\vec{\nabla}_p f_0 = 0$$

vanishes in virtue of collinearity of the vectors $\vec{v}$ and $\vec{\nabla}_p f_0$.

Using (3.38), it is easy obtaining that the intensity vector $\vec{E}(\vec{r},t)$ of the resultant electric field satisfies some equation. We find the divergence of the left and right sides of (3.38). Given that

$$\operatorname{div} \vec{\nabla}_r \left(\frac{1}{|\vec{r}-\vec{r}'|}\right) = \triangle\left(\frac{1}{|\vec{r}-\vec{r}'|}\right) = -4\pi\delta(\vec{r}-\vec{r}'),$$

expression (3.38) yields one of the well-known Maxwell equations:

$$\operatorname{div} \vec{E}(t,\vec{r}) = -4\pi e n \int d\vec{v} f(t,\vec{v},\vec{r}), \tag{3.39}$$

where $\triangle$ is the Laplace operator.

Using the definition (3.38), we write down the linearized Vlasov equation (3.36) for the given case,

$$\frac{\partial f(t,\vec{v},\vec{r})}{\partial t} + \frac{\vec{p}}{m}\frac{\partial f(t,\vec{v},\vec{r})}{\partial \vec{r}} - e\vec{E}(t,\vec{r})\frac{\partial f_0}{\partial \vec{v}} = 0. \tag{3.40}$$

In this case, the linearization consists in use only of the linear terms in equation (3.40) relating to the small quantity of $f(t,\vec{v},\vec{r})$. Since the electric field $\vec{E}(t,\vec{r})$ of the latter summand in (3.40) is linear with respect to this increment, the gradient $\vec{\nabla}_p F_1$ can be replaced by $\vec{\nabla}_p f_0$, i. e. by the quantity being estimated over the equilibrium distribution.

The set of equations (3.39) and (3.40) allows one solve the problem of determining the spectrum of longitudinal oscillations of the electron plasma. As mentioned above, the Vlasov equation is time reversible. This reversibility leads to the solution degeneration regarding the time-reversal operation. The degeneration can be removed by adding an infinitely small source into the right side of equation (3.40), which is similar to the collision integral (3.32) for recording the kinetic equation in the relaxation time approximation:

$$\left(\frac{\partial F_1}{\partial t}\right)_{\text{col}} = \varepsilon f(t,\vec{v},\vec{r}).$$

In this formula ε is a small quantity which should be trended towards zero after performing the thermodynamic limit $N \to \infty$, $V \to \infty$, $N/V = n = \text{const}$. For further calculations we will use the "corrected" Vlasov equation,

$$\frac{\partial f(t,\vec{v},\vec{r})}{\partial t} + \frac{\vec{p}}{m}\frac{\partial f(t,\vec{v},\vec{r})}{\partial \vec{r}} - e\vec{E}(t,\vec{r})\frac{\partial f_0}{\partial \vec{p}} = -\varepsilon f(t,\vec{v},\vec{r}), \tag{3.41}$$

giving solutions of retarding type.

Since we are interested in the longitudinal vibrations, instead of the Fourier transform equations (3.39), (3.41) one can simply look for a solution in the form

$$f(t,\vec{v},\vec{r}) = f_{k,\omega}(\vec{v})e^{i(kx-\omega t)}, \quad \vec{E}(t,\vec{r}) = E(k,\omega)e^{i(kx-\omega t)}.$$

Substituting these expressions into (3.39), (3.41), we have equations for the Fourier components of $f_{k,\omega}(\vec{v})$ and $E(k,\omega)$:

$$\begin{aligned} &-i(\omega - v_x k + i\varepsilon)f_{k,\omega}(\vec{v}) - \frac{e}{m}\frac{\partial f_0}{\partial v_x}E(k,\omega) = 0, \\ &ikE(k,\omega) = -4\pi e n \int d\vec{v} f_{k,\omega}(\vec{v}). \end{aligned} \tag{3.42}$$

From the first equation (3.42) we find

$$f_{k,\omega}(\vec{v}) = i\frac{e}{m}\frac{\partial f_0}{\partial v_x}\frac{E(k,\omega)}{\omega - v_x k + i\varepsilon}.$$

Now we substitute the found value into the right side of the second of equations (3.42). This results in obtaining the equation for the Fourier components of the field:

$$kE(k,\omega) = -E(k,\omega)\frac{4\pi e^2 n}{m}\int d\vec{v}\frac{\partial f_0}{\partial v_x}\frac{1}{\omega - v_x k + i\varepsilon}. \tag{3.43}$$

Equation (3.43) has the trivial solution $E(k,\omega) = 0$. If there is a nontrivial solution and $E(k,\omega) \neq 0$, the following condition must be fulfilled:

$$1 + \frac{4\pi e^2 n}{mk} \int d\vec{v} \frac{\partial f_0}{\partial v_x} \frac{1}{\omega - v_x k + i\varepsilon} = 0. \tag{3.44}$$

This equation is the required dispersion relation, expressing the dependence of the electrons oscillation frequency ω on their wave vector k (we recall that longitudinal oscillations of electron plasma along the axis X are being considered). To obtain the explicit form of the dependence $\omega(k)$, one needs to calculate the integral in formula (3.44).

As for the above example, the main objective was to illustrate an application of the Vlasov equation for solving physical kinetics problems and we have accomplished this problem, having obtained equation (3.44). Then we can omit the details in the following calculations, referring the reader to the literature [17].

To find an explicit dispersion law, we introduce the plasma frequency ω_0, having defined it by the relationship

$$\omega_0^2 = \frac{4\pi e^2 n}{m}.$$

Now we integrate over the velocity components v_y and v_z. Given that

$$\frac{\partial f_0}{\partial v_x} = -\frac{m v_x}{k_B T} f_0, \quad \left(\frac{m}{2\pi k_B T}\right) \int dv_y\, dv_z \exp\left(-\frac{m(v_y^2 + v_z^2)}{2k_B T}\right) = 1,$$

one has

$$1 - \frac{\omega_0^2}{k} \frac{m}{k_B T} \left(\frac{m}{2\pi k_B T}\right)^{1/2} \int_{-\infty}^{\infty} dv_x \exp\left(-\frac{m v_x^2}{2k_B T}\right) \frac{v_x}{\omega - v_x k + i\varepsilon} = 0. \tag{3.45}$$

The integral in the expression (3.45) holds a singularity at $v_x = \omega/k$. To calculate it we apply the well-known relation

$$\lim_{\varepsilon \to 0} \frac{1}{x + i\varepsilon} = P\frac{1}{x} - i\pi\delta(x),$$

where P is the principal value of the function, the punctured singular point. This results in writing the dispersion relation as follows:

$$1 - \operatorname{Re} I + i \operatorname{Im} I = 0, \tag{3.46}$$

$$\operatorname{Re} I = \frac{\omega_0^2}{k} \frac{m}{k_B T} \left(\frac{m}{2\pi k_B T}\right)^{1/2} \int_{-\infty}^{\infty} dv_x \exp\left(-\frac{m v_x^2}{2k_B T}\right) \frac{v_x}{\omega - v_x k}, \tag{3.47}$$

$$\operatorname{Im} I = \pi \frac{\omega_0^2}{k} \frac{\omega}{k^2} \frac{m}{k_B T} \left(\frac{m}{2\pi k_B T}\right)^{1/2} \exp\left(-\frac{m(\omega/k)^2}{2k_B T}\right). \tag{3.48}$$

When integrating over v_x there has been used the identity $\delta(\omega - kv_x) = 1/k\delta(\omega/k - v_x)$.

Unfortunately, the integral cannot be exactly calculated according to the formula (3.47), therefore, we consider only long-wave approximation $kv_x/\omega \ll 1$ and expand the fraction in the integrand in this small parameter, retaining only the first few terms:

$$\frac{v_x}{\omega - v_x k} = \frac{1}{\omega}\left(v_x + \frac{v_x^2 k}{\omega} + \frac{v_x^3 k^2}{\omega^2} + \frac{v_x^4 k^3}{\omega^3} + \cdots\right).$$

Given that the odd Maxwell–Boltzmann moments are zero, while the second and fourth moments are easily calculated: $\overline{v_x^2} = k_\mathrm{B}T/m$, $\overline{v_x^4} = 3(K_BT/m)^2$, where

$$\overline{v_x^n} = \left(\frac{m}{2\pi k_\mathrm{B}T}\right)^{1/2} \int\limits_{-\infty}^{\infty} dv_x v_x^n \exp\left(-\frac{mv_x^2}{2k_\mathrm{B}T}\right),$$

one obtains an approximate expression for the right side of (3.47)

$$\mathrm{Re}\, I = \frac{\omega_0^2}{\omega^2}\left(1 + 3\frac{k^2}{\omega^2}\frac{k_\mathrm{B}T}{m}\right). \tag{3.49}$$

Now we return to an analysis of the dispersion relation (3.46). Discarding the imaginary part for a while, we find the dispersion relation of the long-wave approximation without damping. For this purpose, we substitute the expression (3.49) into (3.46) and then the squared electron oscillation frequency can be written as follows:

$$\omega^2 = \omega_0^2\left(1 + 3\frac{k^2}{\omega^2}\frac{k_\mathrm{B}T}{m}\right). \tag{3.50}$$

This yields the equality $\omega = \omega_0$ in the zero approximation in the small parameter

$$\frac{k^2}{\omega^2}\frac{k_\mathrm{B}T}{m}.$$

Substituting this result into the right side of (3.50) and extracting the square root of the left and right sides, in the first approximation in k^2, we have

$$\omega = \omega_0\left(1 + \frac{3}{2}\frac{k_\mathrm{B}T}{m}\frac{k^2}{\omega_0^2}\right). \tag{3.51}$$

The damping of plasma oscillations can be sought in the zero approximation over the parameter kv_x/ω, assuming that $\omega = \Omega + i\gamma$, where Ω is the real part of the plasma frequency ω, and γ is the imaginary part of the plasma frequency ω.

To find the relationship between γ and $\mathrm{Im}\, I$, one should return to the dispersion equation (3.46) by writing it as

$$1 - \frac{\omega_0^2}{(\Omega + i\gamma)^2} = 0.$$

Assuming that the damping is weak and $\gamma/\Omega \ll 1$, we expand the denominator in the last expression up to linear terms in γ/Ω:

$$1 - \frac{\omega_0^2}{\Omega^2}\left(1 - i\frac{2\gamma}{\Omega}\right) = 0.$$

Upon comparing the above expression with equation (3.46) in zero order over k ($\Omega = \omega_0$), one can find the following expression for the damping constant:

$$\gamma = \sqrt{\pi}\frac{\omega_0^4}{k^3}\left(\frac{m}{2k_B T}\right)^{3/2} \exp\left(-\frac{m(\omega_0/k)^2}{2k_B T}\right). \tag{3.52}$$

The damping of the longitudinal plasma wave, being defined by the expression (3.52), was found by Landau in 1946. One should pay attention to the fact that this damping was obtained without regard of electron collisions with scatterers. In this case, the electric field plays the role of an elastic force. The damping is determined only by part of the electrons, for which the velocity along the X-axis coincides with the phase velocity of the wave, equal to ω/k. These electrons are the most effectively periodically accelerated and then decelerated by the electric field. After acceleration in the electric field, the number of electrons, for which the condition $v_x = \omega/k$ is met, are more than after slowing-down. Therefore, the inhibitory effect of the electric field is more effective than the accelerating effect and electrons of such a group lose on average part of their energy for an oscillation cycle. It should be noted that there are other mechanisms besides the Landau damping without collisions in damping of the plasma oscillations, for example, accelerated moving electrons radiate energy.

3.2.5 The Boltzmann equation for a low-density gas

One of the main results of the kinetic theory is a kinetic equation for a one-particle distribution function. It was obtained by Ludwig Boltzmann in 1872. Consider two different approaches allowing deriving this equation: qualitative method which L. Boltzmann preferred and a first-principles derivation by using a chain of coupled equations of motion for the Bogoliubov distribution functions. Before proceeding to the direct derivation of the kinetic equation, one should analyze applicability conditions for the approach at hand.

First, only pair collisions should be considered because a two-body scattering problem has an analytic solution. However, for the three-body problem (and, more generally, the n-body problem for $n \geq 3$) only numerical solutions are possible, so they cannot be represented in an analytical form. For this reason, we confine our attention to spatially homogeneous systems. In fact, the requirement of spatial homogeneity is not a severe restriction. A distribution function does not have to change significantly at distances $r \sim \lambda$ (of order of the free path of particles). The distribution function, however, may depend on coordinates as a parameter. This turns out to be sufficient for most physical applications.

Second, we consider a system of particles with potential interaction of repulsion type. In this case, bound states cannot arise while scattering.

Third, we assume that the radius r_0 of the interaction forces is much less than the average distance $a_0 = v^{1/3}$, between particles, where v is volume per one particle. Parameter r_0^3/v is small in comparison with other spatial scales of the system.

The existence of two different spatial scales leads to the appearance of two different temporal scales. If one imagines that a system consists of one-sort particles, having the average speed $\overline{v}$, then one can introduce the characteristic time of particle interaction $\tau_0 = r_0/\overline{v}$ and the characteristic time of the particle mean free path $\tau = \lambda/\overline{v}$, where λ is the free path. It is obvious that the condition $\tau_0 \ll \tau$ is fulfilled.

Indeed, there are numerical estimates for a gas under normal conditions: $N/V \sim 3 \cdot 10^{19}$ particles/cm^3; the volume per particle is equal to $v_0 \sim 3 \cdot 10^{-20}$ cm^3; the average distance between particles is equal to $a_0 = 3 \cdot 10^{-7}$ cm, the characteristic radius of interaction forces between particles is $r_0 \sim 10^{-8}$ cm; the free path is $\lambda \sim 10^{-5}$ cm; the thermal velocity of the molecules is $\overline{v} = 10^5$ cm/s. This yields the following data: $\tau_0 \sim 10^{-13}$ s, and $\tau \sim 10^{-10}$ s. The presence of different temporal and spatial scales allows one to roughen the description and go over from the dynamical description to the statistical one.

3.2.6 Qualitative derivation of the Boltzmann equation

Each particle of a gas can be deemed as a closed subsystem provided that collisions between molecules are not taken into account at all. Then the Liouville theorem holds for the one-particle distribution function $F_1(t, \vec{p}, \vec{r})$ and it can be written as

$$\frac{dF_1}{dt} = 0;$$

or

$$\frac{\partial F_1}{\partial t} + \frac{\vec{p}}{m}\vec{\nabla}_r F_1 + \vec{F}\vec{\nabla}_p F_1 = 0.$$

The allowance for the collisions leads to the fact that the distribution function will suffer a change as the particle moves along a phase trajectory and collides with other particles. This part of the change in the distribution function, as mentioned above, is called the collision integral. The merit of the Boltzmann approach is that he succeeded in constructing the collision integral for a low-density gas.

Although the kinematics of collision processes is necessary to take into account while constructing the collision integral, the qualitative derivation of the Boltzmann equation cannot be purely dynamic. If the system's behavior over periods $\tau_0 \ll t \le \tau$

is of interest to us, there is no need to describe precisely the particle collision process. It suffices to know only the asymptotic behavior of the system. In other words, it is enough to find a relationship between states long before and long after the collision.

When elastic collisions of two particles take place, there must be satisfied laws of conservation of momentum and energy:

$$\vec{p} + \vec{p}_1 = \vec{p}' + \vec{p}_1' = \vec{P},$$
$$p^2 + p_1^2 = p'^2 + p_1'^2, \tag{3.53}$$

where $\vec{p}, \vec{p}_1$ are for momentum of the particles before collision $\vec{p}', \vec{p}_1'$ for the momentum of the particles after collision; $\vec{P}$ is the total momentum of the system consisting of two particles.

The velocities of the relative motion $\vec{u} = (\vec{p}_1 - \vec{p})/m$ and $\vec{u}' = (\vec{p}_1' - \vec{p}')/m$ before the collision and after it are equal in value and have opposite arrows: $\vec{u} = -\vec{u}'$.

It is obvious that the particle momentum before and after the collision can be expressed via two quantities by using equations (3.53), namely, the total momentum of the particles $\vec{P}$ and the relative velocity $\vec{u}$:

$$\vec{p} = \frac{\vec{P}}{2} - \frac{m\vec{u}}{2}, \quad \vec{p}_1 = \frac{\vec{P}}{2} + \frac{m\vec{u}}{2},$$
$$\vec{p}' = \frac{\vec{P}}{2} - \frac{m\vec{u}'}{2}, \quad \vec{p}_1' = \frac{\vec{P}}{2} + \frac{m\vec{u}'}{2}. \tag{3.54}$$

Going over to constructing the collision integral, we introduce the following notations for simplification of records:

$$F_1(t, \vec{r}, \vec{p}) = f, \quad F_1(t, \vec{r}, \vec{p}_1) = f_1,$$
$$F_1(t, \vec{r}, \vec{p}') = f', \quad F_1(t, \vec{r}, \vec{p}_1') = f_1'. \tag{3.55}$$

At some moment in time t particles $nfd\vec{r}d\vec{p}$ will be found in the volume element $d\vec{r}d\vec{p}$ of the phase space, where $n = 1/v$ is the number of particles per unit volume (concentration). The collision integral determines the rate of change of particles located in the volume element $d\vec{r}d\vec{p}$ of the phase space in the vicinity of the point $\vec{r}$, $\vec{p}$. In order to find the rate of change, it is necessary to count up the number of particles leaving and entering the volume per unit time. As far as all the scattering events are independent and the scattered particles get fully thermalized before the next act of scattering, each scattering event can be also considered as independent.

Let us consider a single particle with coordinates $\vec{r}$, $\vec{p}$ then we will stop it, i. e. we will move over to the coordinate system associated with the particle. Now we take a model of hard spheres as a model of the interaction between the particles, assuming that each particle has radius r_0. We surround the selected particle by sphere of the particle interaction with radius equal to $2r_0$, the center of it to be an origin of cylindrical coordinate system. The Z-axis of the system will be directed along the relative velocity vector $\vec{u}$. The selected-axis scheme of the axes is shown in Figure 3.5.

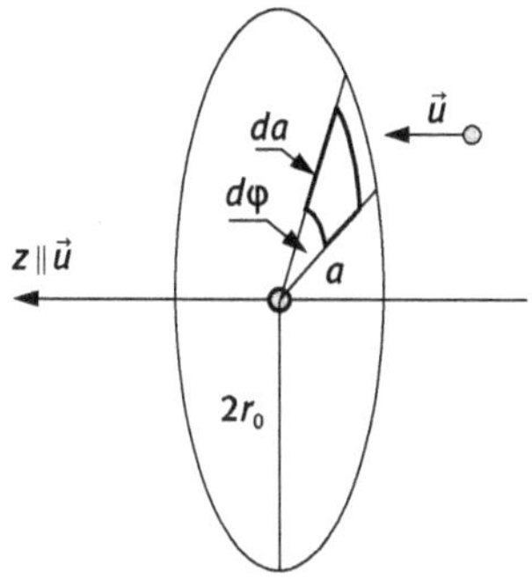

Figure 3.5: The coordinates choice for an analysis of pair collisions of particles.

Coordinate, setting a polar angle can be denoted as φ and letter a can be assigned to a radial variable. Besides, we denote the infinitely small element of a cross-section area of the sphere $ada d\varphi$ as $d\Omega$. In Figure 3.5 this area element is highlighted by a bolder line. An average number of particles with momentum from $\vec{p}_1$ to $\vec{p}_1 + d\vec{p}_1$, falling on this area per unit time, is equal to $nf_1 ud\Omega d\vec{p}_1$. Then the average number of particle collisions, located into the phase volume element $d\vec{r}d\vec{p}$ and particles, possessing momentum from $\vec{p}_1$ to $\vec{p}_1 + d\vec{p}_1$, will be determined by the expression

$$nfd\vec{r}d\vec{p}nf_1 ud\Omega d\vec{p}_1. \tag{3.56}$$

As far as one particle, changing its momentum, leaves the volume element of the phase space $d\vec{r}d\vec{p}$ owing to each of these collisions, the total number of such collisions can be found by integrating both over all possible values of momentum $\vec{p}_1$ and over the cross-section area of the scattering sphere. As a result, we get the number of particles leaving the phase volume element $d\vec{r}d\vec{p}$ per unit time:

$$n\frac{\partial f}{\partial t}\bigg|_- d\vec{r}d\vec{p} = -n^2 \int_0^{2\pi} d\varphi \int_0^{2r_0} a\,da \int ff_1 u\,d\vec{p}_1\,d\vec{r}\,d\vec{p}. \tag{3.57}$$

To find the number of particles entering the volume element of phase space per unit time $d\vec{r}d\vec{p}$ is suffices to note that particles having momenta $\vec{p}$ and $\vec{p}_1$ before the scattering event will possess the momentum $\vec{p}'$ and $\vec{p}_1'$ after the scattering. Consequently, the scattering process reversal is needed for finding the number of particles entering the volume element of the phase space $d\vec{r}d\vec{p}$. To do so, one should make the following replacements in formula (3.57): $f \to f', f_1 \to f_1', d\vec{p} \to d\vec{p}', d\vec{p}_1 \to d\vec{p}_1', \vec{u} \to -\vec{u}$. This yields the expression

$$n\frac{\partial f}{\partial t}\bigg|_+ d\vec{r}d\vec{p}' = n^2 \int_0^{2\pi} d\varphi \int_0^{2r_0} a\,da \int f'f_1' u\,d\vec{p}_1'\,d\vec{r}\,d\vec{p}'. \tag{3.58}$$

Using the kinematics of scattering laws, it is easy to prove that $d\vec{p}d\vec{p}_1 = d\vec{p}'d\vec{p}_1'$. Actually, as known, law of the transition from one coordinate system to another is given with the aid of the Jacobian transformation $d\vec{p}d\vec{p}_1 = |D|d\vec{p}'d\vec{p}_1'$, where the Jacobian transformation D (functional determinant) is determined by the expression

$$D = \frac{\partial(\vec{p}',\vec{p}_1')}{\partial(\vec{p},\vec{p}_1)} = \frac{\partial(\vec{p}',\vec{p}_1')}{\partial(\vec{P},\vec{u}')}\frac{\partial(\vec{P},\vec{u}')}{\partial(\vec{P},\vec{u})}\frac{\partial(\vec{P},\vec{u})}{\partial(\vec{p},\vec{p}_1)}.$$

Upon analyzing equation (3.54), one finds that

$$\frac{\partial(\vec{p}',\vec{p}_1')}{\partial(\vec{P},\vec{u}')} = \frac{\partial(\vec{p},\vec{p}_1)}{\partial(\vec{P},\vec{u})},$$

for the relationship between the variables $\vec{p}'$, $\vec{p}_1'$ and $\vec{P}$, $\vec{u}'$ is exactly the same as for the variables $\vec{p}$, $\vec{p}_1$ and $\vec{P}$, $\vec{u}$. It follows that [1, 3]

$$\frac{\partial(\vec{p}',\vec{p}_1')}{\partial(\vec{P},\vec{u}')}\frac{\partial(\vec{P},\vec{u})}{\partial(\vec{p},\vec{p}_1)} = 1, \quad D = \frac{\partial(\vec{P},\vec{u}')}{\partial(\vec{P},\vec{u})} = \left.\frac{\partial(\vec{u})}{\partial(\vec{u}')}\right|_{\vec{P}=\text{const}}.$$

To find the functional determinant D, we substitute values of the relative velocity projections before the pair collision and after it: $\vec{u} = (u_x, u_y, u_z)$, $\vec{u}' = (u_x, u_y, -u_z)$. Since in the chosen coordinate system, intended for the consideration of elastic scattering event, one changes only the component of the velocity u_z and the velocity components u_x and u_y remain constant, the functional determinant, using the properties [1, 3] can be simplified further:

$$D = \frac{\partial(\vec{u})}{\partial(\vec{u}')} = \frac{\partial(u_x, u_y, u_z)}{\partial(u_x, u_y, -u_z)} = \left.\frac{\partial(u_z)}{\partial(-u_z)}\right|_{\substack{u_x=\text{const}\\ u_y=\text{const}}} = -1.$$

Thus, we have proved that $|D| = 1$ and $d\vec{p}d\vec{p}_1 = d\vec{p}'d\vec{p}_1'$. This result allows one to combine the terms describing the entering and escape of particles from the volume element $d\vec{r}d\vec{p}$ under a general integral sign and make the reduction of the same terms in the left and right sides of equations (3.57) and (3.58). Having combined data obtained, we write down the Boltzmann equation for a low-density gas with the collision integral on the right side:

$$\frac{\partial f}{\partial t} + \frac{\vec{p}}{m}\vec{\nabla}_r f + \vec{F}\vec{\nabla}_p f = n\int_0^{2\pi} d\varphi \int_0^{2r_0} a\,da \int (f'f_1' - ff_1)u\,d\vec{p}_1. \tag{3.59}$$

As for practical calculations in the formula (3.59), the variables $\vec{p}'$ and $\vec{p}_1'$, which control the functions f' and f_1', one should express via the variable $\vec{p}$ and $\vec{p}_1$, using relations (3.53), (3.54).

Let us consider alternative ways to write the collision integral in the case of pair collisions in a central field. It is worth recalling that a force field whose potential depends only on the distance from the force center is referred to as a central field. Under the central field during collision process, the angular momentum is also preserved in addition to energy and momentum. This leads to the fact that each elementary scattering act occurs in a plane perpendicular to the angular momentum vector. The above considered case of the collision between elastic balls is a particular example of the scattering event under the central field. Scattering by a central field is usually described in terms of the scattering cross-section.

Let a homogeneous beam of particles fall on a stationary scattering center with the constant velocity $\vec{u}$, then *the scattering cross-section* $\sigma(\Omega, u)$ is called a proportionality factor between the magnitude of the density flux of falling particles I and the number of particles dN, scattered into the solid angle $d\Omega = \sin\theta d\theta d\varphi$ per unit time:

$$dN = I\sigma(\Omega, u)d\Omega, \tag{3.60}$$

where θ is the so-called scattering angle, i. e. the angle between the relative velocity vectors $\vec{u}$ and $\vec{u}'$ before and after the scattering event, respectively.

The geometric meaning of the parameters describing the collision of particles in the central field is shown in Figure 3.6. The example of the collision between elastic balls with radius r_0 is illustrated in this figure. All particles with impact-parameter from b to $b + db$ will fall into the spherical belt, located on the scattering sphere, and will have scattering angles from θ to $\theta + d\theta$. The belt is depicted in Figure 3.6(a). This implies that all particles entering the surface element $bdbd\varphi$ of the scattering sphere will be scattered into the solid angle $d\Omega$. Consequently, we have the following equation $dN = Ibdbd\varphi$. Upon comparing this expression with (3.60), we find

$$\sigma(\Omega, u) = \frac{b}{\sin\theta}\left|\frac{db}{d\theta}\right|. \tag{3.61}$$

The derivative $db/d\theta$ here has been taken with respect to the modulus, because it increases with diminishing the impact-parameter while determining the scattering angle.

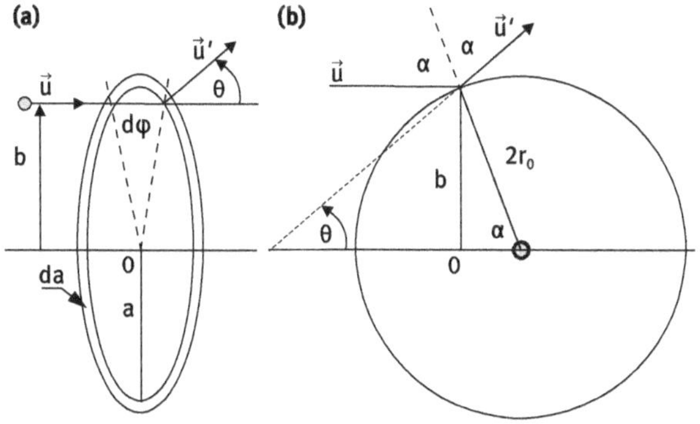

Figure 3.6: The kinematics of an elastic scattering: (a) area cross-section of the interaction between a plane perpendicular to $\vec{u}$; (b) scheme, allowing to find the interrelation of the impact-parameter b with the scattering angle θ.

Functional connection between the impact-parameter b and the scattering angle θ can be found by using the building in Figure 3.6(b). As follows from the figure $b = 2r_0 \sin\alpha$. The angle α is related with the scattering angle θ by the simple relation $\alpha = \pi/2 - \theta/2$. Therefore $b = 2r_0 \cos\theta/2$. Substituting this result into the formula (3.61), one obtains an expression for the scattering cross-section for an elastic collision of particles of radius r_0:

$$\sigma(\Omega, u) = \frac{2r_0^2 \cos\theta/2 \sin\theta/2}{\sin\theta} = r_0^2. \tag{3.62}$$

The total scattering cross-section σ_T can be found by integrating over the entire solid angle

$$\sigma_T = \int \sigma(\Omega, u)\, d\Omega = \int_0^{2\pi} d\varphi \int_0^{2r_0} b\, db = 4\pi r_0^2. \tag{3.63}$$

The second equality is obtained by using the definition of (3.61).

Comparing (3.63) and right side of formula (3.59) it is easy to reveal that the collision integral in the kinetic Boltzmann equation can be written, using the concept of the scattering cross-section. In such a case, instead of (3.59) we have

$$\frac{\partial f}{\partial t} + \frac{\vec{p}}{m}\vec{\nabla}_r f + \vec{F}\vec{\nabla}_p f = n \int u\sigma(\Omega, u)(f' f_1' - f f_1)\, d\vec{p}_1\, d\Omega. \tag{3.64}$$

For practical purposes it is convenient to write the collision integral in such a manner that it would explicitly obey the energy and momentum conservation laws. To do this, in the collision integral there is need to add the integration both over momentum $\vec{p}_1'$ and over energy E_1' of incident particles after scattering. Also one should add the appropriate δ-functions, expressing the energy conservation law:

$$\begin{aligned} \left.\frac{\partial f}{\partial t}\right|_{\text{col}} &= n \int d\vec{p}_1\, d\vec{p}_1'\, dE_1' u\sigma\, d\Omega [f(\vec{p}\,')f(\vec{p}_1') - f(\vec{p})f(\vec{p}_1)] \\ &\quad \times \delta(\vec{p} + \vec{p}_1 - \vec{p}\,' - \vec{p}_1')\delta(E + E_1 - E' - E_1'). \end{aligned} \tag{3.65}$$

Considering the particle scattering as a system transition from state $\vec{p}, \vec{p}_1$ in state $\vec{p}\,'\vec{p}_1'$, it should be introduced the concept of the transition probability $W(\vec{p}, \vec{p}_1; \vec{p}\,', \vec{p}_1')$ determining its by relationship

$$dE_1' u\sigma d\Omega = d\vec{p}\,' W(\vec{p}, \vec{p}_1; \vec{p}\,', \vec{p}_1').$$

This yields the following entry of the collision integral:

$$\begin{aligned} \left.\frac{\partial f}{\partial t}\right|_{\text{col}} &= n \int d\vec{p}_1\, d\vec{p}\,'\, d\vec{p}_1' W(\vec{p}, \vec{p}_1; \vec{p}\,', \vec{p}_1')[f(\vec{p}\,')f(\vec{p}_1') \\ &\quad - f(\vec{p})f(\vec{p}_1)]\delta(\vec{p} + \vec{p}_1 - \vec{p}\,' - \vec{p}_1')\delta(E + E_1 - E' - E_1'). \end{aligned} \tag{3.66}$$

Upon analyzing the collision integral structure (3.66), it is easy to note that it is falling into two contributions, describing an arrival of particles in the state with momentum $\vec{p}$ and escape them from this state. In order to such a representation would be possible, the transition probability $W(\vec{p},\vec{p}_1;\vec{p}',\vec{p}_1')$ must satisfy the condition

$$W(\vec{p},\vec{p}_1;\vec{p}',\vec{p}_1') = W(\vec{p}',\vec{p}_1';\vec{p},\vec{p}_1). \tag{3.67}$$

The relation (3.67) is a special case for demonstrating the detailed balancing principle, which, in the given case, is reduced simply to the fact that the mechanical (quantum-mechanical) transition probabilities between states in forward and reverse directions are equal.

3.2.7 Derivation of the Boltzmann equation from the Bogoliubov equations chain

We write the first two equations of the BBGKY equations chain for the distribution functions F_1 and F_2 in the linear approximation for the parameter r_0^3/v. Having opened the Poisson brackets in (3.29) and (3.23), we get the following expressions:

$$\left(\frac{\partial}{\partial t} + \vec{F}(\vec{r})\vec{\nabla}_p + \frac{\vec{p}}{m}\vec{\nabla}_r\right)F_1 = \frac{1}{v}\int d\vec{r}_1\, d\vec{p}_1 \frac{\partial\Phi(|\vec{r}-\vec{r}_1|)}{\partial\vec{r}}\frac{\partial F_2}{\partial\vec{p}}, \tag{3.68}$$

$$\left(\frac{\partial}{\partial t} + \frac{\vec{p}}{m}\vec{\nabla}_r + \frac{\vec{p}_1}{m}\vec{\nabla}_{r_1}\right)F_2 - \frac{\partial(\Phi(|\vec{r}-\vec{r}_1|) + U(\vec{r}))}{\partial\vec{r}}\frac{\partial F_2}{\partial\vec{p}} - \frac{\partial(\Phi(|\vec{r}-\vec{r}_1|) + U(\vec{r}_1))}{\partial\vec{r}_1}\frac{\partial F_2}{\partial\vec{p}_1} = 0. \tag{3.69}$$

Since the integral term in the right side of (3.68) already has the first power of the small parameter r_0^3/v, we have omitted the integral term on the right side of equation (3.69), containing the first power of the small r_0^3/v parameter (the interaction potential differs from zero only inside a sphere of radius r_0). Now, the equation for the distribution function F_2 turns into the Liouville equation for the two-particle distribution function.

In looking for a solution of the set of equations (3.68) and (3.69) satisfying the principle of spatial correlation attenuation, which drops if particles are sufficiently far away from each other, the pair correlation function can be written as a product of the one-particle functions:

$$F_2(t,\vec{p},\vec{r},\vec{p}_1,\vec{r}_1)|_{|\vec{r}-\vec{r}_1|\to\infty} = F_1(t,\vec{r},\vec{p})F_1(t,\vec{r}_1,\vec{p}_1). \tag{3.70}$$

Expression (3.70) can be regarded as a boundary condition for the distribution function; consequently it allows for finding a physically meaningful solution.

Since equation (3.69) is the Liouville equation for the two-particle distribution function upon full neglecting collisions with other particles, the function that remains constant during the motion along the phase trajectory will be a solution of this equation:

$$
\begin{aligned}
F_2(t, x(t, x_0), x_1(t, x_0)) &= F_2(t-\tau, x(t-\tau, x_0), x_1(t-\tau, x_0)) \\
&= S_{-\tau}(x, x_1) F_2(t-\tau, x(t, x_0), x_1(t, x_0)).
\end{aligned}
\tag{3.71}
$$

The quantities x, x_1 in formula (3.71) are used to denote the aggregate of the coordinate and momentum of a particle. The quantity x_0 denotes a set of coordinates and momentum of two particles at initial time. The entry $x(t, x_0)$ implies that coordinate and momentum of the particle are estimated for solving the mechanical problem with the initial condition $\{x, x_1\} = x_0$. In writing the second part of (3.71) there has been used the $S_{-\tau}(x, x_1)$ operator, which shifts the particles along a phase trajectory for the time τ interval:

$$
S_{-\tau}(x, x_1) = e^{-iL_2\tau}, \quad iL_2 A = [A, H],
$$

where iL_2 is for the Liouville operator of the two particles.

Assume now that the τ time is so large and the two particles are moving away from each other to a distance exceeding the characteristic correlation radius. In this case, the two-particle distribution function is falling into the product of the one-particle distribution functions. Then, having continued the chain of equalities (3.71), the following expression appears:

$$
\begin{aligned}
F_2(t, x, x_1) &= S_{-\tau}(x, x_1) F_2(t-\tau, x, x_1) \\
&= S_{-\tau}(x, x_1) F_1(t-\tau, x) F_1(t-\tau, x_1) \\
&= S_{-\tau}(x, x_1) S_\tau(x) S_\tau(x_1) F_1(t, x) F_1(t, x_1),
\end{aligned}
\tag{3.72}
$$

where $S_\tau(x)$, $S_\tau(x_1)$ are the one-particle operators of evolution.

Expression (3.72) defines the relationship between the one-particle and two-particle distribution functions, taken at the same moment of time. This equation is valid in the approximation of a low-density gas when $r_0^3/v \ll 1$ for mechanical systems, in which the spatial correlation attenuation is implemented.

We will be concerned with a spatially homogeneous case. Then the dependence of the one-particle function F_1 on coordinates can only be parametric, i. e. related with graded-change in external conditions, for example such as the presence of a temperature gradient. Besides, the function is independent on coordinates at distances of order of the free path. Therefore, the relation (3.72) can be simplified further:

$$
F_2(t, x, x_1) = S_{-\tau}(x, x_1) S_\tau(\vec{p}) S_\tau(\vec{p}_1) F_1(t, \vec{p}) F_1(t, \vec{p}_1).
\tag{3.73}
$$

This representation for the two-particle distribution functions is formally exact, if the system is spatially homogeneous and the correlation attenuation principle holds.

Upon analyzing equation (3.69) the partial derivative over time in the given approximation can be omitted. Indeed, explicit dependence of the distribution function on time can only occur due to external interactions with respect to the chosen system consisting of two particles. These interactions can be collisions with other particles or interactions with alternating external field which has a characteristic frequency ω. Assume that the following condition is met:

$$\frac{\partial F_2}{\partial t} \sim \omega F_2 \ll \left.\frac{\partial F_2}{\partial t}\right|_{\text{col}} \sim \frac{F_2}{\tau}.$$

The restriction imposed is not very stringent, since $\tau \sim 10^{-14}$ s and the condition $\omega \ll 1/\tau$ is well satisfied up to frequencies of optical range.

In this case, the partial derivative $\partial F_2/\partial t$ and the collision integral have the same order and, consequently, the order of r_0^3/v. Since we are constructing a kinetic equation in the first approximation for this parameter while the right side of (3.68) it already has, then linear terms for the function F_2 in equation (3.69) can be omitted. In essence, the rejection of the private derivative in equation (3.69) is equivalent to assuming that the collision of two particles occurs at stationary conditions.

We integrate (3.69) over $\vec{r}_1$ and $\vec{p}_1$. Introducing the relative coordinate $\vec{R} = \vec{r}_1 - \vec{r}$, the following expression appears as

$$\int \frac{\vec{p}_1 - \vec{p}}{m} \frac{\partial F_2(\vec{R}, \vec{p}, \vec{p}_1)}{\partial \vec{R}}\, d\vec{p}_1\, d\vec{R} = \int \frac{\partial \Phi(|\vec{r} - \vec{r}_1|)}{\partial \vec{r}} \frac{\partial F_2}{\partial \vec{p}}\, d\vec{r}_1\, d\vec{p}_1. \tag{3.74}$$

The last entry takes into account that

$$\int \frac{\partial \Phi(|\vec{r} - \vec{r}_1|)}{\partial \vec{r}_1} \frac{\partial F_2}{\partial \vec{p}_1}\, d\vec{r}_1\, d\vec{p}_1 = 0$$

in virtue of the condition $F_2|_{p_1=\pm\infty} = 0$ and $U(\vec{r}) = 0$.

The right side of (3.74) coincides with the collision integral for the one-particle distribution function up to the multiplier and therefore the right side in formula (3.68) can be presented as

$$\left.\frac{\partial F_1}{\partial t}\right|_{\text{col}} = \frac{1}{V} \int \frac{\vec{p}_1 - \vec{p}}{m} \frac{\partial F_2(\vec{R}, \vec{p}, \vec{p}_1)}{\partial \vec{R}}\, d\vec{p}_1\, d\vec{R}. \tag{3.75}$$

In integrating the right-hand side of (3.75) over R we pass to a polar coordinate system having selected an origin at the point $\vec{r}$, where one of the particles is located. Let the Z-axis be directed along the relative velocity vector $\vec{u} = (\vec{p}_1 - \vec{p})/m$ and polar coordinates will be denoted through the letters a and φ (see Figure 3.2). Besides, the expressing of the function F_2 via the one-particle distribution functions (3.73) is needed to perform. We assume that the particles are elastic balls of radius r_0, so the interaction region will be sphere of radius $2r_0$. This results in the following expression:

$$\left.\frac{\partial F_1}{\partial t}\right|_{\text{col}} = \frac{1}{V} \int d\vec{p}_1 \int_0^{\infty} a\, da \int_0^{2\pi} d\varphi u \int_{-\infty}^{\infty} dz \frac{d}{dz} S_{-\tau}(x, x_1) F_1(t, \vec{p}) F_1(t, \vec{p}_1). \tag{3.76}$$

In writing this expression the evolution operators for one-particle distribution function have been omitted because the momentum of each particle is not altered when a particle moves along the phase trajectory. Therefore,

$$S_\tau(\vec{p})S_\tau(\vec{p}_1)F_1(t,\vec{p})F_1(t,\vec{p}_1) = F_1(t,\vec{p})F_1(t,\vec{p}_1).$$

Now we carry out integrating over z in the right side of (3.76):

$$\left.\frac{\partial F_1}{\partial t}\right|_{col} = \frac{1}{v}\int d\vec{p}_1 \int_0^\infty a\,da \int_0^{2\pi} d\varphi u S_{-\tau}(x,x_1)F_1(t,\vec{p})F_1(t,\vec{p}_1)|_{-\infty}^\infty. \qquad (3.77)$$

When substituting the lower limit $z = -\infty$, the particles are already separated by a sufficiently large distance and no interaction exists between them. These particles are moving away farther apart from each other with the aid of the evolution operator $S_{-\tau}(x,x_1)$ and therefore the particle momentum is not changed. As a result, we get

$$S_{-\tau}(x,x_1)F_1(t,\vec{p})F_1(t,\vec{p}_1)\big|_{-\infty} = -F_1(t,\vec{p})F_1(t,\vec{p}_1).$$

When substituting the upper limit $z = \infty$, the particles also prove to be separated by a sufficiently large distance and therefore do not interact with each other. But the evolution operator $S_{-\tau}(x,x_1)$, shifting the particles along the phase trajectory, makes them interact. Therefore, if the condition $a < 2r_0$ for the impact-parameter is met, one obtains

$$S_{-\tau}(x,x_1)F_1(t,\vec{p})F_1(t,\vec{p}_1)\big|^\infty = F_1(t,\vec{p}')F_1(t,\vec{p}_1').$$

The impact-parameter being $a > 2r_0$, the particles will not collide with each other and their momenta remain unchanged as a result of the evolution operator action:

$$S_{-\tau}(x,x_1)F_1(t,\vec{p})F_1(t,\vec{p}_1)\big|^\infty = F_1(t,\vec{p})F_1(t,\vec{p}_1).$$

Given the above results, the collision integral in the case of the elastic ball interaction can be written as

$$\left.\frac{\partial F_1}{\partial t}\right|_{col} = \frac{1}{v}\int d\vec{p}_1 \int_0^{2r_0} a\,da \int_0^{2\pi} d\varphi u[F_1(t,\vec{p}')F_1(t,\vec{p}_1') - F_1(t,\vec{p})F_1(t,\vec{p}_1)]. \qquad (3.78)$$

It is easy to see that the expression obtained for the collision integral is identical with the collision integral in the Boltzmann equation (3.59).

3.2.8 The Fokker–Planck equation

Processes, in which a change of the parameters of the distribution function of each elementary scattering act in comparison with their characteristic values is small, make up a significant part of transport phenomena. The momentum relaxation of a heavy particle in a gas consisting of light particles may serve as a typical case in point of such a problem. The concentration of heavy particles is assumed small, and therefore collisions between heavy particles can be neglected. When the heavy particle collides with a light particle, momentum of the heavy particle changes slightly as in magnitude and direction. We denote transmission momentum in an elementary scattering act by the letter $\vec{q}$, $\vec{p} \gg \vec{q}$ and find an equation which holds for the one-particle distribution function $f(t,\vec{p})$. Here and below we replace the notation F_1 for the one-particle distribution function for the sake of simplification.

Assign the notation $w(\vec{p},\vec{q})d\vec{q}$ for the number of transitions of heavy particles from the state with the momentum $\vec{p}$ in the state with momentum $\vec{p}-\vec{q}$ per unit time. Then the quantity $w(\vec{p}+\vec{q},\vec{q})d\vec{q}$ is equal to the transition rate from the state $\vec{p}+\vec{q}$ in the state with momentum $\vec{p}$. As shown above, the collision integral in the kinetic equation can be written as the difference between two terms, one of which describes the particle transition rate in state with momentum $\vec{p}$ and another describes the particle escape rate from this state. Applying this principle, we can construct the collision integral for the heavy particle in a light gas [21]:

$$\left.\frac{\partial f}{\partial t}\right|_{\text{col}} = \int [w(\vec{p}+\vec{q},\vec{q})f(t,\vec{p}+\vec{q}) - w(\vec{p},\vec{q})f(t,\vec{p})]d\vec{q}. \tag{3.79}$$

According to assumptions made above, the quantity $w(\vec{p},\vec{q})$ rapidly decreases with increasing $\vec{q}$ (momentum transmission is small). Therefore, the quantity $\vec{q}$ is small compared with the momentum of particles $\vec{p}$. This circumstance allows for expanding the integral collision (3.79),

$$\begin{aligned} w(\vec{p}+\vec{q},\vec{q})f(t,\vec{p}+\vec{q}) \simeq{} & w(\vec{p},\vec{q})f(t,\vec{p}) + \vec{q}\frac{\partial}{\partial\vec{p}}[w(\vec{p},\vec{q})f(t,\vec{p})] \\ & + \frac{1}{2}q^\alpha q^\beta \frac{\partial^2}{\partial p^\alpha \partial p^\beta}[w(\vec{p},\vec{q})f(t,\vec{p})]. \end{aligned} \tag{3.80}$$

Substituting (3.80) into (3.79), the following expression can be written:

$$\left.\frac{\partial f}{\partial t}\right|_{\text{col}} = \frac{\partial}{\partial p^\alpha}\left[A'_\alpha f(t,\vec{p}) + \frac{\partial}{\partial p^\beta}B_{\alpha\beta}f(t,\vec{p})\right]; \tag{3.81}$$

$$A'_\alpha = \int q^\alpha w(\vec{p},\vec{q})\,d\vec{q}; \quad B_{\alpha\beta} = \frac{1}{2}\int q^\alpha q^\beta w(\vec{p},\vec{q})\,d\vec{q}. \tag{3.82}$$

The main distinguishing trait of the kinetic Fokker–Planck equation is to express the collision integral through averaged characteristics of a single scattering event. Equations (3.81) and (3.82) confirm this fact. The collision integral contains only the average characteristics of the scattering process and they can be expressed through constant A'_α and $B_{\alpha\beta}$. As shown below, in many practically important cases the number of constants can be reduced to only one constant.

Let us draw reader's attention to one other feature of the resulting collision integral. The right side of (3.81) presents the divergence of the particle number flux vector s_α in momentum space:

$$s_\alpha = -A'_\alpha f(t,\vec{p}) - \frac{\partial}{\partial p^\beta} B_{\alpha\beta} f(t,\vec{p}).$$

Similar principles have also been used in constructing the Fokker–Planck equation to describe the motion of Brownian particles (see formulae (2.29)–(2.33)), although there has been there the case of the particle flux in momentum space but not in coordinate one.

As far as the quantities A'_α and $B_{\alpha\beta}$ are just some constants, it is convenient to introduce a new constant A_α instead of the constant A'_α for further discussion, having defined it by relation

$$A_\alpha = A'_\alpha + \frac{\partial}{\partial p^\beta} B_{\alpha\beta}.$$

At the same time expression for the flux s_α can be significantly simplified:

$$s_\alpha = -A_\alpha f(t,\vec{p}) - B_{\alpha\beta} \frac{\partial}{\partial p^\beta} f(t,\vec{p}).$$

Indeed, the constants A_α and $B_{\alpha\beta}$ are not independent. It is easy to create a relationship between them, if equilibrium case will be considered. In equilibrium, the distribution function is known: it is the distribution function of Maxwell–Boltzmann. Further, the distribution functions will be normalized for a total number of particles in the sample n. It is worth recalling that earlier the concentration of particles was denoted by the letter n. This discrepancy in the notation is not principle, because for any estimates the sample volume is always assumed to be equal to unity:

$$f(t,\vec{p}) = \frac{n}{Z} \exp\left(-\frac{p^2}{2mk_\mathrm{B}T}\right), \quad Z = \int_{-\infty}^{\infty} d\vec{p} \exp\left(-\frac{p^2}{2mk_\mathrm{B}T}\right).$$

Besides, in equilibrium the flux s_α^0 must vanish. Hence, performing the necessary calculations, one obtains

$$s_\alpha^0 = -A_\alpha + B_{\alpha\beta} \frac{p^\beta}{mk_\mathrm{B}T} = 0, \quad A_\alpha = \frac{B_{\alpha\beta} p^\beta}{mk_\mathrm{B}T}.$$

If the transition probability $w(\vec{p},\vec{q})$ depends only on modulus of the vector $\vec{q}$, then $B_{\alpha\beta} = B\delta_{\alpha\beta}$ as follows from the definition of coefficients $B_{\alpha\beta}$ (3.82) by virtue of the symmetry conditions. In particular, the velocity of heavy particles being neglected in comparison with speed of light particles, such a situation can be realized. Then one can run the scattering analysis as the velocity of heavy particles is equal to $\vec{p}/m \simeq 0$.

In this case, entry of the collision integral for the Fokker–Planck equation turns out to be the simplest:

$$\left.\frac{\partial f}{\partial t}\right|_{\text{col}} = B\frac{\partial}{\partial p^\alpha}\left[\frac{p^\alpha}{mk_{\text{B}}T}f(t,\vec{p}) + \frac{\partial}{\partial p^\alpha}f(t,\vec{p})\right]. \tag{3.83}$$

Problem 3.3. Determine mobility μ of a heavy particle into a light gas with the help of the Fokker–Planck equation.

Solution. Let a force $\vec{F} = e\vec{E}$ act on the charged heavy particle where e is the particle charge in the presence of an external electric field, defined by the electrical vector $\vec{E}$. Assume that the electric field is homogeneous and constant. Under these conditions, a distribution function depends only on momentum and does not depend on coordinates and time. Then, considering (3.59), a kinetic equation with the collision integral for the distribution function $f(\vec{p})$ can be written in the form of (3.83):

$$eE^\alpha\frac{\partial f}{\partial p^\alpha} = B\frac{\partial}{\partial p^\alpha}\left[\frac{p^\alpha}{mk_{\text{B}}T}f(t,\vec{p}) + \frac{\partial}{\partial p^\alpha}f(t,\vec{p})\right]. \tag{3.84}$$

As far as the left and right sides of equation (3.84) contain identical derivatives $\partial/\partial p^\alpha$, the following equality should be fulfilled up to a negligible constant:

$$eE^\alpha f(\vec{p}) = B\left[\frac{p^\alpha}{mk_{\text{B}}T}f(t,\vec{p}) + \frac{\partial}{\partial p^\alpha}f(t,\vec{p})\right]. \tag{3.85}$$

In looking for the solution of equation (3.85) in the linear approximation over external forces, one can write down the non-equilibrium distribution function as the sum of the equilibrium distribution function f_0 and small correction $\delta f : f = f_0 + \delta f$. Since there exists a first order in the external force on the left side of equation (3.85), here we replace f by f_0. Substituting the equilibrium distribution function f_0 in the collision integral gives a zero. Therefore, in the linear approximation over the external forces, a simple equation for the correction of the distribution function appears as

$$\frac{\partial}{\partial p^\alpha}\delta f + \frac{p^\alpha}{mk_{\text{B}}T}\delta f = \frac{eE^\alpha}{B}f_0. \tag{3.86}$$

The equation found is a linear inhomogeneous equation. It is easy to verify that the general solution of a homogeneous equation is the equilibrium distribution function f_0. The particular solution of inhomogeneous equation will be sought in the form

$$\delta f = C^\alpha p^\alpha f_0,$$

where C^α are unknown factors.

Substituting the trial solution of equation (3.86), one finds values of the coefficients C^α and an explicit form of correction to the distribution function δf:

$$C^\alpha = \frac{eE^\alpha}{B}; \quad \delta f = \frac{eE^\alpha}{B} p^\alpha f_0.$$

In the case of one type of charge carriers, the electrical conductivity σ and the mobility μ are determined from the phenomenological expressions

$$\vec{J} = \sigma\vec{E} = en\mu\vec{E}, \quad \sigma = en\mu. \tag{3.87}$$

The total current $\vec{J}$ can be expressed as an average velocity of charge carrier drift in an electric field $\vec{v}_{dr}$:

$$\vec{J} = en\vec{v}_{dr}.$$

Upon comparing this result with (3.87), one obtains

$$\mu = \frac{v_{dr}}{E}.$$

Thus, mobility is numerically equal to an average velocity of charge carrier drift if the electrical intensity E is equal to unity.

The total electric current in the sample is determined by a correction to the distribution function δf. Substituting this value in the definition of the charge flux, we have

$$J^\alpha = en\mu E^\alpha = \frac{e^2}{B} \int v^\alpha E^\beta p^\beta f_0(\vec{p})\, d\vec{p}. \tag{3.88}$$

Hence it follows that the expression for the mobility μ appears as

$$\mu = \frac{2e}{3nB} \int \frac{p^2}{2m} f_0(\vec{p})\, d\vec{p}. \tag{3.89}$$

In writing the expression (3.89) we have relied on the fact that the function actually depends on modulus of this vector, and therefore, integrating the expression (3.88) one can write down the following relation:

$$v^\alpha p^\beta = \frac{1}{3}\frac{p^2}{m}\delta_{\alpha\beta},$$

where $\delta_{\alpha\beta}$ is the Kronecker symbol.

Integral on the right side of (3.89) represents the average energy of particles. Indeed, going over to the integration in a spherical coordinate system, taking into account the conditions of normalization of the functions $f(\vec{p})$ and setting $\varepsilon = p^2/2m$, one finds the following:

$$\frac{4\pi n \int_0^\infty \varepsilon \exp(-\varepsilon/k_B T) p\, dp^2}{4\pi \int_0^\infty \exp(-\varepsilon/k_B T) p\, dp^2} = \frac{n \int_0^\infty \varepsilon^{3/2} \exp(-\varepsilon/k_B T)\, d\varepsilon}{\int_0^\infty \varepsilon^{1/2} \exp(-\varepsilon/k_B T)\, d\varepsilon} = nk_B T \frac{\Gamma(5/2)}{\Gamma(3/2)}. \tag{3.90}$$

Using the well-known relations for the gamma-function

$$\Gamma(n) = \int_0^\infty x^{n-1}e^{-x}\,dx; \quad \Gamma(5/2) = 3/2 \cdot \Gamma(3/2),$$

one obtains a simple formula for the mobility of heavy particles into a light gas

$$\mu = \frac{ek_B T}{B}.$$

In this expression the constant B can be regarded as a phenomenological parameter to be found from experiment or estimated from the first principles, having defined the explicit form of expression for the transition probability in formula (3.82).

3.3 Solving for kinetic equations

3.3.1 The solution of the Boltzmann equation for the equilibrium state

The analysis of the problem to solve the kinetic Boltzmann equation should start with the simplest case of a system's equilibrium state. In equilibrium, a distribution function in the absence of external forces driving the system out of its equilibrium contains no explicit dependence both on coordinates and on time. Consequently, the left-hand side of the expression (3.64) vanishes and a kinetic equation for the equilibrium state is reduced to the vanishing collision integral:

$$\int u\sigma(\Omega, u)(f'f_1' - ff_1)\,d\vec{p}_1 d\Omega = 0. \tag{3.91}$$

Note that we have still been writing the kinetic equation for the one-particle functions F_1, and it is this function that is involved in formula (3.59). It is worth pointing out that the notation $F_1(t, \vec{r}, \vec{p}) = f$ introduced by (3.55) must not be misleading. It is convenient to pass on to the more usual definition of the distribution function which is normalized to the concentration. Since the one-particle distribution function $F_1(t, \vec{r}, \vec{p})$ is related to the function $f(t, \vec{r}, \vec{p})$, normalized to the concentration, by the simple relation $f(t, \vec{r}, \vec{p}) = nF_1(t, \vec{r}, \vec{p})$, it suffices to omit the expression for the concentration in the collision integral (3.59) in order to pass on to the new notations for writing the kinetic equation. The functions appearing in the kinetic equation are assumed to be normalized to concentration and the above changeover has been already achieved. That is why when writing the collision integral (3.91) the expression for the concentration n in front of the integral has been omitted.

It is obvious that the vanishing of (3.91) is attained by fulfilling the condition

$$f(\vec{p}\,')f(\vec{p}_1') = f(\vec{p})f(\vec{p}_1).$$

Also, the equality holds for taking the logarithm:

$$\ln f(\vec{p}') + \ln f(\vec{p}_1') = \ln f(\vec{p}) + \ln f(\vec{p}_1). \tag{3.92}$$

Equation (3.92) can be interpreted as some conservation law: the sum of logarithms of the distribution function of particles before the collision is equal to the sum of logarithms of the distribution function of particles after the collision. It is known that the pair elastic collisions are characterized by the presence of additive conservation laws of momentum, energy and particle number or mass. The conservation law of the quantity A is called *additive*, if this quantity can be represented as the sum of A_i for all parts of the system provided no interaction exists between them. No other additive laws of conservation are in this problem. All in all, angular momentum is also an additive integral of motion, but in the event of the molecular rotation and change momentum in the collision not being taken into account, the motion integral can be ignored. Therefore, the distribution function logarithm can only depend on the above five additive invariants of the collision:

$$\ln f(\vec{p}) = A\frac{p^2}{2m} + \vec{B}\vec{p} + C, \tag{3.93}$$

where A, $\vec{B}$ and C are some constants. We choose these constants so that the momenta of the distribution function would have meaningful physical values:

$$\int d\vec{p} f(\vec{p}) = n, \tag{3.94}$$

$$\int d\vec{p} f(\vec{p})\vec{p} = nm\vec{v}_0, \tag{3.95}$$

$$\int d\vec{p} f(\vec{p})\frac{(\vec{p} - m\vec{v}_0)^2}{2m} = \frac{3}{2}k_B Tn. \tag{3.96}$$

The momentum of zero order of (3.94) is a condition of the distribution function normalization; n is the total number (or concentration) of the particles in the sample. The first momentum (3.95) is a total momentum of the particle system; $\vec{v}_0$ is the average drift speed while the second momentum of (3.96) is equal to the total energy of chaotic motion of particles. It is easy to see that, this being so with respect to such a choice of the constants A, $\vec{B}$ and C, the distribution function has the form

$$f(\vec{p}) = \frac{n}{(2\pi m k_B T)^{3/2}} \exp\left(-\frac{(\vec{p} - m\vec{v}_0)^2}{2mk_B T}\right). \tag{3.97}$$

Thus, for the equilibrium case a solution of the kinetic equation (3.91) is to be the well-known Maxwell–Boltzmann distribution function.

The results of (3.94)–(3.97) can be summarized as follows. First, the previous consideration can be applied to local-equilibrium state. In this case the distribution function will depend parametrically on coordinates and time through the local concentration $n(\vec{r}, t)$, the local temperature $T(\vec{r}, t)$, and the drift velocity $\vec{v}_0(\vec{r}, t)$. Such an approach allows for making use of the Boltzmann equation to derive the hydrodynamic equations of balance. In the next chapter, equations of momentum balance, energy and particle number for a system of hot electrons in conductive crystals will be obtained using this method.

Second, it is not hard to generalize the results (3.94)–(3.97) both for the case when particles of a gas are in a stationary potential field of force $U(\vec{r})$ and for the case of the inelastic particle scattering. These results can be found in [22].

3.3.2 The Boltzmann *H*-theorem

In contrast to the dynamic equations, which are time reversible, the Boltzmann kinetic equation is noninvariant with respect to time-reversal operation. In order to verify this, we apply the time-reversal operation ($t \to -t$, $\vec{p} \to -\vec{p}$, $\vec{r} \to \vec{r}$) to the kinetic equation (3.59). Having adopted the notation $\hat{f} = f(-t, -\vec{p}, \vec{r})$, we get the triple integral,

$$-\frac{\partial \hat{f}}{\partial t} - \frac{\vec{p}}{m}\vec{\nabla}_r \hat{f} - \vec{F}\vec{\nabla}_p \hat{f} = \int_0^{2\pi} d\varphi \int_0^{2r_0} a\,da \int (\hat{f}'\hat{f}_1' - \hat{f}\hat{f}_1) u\, d\vec{p}_1. \tag{3.98}$$

After completing the time-reversal operation, we see that the left-hand member of equation (3.98) has changed sign for the function $\hat{f} = f(-t, -\vec{p}, \vec{r})$, while the right-hand member has not. Irreversibility of the Boltzmann equation is accounted for by the fact that only those solutions, which satisfy the correlation attenuation principle are selected from all possible solutions of the Bogoliubov hierarchy. Boltzmann's contemporaries made a slashing criticism of him for his breakaway from the ideas of determinism. From the standpoint of contemporary knowledge, as mentioned in Section 3.1.2, an exact solution of the dynamic problem in systems demonstrating dynamic chaos, is completely pointless. Therefore, it is necessary to go over to a statistical description to get results making practical sense. It is this idea that was implemented by Boltzmann.

The irreversible behavior of a system, whose description is produced by language of the distribution functions satisfying the Boltzmann equation, becomes obvious if the quantity of H will be determined due to Boltzmann. In other words, it is required to find the Lyapunov function (see (1.97)):

$$H(t) = \int d\vec{p} f(\vec{p}, t) \ln f(\vec{p}, t). \tag{3.99}$$

One must mention that the above function is a nonincreasing function of time. It is obvious that non-decreasing quantity of $S(t) = -H(t)$ can be also ascertained, it will coincide with the system entropy up to a dimension factor. The nonincreasing function $H(t)$ being determined by the integral of (3.99) is commonly referred to as Boltzmann's H-theorem.

We give a proof of this theorem for the case of spatially homogeneous distribution of a gas in the absence of external forces. Then the kinetic equation describes the relaxation of the gas relating to the equilibrium state and has the simplest form:

$$\frac{\partial f(\vec{p})}{\partial t} = \int u\sigma(\Omega, u)[f(\vec{p}\,')f(\vec{p}_1') - f(\vec{p})f(\vec{p}_1)]\, d\vec{p}_1\, d\Omega. \tag{3.100}$$

Now it would be advisable to find a derivative of the function $H(t)$ over time and to show that $H(t)$ is always nonpositive. Upon carrying out the time-differentiation in formula (3.99), the following expression appears:

$$\frac{\partial H(t)}{\partial t} = \int d\vec{p}[1 + \ln f(\vec{p})]\frac{\partial f(\vec{p})}{\partial t}.$$

We substitute the value of the derivative of the distribution function from the kinetic equation (3.100) in this expression, then we obtain

$$\frac{\partial H(t)}{\partial t} = \int u\sigma(\Omega, u)[f(\vec{p}\,')f(\vec{p}_1') - f(\vec{p})f(\vec{p}_1)] \times [1 + \ln f(\vec{p})]\, d\vec{p}\, d\vec{p}_1\, d\Omega. \tag{3.101}$$

As far as the integration is performed over $\vec{p}$ and $\vec{p}_1$ at the same limits, the expression (3.101) can be symmetrized by writing it as

$$\frac{\partial H(t)}{\partial t} = \frac{1}{2}\int u\sigma(\Omega, u)[f(\vec{p}\,')f(\vec{p}_1') - f(\vec{p})f(\vec{p}_1)] \times [2 + \ln(f(\vec{p})f(\vec{p}_1))]\, d\vec{p}\, d\vec{p}_1\, d\Omega. \tag{3.102}$$

The result found can be undergone by further symmetrization, because $d\vec{p}d\vec{p}_1 = d\vec{p}\,'d\vec{p}_1'$, $u = -u'$. Therefore, the following expression holds:

$$\frac{\partial H(t)}{\partial t} = \frac{1}{4}\int u\sigma(\Omega, u)[f(\vec{p}\,')f(\vec{p}_1') - f(\vec{p})f(\vec{p}_1)] \times [2 + \ln(f(\vec{p})f(\vec{p}_1)) - 2 - \ln(f(\vec{p}\,')f(\vec{p}_1'))]\, d\vec{p}\, d\vec{p}_1\, d\Omega. \tag{3.103}$$

Only the expressions in square brackets being considered, one may observe that the integrand on the right-hand side of (3.103) can be presented as

$$f(x, y) = (x - y)\ln\frac{y}{x},$$

where $x = f(\vec{p}\,')f(\vec{p}_1')$, $y = f(\vec{p})f(\vec{p}_1)$. Now, it is clear that the function $f(x, y)$ is negative for all values of $x \neq y$. The vanishing is achieved only when the $x = y$ equality is valid. Since the u relative velocity of the particles before the collision and the $\sigma(\Omega, u)$ scattering cross-section are positive, the integrand on the right-hand member of equation (3.103) is a nonpositive magnitude over the entire integrated interval:

$$\frac{\partial H(t)}{\partial t} \leq 0.$$

This brings in evidence the Boltzmann H-theorem.

Note that the Boltzmann H-theorem proved above is equivalent to the second law of thermodynamics. It reads: a system's entropy cannot decrease. In fact the Boltzmann H-theorem is even a more general statement, since it also holds for systems being far away from their equilibrium state. This theorem allows one to claim that the Lyapunov function which is in some sense equivalent to the entropy can be defined for a non-equilibrium state. The special literature [23, 24] discusses both other formulations of the proof of the H-theorem and difficulties related to irreversibility of solutions of the Boltzmann equation.

3.3.3 The Hilbert expansion

The Boltzmann kinetic equation (3.64) is a non-linear integro-differential equation; therefore, the process of finding solutions satisfying initial and boundary conditions is an extremely complex problem. It is no wonder that there is still no complete analysis of the existence and uniqueness of the solutions in a general form. At present, to the best of our knowledge, results relating to this issue are modest enough, and the books [23, 24] discuss most of them. The main direction of a practical use of the Boltzmann equation for solving problems of physical kinetics is to construct a perturbation theory.

The method of a linearized collision integral is the simplest and physically clear approach. In this case the perturbation theory is built in powers of a system's deviation from an equilibrium state, but a solution of the kinetic equation $f(\vec{p}, t)$ needs to be sought in the form of the equilibrium distribution function $f_0(\vec{p})$ and a small correction $\delta f(\vec{p}, t)$. The linearization of the collision integral is due to retaining only linear terms in $\delta f(\vec{p}, t)$. For the linearized Boltzmann equation, there are a number of strict results relating to the existence and uniqueness of problem solutions with initial and boundary conditions [24]. A disadvantage of this approach is that the analysis holds only for weakly non-equilibrium states.

Other methods of the perturbation theory consist in expanding the distribution function into a power-series of some small parameter and in constructing an iterative scheme for the successive determination of expansion coefficients. In 1912 German mathematician Hilbert was first to apply this technique to an analysis of solutions of the Boltzmann equation. In a nutshell, consider a description of its nature and a result of the Hilbert expansion.

Let us first estimate the order of different terms in the Boltzmann equation. If one adopts that ω is a characteristic frequency of external effects, v corresponds to a characteristic velocity of particles, d denotes a characteristic size of spatial non-homogeneity of a system, l is mean free path of a particle, the time of free path of the particles is designated as $l/v = \tau$, one may estimate the order of various members in the Boltzmann equation:

$$\frac{\partial f}{\partial t} \sim \omega f, \quad \vec{v}\vec{\nabla}_r f \sim \frac{v}{d}f, \quad \left.\frac{\partial f}{\partial t}\right|_{col} \sim \frac{f}{\tau} \sim \frac{v}{l}f.$$

It follows that there may be introduced two dimensionless parameters characterizing a relative magnitude of collision integral compared with contributions of the summands on the left-hand side of the equation, $\omega\tau$ and l/d. One can assume that these parameters as a first approximation are close in value, consequently, the relative contribution of the collision integral is determined by only one parameter, $\mathrm{Kn} = l/d$, which is called the *Knudsen number*. For small values of the Knudsen number, the mean free path is small and collisions occur rather often and the contribution of the collision integral is large. At large values of the Knudsen number $\mathrm{Kn} \gg 1$, the free-molecular-flow regime in a gas is possible, thus the collision integral in the kinetic equation can be omitted. This analysis is a point of departure of the perturbation theory which may be constructed for the kinetic equations for two different limiting cases, i. e. when the Knudsen number $\mathrm{Kn} \to 0$, and when this number is large and $\mathrm{Kn} \to \infty$.

The Hilbert expansion corresponds to the first case where the Knudsen number $\mathrm{Kn} = \epsilon$ is a small parameter (dense gases). We write down the kinetic equation (3.64), introducing $I(f,f)$ as a symbolic notation for the collision integral:

$$\epsilon\left[\frac{\partial f}{\partial t} + \frac{\vec{p}}{m}\vec{\nabla}_r f + \vec{F}\vec{\nabla}_p f\right] = I(f,f). \tag{3.104}$$

The ϵ quantity has been introduced on the left-hand side of (3.104) in constructing an iterative procedure for selecting members of the same order of the ϵ parameter.

A solution of the kinetic equation f should be expanded into an infinite series form in powers of ϵ:

$$f = f^{(0)} + \epsilon f^{(1)} + \epsilon^2 f^{(2)} + \cdots. \tag{3.105}$$

After finding all terms of the expansion set the ϵ parameter should be set equal to unity and plugged back into the original definitions. Following this line of reasoning, the expansion is a formal subterfuge for selecting members of the same magnitude order when appropriate.

We substitute the expansion (3.105) into equation (3.104) and equate the terms containing zero, the first, the second, etc. orders of the ϵ parameter on the left and right sides of equation (3.104). As a result, an infinite sequence of the equations, enabling one to determine the expansion coefficients $f^{(i)}$ can be written:

$$0 = I(f^{(0)},f^{(0)}), \tag{3.106}$$

$$\frac{\partial f^{(0)}}{\partial t} + \frac{\vec{p}}{m}\vec{\nabla}_r f^{(0)} + \vec{F}\vec{\nabla}_p f^{(0)} = I(f^{(1)},f^{(0)}) + I(f^{(0)},f^{(1)}), \tag{3.107}$$

$$\frac{\partial f^{(1)}}{\partial t} + \frac{\vec{p}}{m}\vec{\nabla}_r f^{(1)} + \vec{F}\vec{\nabla}_p f^{(1)} = I(f^{(0)},f^{(2)}) + I(f^{(2)},f^{(0)}) + I(f^{(1)},f^{(1)}), \tag{3.108}$$

$$\cdots.$$

Equation (3.106) allows the $f^{(0)}$ to be determined. It is easy to see that, in fact, this equation coincides with equation (3.91). A solution of (3.106) is the quasi-equilibrium distribution function (3.97):

$$f^{(0)}(\vec{p},\vec{r},t) = \frac{n}{(2\pi m k_B T)^{3/2}} \exp\left(-\frac{(\vec{p}-m\vec{v}_0)^2}{2mk_B T}\right), \tag{3.109}$$

where the n, v_0, T parameters are local-equilibrium quantities, depending on coordinates and time.

Let us analyze the structure of equations (3.107) and (3.108). Further calculations are rather cumbersome. As far as only fundamental aspects of the method are interesting rather than the applied ones, the term proportional to an external force in equations (3.107), (3.108) can be omitted without any loss. Each of these equations enables determining the next correction in the expansion (3.105). Thus, in principle, all terms of the expansion (3.105) can be found, but to do this, a linear inhomogeneous integral equation has to be solved at each step of the procedure. In looking for the next correction the integral equation structure $f^{(n)} = f^{(0)}h^{(n)}$ remains the same and can be written in the symbolic form

$$\left[\frac{\partial}{\partial t} + \frac{\vec{p}}{m}\vec{\nabla}_r\right] f^{(0)}h^{(n-1)} = Lh^{(n)} + S^{(n)}, \quad n = 1, 2\ldots, \tag{3.110}$$

$$Lh^{(n)} = I(f^{(0)}, f^{(0)}h^{(n)}) + I(f^{(0)}h^{(n)}, f^{(0)}), \quad h^{(0)} = 1, \tag{3.111}$$

$$S^{(1)} = 0, \quad S^{(n)} = \sum_{k=1}^{n-1} I(f^{(0)}h^{(k)}, f^{(0)}h^{(n-k)}), \quad n = 2, 3\ldots, \tag{3.112}$$

where L is the linear integral operator and $S^{(n)}$ is some function, whose explicit form is known, if previous terms of the expansion (3.105) are found. As far as the quantities of $f^{(0)}$ are known, one can go over to a set of equations for finding the functions $h^{(n)}$, whose structure is identical and at each step is a nonhomogeneous linear integral Fredholm's equation of the second kind:

$$Lh^{(n)} = g^{(n)}. \tag{3.113}$$

Solutions of the homogeneous equation $Lh^{(n)} = 0$ for the case of an elastic scattering are additive invariants of the collision ψ_α, $\alpha = 1, 2, \ldots, 5$, i. e. a constant, three momentum components and kinetic energy. Note that these quantities are eigenfunctions of the equation $Lh = \lambda h$, corresponding to the eigenvalue $\lambda = 0$. A solution of the nonhomogeneous equation (3.113) is equivalent to finding the L^{-1} inverse linear operator, which in the general case is impossible, since $\lambda = 0$ is one of the possible eigenvalues of L. Therefore, in addition, it is required that the $g^{(n)}$ vector, defining the nonhomogeneity, would be orthogonal to ψ_α. Now, finding the solution of such a class of functions is becoming possible. However, one may refer to a more general statement, which tells that a solution of the inhomogeneous Fredholm equation of the second kind exists if and only if its right-hand side (nonhomogeneity) is orthogonal to all of its solutions. As a result, one obtains the very important condition that allows the transport equations ($n = 1, 2, \ldots$) to be found:

$$\int \psi_\alpha \left[\left(\frac{\partial}{\partial t} + \frac{\vec{p}}{m} \vec{\nabla}_r \right) f^{(0)} h^{(n-1)} - S^{(n)} \right] d\vec{p} = 0; \quad \alpha = 1, 2, \ldots, 5. \tag{3.114}$$

In the formula (3.114) ψ_α is a vector whose components are invariants of a collision (a set of eigenfunctions of homogeneous equation $Lh^{(n-1)} = 0$ at step $n - 1$).

A general solution of the homogeneous equation (3.113) is the sum of the particular solution $\hat{h}^{(n)}$ of the nonhomogeneous equation and the general solution of the homogeneous equation:

$$h^{(n)} = \hat{h}^{(n)} + C_\alpha^{(n)} \psi_\alpha, \quad \alpha = 1, 2, \ldots, 5. \tag{3.115}$$

Here $\hat{h}^{(n)}$ denotes any particular solution of equation (3.113), $C_\alpha^{(n)}$ presents five variables which are similar to the A, $\vec{B}$ and C coefficients in equation (3.93). These quantities depend on coordinates and time that are to be determined at each iteration step.

An unambiguous choice of the functions $\hat{h}^{(n)}$ calls for imposing five additional conditions:

$$\int \psi_\alpha \hat{h}^{(n)} f^{(0)} \, d\vec{p} = 0, \quad \alpha = 1, 2, \ldots, 5. \tag{3.116}$$

In mathematical terms, the necessity of these conditions is related with finding a solution of the inhomogeneous equation inside a multitude of functions orthogonal to basis of eigenfunctions of the homogeneous equation $Lh^{(n)} = 0$, which ensures the existence of the L^{-1} operator.

Thus, the function $f(0)$ appears from equation (3.106) and it coincides with the equilibrium distribution function f_0. The corrections

$$f^{(n)} = f^{(0)} (\hat{h}^{(n)} + C_\alpha^{(n)} \psi_\alpha) \tag{3.117}$$

for $n = 1, 2, \ldots$ contain unknown functions $C_\alpha^{(n)}$ of coordinates and time and unknown functions $\hat{h}^{(n)}$. These functions should be found from the conditions (3.114), (3.110), ensuring the existence of a correction at the $n + 1$ step. Thus, at least in principle, there is a possibility to construct an iterative scheme for determining all terms in the expansion (3.105).

Now, we embody the above scheme for the case $n = 1$. As far as the integrals

$$\int \psi_\alpha S^{(n)} \, d\vec{p} = 0$$

for all n provided that the ψ_α quantities are collision invariants [23], the orthogonality conditions (3.114) are reduced to the five equations of Euler for a nonviscous medium

$$\int \psi_\alpha \left[\frac{\partial}{\partial t} + \frac{\vec{p}}{m} \vec{\nabla}_r \right] f^{(0)} \, d\vec{p} = 0, \quad \alpha = 1, 2, \ldots, 5. \tag{3.118}$$

It is worth recollecting that five invariants of the collisions should be taken as the values ψ_α, the mass m of a particle, three components $m\vec{v}$ of the particle momentum and the kinetic energy $mv^2/2$. Taking $\psi_1 = m$, from (3.118) one obtains the continuity equation

$$\frac{\partial \rho}{\partial t} + \frac{\partial \rho v_{0i}}{\partial r_i} = 0, \tag{3.119}$$

$$\rho = mn = \int mf^{(0)}\, d\vec{p}, \quad v_{0i} = \frac{1}{n}\int v_i f^{(0)}\, d\vec{p}. \tag{3.120}$$

For $\psi_{2,3,4} = m\vec{v}$ from the expression (3.118) we get the momentum balance equation:

$$\frac{\partial \rho v_{0i}}{\partial t} + \frac{\partial}{\partial r_j}(P_{ij} + \rho v_{0i}v_{0j}) = 0, \tag{3.121}$$

$$P_{ij} = m\int (v_i - v_{0i})(v_j - v_{0j})f^{(0)}\, d\vec{p} = m\int c_i c_j f^{(0)}\, d\vec{p}, \tag{3.122}$$

where P_{ij} are the stress tensor components, and $c_i = v_i - v_{0i}$ the velocity components of the thermal motion.

In deriving the formula (3.121) the particle's velocity v_i should be written down as the sum of the velocity c_i of a thermal motion and the drift velocity v_{0i}. Then, considering that the average rate of the heat motion is equal to zero, the following expressions appear:

$$\begin{aligned} m\int v_i v_j f^{(0)}\, d\vec{p} &= m\int (c_i + v_{0i})(c_j + v_{0j})f^{(0)}\, d\vec{p} \\ &= m\int c_i c_j f^{(0)}\, d\vec{p} + mv_{0i}v_{0j}\int f^{(0)}\, d\vec{p} = P_{ij} + \rho v_{0i}v_{0j}. \end{aligned}$$

Finally, substituting the kinetic energy of a particle $mv^2/2$ as the fifth parameter ψ_5, the macroscopic energy balance equation reads

$$\frac{\partial}{\partial t}\rho\left(\frac{3}{2}\frac{k_B T}{m} + \frac{1}{2}v_0^2\right) + \frac{\partial}{\partial r_j}\left[\rho v_{0j}\left(\frac{3}{2}\frac{k_B T}{m} + \frac{1}{2}v_0^2\right) + v_{0i}P_{ij} + q_j\right] = 0. \tag{3.123}$$

The derivation of (3.123) is not difficult, given the relations (3.96) and the definition of the stress tensor (3.122). The quantity q_j is the heat flux

$$q_j = \frac{m}{2}\int c_i^2 c_j f^{(0)}\, d\vec{p}.$$

Equations (3.119), (3.121), (3.123) are the Euler equations for the five macroscopic quantities involved in $f^{(0)}$, i. e. $n(\vec{r}, t)$ or $\rho(\vec{r}, t)$, $\vec{v}_0(\vec{r}, t)$ and $T(\vec{r}, t)$. However, the hydrodynamic parameters found from solving these equations are not yet the true density, speed and temperature. They can be regarded as a first approximation to the true parameters. To find the next correction one should go over to the next step of the iteration.

We can generalize the results of (3.119), (3.121), (3.123) for the case of arbitrary values of $n \geq 1$. It is obvious that the combination of five hydrodynamic equations appearing from the condition of (3.114) can always be written as

$$\frac{\partial \rho_\alpha^{(n)}}{\partial t} + \operatorname{div} \vec{J}_\alpha^{(n)} = 0, \quad \alpha = 1, 2, \dots, 5, \quad n = 0, 1 \dots, \tag{3.124}$$

$$\rho_\alpha^{(n)} = \int \psi_\alpha f^{(n)} \, d\vec{p}, \quad \vec{J}_\alpha^{(n)} = \int \psi_\alpha \vec{v} f^{(n)} \, d\vec{p}. \tag{3.125}$$

It is worth understanding that it is necessary to solve the set of equations (3.124) at each iteration step and to find the unknown coefficients $C_\alpha^{(n)}$, involved in the corrections $f^{(n)}$. Further, the particular solution (quantities of $\hat{h}^{(n)}$) of the corresponding homogeneous equation (3.110) need also be found. Thus, it is clear that the planned program is difficult to realize, and a main merit of the Hilbert expansion is not to be practical method of finding solutions of the Boltzmann equation but one is to prove the existence and uniqueness of a solution. In addition, the Hilbert expansion allows the one-to-one correspondence between the distribution function $f(\vec{p}, \vec{r}, t)$ and its first moments of $n(\vec{r}, t)$, $\vec{v}_0(\vec{r}, t)$ and $T(\vec{r}, t)$ to be established. In other words, the Hilbert expansion enables one to prove that the Boltzmann kinetic equation uniquely determines the distribution function $f(\vec{p}, \vec{r}, t)$, if the first five moments of the distribution function are given at the initial time.

To prove this statement it suffices to substitute the expression for $f^{(n)}$ from (3.117) into the expression (3.125) for the density $\rho_\alpha^{(n)}$. This results in obtaining the equation of the interrelationship of the $\rho_\alpha^{(n)}$ quantities and the $C_\alpha^{(n)}$ coefficients:

$$\rho_\alpha^{(n)} = \int \psi_\alpha f^{(0)} \hat{h}^{(n)} \, d\vec{p} + \sum_{\beta=1}^{5} C_\beta^{(n)} \int \psi_\alpha f^{(0)} \psi_\beta \, d\vec{p}. \tag{3.126}$$

In virtue of (3.116) the first integral in the expression (3.126) is zero, and hence there follow five equations expressing the $C_\alpha^{(n)}$ coefficients via $\rho_\alpha^{(n)}$. As far as the $C_\alpha^{(n)}$ quantities are found by solving differential equations, it is necessary to specify initial conditions for their unambiguous determination. Having recently shown the exact correspondence between the $C_\alpha^{(n)}$ and $\rho_\alpha^{(n)}$ quantities, then the initial conditions at each iteration step can be defined not for $C_\alpha^{(n)}$ but for $\rho_\alpha^{(n)}$. Thus, all corrections to the distribution function will be found from the Boltzmann equation, if the $\rho_\alpha^{(n)}$ quantities will be given at the initial time. In other words, the distribution function $f(\vec{p}, \vec{r}, t)$ is uniquely determined by the five parameters $n(\vec{r}, t)$, $\vec{v}_0(\vec{r}, t)$ and $T(\vec{r}, t)$. As far as one may pick any initial moment of time, one can claim that the correspondence between distribution function $f(\vec{p}, \vec{r}, t)$ and the vector of its first five moments, given at arbitrary time, occurs. Consequently, there is a possibility to substantiate applicability of hydrodynamic equations to describe evolution of a system.

The $\hat{h}^{(n)}$ quantities, of course, also should be determined as partial solutions of equations (3.110) at each iteration step. But equations (3.110) do not require specifying initial conditions and contain the $C_\alpha^{(n-1)}$ found already within previous step. Therefore, the problem of finding the $\hat{h}^{(n)}$ quantities does not affect the above conclusion that the assignments of the first five moments for the distribution function at the initial instant of time uniquely determine the solution of the Boltzmann equation.

Thus, Hilbert proved the existence and uniqueness of the solution of the Boltzmann equation in the class of solutions which can be presented in the expansion form (3.105). Unfortunately, one has so far not yet succeeded in proving the possibility of such an expansion and even more in ascertaining its convergence. Nevertheless, the Hilbert expansion serves as a theoretical basis for most practically applicable methods for solving the Boltzmann equation; in particular, the Enskog–Chapman method whose main ideas are outlined below.

3.3.4 The Enskog–Chapman method. Derivation of hydrodynamic equations

As shown in the previous section a solution of the Boltzmann equation can be constructed as an expansion in small parameter (the Knudsen number), which is completely determined by assigning hydrodynamic quantities at an initial instant of time. However, if the distribution function $f(\vec{p}, \vec{r}, t)$ at any time t is expressed via the hydrodynamic quantities at the initial moment of time, then these hydrodynamic quantities at any given time must be expressed through the initial values of the hydrodynamic parameters. Consequently, we can exclude the distribution function and establish a direct link between the hydrodynamic quantities at different time. This result of Hilbert's theory can substantiate the use of the hydrodynamic equations to describe the gas dynamics.

The set of the hydrodynamic equations (3.119), (3.121), (3.123) presents five independent equations to determine *thirteen* unknowns. The unknown quantities are the density ρ, three components of the average speed $\vec{v}_0$, six components of the symmetric stress tensor P_{ij} and three components of the heat flux $\vec{q}$. The temperature T can easily be expressed in terms of the diagonal components of the stress tensor. Indeed, determining the pressure by the relation

$$p = \frac{1}{3}(P_{11} + P_{22} + P_{33}),$$

where the components of P_{ij} are defined by (3.122), and recollecting the condition (3.96), we obtain the well-known relation $p = nk_B T$ where the temperature can be determined through other hydrodynamic parameters.

Thus, the set of the hydrodynamic equations is not closed. To close it, the P_{ij} and q_i quantities need be expressed via the hydrodynamic quantities $n, \vec{v}_0$, p or T. Then the set of the hydrodynamic equations will be closed and we have five independent equations to determinate the five hydrodynamic parameters at each iteration step.

The purpose of the Enskog–Chapman method is to establish the above connection and to obtain a closed set of hydrodynamic balance equations. The Enskog–Chapman method improves the Hilbert method by means of modification of the Hilbert expansion in powers of the small parameter ϵ (the Knudsen number) for the distribution function $f(\vec{p},\vec{r},t)$ [24]. Such a reconstruction is necessary because the Hilbert expansion in any order of ϵ provides only the hydrodynamic equations of a nonviscous fluid. In physics, there are a lot of examples where, in any order of the perturbation theory, theoretical results are not in agreement with experiment and the perturbation theory requires a rearrangement of a series. Note that such a rearrangement is often equivalent to the summation of some infinite sequence of terms of a series in the perturbation theory. An application both of the diagram technique and the mass operator method in the problems of solid physics may serve as a case in point of such an approach.

Since there is no opportunity to set forth all details of the original Enskog–Chapman method, we will confine ourselves only to a discussion of principles that enable one to obtain closed equations of hydrodynamics suitable for describing a viscous fluid. These equations are referred to as the Navier–Stokes equations.

The initial steps of building the Enskog–Chapman expansion coincides completely with the Hilbert expansion. Thus, similar to line of reasoning in the previous paragraph, we arrive at equations (3.105)–(3.108). For simplicity, we restrict ourselves to the case when an external force $\vec{F} = 0$.

The function (3.109) is a solution of (3.106). The n, $\vec{v}_0$, T parameters present the local density of particles, their average speed and temperature, respectively, and in the general case they are arbitrary functions of coordinates and time. Strictly speaking, the $n^{(0)}$, $\vec{v}_0^{(0)}$, $T^{(0)}$ quantities, i. e. the hydrodynamic parameters of zero approximation must be involved in equation (3.109). However, if the n, $\vec{v}_0$, T parameters are assumed to meet the equations below, the theory becomes considerably more elegant, and results are easily interpreted;

$$\int d\vec{p} f^{(0)}(\vec{p},\vec{r},t) = n, \tag{3.127}$$

$$\int d\vec{p} f^{(0)}(\vec{p},\vec{r},t)\vec{p} = nm\vec{v}_0, \tag{3.128}$$

$$\int d\vec{p} f^{(0)}(\vec{p},\vec{r},t)\frac{(\vec{p}-m\vec{v}_0)^2}{2m} = \frac{3}{2}k_B T n. \tag{3.129}$$

Here, $f^{(0)}$ is a function of the distributions (3.109). Then the corrections $f^{(n)}$, $n = 1, 2, \ldots$, must satisfy the system of definitions

$$\int d\vec{p} f^{(n)}(\vec{p},\vec{r},t) = 0, \tag{3.130}$$

$$\int d\vec{p} f^{(n)}(\vec{p},\vec{r},t)\vec{p} = 0, \tag{3.131}$$

$$\int d\vec{p} f^{(n)}(\vec{p},\vec{r},t)\frac{(\vec{p}-m\vec{v}_0)^2}{2m} = 0. \tag{3.132}$$

The five equations (3.130)–(3.132) are analogs of equations (3.116). How these equations should be used in constructing the hydrodynamic equations, we will discuss later.

If one confines oneself to the first term into the expansion (3.105) and supposes that $f = f^{(0)}$, then the Euler equations of hydrodynamics such as (3.119), (3.121) and (3.123) can be obtained. It is easy to note that in this case the stress tensor can be written as $P_{ij} = p\delta_{ij}$ and the heat flux $\vec{q}$ is equal to zero. Then the set of the hydrodynamic equations is closed. This result coincides with the Hilbert expansion result.

Considering now the correction $f^{(1)}$ in the expansion (3.105); taking

$$f = f^{(0)} + \epsilon f^{(1)} = f^{(0)}(1 + \epsilon h^{(1)}) \tag{3.133}$$

one can write down the integral equation for $h^{(1)}$,

$$\left[\frac{\partial}{\partial t} + \frac{\vec{p}}{m}\vec{\nabla}_r\right] f^{(0)} = I(f^{(0)}, f^{(0)}h^{(1)}) + I(f^{(0)}h^{(1)}, f^{(0)}). \tag{3.134}$$

The inhomogeneous integral equation (3.134) to determine $h^{(1)}$ can be found by setting n equal to unity in equations (3.110)–(3.112). An analysis of equation (3.134) in the frame of the Enskog–Chapman method differs radically from Hilbert's analysis. As already mentioned, the main goal of the Enskog–Chapman method is to derive hydrodynamic equations. By virtue of the fact that the Hilbert expansion in any order of ϵ does not permit obtaining motion equations for a viscous fluid, the expansion should be rearranged. This rearrangement is based on results obtained by Hilbert. As far as a solution of the Boltzmann equation is uniquely determined by the first five moments of the distribution function, the derivative over time in equation (3.134) can be expressed through these moments.

To implement this program, we substitute the function $f^{(0)}$, being determined by the expression (3.109), into the left-hand side of equation (3.134) and perform the differentiation concerning coordinates and time, assuming that the hydrodynamic parameters n, $\vec{v}_0$, T are functions of the coordinates and time. As a result of simple calculations, one finds the following equality:

$$\begin{aligned}\left[\frac{\partial}{\partial t} + \frac{\vec{p}}{m}\vec{\nabla}_r\right] f^{(0)} = f^{(0)}\bigg\{&\frac{1}{n}\frac{\partial n}{\partial t} + \frac{1}{T}\frac{\partial T}{\partial t}\left[\frac{(\vec{p} - m\vec{v}_0)^2}{2mk_B T} - \frac{3}{2}\right] \\ &+ \frac{\vec{p} - m\vec{v}_0}{k_B T}\frac{\partial \vec{v}_0}{\partial t} + \frac{1}{n}\vec{v}\frac{\partial n}{\partial \vec{r}} + \frac{1}{T}\vec{v}\frac{\partial T}{\partial \vec{r}}\left[\frac{(\vec{p} - m\vec{v}_0)^2}{2mk_B T} - \frac{3}{2}\right] \\ &+ ((\vec{v}\vec{\nabla}_r)\vec{v}_0)\frac{\vec{p} - m\vec{v}_0}{k_B T}\bigg\}.\end{aligned} \tag{3.135}$$

All derivatives over time in the right-hand side of (3.135) can be excluded with the help of the hydrodynamic equations (3.119), (3.121) and (3.123).

If one takes $P_{ij} = p\delta_{ij}$, $p = nk_BT$, $\vec{q} = 0$, $\rho = nm$, these hydrodynamic equations can be written in a simpler form, which is represented by

$$\frac{\partial n}{\partial t} + \operatorname{div} n\vec{v}_0 = 0, \tag{3.136}$$

$$\frac{\partial \vec{v}_0}{\partial t} + ((\vec{v}_0\vec{\nabla}_r)\vec{v}_0) = -\frac{1}{\rho}\vec{\nabla}_r p, \tag{3.137}$$

$$\frac{\partial T}{\partial t} + \vec{v}_0\vec{\nabla}_r T + \frac{2}{3}T \operatorname{div}\vec{v}_0 = 0. \tag{3.138}$$

Before deriving the last equation, the expression (3.123) should be converted by using the conservation laws of (3.136) and (3.137). After excluding the time-derivatives, as a result of simple but rather cumbersome transformations, the right-hand side of (3.135) can be written as [22]

$$f^{(0)}\left\{\left[\frac{\vec{v}-\vec{v}_0}{T}\vec{\nabla}_r T\right]\left[\frac{m(\vec{v}-\vec{v}_0)^2}{2k_BT} - \frac{5}{2}\right] - \frac{1}{3}\frac{m}{k_BT}(\vec{v}-\vec{v}_0)^2 \right.$$
$$\left. \times \operatorname{div}\vec{v}_0 + \frac{m}{k_BT}(\vec{v}-\vec{v}_0)_i(\vec{v}-\vec{v}_0)_j\frac{\partial v_{0i}}{\partial r_j}\right\}. \tag{3.139}$$

We now write down the integral equation (3.134) using the result found above. For simplicity, as before, we will use the thermal velocity $\vec{c} = \vec{v} - \vec{v}_0$. For the collision integral we use the expression in the right-hand side of (3.100) and substitute the expansion (3.133) for the distribution function f. Then, given the energy conservation law, one can get

$$f^{(0)}\left[\frac{\vec{c}}{T}\vec{\nabla}_r T\left(\frac{m\vec{c}^2}{2k_BT} - \frac{5}{2}\right) + \frac{m}{k_BT}\left(\vec{c}_i\vec{c}_j - \frac{1}{3}\vec{c}^2\delta_{ij}\right)\frac{\partial v_{0i}}{\partial r_j}\right]$$
$$= \int u\sigma(\Omega, u) f^{(0)} f_1^{(0)} [h^{(1)\prime} + h_1^{(1)\prime} - h^{(1)} - h_1^{(1)}]\, d\vec{p}_1\, d\Omega. \tag{3.140}$$

Equation (3.140) is an inhomogeneous Fredholm equation and its solution is the superposition of a general solution of the homogeneous equation and a particular solution of the nonhomogeneous equation. It allows a correction to the distribution function of the first order of ϵ to be found. The technique of solving equation (3.140) is presented in detail in the monograph by Kogan [25]. Not going into the details of the calculations, we point out that the particular solution of the integral equation (3.140) is sought in the form

$$h^{(1)} = -Ac_i\frac{\partial T}{\partial r_i} - B\left(c_ic_j - \frac{1}{3}c^2\delta_{ij}\right)\frac{\partial v_{0i}}{\partial r_j}, \tag{3.141}$$

where the scalar A and B quantities are assumed to depend on a velocity modulus of a thermal motion, density and temperature. To determine these constants, the expression (3.141) should be substituted into equation (3.140). As a result, it falls into two equations: the equation for finding the parameter A and the equation for determining the parameter B. Then these equations are to be solved.

The distribution function can be written up to first order of ϵ as

$$f = f^{(0)}\left\{1 - A^* c_i \frac{\partial T}{\partial r_i} - B\left(c_i c_j - \frac{1}{3}c^2 \delta_{ij}\right)\frac{\partial v_{0i}}{\partial r_j}\right\},\qquad (3.142)$$

where A^* is a renormalized scalar quantity A. This renormalization appears when the solution of the homogeneous equation [25] is taken into account.

The expression for the distribution function (3.142) allows one to seek the heat flux and a more exact expression for the stress tensor. Substituting (3.142) in the definitions of q_j the flux density and P_{ij} stress tensor

$$q_j = \frac{m}{2}\int c_i^2 c_j f\, d\vec{p},\quad P_{ij} = m\int c_i c_j f\, d\vec{p},$$

we have

$$q_j = -\lambda\frac{\partial T}{\partial r_j},\quad P_{ij} = p\delta_{ij} - \mu\left(\frac{\partial v_{0i}}{\partial r_j} + \frac{\partial v_{0j}}{\partial r_i} - \frac{2}{3}\delta_{ij}\,\text{div}\,\vec{v}_0\right);\qquad (3.143)$$

$$\lambda = \frac{m}{6}\int A^* c^4 f^{(0)}\, d\vec{p},\quad \mu = \frac{m}{15}\int B f^{(0)} c^4\, d\vec{p}.\qquad (3.144)$$

The λ and μ constants, contained in the expression (3.144), are to be found by solving equation (3.140). For this purpose, the $A(c)$ and $B(c)$ functions are expanded in a series of Sonin polynomials [25]. The expansion procedure is cumbersome enough, as the result depends on the model type of the particle interaction. We give here the result only for the case when the particles are elastic balls with diameter d and in the case that only the first term of the Sonin expansion of polynomials is left [25]. In this case

$$\mu = \frac{15}{16}\frac{\sqrt{mk_B T}}{\sqrt{\pi}d^2},\quad \lambda = \frac{5}{2}C_v\mu.$$

In the last formula C_v is the specific heat of a gas at a constant volume.

Thus, the correction $h^{(1)}$ to the distribution function allows instead of (3.136)–(3.138) for obtaining a new closed set of hydrodynamic balance equations with a renormalized value of the stress tensor and non-zero heat flux. Renormalization of the stress tensor is associated with taking into account the irreversible (viscous) momentum transfer in a gas. The coefficient μ is called the medium viscosity coefficient, λ is the thermal conductivity coefficient. It is important to note that the viscosity and thermal conductivity coefficients are not phenomenological parameters but are calculated from first principles.

In conclusion, we would like to point out that although the step-by-step procedure for finding the expansion coefficients of (3.133) can be continued in the frame of the Enskog–Chapman method, one fails to obtain the corrections of a higher order than the second to the distribution function due to computation difficulties. Also, one has so far not succeeded in proving convergence of procedure of the expansion (3.133) in a general form. Therefore, although the Enskog–Chapman method is widely used in practice, the scope of its applicability does remain not fully investigated.

3.3.5 The method of moments

The most universal technique to build a closed set of hydrodynamic balance equations for arbitrary Knudsen numbers is through the method of moments. In essence, the above hydrodynamic variables are moments of the distribution function:

$$n(\vec{r},t) = M^{(0)} = \int f\,d\vec{p}, \tag{3.145}$$

$$nv_{0i}(\vec{r},t) = M_i^{(1)} = \int \frac{p_i}{m} f\,d\vec{p}, \tag{3.146}$$

$$P_{ij}(\vec{r},t) = m\mathcal{M}_{ij}^{(2)} = m\int c_i c_j f\,d\vec{p}, \tag{3.147}$$

$$q_i(\vec{r},t) = \frac{m}{2}\mathcal{M}_{ijj}^{(3)} = \frac{m}{2}\int c_i c^2 f\,d\vec{p}. \tag{3.148}$$

In formulae (3.145)–(3.148) the indices i, j are taking the values 1, 2, 3. The moments $\mathcal{M}$ are referred to as central and identified for speed deviations about the mean. The moments M of the distribution function and the central moments $\mathcal{M}$ are obviously related to each other and can be easily expressed via each other.

The main idea of the method of moments is to express the distribution function through its moments

$$f(\vec{p},\vec{r},t) = f(\vec{p}, M^{(0)}, M^{(1)}, \dots), \tag{3.149}$$

where the moments $M^{(k)}$ are a function of coordinates and time. Then a set of equations for finding the moments of the distribution function is obtained by inserting the distribution function written in such a way into the Boltzmann kinetic equation. In the general case, Boltzmann's kinetic equation is equivalent to an infinite set of equations for the moments, but in most practically important cases one may restrict oneself to the few first moments.

Grad was the first to use in 1949 the method of moments for solving the kinetic equation. Following Grad, we expand the distribution function in a series of three-dimensional Hermite polynomials:

$$f = f^{(0)}\left(a^{(0)}H^{(0)} + a_i^{(1)}H_i^{(1)} + \frac{1}{2!}a_{ij}^{(2)}H_{ij}^{(2)} + \frac{1}{3!}a_{ijk}^{(3)}H_{ijk}^{(3)} + \cdots\right). \tag{3.150}$$

In this formula, the $a_{i,j,\dots}^{(N)}$ coefficients are functions of the coordinates and time. The Hermite polynomials are functions of the dimensionless relative velocity,

$$\vec{\xi} = \frac{\vec{p} - m\vec{v}_0}{\sqrt{mk_B T}},$$

and their explicit form can be obtained with the aid of the following formula:

$$H^{(N)}_{ij\ldots k} = (-1)^N \exp\left(\frac{\xi^2}{2}\right)\frac{\partial^N}{\partial\xi_i\partial\xi_j\ldots\partial\xi_k}\exp\left(-\frac{\xi^2}{2}\right). \tag{3.151}$$

Using the formula (3.151) one can easily calculate the explicit form of the Hermite polynomial of any order. In practice, it is only required polynomials of lower orders, some of them are given by

$$\begin{aligned} &H^{(0)} = 1, \quad H^{(1)}_i = \xi_i, \quad H^{(2)}_{ij} = \xi_i\xi_j - \delta_{ij}, \\ &H^{(3)}_{ijk} = \xi_i\xi_j\xi_k - (\xi_i\delta_{jk} + \xi_j\delta_{ik} + \xi_k\delta_{ij}). \end{aligned} \tag{3.152}$$

From the definition of the Hermite polynomials (3.151) it follows that all polynomials, which differ by interchanging the indices, are identically equal. The Hermite polynomials (3.151) are orthogonal to some weighting function:

$$\frac{1}{(2\pi)^{3/2}}\int \exp\left(-\frac{\xi^2}{2}\right)H^{(n)}_\alpha H^{(m)}_\beta\, d\vec{\xi} = \delta_{nm}\delta_{\alpha\beta}. \tag{3.153}$$

The function $f^{(0)}$ in formula (3.150) is defined by the relation

$$f^{(0)} = \frac{n}{(2\pi mk_{\mathrm{B}}T)^{3/2}}\exp\left(-\frac{\xi^2}{2}\right). \tag{3.154}$$

Using the orthogonality of the Hermite polynomials, the expansion coefficients can be expressed in terms of hydrodynamic parameters or moments of the distribution function:

$$a^{(N)}_\alpha = \frac{(mk_{\mathrm{B}}T)^{3/2}}{n(\vec{r},t)}\int fH^{(N)}_\alpha\, d\vec{\xi}. \tag{3.155}$$

Let us give explicit expressions for a few first coefficients of the expansion:

$$\begin{aligned} &a^{(0)} = M^{(0)} = 1, \quad a^{(1)}_i = \mathcal{M}^{(1)}_i = 0, \quad a^{(2)}_{ij} = \frac{P_{ij} - p\delta_{ij}}{p}, \\ &a^{(3)}_{ijk} = \frac{m\mathcal{M}^{(3)}_{ijk}}{p}\sqrt{\frac{m}{k_{\mathrm{B}}T}}. \end{aligned} \tag{3.156}$$

As far as the $a^{(N)}_\alpha$ expansion coefficients are expressed via moments of the distribution function, and those, in turn, are the hydrodynamic quantities in question, the problem of finding the hydrodynamic balance equations in the frame of the method of moments is reduced to the problem of finding the equations for the coefficients in the expansion (3.150). The equations of motion for the coefficients $a^{(N)}_\alpha$ can be found with the help of the Boltzmann kinetic equation. To do this, the distribution function (3.150) needs be substituted in the kinetic equation. Then it is necessary to multiply both sides of equation by the corresponding Hermite polynomial with weighting function and integrate it over the relative speed. The orthogonality condition of the Hermite polynomials permits the number of terms in each of equations to be limited significantly. Although this procedure is rather simple but at the same time extremely cumbersome, the derivation of these equations should be omitted, referring the reader to the literature [25].

From a practical point of view, it would be desirable to obtain equations for the moments (hydrodynamic quantities), which are measurable and have clear physical meaning. As noted above, there are 13 such moments: the concentration n, three components of the drift velocity v_{0i}, the temperature T, six components of the symmetric stress tensor p_{ij} and three components of the heat flux q_i. To obtain the hydrodynamic equations for these variables, it suffices to approximate the distribution function (3.150) by the expression

$$f = f^{(0)}\left(1 + \frac{1}{2}a_{ij}^{(2)}H_{ij}^{(2)} + \frac{1}{10}a_{ijj}^{(3)}H_{ikk}^{(3)}\right), \tag{3.157}$$

there remaining in it only the first three terms of the expansion. This approximation of the distribution function is known in the literature as Grad's 13-moment approximation. The results obtained in this approximation are in full agreement with the results of the Enskog–Chapman method. The complete derivation of the hydrodynamic equations corresponding to the 13-moment Grad approximation can be found in the aforementioned book [25].

It should be noted that in the frame of the method of moments an approximation of the distribution function can be virtually arbitrary by using a combination of hydrodynamic parameters. A certain form of the approximating function depends on the problem posed and peculiarities of the physical phenomenon involved. In the next chapter, the method of moments will be applied to obtain closed hydrodynamic equations for a system of hot electrons in conducting crystals.

3.4 Problems to Chapter 3

3.1. Assume that the exact solution to classical mechanics equations for a system consisting of N particles is known, and the time dependence between their coordinates $\vec{r}_i(t)$ and momenta $\vec{p}_i(t)$ is defined. Write down explicitly the N-particle distribution density function

$$\rho(\vec{r}_1, \vec{r}_2, \dots, \vec{r}_N, \vec{p}_1, \vec{p}_2, \dots, \vec{p}_N, t),$$

for the system in the so-called N "quasiparticle" representation to satisfy both the Liouville equation and the normalization condition.

3.2. Write down the Liouville equation for a material particle with mass m, located in a gravitational field with potential $U = mqx$, where q is the acceleration of gravity, x the coordinate of the particle.

3.3. Let us consider a one-dimensional harmonic oscillator with mass m and the elastic coupling constant k. Find the general solution to the Liouville equation for this system.
To solve the problem, it is necessary to determine the particle's coordinate and momentum as functions of time and also constants of integration of Hamilton's equations of motion C_1, C_2. Express the constants in terms of the system's coordinate and momentum. An arbitrary function of the constants $\rho(C_1, C_2)$ will be the general solution of the Liouville equation, satisfying the condition

$$\frac{d\rho}{dt} = 0.$$

Construct the phase portrait of the system.

3.4. Ascertain that if a system has the number of particles $N \approx 10^{20}$, volume $V \approx 1\,\text{cm}^3$, the mean free path of the particles $\lambda \approx 10^{-5}$ cm, and the mean free time $\tau \approx 10^{-10}$ s, it returns to its initial state (the Poincaré recurrence time) after this period of time $T \approx T_1^N \approx 10^{9^{20}}$ s, where T_1 is the time to return to its "cube" with volume λ^3 of a single particle. Note that here the coordinate space should be taken as the phase space only, but not take into account the momentum space.

3.5. Write down an equation in the relaxation time approximation for a system of charged particles with mass m and charge e, located in an external uniform electric field $\vec{E}$ and a constant magnetic field $\vec{B}$. (One should not take into account quantization of orbital motion in the magnetic field.)

3.6. Find the solution to the Boltzmann equation for a local-equilibrium system in an external static field with potential $U(\vec{r})$. To solve the problem, use the fact that $\ln f(\vec{r}, \vec{p})$ in the local-equilibrium state is a function of five additive collision invariants only (see (3.93)).

3.7. Prove that the collision integral in the right-hand side of (3.59) does not contribute to the change in density of quantities conserving in the collision process (the number of particles, momentum, and energy):

$$n \int_0^{2\pi} d\varphi \int_0^{2r_0} a\,da \int (f'f_1' - ff_1) u \psi(\vec{p})\, d\vec{p}\, d\vec{p}_1 = 0,$$

provided that the function ψ is an arbitrary function of collision invariants.

3.8. Derive an Euler hydrodynamics equation for an ideal fluid using equation (3.64) and the Chapman–Enskog method.

3.9. Find the solution to equation (3.32) for a weakly non-equilibrium gas of classical particles by the Chapman–Enskog method. The non-equilibrium is due to a temperature gradient field and external static potential field. Having obtained the solution, it is necessary to eliminate all time-derivatives using the continuity equations.

3.10. With the aid of the correction obtained in the previous problem to the distribution function, prove that the coefficient of thermal conductivity κ in a linear approximation is calculated by the simple formula

$$\kappa = \frac{5}{2}\frac{n}{m}k_B^2 T\tau.$$

4 Kinetic equation for electrons and phonons in conducting crystals

4.1 Kinetic coefficients in the relaxation time approximation

4.1.1 Kinetic equation for electrons and its solution to the relaxation time approximation

Consider the simplest model of the conductor, according to which quasi-free electrons or holes are current carriers. The current carriers are interacting with crystal lattice defects or phonons as a result of the collision processes. For simplicity, the dispersion law of electrons (holes) is assumed to be parabolic and to have the form

$$\varepsilon_{\vec{p}} = \frac{p^2}{2m}, \tag{4.1}$$

where $\vec{p}$ is for the current carrier momentum vector, m is for the current carrier mass.

The assumption of the parabolic nature of the dispersion law is not crucial for the elementary theory of transport phenomena under consideration in this chapter. All the results can be generalized to the case of a spherically symmetric conduction band, the electron energy being dependent arbitrarily on modulus of a wave vector $\vec{k}$.

In thermodynamic equilibrium, properties of an electron gas are determined by the Fermi–Dirac distribution function:

$$f_0(\varepsilon_{\vec{p}}) = \left\{\exp\left(\frac{\varepsilon_{\vec{p}} - \zeta}{k_B T}\right) + 1\right\}^{-1}, \tag{4.2}$$

where k_B is the Boltzmann constant.

In the non-equilibrium case, one can also introduce a non-equilibrium distribution function $f(\vec{r}, \vec{p}, t)$ that depends on the coordinates $\vec{r}$, momentum $\vec{p}$ and time t, and satisfies the normalization condition:

$$\sum_{\sigma} \frac{V}{(2\pi\hbar)^3} \int d\vec{p}\, d\vec{r} f(\vec{r}, \vec{p}, t) = n, \tag{4.3}$$

where σ enumerates the projection of the electron spin on the Z-axis ($\sigma = \pm 1/2$), n being the number of electrons in the specimen. Furthermore, the specimen volume is assumed to be equal to unity and the value of n will have the meaning of electron concentration. The multiplier

$$\frac{V}{(2\pi\hbar)^3} = \frac{L_x L_y L_z}{(2\pi\hbar)^3}$$

is the density number of the electrons in the momentum space. The appearance of the multiplier in the formula (4.3) is due to quantized electron states:

$$p_x = \pm\frac{2\pi\hbar n_1}{L_x}, \quad p_y = \pm\frac{2\pi\hbar n_2}{L_y}, \quad p_z = \pm\frac{2\pi\hbar n_3}{L_z},$$

https://doi.org/10.1515/9783110727197-004

where the quantities n_1, n_2, n_3 are integers running from zero to infinity. So when counted in the formula (4.3), it is necessary to sum the number of states over discrete states of the electrons with different values of momentum components. To avoid complicated calculations, the summation is usually replaced by integration. In this case, the integral is multiplied by dimension factor having the meaning of density of states in the momentum space. It follows from (4.3) that the expression

$$\frac{2V}{(2\pi h)^3} d\vec{p}d\vec{r}f(\vec{r},\vec{p},t)$$

gives the number of the electrons with the momentum $\vec{p}$ and the coordinate $\vec{r}$ trapped by the phase volume element $d\vec{p}d\vec{r}$ at the moment in time t.

If one assumes that the electrons are noninteracting particles, then every electron can be considered as an isolated system. Phase points corresponding to possible different states of the particle just migrate from one region of the phase space to another not vanishing and appearing again. We do not discuss processes of birth and destruction of particles here. The phase points' motion pattern is schematically depicted in Figure 4.1.

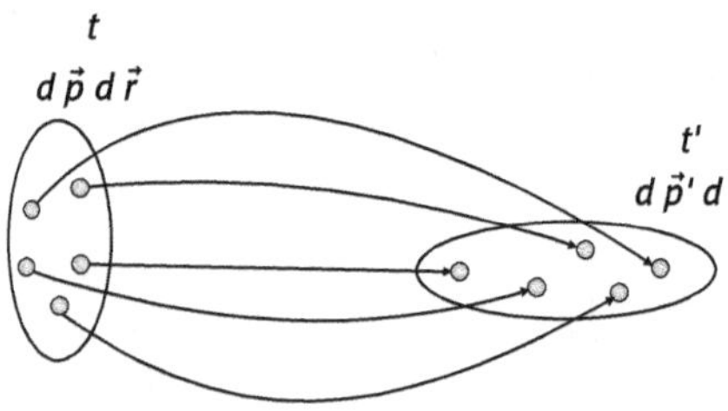

Figure 4.1: Scheme of migration of phase points in the phase space.

Each phase point of the phase space being in the region $d\vec{p}d\vec{r}$ at time t will have moved to some region by the moment in time t', as shown in Figure 4.1. Therefore, one can write down the equality:

$$f(\vec{r},\vec{p},t)d\vec{p}d\vec{r} = f(\vec{r}',\vec{p}',t')d\vec{p}'d\vec{r}'. \tag{4.4}$$

As already mentioned in Section 3.2.6, the phase flux preserves the phase volume of the system according to the Liouville theorem, and therefore the equality $d\vec{p}d\vec{r} = d\vec{p}'d\vec{r}'$ takes place. Then an important result follows from the formula (4.4),

$$f(\vec{r},\vec{p},t) = f(\vec{r}',\vec{p}',t'). \tag{4.5}$$

In accordance with the result obtained above, the non-equilibrium distribution function of the noninteracting electrons is an integral of motion and its total time-derivative must be zero. If, nevertheless, there is an interaction between them, the total derivative is equal to not zero, but to a change in the distribution function due to the interaction (collisions) with scatterers such as lattice defects or phonons. Thus, we get

$$\frac{\partial}{\partial t}f(\vec{r},\vec{p},t) + \dot{\vec{r}}\vec{\nabla}_{\vec{r}}f(\vec{r},\vec{p},t) + \dot{\vec{p}}\vec{\nabla}_{\vec{p}}f(\vec{r},\vec{p},t) = -\left(\frac{\partial f}{\partial t}\right)_{\text{col}}. \tag{4.6}$$

The left-hand side of the expression (4.6) describes the change in the phase-space distribution function under its evolution, while the right-hand side is for the change in the distribution function due to collisions. In general, it follows from the material presented in the previous chapter, the collision term on the right-hand side of (4.6) is a nonlinear functional whose kernel contains the distribution function and depends on a particular mechanism of interaction between electrons and subsystems in a crystal. Equation (4.6) represents a kinetic equation for mobile charge carriers (electrons or holes) in a quasi-classical approximation. Conditions of applicability of the quasi-classical description of the electrons' motion in the crystal will be considered later.

As noted above, an attempt to solve strictly the kinetic equation for even the simplest interaction potentials encounters serious computational difficulties. However, it is well known that many peculiarities of kinetic phenomena in metals and semiconductors can be apprehensible within the framework of the relaxation time approximation when the collision integral is approximated by the expression

$$\left(\frac{\partial f}{\partial t}\right)_{\text{col}} = \frac{f - f_0}{\tau_{\vec{p}}}. \tag{4.7}$$

Therefore, it would be proper to start an acquaintance with the theory of transport phenomena in conducting crystals from this simple approximation.

For further simplification of the kinetic equation (4.6), it is worth emphasizing that

$$\begin{aligned} \frac{\partial f(\vec{r},\vec{p},t)}{\partial t} &\simeq \omega f_1(\vec{r},\vec{p},t) \ll \frac{f_1(\vec{r},\vec{p},t)}{\tau_{\vec{p}}}, \\ f_1(\vec{r},\vec{p},t) &= f(\vec{r},\vec{p},t) - f_0(\varepsilon_{\vec{p}}). \end{aligned} \tag{4.8}$$

In the expression (4.8) $f_1(\vec{r},\vec{p},t)$ is a correction to the equilibrium distribution function, which emerges under the action of external thermodynamic forces. The parameter $\omega\tau_{\vec{p}} \ll 1$ up to an optical frequency range, and therefore in the kinetic equation (4.6) the time-derivative can be omitted. In other words, as far as the momentum relaxation time is sufficient small ($\tau_{\vec{p}} \simeq 10^{-13}$ s), the system of the electrons manages to tune to an external field. Then the alternating electric field can be considered as static at every moment in time. This means that the apparent dependence of the nonequilibrium distribution function on the time in the kinetic equation (4.6) can be neglected.

Another essential simplification is due to the fact that the non-equilibrium distribution function in locally-equilibrium state depends on coordinates only parametrically via a thermodynamic parameters, such as the temperature and chemical potential:

$$\vec{\nabla}_{\vec{r}} f = \frac{\partial f}{\partial T}\vec{\nabla} T + \frac{\partial f}{\partial \zeta}\vec{\nabla}\zeta. \tag{4.9}$$

If one restricts oneself to a linear approximation over the thermodynamic forces $\vec{\nabla} T, \vec{\nabla}\zeta$ in the kinetic equation (4.6), assuming that non-equilibrium is weak ($f_0 \gg f_1$), then the non-equilibrium distribution function on the right-hand side of the expression (4.9) can be replaced by the equilibrium function $f_0(\varepsilon_{\vec{p}})$. As a result of simple calculations, one obtains

$$\vec{\nabla}_{\vec{r}} f = -\frac{\partial f_0}{\partial \varepsilon_{\vec{p}}}\left[\vec{\nabla}\zeta + \frac{\varepsilon_{\vec{p}} - \zeta}{T}\vec{\nabla} T\right]. \tag{4.10}$$

In the presence of an external electric field $\vec{E}$ and an induced magnetic field $\vec{H}$ we have

$$\dot{\vec{p}} = e\vec{E} + \frac{e}{c}[\vec{v} \times \vec{H}]. \tag{4.11}$$

Therefore, in the linear approximation in the thermodynamic forces (in the given case, the magnetic field is not the thermodynamic force that gives rise to a deviation from equilibrium) we arrive at

$$\dot{\vec{p}}\vec{\nabla}_{\vec{p}} f = \frac{\partial f_0}{\partial \varepsilon_{\vec{p}}} e\vec{v}\vec{E} + \vec{\nabla}_{\vec{p}} f_1 \frac{e}{c}[\vec{v} \times \vec{H}]. \tag{4.12}$$

In obtaining the result (4.12) we have taken into account that $\vec{\nabla}_{\vec{p}} f_0 \sim \vec{v}$, and the contribution

$$\vec{\nabla}_{\vec{p}} f_0 \frac{e}{c}[\vec{v} \times \vec{H}] = 0,$$

since $\vec{v}[\vec{v} \times \vec{H}] = 0$.

Having substituted the results (4.7), (4.10) and (4.12) into the kinetic equation (4.6), we get

$$-\frac{\partial f_0}{\partial \varepsilon_{\vec{p}}}\vec{v}\left(e\vec{\varepsilon} - \frac{\varepsilon_{\vec{p}} - \zeta}{T}\vec{\nabla} T\right) = \frac{f_1}{\tau_{\vec{p}}} + \vec{\nabla}_{\vec{p}} f_1 \frac{e}{c}[\vec{v} \times \vec{H}]. \tag{4.13}$$

In the case when the magnetic field is zero, the expression (4.13) allows one to immediately determine the correction to the distribution function f_1 that is linear both in the electrochemical potential gradient $\vec{\varepsilon} = -\vec{\nabla}(\varphi + 1/e\zeta)$ (1.14) and in the temperature gradient T:

$$f_1 = \tau_{\vec{p}}\left(-\frac{\partial f_0}{\partial \varepsilon_{\vec{p}}}\right)\vec{v}\left(e\vec{\varepsilon} - \frac{\varepsilon_{\vec{p}} - \zeta}{T}\vec{\nabla} T\right). \tag{4.14}$$

If the magnetic field is nonzero, to solve the equation (4.13) the correction f_1 needs to be sought in the form

$$f_1 = -\frac{\partial f_0}{\partial \varepsilon_{\vec{p}}}(\vec{v}\vec{\chi}(\varepsilon_{\vec{p}})), \tag{4.15}$$

where $\vec{\chi}(\varepsilon_{\vec{p}})$ is the unknown vector function depending only on energy.

Let us calculate the gradient of functions f_1 in the momentum space. Using the definition (4.15), one finds

$$\begin{aligned} \vec{\nabla}_{\vec{p}} f_1 &= -\frac{\partial^2 f_0}{\partial \varepsilon_{\vec{p}}^2}\vec{v}(\vec{v}\vec{\chi}(\varepsilon_{\vec{p}})) - \frac{\partial f_0}{\partial \varepsilon_{\vec{p}}}\vec{\nabla}_{\vec{p}}(\vec{v}\vec{\chi}(\varepsilon_{\vec{p}})), \\ \vec{\nabla}_{\vec{p}}(\vec{v}\vec{\chi}(\varepsilon_{\vec{p}})) &= v_i\vec{\nabla}_{\vec{p}}\chi_i(\varepsilon_{\vec{p}}) + \chi_i(\varepsilon_{\vec{p}})\vec{\nabla}_{\vec{p}}v_i, \\ \vec{\nabla}_{\vec{p}}\chi_i(\varepsilon_{\vec{p}}) &= \frac{\partial \chi_i(\varepsilon_{\vec{p}})}{\partial \varepsilon_{\vec{p}}}\vec{v}, \quad \chi_i(\varepsilon_{\vec{p}})\vec{\nabla}_{\vec{p}}v_i = \frac{\vec{\chi}(\varepsilon_{\vec{p}})}{m}. \end{aligned} \tag{4.16}$$

Substituting the results obtained in the last summand on the right-hand side of the expression (4.13) and considering that the terms proportional to the velocity vector $\vec{v}$ do not contribute, one may obtain a simple expression:

$$\vec{\nabla}_{\vec{p}} f_1 \frac{e}{c}[\vec{v}\times\vec{H}] = -\frac{\partial f_0}{\partial \varepsilon_{\vec{p}}}\omega_0\vec{v}[\vec{h}\times\vec{\chi}(\varepsilon_{\vec{p}})]. \tag{4.17}$$

In deriving the formula (4.17), we have used the definition of the Larmor precession frequency ω_0 of electrons in the magnetic field:

$$\omega_0 = \frac{eH}{mc}, \tag{4.18}$$

and introduced the unit vector $\vec{h}$ oriented along the direction of the magnetic field vector ($\vec{H} = \vec{h}H$), H being the modulus of the magnetic field. Then we have performed a rearrangement in the order of the vectors of the vector-scalar product.

Having substituted the results (4.15), (4.17) into the formula (4.13) and produced necessary reductions, we obtain the vector equation for the function $\vec{\chi}(\varepsilon_{\vec{p}})$:

$$\vec{\chi}(\varepsilon_{\vec{p}}) + [\vec{a}\times\vec{\chi}(\varepsilon_{\vec{p}})] = \vec{b}; \tag{4.19}$$

$$\vec{a} = \omega_0\tau_{\vec{p}}\vec{h}; \quad \vec{b} = \tau_{\vec{p}}\left(e\vec{\mathcal{E}} - \frac{\varepsilon_{\vec{p}} - \zeta}{T}\vec{\nabla}T\right). \tag{4.20}$$

To solve the equation (4.19), we scalarwise multiply it once by the vector $\vec{a}$ and a second time vectorially by the vector $\vec{a}$. Simple algebraic manipulations result in expressing the vector $\vec{\chi}(\varepsilon_{\vec{p}})$ through the vectors $\vec{a}$ and $\vec{b}$:

$$\vec{\chi}(\varepsilon_{\vec{p}}) = \frac{\vec{b} + \vec{a}(\vec{a}\vec{b}) - [\vec{a}\times\vec{b}]}{1 + a^2}. \tag{4.21}$$

For the solution structure to be a more convenient we make use of the identity

$$\vec{b} = \frac{1}{a^2}\{\vec{a}(\vec{a}\vec{b}) - [\vec{a} \times [\vec{a} \times \vec{b}]]\} \tag{4.22}$$

and substitute this expression for $\vec{b}$ in the solution of (4.21). As a result, the representation for the function $\vec{\chi}(\varepsilon_{\vec{p}})$ expressed via the $\vec{a}$ and $\vec{b}$ vectors appears as

$$\vec{\chi}(\varepsilon_{\vec{p}}) = \frac{1}{a^2}\vec{a}(\vec{a}\vec{b}) - \frac{[\vec{a} \times \vec{b}] + 1/a^2[\vec{a} \times [\vec{a} \times \vec{b}]]}{1 + a^2}. \tag{4.23}$$

Finally, considering the explicit form of the $\vec{a}$ and $\vec{b}$ vectors and substituting their values of (4.20) into formula (4.23), the vector $\vec{\chi}(\varepsilon_{\vec{p}})$ is

$$\begin{aligned}\vec{\chi}(\varepsilon_{\vec{p}}) &= e\tau_{\vec{p}}\left\{(\vec{h}\vec{\varepsilon})\vec{h} - \frac{\omega_0\tau_{\vec{p}}[\vec{h} \times \vec{\varepsilon}] + [\vec{h} \times [\vec{h} \times \vec{\varepsilon}]]}{1 + (\omega_0\tau_{\vec{p}})^2}\right\} \\ &+ \tau_{\vec{p}}\frac{\zeta - \varepsilon_{\vec{p}}}{T}\left\{(\vec{h}\vec{\nabla}T)\vec{h} - \frac{\omega_0\tau_{\vec{p}}[\vec{h} \times \vec{\nabla}T] + [\vec{h} \times [\vec{h} \times \vec{\nabla}T]]}{1 + (\omega_0\tau_{\vec{p}})^2}\right\}.\end{aligned} \tag{4.24}$$

We will use the formulas (4.14), (4.15) and (4.24) to determine fluxes of charge and heat and also to calculate kinetic coefficients for describing both thermomagnetic and the galvanomagnetic phenomena in conducting crystals.

4.1.2 Conditions of applicability for the quasi-classical description of electrons in conducting crystals

The kinetic equation (4.6), or (4.13) written in the previous section is quasi-classical. So far as, it is well known that electrons in a crystal are quantum objects and there are a lot of convincing effects where electrons exhibit quantum nature (for example, diffraction of electrons in crystals), there arises an issue concerning applicability of such a description. In fact, the quasi-classical description imposes some restrictions on the conditions required to conduct a physical experiment. However, one can show that the quasi-classical description is perfectly justified in most realistic situations, in which measurements of kinetic phenomena in solids are made. Below, basic conditions of applicability for the quasi-classical kinetic equation are formulated to describe kinetic phenomena in conducting crystals in a static external magnetic field and in the absence of it.

These conditions include three main restrictions. First, the electron wavelength λ must be less than other characteristic spatial scales of the problem what allows considering an electron as a point object. In the absence of a magnetic field, the mean free path l is a natural parameter of the length dimension. Therefore, the quasi-classical description is possible if the condition is met that

$$\lambda \ll l.$$

Second, the electron $\triangle E$ energy uncertainty, which is a consequence of quantum-mechanical uncertainty principle, must be small compared to the average electron energy $\overline{\varepsilon}$ (the average energy $\overline{\varepsilon} \simeq k_B T$ for nondegenerate case, $\overline{\varepsilon} \simeq \zeta$ under conditions of degeneration)

$$\frac{\hbar}{\tau_0} \ll \overline{\varepsilon},$$

where τ_0 is a characteristic time of interaction between an electron and other subsystems of a crystal. Therefore, the time of the electron–scatterer interaction τ_0 must be sufficiently large. In this case, collisional broadening of energy levels is thought to be negligible and temperature remains the only parameter that leads to chaotization of motion of charge carriers. This condition provides a basis to describe the electron system in terms of a distribution function. Moreover, the characteristic time of collisions must be substantially less than a time between two successive collisions because each of them is considered as an independent process and a non-equilibrium distribution in the system is deemed to be established after each collision. Therefore, one may take the time between two successive collisions as an upper estimate of τ_0 and assume that $\tau_0 \approx \tau_{\bar{p}}$. This is even more justified, since the quantity τ_p yields easily an experimental evaluation. In this case, the condition

$$\frac{\hbar}{\tau_0} \approx \frac{\hbar}{\tau_{\bar{p}}} \ll \overline{\varepsilon}$$

is reduced to the condition $\lambda \ll l$. It is easily seen that the given condition holds, if the previous condition will be written in the form

$$\frac{\overline{v}\hbar}{\overline{\varepsilon}} \ll \overline{v}\tau_{\bar{p}},$$

multiplying both sides of the inequality by the average velocity $\overline{v}$ of the electrons. Then the left-hand side of the inequality has the quantity $\hbar/p \approx \lambda$ as the right-hand side contains the mean free path $l = \overline{v}\tau_{\bar{p}}$. Thus, in the absence of a magnetic field the first and second restrictions lead to the same condition, $\lambda \ll l$.

Another limitation occurs when an electron is in effective area of external force fields. Then, if the electron is regarded as a point object, a change in its energy on the length scale of the de Broglie wavelength should be much less than the average electron energy. In other words, if, for example, the electron moves in an external electric field applied, then, in a nondegenerate case, the condition $\lambda eE \ll \overline{\varepsilon} \simeq k_B T$ must be met. This is not a very significant restriction, since simple estimations give the limit $E < 10^6$ V/m, which is acceptable in most experimental situations.

Consider conditions of applicability for a kinetic equation to describe non-equilibrium current carriers in a magnetic field. In this case, there are three characteristic parameters with dimension length: l_H, the so-called "magnetic length" or characteristic size of the electron Larmor orbit,

$$l_H = \left(\frac{\hbar c}{eH}\right)^{1/2},$$

the de Broglie wavelength of the electron,

$$\lambda = \frac{2\pi\hbar}{\sqrt{2m\bar{\varepsilon}}},$$

and the mean free path of the electron between two successive collisions l.

In these conditions, the first criterion of the quasi-classical description is to be the following condition:

$$\frac{2\pi\hbar}{\sqrt{2m\bar{\varepsilon}}} \ll \left(\frac{\hbar c}{eH}\right)^{1/2}.$$

Putting that $\bar{\varepsilon} \simeq k_B T$, we arrive at the condition of applicability for the quasi-classical description of the electrons in the magnetic field:

$$\hbar\omega_0 \ll k_B T. \tag{4.25}$$

This condition of applicability is well known in the literature and it admits a simple interpretation: in the quasi-classical description, the distance between the quantized energy levels of electrons in a magnetic field must be small, compared with the average energy of a thermal electron motion.

Another condition for the applicability of the kinetic equation in a magnetic field is also associated with the influence of the magnetic field on the orbital electron motion. The distance between the Landau levels in the magnetic field $\hbar\omega_0$ must be substantially less than collisional broadening of the level of $\sim \hbar/\tau_{\vec{p}}$ caused by electron scattering by crystal lattice defects or phonons. This condition is usually written as

$$\omega_0 \tau_{\vec{p}} \ll 1. \tag{4.26}$$

The condition (4.26) may have another interpretation: in order for the quasi-classical description to be applicable, it is necessary that the electron moving along a cyclotron orbit between two scattering events should be able to cover only a small fraction of period T of the circular trajectory:

$$T \simeq \frac{2\pi l_H}{\bar{v}}, \quad \frac{2\pi}{T} = \omega_0, \quad \frac{l_H}{\bar{v}} = \frac{1}{\omega_0},$$

where $\bar{v}$ is for the electron velocity at the Fermi surface for the degenerate case. It is worth noting that in the nondegenerate case, the electron velocity at the Fermi surface should be substituted for the average thermal velocity. Since the momentum relaxation time is expressed as

$$\tau_{\vec{p}} \simeq \frac{l}{\bar{v}},$$

the condition $\omega_0\tau_{\vec{p}} \ll 1$ can also be written in the form $l_H \gg l$. In other words, the radius of the cyclotron orbit must be much greater that the electron free path.

The inequalities obtained allow one to distinguish three regions of change in the external magnetic field: a weak field, a strong field and a quantizing magnetic field.

If the inequality $l_H \gg l$ is fulfilled, which is equivalent to $\omega_0\tau_{\vec{p}} \ll 1$, the magnetic fields are called weak.

If the inequality $l_H \ll l$ and, consequently $\omega_0\tau_{\vec{p}} \gg 1$ holds true, magnetic fields are called strong. In this case, the magnetic field essentially distorts the electron path. However, if its influence can be ignored in calculating scattering probabilities, the kinetic equation is applicable to describing the transport phenomena in the magnetic field even under such conditions as $\omega_0\tau_{\vec{p}} \gg 1$. Naturally, the condition $\lambda \ll l$ should remain valid.

With further increase of a magnetic field the condition (4.25) is infringed, and the magnetic field becomes quantizing one. In this case, the spectrum of charge carriers in a magnetic field is rearranged completely and an influence of the magnetic field should be considered not only when an analysis of the orbital motion of electrons runs, but in calculating the scattering probabilities for each elementary act of the collision.

4.1.3 How to determine charge and heat fluxes and calculate kinetic coefficients when $H = 0$

Summarizing the simplest expression for the flux of charged particles $\vec{J} = en\vec{v}$, where n is the number of particles with velocity $\vec{v}$, we get the expressions for the charge and heat flux densities:

$$\vec{J} = \sum_{\sigma} \frac{e}{(2\pi\hbar)^3} \int d\vec{p} f(\vec{p})\vec{v}, \tag{4.27}$$

$$\vec{J}_Q = \sum_{\sigma} \frac{1}{(2\pi\hbar)^3} \int d\vec{p}(\varepsilon_{\vec{p}} - \zeta) f(\vec{p})\vec{v}. \tag{4.28}$$

The summation over the spin quantum number in the formulas (4.27), and (4.28) gives the numerical multiplier equal to two, because the spin-splitting of levels is not considered. In writing the expression (4.28) we have used the definition (1.12).

From physical reasoning it is easy to see that only non-equilibrium correction to the distribution function $f_1(\vec{p})$ makes a non-zero contribution in the formulas (4.27) and (4.28). The correction is defined either by the expression (4.14) in the absence of a magnetic field or by the expressions (4.15) and (4.24) in the presence of an external magnetic field.

Consider first transport phenomena in the absence of the external magnetic field. In this case, the kinetic coefficients are scalar quantities. Then, substituting the expression (4.14) into the formulas determining the charge and heat fluxes, we have

$$\vec{j} = e^2 K_0 \vec{\varepsilon} - \frac{e}{T} K_1 \vec{\nabla} T, \tag{4.29}$$

$$\vec{j}_Q = e K_1 \vec{\varepsilon} - \frac{1}{T} K_2 \vec{\nabla} T, \tag{4.30}$$

where the integrals K_l, $l = 0, 1, 2$, are defined by the relation

$$K_l = \frac{2}{(2\pi\hbar)^3} \frac{1}{3} \int d\vec{p} \left(-\frac{\partial f_0}{\partial \varepsilon_{\vec{p}}} \right) \tau_{\vec{p}} v^2 (\varepsilon_{\vec{p}} - \zeta)^l. \tag{4.31}$$

While deriving the formulas (4.29)–(4.31) we have taken into account the fact that the following representation holds for the arbitrary function $\Phi(\varepsilon_{\vec{p}})$ of the modulus of the electron quasi-momentum:

$$\int d\vec{p} \Phi(\varepsilon_{\vec{p}}) v_i v_j = \frac{1}{3} \int d\vec{p} \Phi(\varepsilon_{\vec{p}}) v^2 \delta_{ij},$$

where δ_{ij} is the Kronecker symbol; $i, j = x, y, z$.

Having compared the formulas (4.29), (4.30) with the corresponding phenomenological results of (1.15), (1.35) and (1.36) we can express the kinetic coefficients, which describe thermal and electrical conductivity, as well as thermoelectric phenomena via the above-introduced integrals K_l:

$$\rho = \frac{1}{e^2 K_0}, \quad \alpha = \frac{K_1}{e T K_0},$$
$$\widetilde{\kappa} = \frac{K_2 K_0 - K_1^2}{T K_0}. \tag{4.32}$$

Thus, to calculate these kinetic coefficients integrals of K_l (4.31) are necessary to calculate too. Passing on to the integration over the energy in formula (4.31), after performing the integration over the polar and azimuthal angles in the spherical coordinate system, we have

$$K_l = \frac{2(2m)^{1/2}}{3\pi^2\hbar^3} \int_0^\infty d\varepsilon_{\vec{p}} \left(-\frac{\partial f_0}{\partial \varepsilon_{\vec{p}}} \right) \tau_{\vec{p}} \varepsilon_{\vec{p}}^{3/2} (\varepsilon_{\vec{p}} - \zeta)^l. \tag{4.33}$$

The energy integrals containing the Fermi function or of its derivatives can be reduced to the so-called the Fermi integrals $F_p(\zeta / k_B T)$ for index p:

$$F_p\left(\frac{\zeta}{k_B T} \right) = \int_0^\infty \frac{x^p}{\exp(x - \zeta / k_B T) + 1} dx. \tag{4.34}$$

These integrals depend on the $\zeta / k_B T$ parameter and various asymptotic representations [8] are well known for them. General expressions which can be obtained at the same time prove to be quite cumbersome, and therefore it would be proper to consider only two extreme cases whose simple estimates can be easily carried out.

The case of strong degeneracy

In this case, the conditions $\zeta > 0$, $\zeta/k_B T \gg 1$ are fulfilled and the derivative over the energy of the distribution function has a sharp peak at $\varepsilon_{\vec{p}} = \zeta$. The dependence of the Fermi–Dirac function and its the first derivative on a dimensionless parameter x is represented in Figure 4.2. We have

$$f_0(x) = \frac{1}{e^x + 1}, \qquad x = \frac{\varepsilon_{\vec{p}} - \zeta}{k_B T}.$$

As can be seen from Figure 4.2(b), the derivative of the distribution function differs from zero only at a small energy interval $\varepsilon_{\vec{p}} \simeq k_B T$. This peculiarity of the derivative is widely used for constructing approximate formulas to calculate integrals containing the product of the smooth function $\Phi(\varepsilon_{\vec{p}})$ and the derivative of the Fermi–Dirac distribution function over energy $\varepsilon_{\vec{p}}$ as integrand.

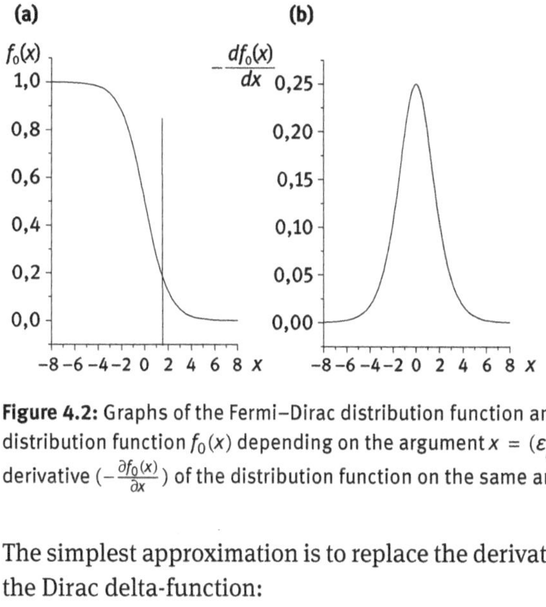

Figure 4.2: Graphs of the Fermi–Dirac distribution function and its derivative: (a) the Fermi–Dirac distribution function $f_0(x)$ depending on the argument $x = (\varepsilon_{\vec{p}} - \zeta)/k_B T$; (b) the dependence of the derivative $(-\frac{\partial f_0(x)}{\partial x})$ of the distribution function on the same argument.

The simplest approximation is to replace the derivative of the distribution function by the Dirac delta-function:

$$-\frac{\partial f_0(\varepsilon_{\vec{p}} - \zeta)}{\partial \varepsilon_{\vec{p}}} \simeq \delta(\varepsilon_{\vec{p}} - \zeta).$$

In general, there takes place a build-up of the expansion of the integrand over the small $k_B T/\zeta$ parameter [26]. As rule, it suffices to keep the first two terms in the expansion:

$$\int_0^\infty d\varepsilon_{\vec{p}} \Phi(\varepsilon_{\vec{p}}) \left(-\frac{\partial f_0}{\partial \varepsilon_{\vec{p}}} \right) \simeq \Phi(\zeta) + \frac{\pi^2}{6} (k_B T)^2 \frac{\partial^2 \Phi(\varepsilon_{\vec{p}})}{\partial \varepsilon_{\vec{p}}^2} \bigg|_{\varepsilon_{\vec{p}} = \zeta} + \cdots . \tag{4.35}$$

Performing simple calculations with the help of formulas (4.33) and (4.35) and keeping the first nonvanishing terms over the small $k_B T/\zeta$ parameter, we get

$$K_0 = \frac{n}{m}\tau_{\vec{p}}(\zeta), \quad K_2 = \frac{\pi^2}{3}(k_B T)^2 K_0,$$
$$K_1 = \frac{\pi^2}{3}\frac{n}{m}\frac{(k_B T)^2}{\zeta}\left[\zeta\frac{\partial \tau_{\vec{p}}(\zeta)}{\partial \zeta} + \frac{3}{2}\tau_{\vec{p}}(\zeta)\right]. \tag{4.36}$$

The numerical value and type of functional dependence of momentum relaxation time on energy $\tau_{\vec{p}}(\varepsilon_{\vec{p}})$ is necessary for practical application of the formulas (4.36). Usually this dependence is deemed to be power-like:

$$\tau_{\vec{p}}(\varepsilon_{\vec{p}}) = \tau_0(\varepsilon_{\vec{p}}/k_B T)^r,$$

where r is for the scattering coefficient, whose value depends on a particular mechanism for the relaxation of the electron momentum, τ_0 is for the a size factor, which depends on the scattering mechanism and temperature. The particular values of τ_0 and r for different scattering mechanisms can be found in the monographs [8, 26, 27].

The expressions for the kinetic coefficients in the limit of high degeneracy can be easily obtained, the expressions (4.36) for the integrals of K_l being substituted in (4.32):

$$\sigma = \frac{1}{\rho} = \frac{e^2 n}{m}\tau_{\vec{p}}(\zeta), \tag{4.37}$$

$$\alpha = \frac{\pi^2}{3}\frac{k_B}{e}\frac{k_B T}{\zeta}(3/2 + r), \tag{4.38}$$

$$\widetilde{\kappa} \simeq \kappa = \frac{\pi^2}{3}k_B^2 T \frac{n}{m}\tau_{\vec{p}}(\zeta). \tag{4.39}$$

In deriving the last relation we have taken into account that $K_0 K_2 \gg K_1^2$.

The results obtained (4.37)–(4.39) are qualitatively correct to describe the behavior of the electrical conductivity σ, the differential thermopower α and thermal conductivity $\widetilde{\kappa}$ in normal metals and highly-degenerate semiconductors. We do not give here values of the thermoelectric coefficients, characterizing the phenomenon of Peltier and Thomson, because, as shown in Chapter 1, they are expressed in the isotropic case via the coefficient of the thermoelectric power. We estimate the order of magnitude of the differential thermopower α, using formula (4.38)

$$\alpha \simeq \frac{\pi^2}{3}\frac{k_B}{e}\frac{k_B T}{\zeta} \simeq 10^{-8} T \text{ (V/K)}, \tag{4.40}$$

where the temperature T is measured in degrees of the Kelvin scale. This evaluation in the order of the magnitude coincides with the known experimental data for the thermopower of most metals ($\alpha = 3 \div 10\,\mu\text{V/K}$). Significant deviations from the formula (4.38) can occur, for example, in the presence of magnetic impurities (the Kondo effect). We will not dwell on this interesting question and refer the reader to special literature [28, 30].

Another important result of the theory in question is to implement the Wiedemann–Franz law for the σ and $\tilde{\kappa}$ coefficients:

$$\tilde{\kappa} = \sigma T L, \quad L = \frac{\pi^2}{3}\frac{k_B^2}{e^2},$$

This result is well confirmed by experiment at sufficiently high temperatures ($T \simeq 300\,\mathrm{K}$).

Problem 4.1. Explain with quantum statistics how an external electric field and a temperature gradient applied to a conductor induce an electric current in it using the expression (4.14) for the correction to the distribution function.

Solution. Let there be an external electric field $\vec{E}$ only. Before switching on the external electric field the equilibrium distribution function was spherically symmetric and depended only on momentum modulus $\vec{p} = \hbar\vec{k}$. Once the external electric field emerges some direction is distinguished and the distribution function becomes non-spherically symmetric. In a stationary non-equilibrium state, the electron momentum has an additive component $\Delta\vec{p} = e\vec{E}\tau_{\vec{p}}$ appearing under the action of the external electric field. In a degenerate case, as seen in Figure 4.2(b) only a small electron layer with width of the order of $k_B T$ near the Fermi surface participates in electromigration. The rest electrons cannot be accelerated by the external electric field because neighboring energy states are occupied. Therefore, only those electrons which are lying on the Fermi surface absorb the additive component to the momentum.

For simplicity, consider the direction that coincides with the external electric field. The electrons moving along this direction are slowed down by the field (an electron is a negatively charged particle) and on the Fermi surface have a smaller speed than the electrons moving in the opposite direction. This situation is schematically shown in Figure 4.3(a).

Thus, upon switching on the electric field there arises a group of electrons near the Fermi surface. These electrons move against the field and have an additional correction to velocity $\Delta\vec{v} = |e|/m\vec{E}\tau_{\vec{p}}(\zeta)$. Another group of electrons moving along the field have less speed which leads to directional motion of electrons upon switching on the electric field.

In other words, the switching on of a constant electric field leads to a shift of the Fermi surface in the momentum space by the magnitude $\Delta\vec{p} = e\vec{E}\tau_{\vec{p}}$. Therefore, the distortion of the distribution function can be found through the equilibrium distribution in a coordinate system shifted by this magnitude:

$$f_0(\varepsilon_{\vec{p}-e\vec{E}\tau_{\vec{p}}}) = f_0\left(\frac{(\vec{p}-e\vec{E}\tau_{\vec{p}})^2}{2m}\right) \simeq f_0\left(\frac{p^2}{2m} - e\vec{v}\vec{E}\tau_{\vec{p}}\right) = f_0 - e\frac{\partial f_0}{\partial \varepsilon_{\vec{p}}}\vec{v}\vec{E}\tau_{\vec{p}}.$$

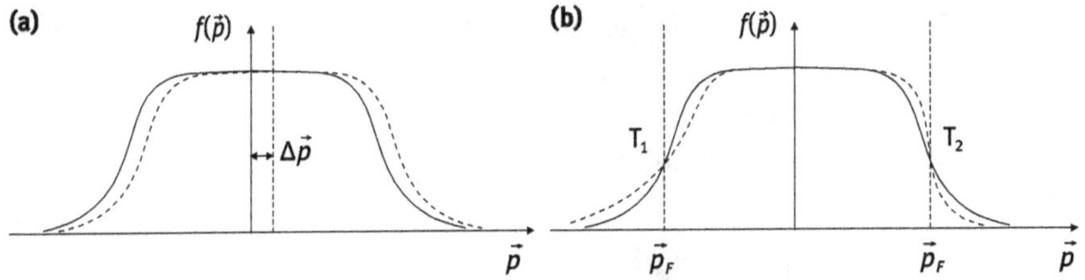

Figure 4.3: Appearance of the distribution function asymmetry in present of an external field and temperature gradient: (a) the Fermi level shift in the presence of an electric field in magnitude of $\Delta\vec{p} = |e|\vec{E}\tau_{\vec{p}}(\zeta)$; (b) a change in shape of the distribution function near the Fermi level in the presence of a temperature gradient.

It is the last expression in this formula that is the correction to the distribution function in an electric field. Although, as already noted, the concept of a shift of the Fermi surface in momentum space is not completely correct, it allows one to deduce an accurate expression for the correction to the distribution function in the electric field.

Now consider the influence of a temperature gradient and again analyze the motion of electrons only along one direction, which coincides with the orientation of the temperature gradient. Let us take two cross-sections of the specimen with a distance between them less than the mean free path. Let T_1 and T_2 be temperatures corresponding to the "hot" and "cold" cross-sections, respectively, i. e. $T_1 > T_2$. Consequently, electrons moving through the given cross-sections of the conductor will have appropriate thermal equilibrium distributions. Both distributions are schematically shown in Figure 4.3(b). As far as electrons located only in a narrow energy layer with width of a few $k_B T$ make a contribution to the transfer, it is important to observe changes in the velocity distributions. Into the "hot" cross-section, the number of electrons having a greater momentum than the Fermi momentum p_F increases compared with their number under equilibrium conditions. On the contrary, the number of electrons into the "cold" cross-section decreases. It is just this change in the shape of the electron distribution that leads to the appearance of an electric current when a conductor is exposed to a temperature gradient. The above reasoning yields also a quantitative evaluation of kinetic coefficients to describe thermoelectric effects.

Nondegenerate electron gas

In the other limiting case of a nondegenerate electron gas obeying the Maxwell – Boltzmann statistics, the conditions $\zeta < 0$, $|\zeta|/k_B T \gg 1$ are fulfilled and the Fermi–Dirac function is approximated by the expression

$$f_0(\varepsilon_{\vec{p}}) = \exp\left(-\frac{\varepsilon_{\vec{p}} - \zeta}{k_B T}\right). \tag{4.41}$$

Consider the computation process to find the integrals of K_l in this limit. Taking into account the expression (4.41) and using the expression (4.33) for the integrals K_l, we get

$$K_0 = \frac{2(2m)^{1/2}e^{\zeta/k_BT}(k_BT)^{3/2}}{3\pi^2\hbar^3}\int\limits_0^\infty dxe^{-x}x^{3/2}\tau_{\vec{p}}(x)$$

$$= \frac{4n}{3\pi^{1/2}m}\int\limits_0^\infty dxe^{-x}x^{3/2}\tau_{\vec{p}}(x) = \frac{n}{m}\langle\tau_{\vec{p}}(x)\rangle, \tag{4.42}$$

$$\langle\tau_{\vec{p}}(x)\rangle = \frac{4}{3\sqrt{\pi}}\int\limits_0^\infty dxe^{-x}x^{3/2}\tau_{\vec{p}}(x); \tag{4.43}$$

$$n = \frac{(2mk_BT)^{3/2}}{4\pi^{3/2}\hbar^3}e^{\zeta/k_BT}. \tag{4.44}$$

The integrals K_1 and K_2 are calculated quite analogously. Summarizing the results, we have

$$K_0 = \frac{n}{m}\langle\tau_{\vec{p}}(x)\rangle,$$
$$K_1 = \frac{n}{m}k_BT\langle\tau_{\vec{p}}(x)(x-\zeta/k_BT)\rangle,$$
$$K_2 = \frac{n}{m}(k_BT)^2\langle\tau_{\vec{p}}(x)(x-\zeta/k_BT)^2\rangle. \tag{4.45}$$

Now, it is necessary to find expressions for the kinetic coefficients which are of interest to us. For this purpose, the values of the integrals K_l should be substituted into the definitions (4.32). As a result, one obtains

$$\rho^{-1} = \sigma = \frac{e^2n}{m}\langle\tau_{\vec{p}}(x)\rangle, \tag{4.46}$$

$$\alpha = \frac{k_B}{e}\left(\frac{\langle\tau_{\vec{p}}(x)x\rangle}{\langle\tau_{\vec{p}}(x)\rangle} - \frac{\zeta}{k_BT}\right), \tag{4.47}$$

$$\tilde{\kappa} = \frac{n}{m}k_B^2T\left(\langle\tau_{\vec{p}}(x)x^2\rangle - \frac{\langle\tau_{\vec{p}}(x)x\rangle^2}{\langle\tau_{\vec{p}}(x)\rangle}\right). \tag{4.48}$$

Now consider the averages of $\langle\tau_{\vec{p}}(x)x^k\rangle$, involved in the expressions (4.46)–(4.48) for the kinetic coefficients. Taking into account the definition (4.43) and the commonly used approximation $\tau_{\vec{p}}(x) = \tau_0x^r$ we can express the averages through the ratio of gamma functions:

$$\langle\tau_{\vec{p}}(x)x^k\rangle = \frac{4}{3\sqrt{\pi}}\int\limits_0^\infty dxe^{-x}\tau_0x^{r+k+3/2} = \tau_0\frac{\Gamma(r+k+5/2)}{\Gamma(5/2)}, \tag{4.49}$$

where $\Gamma(p)$ denotes the gamma function defined in a standard manner:

$$\Gamma(p) = \int\limits_0^\infty dxe^{-x}x^{p-1}, \quad \Gamma(5/2) = \frac{3\sqrt{\pi}}{4}.$$

Plugging the expression (4.49) into the expressions (4.46)–(4.48) for finding the kinetic coefficients for the nondegenerate case, we have

$$\rho^{-1} = \sigma = \frac{e^2 n}{m} \tau_0 \frac{\Gamma(5/2 + r)}{\Gamma(5/2)}, \tag{4.50}$$

$$\alpha = \frac{k_B}{e} \left[\frac{\Gamma(5/2 + r + 1)}{\Gamma(5/2 + r)} - \frac{\zeta}{k_B T} \right] = \frac{k_B}{e} \left[\frac{5}{2} + r - \frac{\zeta}{k_B T} \right], \tag{4.51}$$

$$\begin{aligned} \tilde{\kappa} &= \frac{n}{m} k_B^2 T \tau_0 \left[\frac{\Gamma(5/2 + r + 2)}{\Gamma(5/2)} - \frac{\Gamma(5/2 + r + 1)^2}{\Gamma(5/2 + r)\Gamma(5/2)} \right] \\ &= \frac{n}{m} k_B^2 T \tau_0 \frac{\Gamma(5/2 + r)}{\Gamma(5/2)} (5/2 + r). \end{aligned} \tag{4.52}$$

In deducing the formulas (4.51), (4.52) the known relation satisfied by the gamma function $\Gamma(p + 1) = p\Gamma(p)$ has been used.

Upon comparing (4.50) and (4.52), it is clear that the electrical conductivity and the electronic component of thermal conductivity, as in the case of strong degeneracy, are related to each other by the Wiedemann–Franz relation:

$$\tilde{\kappa} = \sigma T L, \quad L = \frac{k_B^2}{e^2} (5/2 + r).$$

It is of interest to compare magnitude of the differential thermopower coefficient for the cases of a strongly degenerate and nondegenerate gas. As to the strongly degenerate electron gas, comparing the formula (4.38) with (4.51), it turns out that the expression for the differential thermopower coefficient contains the additional small parameter $k_B T/\zeta \simeq 10^{-2}$. For this reason, the thermoelectric power of typical metals is significantly less than that of typical semiconductors.

Note also that the electron's charge e appearing in the formula (4.51) is negative and therefore the coefficient α is negative too provided that charge carriers are electrons. In the case of a hole conductivity, the expression (4.51) for the differential thermopower coefficient remains true afterward if the value e is substituted for $|e|$ and the chemical potential of electrons ζ is replaced by the chemical potential of holes:

$$\zeta_p = -E_g - \zeta. \tag{4.53}$$

Thus, if charge carriers are holes, then the thermopower coefficient is positive, what can be used in an experiment to determine a type of the charge carriers in the crystal.

We have just given expressions for the kinetic coefficients for both limits concerning strongly degenerate and nondegenerate statistics. In principle, there are formulas to evaluate the Fermi integrals (4.34) and the error in estimate does not exceed 1.2 % for actual values of the index p at all values of $x = \zeta/k_B T$ [29].

Problem 4.2. Deduce a formula for calculating the differential thermal power coefficient in the case of mixed electron-hole conductivity.

Solution. Consider the simplest case of a nondegenerate eigen conductor with valence band that is nearly filled by electrons. If a width of the band gap E_g is not very large, due to thermal excitation of valence electrons, conduction electrons may be excited into the conduction band. Moreover, in the valence band there appear empty, unfilled states, which are called holes. The hole concept is convenient and greatly simplifies a description of transport phenomena where the valence band electrons take part. Consider first equilibrium statistical properties of an electron-hole system.

Electrons in conduction and valence bands are an aggregation of particles as a whole and are characterized by the same thermodynamic potential. We choose the lowest conduction band edge as an origin point of the energy scale for electron states in the valence and conduction bands. Bearing in mind a parabolic dispersion law in the conduction and valence bands, we have

$$\varepsilon_c = \frac{p^2}{2m_c}, \quad \varepsilon_v = -E_g - \frac{p^2}{2m_v}, \tag{4.54}$$

where m_c and m_v is for the effective mass of electrons in the conduction and valence bands, respectively.

The chemical potential of electrons can be determined by using the law of conservation of particles: the number of electrons n in the conduction band must coincide with the number of empty places p (holes) in the valence band. As a result in equilibrium we obtain the obvious relation for determining the chemical potential:

$$\sum_{\sigma,\vec{p}} \left[\exp\left(\frac{\varepsilon_c - \zeta}{k_B T}\right) + 1\right]^{-1} = \sum_{\sigma,\vec{p}} \left\{1 - \left[\exp\left(\frac{\varepsilon_v - \zeta}{k_B T}\right) + 1\right]^{-1}\right\}. \tag{4.55}$$

We transform the brace on the right-hand side of the expression (4.55)

$$1 - \left[\exp\left(\frac{\varepsilon_v - \zeta}{k_B T}\right) + 1\right]^{-1} = \left[\exp\left(\frac{-\varepsilon_v + \zeta}{k_B T}\right) + 1\right]^{-1} = \left[\exp\left(\frac{\varepsilon_p - \zeta_p}{k_B T}\right) + 1\right]^{-1}.$$

We have used the definition ε_v (4.54) for getting the last result and introduced the ε_p and ζ_p notations for the energy and for the hole chemical potential, respectively:

$$\varepsilon_p = \frac{p^2}{2m_v}, \quad \zeta_p = -E_g - \zeta.$$

Passing on from the summation over the quasi-momentum in the expression (4.55) to the integration over the quasi-momentum in the spherical coordinate system, and then going over to the integration over the $x = \varepsilon_c/k_B T$ and $x = \varepsilon_p/k_B T$ dimensionless variables for electrons and for holes, respectively, one finds

$$n = \frac{(2m_c k_B T)^{3/2}}{2\pi^2\hbar^3} F_{1/2}\left(\frac{\zeta}{k_B T}\right), \quad p = \frac{(2m_v k_B T)^{3/2}}{2\pi^2\hbar^3} F_{1/2}\left(\frac{-E_g - \zeta}{k_B T}\right), \tag{4.56}$$

$$F_{1/2}\left(\frac{\zeta}{k_B T}\right) = \begin{cases} \int_0^{\zeta/k_B T} x^{1/2}\,dx = \frac{2}{3}\left(\frac{\zeta}{k_B T}\right)^{3/2}; \\ \int_0^{\infty} e^{\zeta/k_B T} e^{-x} x^{1/2}\,dx = e^{\zeta/k_B T}\Gamma(3/2). \end{cases} \tag{4.57}$$

Using the formulas (4.55)–(4.57), it is easy to find an expression for the chemical potential in an eigen semiconductor:

$$\zeta = -\frac{E_g}{2} + \frac{3}{4}\ln\left(\frac{m_v}{m_c}\right). \tag{4.58}$$

Similarly, using the hole concept, an expression to compute equilibrium thermodynamic potentials in the presence of donor and acceptor impurity centers in semiconductors may be also derived.

Now consider how to calculate a contribution of valence electrons to the electron transfer using the hole concept. Applying the formula (4.27), we write down expressions for the contribution of the valence band electrons to an electric current:

$$\vec{J} = e\sum_{\sigma,\vec{p}} \vec{v} f\left(\frac{\varepsilon_v - \zeta}{k_B T}\right) = |e|\sum_{\sigma,\vec{p}} \vec{v}\left\{1 - f\left(\frac{\varepsilon_v - \zeta}{k_B T}\right)\right\}. \tag{4.59}$$

In writing the second equality in the expression (4.59) we have taken into account the fact that the contribution to the electric current is equal to zero when the valence band is completely filled. The expression in the braces in the last formula represents a hole distribution function,

$$\left\{1 - f\left(\frac{\varepsilon_v - \zeta}{k_B T}\right)\right\} = f\left(\frac{\varepsilon_p - \zeta_p}{k_B T}\right).$$

Therefore, the contribution of the valence electrons to the electric current can be represented as a current of positively charged quasi-particles having a positive mass (a second derivative over momentum of the quasi-particle energy $\varepsilon_p = p^2/2m$) is positive:

$$\vec{J} = |e|\sum_{\sigma,\vec{p}} \vec{v} f\left(\frac{\varepsilon_p - \zeta_p}{k_B T}\right). \tag{4.60}$$

Under external disturbances such as an external electric field and a temperature gradient the equilibrium hole distribution function suffers from distortions that can be calculated in the same way as distortions of the electron distribution function in Problem 4.1. As to the holes, following the reasoning mentioned above in Problem 4.1, we can see that the hole drift velocity is directed along the electric field, and the holes will contribute to the current flow, causing it to increase.

The distortions of the hole distribution function and electron distribution function are exactly identical. Therefore, under the influence of a temperature gradient, holes drift in the same direction as the electron flow. Summarizing these results, we can write down a phenomenological equation governing charge flow in a semiconductor with the mixed-type conductivity in the presence of an electric field and a temperature gradient:

$$\vec{J} = (\sigma_n + \sigma_p)\vec{\varepsilon} - (\beta_n - \beta_p)\vec{\nabla}T, \tag{4.61}$$

which is a generalization of the first equation of the set of the phenomenological transport equations (1.15) in the case of a mixed-type conductivity; σ_n, σ_p and β_n, β_p being the electrical conductivity coefficients and thermoelectric coefficients of the electron and hole subsystems, respectively. Now, using the relations (1.35) and (1.36) we get an expression for the field $\vec{E}$ inside a homogeneous conductor:

$$\vec{E} = \frac{1}{\sigma_n + \sigma_p}\vec{J} + \left[\frac{\beta_n}{\sigma_n + \sigma_p} - \frac{\beta_p}{\sigma_n + \sigma_p}\right]\vec{\nabla}T. \tag{4.62}$$

Introducing the coefficients of the differential thermopower $\alpha_n = \sigma_n^{-1}\beta_n$ and $\alpha_p = \sigma_p^{-1}\beta_p$ for electrons and holes, respectively and considering the fact that the thermopower differential coefficient for electrons is determined by the formula (4.51) (a similar formula should be written for holes), we get

$$\alpha = \frac{k_B}{e\sigma}\left[\sigma_n\left(5/2 + r - \frac{\zeta}{k_B T}\right) - \sigma_p\left(5/2 + r' - \frac{\zeta_p}{k_B T}\right)\right]. \tag{4.63}$$

In this formula $\sigma = \sigma_n + \sigma_p$ represents the complete electrical conductivity, r' being the index of the scattering for the holes.

It should not be supposed that this simple theory of kinetic coefficients based on the parabolic dispersion law for electrons and holes is well consistent with experimental data. For example, magnitude of the thermopower for the typical metals such as lithium, copper, silver, gold coincides with magnitude of a simple evaluation but has a positive sign (it is typical for hole material) over a very wide temperature range up to their melting point. There have been quite a lot of attempts to account for this anomaly. The obvious idea that comes to mind is to explain the effect by the influence both of dispersion law non-parabolicity and of the complex Fermi surface shape. However, we are forced to discard it immediately because the sign of the Hall effect in these materials is typical for charge carriers such as electrons.

The effect can be accounted for by an anomalously sharp energy-dependence of the electron momentum relaxation time [30]. Indeed, it follows from the formula (4.33) that the integral sign is determined by what kind of electrons will make a greater contribution to the integral. Of course, one should keep in mind that l needs to be put equal to unity. The integral can be divided into two pieces, consequently, the contribution of the electrons may be regarded in terms of charge carriers possessing by energy less and greater than ζ. The electrons with the energy $\varepsilon_{\vec{p}} < \zeta$ will give a negative contribution, and the electrons with the kinetic energy $\varepsilon_{\vec{p}} > \zeta$ a positive one. If the contribution of the electrons with the kinetic energy is suppressed due to a sharp decrease in the relaxation time, then a resultant value of the integral K_1 will be negative, and the thermoelectric power will have a positive sign. It is curious to note that the positive sign of the thermopower for electrons means that they will diffuse along the temperature gradient towards higher temperatures.

4.1.4 Scattering of electrons by lattice vibrations

A relaxation time approximation gives good enough results in describing thermoelectric phenomena in conducting crystals. But, firstly, this approximation itself needs to be substantiated, and secondly, there are a number of effects that require going beyond the relaxation time approximation. The phenomenon of electron–phonon drag may serve as an example. In the case, the value of the differential thermopower coefficient is changed strongly at low temperatures. Another argument in favor of a more detailed study of electron scattering processes in the crystal is to estimate independently the relaxation time τ from first principles and to calculate a temperature dependence of the relaxation time.

In order to create a theory that allows posed problems to be solved, it is necessary to derive an explicit form of the electron–scatterer interaction Hamiltonian, to write down the corresponding collision integral, and then to resolve the kinetic equation and determine the thermoelectric coefficients. There are many different mechanisms of the electron–scatterer interaction, and even a brief review would take up too much space. More detailed information can be found in monographs [26, 27, 31]. Here, we consider only two types of interaction: interaction of electrons with longitudinal acoustic vibrations and interaction of electrons with charged impurity centers.

For the electron–phonon interaction Hamiltonian to be derived, it is necessary to write an expression for a shift of atoms of a crystal lattice when small (obeying a harmonic law) thermal vibrations of the atoms take place. In the simplest case of a one-atom lattice, the kinetic energy of the vibrations can be written as

$$E_k = \frac{1}{2} \sum_i M \dot{\vec{u}}_i^2, \tag{4.64}$$

where M is the atomic mass and $\vec{u}_i$ the displacement vector of the i-th atom from the equilibrium position.

One can introduce a smooth function of the displacement of an atom $\vec{u}(\vec{r})$ at the point $\vec{r}$ for the sufficiently long-wavelength vibrations and write down the kinetic energy in the continual form,

$$E_k = \frac{\rho}{2} \int \dot{\vec{u}}^2(\vec{r})\, d\vec{r}, \tag{4.65}$$

where ρ is for the density of the crystal.

The integration is being performed over total crystal volume. For the transition from the classical description of atomic vibrations of the crystal lattice to quantum one, it suffices to introduce quantization rules for coordinates and momenta:

$$M[\dot{u}_i^\alpha, u_j^\beta] = -i\hbar\delta_{ij}\delta_{\alpha\beta}. \tag{4.66}$$

This expression in the continual form can be summarized as follows:

$$\rho[\dot{u}^\alpha(\vec{r}), u^\beta(\vec{r}\,')] = -i\hbar\delta_{\alpha\beta}\delta(\vec{r} - \vec{r}\,'). \tag{4.67}$$

It is easy to verify that such a representation is faithful: it is necessary to sum up both sides of (4.66) over all atoms as both sides of (4.67) to integrate over the whole volume. Then the right-hand sides of the expressions obtained will be equal to $-i\hbar\delta_{\alpha\beta}$, and the left-hand sides will present the same quantity—the commutator of the total momentum of the lattice and of the displacement in one of points of the crystal.

Consider the longitudinal vibrations and expand the displacement operator $\vec{u}(\vec{r})$ in a Fourier series (in fact, the shift can be represented as a superposition of normal coordinates). As far as the displacements $\vec{u}(\vec{r})$ are real quantities, then this expansion should be written in such a manner that the operator $\vec{u}(\vec{r})$ would possess the property of being self-adjoint:

$$u(\vec{r}) = \frac{1}{V^{1/2}} \sum_{\vec{q}} \{u_{\vec{q}} e^{i\vec{q}\vec{r} - i\Omega_{\vec{q}} t} + u_{\vec{q}}^+ e^{-i\vec{q}\vec{r} + i\Omega_{\vec{q}} t}\}. \tag{4.68}$$

where $\vec{q}$ is the wave vector, $\Omega_{\vec{q}}$ the frequency of normal excitations.

After inserting the expansion (4.68) in the quantization condition (4.67), we arrive at the commutation relations for the amplitudes $u_{\vec{q}}$ and $u_{\vec{q}}^+$ of normal vibrations:

$$[u_{\vec{q}}, u_{\vec{q}\,'}^+] = \frac{\hbar}{2\rho\Omega_{\vec{q}}}\delta_{\vec{q}\vec{q}\,'}, \quad [u_{\vec{q}}, u_{\vec{q}\,'}] = 0, \quad [u_{\vec{q}}^+, u_{\vec{q}\,'}^+] = 0. \tag{4.69}$$

We introduce the creation and annihilation operators of phonons with the wave vector $\vec{q}$:

$$b_{\vec{q}}^+ = \left(\frac{2\rho\Omega_{\vec{q}}}{\hbar}\right)^{1/2} u_{\vec{q}}^+, \quad b_{\vec{q}} = \left(\frac{2\rho\Omega_{\vec{q}}}{\hbar}\right)^{1/2} u_{\vec{q}}.$$

Obviously, these operators satisfy the simple commutation relations:

$$[b_{\vec{q}}, b_{\vec{q}\,'}] = 0, \quad [b_{\vec{q}}^+, b_{\vec{q}\,'}^+] = 0, \quad [b_{\vec{q}}, b_{\vec{q}\,'}^+] = \delta_{\vec{q}\vec{q}\,'}. \tag{4.70}$$

Performing the commutation relations (4.70) requires fulfilling the following conditions in the second-quantization representation of the wave function:

$$b_{\vec{q}}^+|N_{\vec{q}}\rangle = \sqrt{N_{\vec{q}} + 1}|N_{\vec{q}} + 1\rangle, \quad b_{\vec{q}}|N_{\vec{q}}\rangle = \sqrt{N_{\vec{q}}}|N_{\vec{q}} - 1\rangle. \tag{4.71}$$

Using the creation and annihilation operators of phonons, we can write down an expression for the shift operator $u(\vec{r})$:

$$u(\vec{r}) = \sum_{\vec{q}} \left(\frac{\hbar}{2\rho\Omega_{\vec{q}}} \right)^{1/2} \{b_{\vec{q}}(t)e^{i\vec{q}\vec{r}} + b^{+}_{\vec{q}}(t)e^{-i\vec{q}\vec{r}}\},$$

$$b_{\vec{q}}(t) = b_{\vec{q}}e^{-i\Omega_{\vec{q}}t}, \quad b^{+}_{\vec{q}}(t) = b^{+}_{\vec{q}}e^{i\Omega_{\vec{q}}t}. \tag{4.72}$$

The expression for the kinetic energy (4.65) allows also for writing down the Hamiltonian for the phonon system in the second quantization representation. It is worth recalling that the average kinetic energy and average potential energy are equal for harmonic oscillations. Therefore, the total energy can be computed by doubling the kinetic energy. Substituting the expansion (4.68) into the expression (4.65) and averaging it over time, we obtain for the average kinetic energy:

$$\overline{E}_k = \frac{\rho}{2} \sum_{\vec{q}} (u_{\vec{q}}u^{+}_{\vec{q}} + u^{+}_{\vec{q}}u_{\vec{q}})\Omega^2_{\vec{q}}. \tag{4.73}$$

Replacing the operators $u_{\vec{q}}, u^{+}_{\vec{q}}$ in this expression by the $b_{\vec{q}}$ and $b^{+}_{\vec{q}}$, taking into account the commutation relations (4.70) we find the formula for the phonon Hamiltonian:

$$H_p = \sum_{\vec{q}} \hbar\Omega_{\vec{q}} \left(b^{+}_{\vec{q}}b_{\vec{q}} + \frac{1}{2} \right). \tag{4.74}$$

The expressions (4.70), (4.72) and (4.74) have been derived as a result of simple qualitative considerations and do not require a strict argument. However, as shown by calculations, a contribution of all three branches of the oscillations of the acoustic phonons and emergence of new optical branches can simply be taken into account, using the above results.

Now we go over to deriving the electron–phonon interaction Hamiltonian. As already mentioned, there are multiple mechanisms causing scattering of electrons by lattice vibrations. Consider the simplest mechanism that consists in the fact that the oscillations of atoms of the crystal lattice lead to a local deformation of the crystal, thereby changing the electron energy.

All properties of the strained crystal are defined by components of the symmetric strain tensor:

$$\epsilon_{ij} = \frac{1}{2} \left(\frac{\partial u_i}{\partial r_j} + \frac{\partial u_j}{\partial r_i} \right).$$

Therefore, the energy of the electrons in the strained crystal is also a function of components of this tensor $\varepsilon(\vec{p}, \epsilon_{ij})$. Expanding the electron energy in such a crystal in a series over the strain tensor components yields

$$\varepsilon(\vec{p}, \epsilon_{ij}) = \varepsilon(\vec{p}) + E_{ij}\epsilon_{ij}.$$

In the isotropic case or in crystals with cubic symmetry, the strain tensor can be represented as $\epsilon_{ij} = \operatorname{div} \vec{u} \delta_{ij}$, and therefore a correction to the electron energy, which has meaning of the interaction Hamiltonian of the electron with lattice vibrations H_{ep}, can be written down as

$$H_{ep} = E_0 \operatorname{div} \vec{u}.$$

The quantity E_0 is usually called the strain potential. Substituting the displacement $\vec{u}(\vec{r})$ (4.72) in the above expression, we get an expression for the interaction Hamiltonian between the electron located at some point $\vec{r}$ of a space, and a phonon field:

$$H_{ep} = i \sum_{\vec{q}} \left(\frac{E_0^2 \hbar}{2\rho \Omega_{\vec{q}}} \right)^{1/2} (\vec{e}_{\vec{q}} \vec{q}) \{ b_{\vec{q}} e^{i\vec{q}\vec{r}} - b_{\vec{q}}^{+} e^{-i\vec{q}\vec{r}} \}, \tag{4.75}$$

where $\vec{e}_{\vec{q}}$ is the unit polarization vector of the sound wave.

One can encounter another definition for Hamiltonian H_{ep} in the literature:

$$H_{ep} = \sum_{\vec{q}\lambda} C_{\vec{q}\lambda} \{ b_{\vec{q}\lambda} e^{i\vec{q}\vec{r}} + b_{\vec{q}\lambda}^{+} e^{-i\vec{q}\vec{r}} \}, \tag{4.76}$$

where $C_{\vec{q}\lambda}$ is the amplitude of the electron–phonon interaction,

$$|C_{\vec{q}\lambda}|^2 = \frac{E_0^2 \hbar}{2\rho s} q^t,$$

s is the sound speed, $\Omega_{\vec{q}\lambda} = sq$, λ is the index of the polarization sound wave.

The power index t varies depending on the mechanism of the electron–phonon interaction ($t = 1$ for scattering by acoustic phonons in the framework of the potential strain method). The Hamiltonian (4.76) in choosing the appropriate constant $C_{\vec{q}\lambda}$ and the exponent t can be used for other mechanisms of the electron–phonon interaction, which are different from the scattering by longitudinal acoustic vibrations.

4.1.5 The Hamiltonian of interaction between electrons and charged impurity centers

Let n be the average concentration of electrons in the crystal, n' be their concentration in the vicinity of the impurity center. If one denotes φ as the total potential of the electrostatic field of an ion located at the origin and the negative charge of electrons $-|e|(n' - n)$, then φ must satisfy the Poisson equation:

$$\nabla^2 \varphi = \frac{4\pi |e|}{\epsilon} (n' - n), \tag{4.77}$$

where ϵ is the high-frequency dielectric permittivity.

In this expression, the electron density n is determined by the formula (4.56), and for the quantity n' an analogous expression can be written, having substituted the chemical potential $\zeta \to \zeta - e\varphi$. Indeed, the electron energy in the resultant electrostatic potential is $\varepsilon_{\vec{p}} + e\varphi$, the electron distribution function being dependent on the argument $\varepsilon_{\vec{p}} + e\varphi - \zeta$. So, the expression for the concentration of the electrons in the vicinity of the impurity center appears as

$$n' = \frac{(2mk_BT)^{3/2}}{2\pi^2\hbar^3} F_{1/2}\left(\frac{\zeta - e\varphi}{k_BT}\right) = n + \frac{(2mk_BT)^{3/2}}{2\pi^2\hbar^3} F'_{1/2}\left(\frac{\zeta}{k_BT}\right)\frac{|e|\varphi}{k_BT}.$$

Substituting the last result in the Poisson equation, one obtains a simple equation to determine the potential φ:

$$\nabla^2\varphi = q_0^2\varphi, \tag{4.78}$$

$$q_0^2 = \frac{2e^2(2m)^{3/2}(k_BT)^{1/2}}{\epsilon\pi\hbar^3} F'_{1/2}\left(\frac{\zeta}{k_BT}\right), \tag{4.79}$$

where the quantity q_0 has meaning of the inverse screening radius of the electrostatic potential of an ion.

The solution of equation (4.78) has a spherical symmetry and satisfies the condition

$$\lim_{r\to\infty} \varphi(r) = 0,$$

it appears as

$$\varphi = \frac{|e|}{\epsilon r} e^{-q_0 r}.$$

The interaction energy E_{ei} between the electron located at the point with the coordinate $\vec{r}$ and the singly ionized impurity center can be written as follows:

$$E_{ei} = -\frac{e^2}{\epsilon\vec{r}} e^{-q_0\vec{r}} = -\sum_{\vec{q}} G_{\vec{q}} e^{i\vec{q}\vec{r}},$$

where $G_{\vec{q}}$ is the Fourier transform of the screened Coulomb potential of a point charge (multiplied by an electron charge):

$$G_{\vec{q}} = \frac{4\pi e^2}{\epsilon(q^2 + q_0^2)}. \tag{4.80}$$

In fact, the crystal has N_i impurities with the coordinates $\vec{R}_j$, and N electrons with coordinates $\vec{r}_i$. If the interaction of the electrons with impurity atoms exhibits an additive nature, it suffices to sum up the expression of E_{ei} over all impurity centers and all electrons for obtaining the interaction Hamiltonian of the interaction between them:

$$H_{ei} = -\sum_i^N \sum_{\vec{q}} G_{\vec{q}}\rho_{-\vec{q}} e^{i\vec{q}\vec{r}_i}; \quad \rho_{\vec{q}} = \sum_{j=1}^{N_i} e^{i\vec{q}\vec{R}_j}. \tag{4.81}$$

For practical applications it is more convenient to write down the Hamiltonian of electron–impurity scattering in the second quantization representation, assuming that electron states are described by a wave function $|\nu, \sigma\rangle$, where ν, σ are quantum numbers to define orbital and spin states, respectively. In this case, using the transition rule to the second quantization representation for the operator A of the additive type [4]:

$$A = \sum_i A_i = \sum_{\nu'\sigma',\nu\sigma} \langle \nu'\sigma'|A|\nu\sigma\rangle a^+_{\nu'\sigma'} a_{\nu\sigma},$$

we get the expression for the Hamiltonian of the electron–impurity scattering:

$$H_{ei} = -\sum_{\vec{q}} G_{\vec{q}} \rho_{-\vec{q}} \langle \nu'\sigma'|e^{i\vec{q}\vec{r}}|\nu\sigma\rangle a^+_{\nu'\sigma'} a_{\nu\sigma}. \tag{4.82}$$

Problem 4.3. Derive the expression (4.80) for the Fourier-transform of the electron–impurity interaction potential.

Solution. One of the best ways for solving the problem is to verify that the inverse Fourier-transform (4.80) leads to an expression for the screened Coulomb potential $\varphi(r)$:

$$\varphi(r) = \frac{1}{(2\pi)^3} \int d\vec{q} \frac{4\pi|e|}{\epsilon(q^2 + q_0^2)} e^{-i\vec{q}\vec{r}}.$$

For that, one should pass on to integration in the spherical coordinate system in the last integral by putting $d\vec{q} = q^2 dq \sin\theta d\theta d\varphi$. Setting $x = \cos\theta$ and after performing the integration over the angle φ, we arrive at the expression for the screened Coulomb potential $\varphi(r)$:

$$\varphi(r) = \frac{|e|}{\pi} \int_0^\infty \frac{q^2\, dq}{\epsilon(q^2 + q_0^2)} \int_{-1}^{1} e^{-iqrx}\, dx = \frac{|e|}{\pi} \int_0^\infty \frac{q\, dq}{\epsilon(q^2 + q_0^2)} \frac{e^{iqr} - e^{-iqr}}{ir}.$$

Next, it is more convenient to perform the integration after substituting the variable $q \to -q$ in the second term of the last integral:

$$\varphi(r) = \frac{|e|}{i\pi r} \int_{-\infty}^{\infty} \frac{q e^{iqr}}{\epsilon(q^2 + q_0^2)} dq. \tag{4.83}$$

The last integral is easily estimated by the theory of residues. It should be recalled that, if the integrand

$$f(z) = \frac{\phi(z)}{\psi(z)}$$

and the function $\phi(z)$ have no poles inside the integration region whereas the function $\psi(z)$ has a simple pole at the point a, then

$$\int_{\Gamma} f(z)\,dz = 2\pi i \frac{\phi(a)}{\psi'(a)}.$$

The integrand in the formula (4.83) has two poles $q = \pm i q_0$. The integration contour in the complex plane should be closed so that the only pole whose potential would tend to zero at infinity should be inside of the contour. Thus, we have proved that

$$\varphi(r) = \frac{|e|}{\epsilon r} e^{-q_0 r}.$$

4.1.6 The collision integral for the electron–phonon interaction

Let us deduce an explicit expression for the collision integral free electrons with the wave vector $\vec{k}$ in the conduction band interacted with a phonons.

We calculate the transition probability $W_{\vec{k}'\vec{k}}$ from a state with the wave vector $\vec{k}$ to a state with the wave vector $\vec{k}'$ under the action of the perturbation, defined by the Hamiltonian (4.76). According to the nonstationary perturbation theory, the transition probability of a system is determined by averaging the squared modulus of the transition amplitude $a_{\vec{k}'\vec{k}}(t)$ over states of the phonon system:

$$W_{\vec{k}'\vec{k}} = \langle |a_{\vec{k}'\vec{k}}(t)|^2 \rangle,$$
$$a_{\vec{k}'\vec{k}}(t) = -\frac{i}{\hbar} \int_0^t dt \langle \vec{k}'|H_{ep}(t)|\vec{k}\rangle e^{\frac{i}{\hbar}(\varepsilon_{\vec{k}'} - \varepsilon_{\vec{k}})t}, \tag{4.84}$$

where $|\vec{k}\rangle$ is a wave function of a free electron in a state with the wave vector $\vec{k}$. The angle brackets $\langle \dots \rangle$ denotes a quantum-statistical averaging over states of the phonon system, the quantity $\varepsilon_{\vec{k}}$ being the energy of an electron with the wave vector $\vec{k}$.

Substituting the explicit form of the Hamiltonian H_{ep} (4.76) into formula (4.84) and considering that the time-dependence of the $b_{\vec{q}\lambda}(t)$ and $b^+_{\vec{q}\lambda}(t)$ Bose-operators is defined by (4.72) and the quantum-statistical averages over the phonon variables for products of the creation–annihilation operators have the form

$$\langle b^+_{\vec{q}\lambda} b_{\vec{q}'\lambda'} \rangle = N_{\vec{q}\lambda} \delta_{\vec{q}\vec{q}'} \delta_{\lambda\lambda'}, \quad \langle b_{\vec{q}\lambda} b^+_{\vec{q}'\lambda'} \rangle = (N_{\vec{q}\lambda} + 1) \delta_{\vec{q}\vec{q}'} \delta_{\lambda\lambda'},$$
$$\langle b_{\vec{q}\lambda} b_{\vec{q}'\lambda'} \rangle = \langle b^+_{\vec{q}\lambda} b^+_{\vec{q}'\lambda'} \rangle = 0,$$
$$N_{\vec{q}\lambda} = \left\{ \exp\left(\frac{\hbar \Omega_{\vec{q}\lambda}}{k_B T} \right) - 1 \right\}^{-1}, \tag{4.85}$$

where $\hbar\Omega_{\vec{q}\lambda}$ is the energy of the phonon with the wave vector $\vec{q}$ and polarization λ, we get

$$W_{\vec{k}'\vec{k}} = \frac{2\pi t}{\hbar} \sum_{\vec{q}\lambda} |C_{\vec{q}\lambda}|^2 \Big\{ \big|\langle \vec{k}'|e^{i\vec{q}\vec{r}}|\vec{k}\rangle\big|^2 N_{\vec{q}\lambda} \delta(\varepsilon_{\vec{k}'} - \varepsilon_{\vec{k}} - \hbar\Omega_{\vec{q}\lambda})$$

$$+ \big|\langle \vec{k}'|e^{-i\vec{q}\vec{r}}|\vec{k}\rangle\big|^2 (N_{\vec{q}\lambda} + 1)\delta(\varepsilon_{\vec{k}'} - \varepsilon_{\vec{k}} + \hbar\Omega_{\vec{q}\lambda}) \Big\}. \quad (4.86)$$

In writing this expression we have taken into account the fact that

$$\left| \frac{e^{i/\hbar(\varepsilon_{\vec{k}'} - \varepsilon_{\vec{k}} + \hbar\Omega_{\vec{q}\lambda})t} - 1}{i/\hbar(\varepsilon_{\vec{k}'} - \varepsilon_{\vec{k}} + \hbar\Omega_{\vec{q}\lambda})} \right|^2 = \frac{4\sin^2[(\varepsilon_{\vec{k}'} - \varepsilon_{\vec{k}} + \hbar\Omega_{\vec{q}\lambda})t/2\hbar]}{1/\hbar^2(\varepsilon_{\vec{k}'} - \varepsilon_{\vec{k}} + \hbar\Omega_{\vec{q}\lambda})^2}$$

$$= 2\pi\hbar t\delta(\varepsilon_{\vec{k}'} - \varepsilon_{\vec{k}} + \hbar\Omega_{\vec{q}\lambda})$$

and availed of the definition of δ-function

$$\delta(x) = \lim_{t\to\infty} \frac{1}{\pi} \frac{\sin^2(xt)}{x^2 t}.$$

Expression (4.86) naturally is divided into two summands, one of that is proportional to $N_{\vec{q}\lambda} + 1$, and describes a transition from the state $\vec{k}$ to the state $\vec{k}'$, generating the phonon. The other one is proportional to $N_{\vec{q}\lambda}$ and describes the processes of the phonon absorption. From the energy conservation law it follows that the transitions from a state with the wave vector $\vec{k}$ and energy $\varepsilon_{\vec{k}}$ to a state with the wave vector $\vec{k}' = \vec{k} \pm \vec{q}$ and energy $\varepsilon_{\vec{k}'} = \varepsilon_{\vec{k}} \pm \hbar\Omega_{\vec{q}\lambda}$ are possible. The two processes lead to a reduction in the number of the electrons in the state $\vec{k}$. In addition, there may be well transitions from a state with the $\vec{k}+\vec{q}$ and $\vec{k}-\vec{q}$ wave vectors to the state $\vec{k}$. These transitions will increase the number of the electrons in the state $\vec{k}$. The possible transitions listed above are schematically illustrated in Figure 4.4.

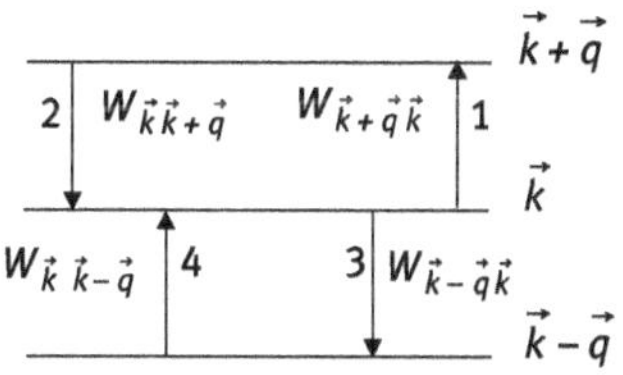

Figure 4.4: Scheme of state-transitions between states of the electrons with energy $\varepsilon_{\vec{k}}$, $\varepsilon_{\vec{k}+\vec{q}}$, $\varepsilon_{\vec{k}-\vec{q}}$. These transitions change the number of the electrons in the state with the wave vector $\vec{k}$.

Using (4.86), we write down expressions for the probability of these transitions:

1. $W_{\vec{k}+\vec{q}\vec{k}} = \dfrac{2\pi t}{\hbar} \sum_{\vec{q}\lambda} |C_{\vec{q}\lambda}|^2 \big|\langle \vec{k}+\vec{q}|e^{i\vec{q}\vec{r}}|\vec{k}\rangle\big|^2 N_{\vec{q}\lambda} \delta(\varepsilon_{\vec{k}+\vec{q}} - \varepsilon_{\vec{k}} - \hbar\Omega_{\vec{q}\lambda})$,
2. $W_{\vec{k}\vec{k}+\vec{q}} = \dfrac{2\pi t}{\hbar} \sum_{\vec{q}\lambda} |C_{\vec{q}\lambda}|^2 \big|\langle \vec{k}|e^{-i\vec{q}\vec{r}}|\vec{k}+\vec{q}\rangle\big|^2 (N_{\vec{q}\lambda} + 1) \times \delta(\varepsilon_{\vec{k}} - \varepsilon_{\vec{k}+\vec{q}} + \hbar\Omega_{\vec{q}\lambda})$,

$$3.\quad W_{\vec{k}-\vec{q}\vec{k}} = \frac{2\pi t}{\hbar}\sum_{\vec{q}\lambda}|C_{\vec{q}\lambda}|^2|\langle\vec{k}-\vec{q}|e^{-i\vec{q}\vec{r}}|\vec{k}\rangle|^2(N_{\vec{q}\lambda}+1)\times\delta(\varepsilon_{\vec{k}-\vec{q}}-\varepsilon_{\vec{k}}+\hbar\Omega_{\vec{q}\lambda}),$$

$$4.\quad W_{\vec{k}\vec{k}-\vec{q}} = \frac{2\pi t}{\hbar}\sum_{\vec{q}\lambda}|C_{\vec{q}\lambda}|^2|\langle\vec{k}|e^{i\vec{q}\vec{r}}|\vec{k}-\vec{q}\rangle|^2 N_{\vec{q}\lambda}\delta(\varepsilon_{\vec{k}}-\varepsilon_{\vec{k}-\vec{q}}-\hbar\Omega_{\vec{q}\lambda}). \tag{4.87}$$

Formulas (4.87) give the quantum-mechanical probability of the transition between the electron states with the $\vec{k}$ and $\vec{k}\pm\vec{q}$ wave vectors and over the time t. This probability is averaged over the phonon states of the system. To calculate the rate of change of the distribution function

$$\left.\frac{\partial f_{\vec{k}}}{\partial t}\right|_{\text{col}},$$

involved in the right-hand side of the kinetic equation (4.6), it is necessary to ascertain the rate of a change in the number of the electrons with the wave vector $\vec{k}$, taking into account both initially filled states and voids in final states. Given that the first and the third transitions lead to a decrease in the number of the electrons in the state $\vec{k}$ as the second and the fourth transitions to an increase in the number of the electrons in this state, we get

$$\begin{aligned}\left.\frac{\partial f_{\vec{k}}}{\partial t}\right|_{\text{col}} = \frac{2\pi}{\hbar}\sum_{\vec{q}\lambda}|C_{\vec{q}\lambda}|^2\Big\{&[(N_{\vec{q}\lambda}+1)f_{\vec{k}+\vec{q}}(1-f_{\vec{k}})-N_{\vec{q}\lambda}f_{\vec{k}}(1-f_{\vec{k}+\vec{q}})]\\ &\times\delta(\varepsilon_{\vec{k}}-\varepsilon_{\vec{k}+\vec{q}}+\hbar\Omega_{\vec{q}\lambda})+[N_{\vec{q}\lambda}f_{\vec{k}-\vec{q}}(1-f_{\vec{k}})-(N_{\vec{q}\lambda}+1)f_{\vec{k}}(1-f_{\vec{k}-\vec{q}})]\\ &\times\delta(\varepsilon_{\vec{k}}-\varepsilon_{\vec{k}-\vec{q}}-\hbar\Omega_{\vec{q}\lambda})\Big\}.\end{aligned} \tag{4.88}$$

In deriving the expression (4.88) we have availed of an evenness of the delta-function $\delta(x) = \delta(-x)$ and taken into account that, by virtue of the normalization requirement,

$$|\langle\vec{k}|e^{-i\vec{q}\vec{r}}|\vec{k}+\vec{q}\rangle|^2 = |\langle\vec{k}+\vec{q}|e^{i\vec{q}\vec{r}}|\vec{k}\rangle|^2 = 1.$$

Using the expression (4.88), we may show that one succeeds in introducing the momentum relaxation time $\tau_{\vec{p}}(\varepsilon_{\vec{p}})$ and in substantiating the representation of the collision integral in the relaxation time approximation (4.7) to describe the elastic scattering of electrons by phonons. Indeed, the scattering is elastic, if the phonon energy $\hbar\Omega_{\vec{q}\lambda}$ is much smaller than the average thermal energy of electrons $\bar{\varepsilon} \simeq k_BT$. In this case, expanding exponent in the denominator of the Planck function $N_{\vec{q}\lambda}$ over the small $\hbar\Omega_{\vec{q}\lambda}/k_BT$ parameter and restricting ourselves to only a linear approximation, we obtain

$$N_{\vec{q}\lambda} = \frac{k_BT}{\hbar\Omega_{\vec{q}\lambda}} \gg 1,\quad N_{\vec{q}\lambda}+1 \simeq N_{\vec{q}\lambda}. \tag{4.89}$$

If one takes into consideration the result (4.89), one can significantly simplify the expression for the collision integral (4.88) by reducing the members, whose function $f_{\vec{k}}$ is quadratically:

$$\left.\frac{\partial f_{\vec{k}}}{\partial t}\right|_{\text{col}} = \frac{2\pi}{\hbar}\sum_{\vec{k}'} V(\vec{k},\vec{k},')(f_{\vec{k}'} - f_{\vec{k}})\delta(\varepsilon_{\vec{k}'} - \varepsilon_{\vec{k}}),$$
$$V(\vec{k},\vec{k},') = \sum_{\vec{q}\lambda} |C_{\vec{q}\lambda}|^2 N_{\vec{q}\lambda}(\delta_{\vec{k}'\vec{k}+\vec{q}} + \delta_{\vec{k}'\vec{k}-\vec{q}}). \tag{4.90}$$

Furthermore, if the non-equilibrium distribution function are written in the form

$$f_{\vec{k}} = f_0(\varepsilon_{\vec{k}}) + f_1(\vec{k}) = f_0(\varepsilon_{\vec{k}}) - \frac{\partial f_0(\varepsilon_{\vec{k}})}{\partial \varepsilon_{\vec{k}}}(\vec{\chi}(\varepsilon_{\vec{k}})\vec{k}), \tag{4.91}$$

where $\vec{\chi}(\varepsilon_{\vec{k}})$ is an unknown vector function, depending on only the electron energy, then the difference of the distribution functions in (4.90) can be expressed via a correction to the distribution function $f_1(\vec{k})$

$$f_{\vec{k}'} - f_{\vec{k}} = \frac{\partial f_0(\varepsilon_{\vec{k}})}{\partial \varepsilon_{\vec{k}}}(\vec{\chi}(\varepsilon_{\vec{k}})\vec{k})\left[1 - \frac{k'_\chi}{k_\chi}\right],$$

where the k'_χ and k_χ quantities are projections of the $\vec{k}'$ and $\vec{k}$ vectors on the vector $\vec{\chi}(\varepsilon_{\vec{k}})$;

$$\left.\frac{\partial f_{\vec{k}}}{\partial t}\right|_{\text{col}} = \frac{f_1(\vec{k})}{\tau_{\vec{k}}(\varepsilon_{\vec{k}})},$$
$$\frac{1}{\tau_{\vec{k}}(\varepsilon_{\vec{k}})} = \frac{2\pi}{\hbar}\sum_{\vec{k}'} V(\vec{k},\vec{k},')\left[1 - \frac{k'_\chi}{k_\chi}\right]\delta(\varepsilon_{\vec{k}'} - \varepsilon_{\vec{k}}). \tag{4.92}$$

In spite of it not being very convenient for practical computations, the result (4.92) justifies the assumption made above introducing the relaxation time to describe the kinetic phenomena in conducting crystals. Also it indicates the limits of the applicability of this approximation.

Quite similarly, the structure of the collision integral at scattering by a screened Coulomb potential and by magnetic impurities can be obtained.

Problem 4.4. Express the average value $\langle c_\nu^+ c_\nu\rangle$ of the product of the c_ν^+ creation and c_ν annihilation operators of bosons (fermions) in the state $|\nu\rangle$ via the bosonic (fermionic) distribution function.

Solution. We show that $\langle c_{\nu'}^+ c_\nu\rangle \equiv \mathrm{Sp}\{\rho_0 c_{\nu'}^+ c_\nu\} = f_\nu \delta_{\nu\nu'}$, where ρ_0 is the equilibrium statistical distribution

$$\rho_0 = \exp\{-\beta(\Phi + H_0)\}, \quad \Phi = \frac{1}{\beta}\ln \mathrm{Sp}\{e^{-\beta H_0}\}, \quad H_0 = \sum_\nu \varepsilon_\nu c_\nu^+ c_\nu, \tag{4.93}$$

where $\beta = 1/k_B T$ is the inverse temperature, f_ν the equilibrium distribution function of bosons or fermions.

If there are other integrals of motion in the system besides the energy, then ρ_0 should be written as

$$\rho_0 = \exp\left\{-\beta\left(\Phi + H_0 + \sum_k P_k F_k\right)\right\}, \tag{4.94}$$

where P_k is for operators, which are conserved quantities (thermodynamic coordinates); the F_k are the thermodynamic forces corresponding to the above coordinates.

For example, in the case of Fermi-particles, the number of particles is often a conserved quantity. Then the operator of the statistical distribution in the second quantization representation should be written as

$$\rho_0 = \exp\left\{-\beta\left[\Phi + \sum_\nu (\varepsilon_\nu - \zeta)\hat{n}_\nu\right]\right\}, \quad \Phi = \frac{1}{\beta} \ln \mathrm{Sp}\{e^{-\beta \sum_\nu (\varepsilon_\nu - \zeta)\hat{n}_\nu}\}. \tag{4.95}$$

In the formula (4.95) $\hat{n}_\nu = c_\nu^+ c_\nu$, the quantity Φ being thermodynamic potential of the system particles. The quantity ζ is a chemical potential for the fermionic system. For the bosonic system the quantity ζ is equal to zero.

The quantity f_μ is the distribution function of quasi-particles. It involves an average value of the operator of the number of particles in some state μ. It is using the thermodynamic potential Φ that is the easiest way to find the quantity f_μ:

$$-\frac{d\Phi}{d\varepsilon_\mu} = \frac{\mathrm{Sp}\{\exp[-\beta \sum_\nu (\varepsilon_\nu - \zeta)\hat{n}_\nu]\hat{n}_\mu\}}{\mathrm{Sp}\{\exp[-\beta \sum_\nu (\varepsilon_\nu - \zeta)\hat{n}_\nu]\}} = \mathrm{Sp}\{\rho_0 \hat{n}_\mu\} = f_\mu. \tag{4.96}$$

Thus, to calculate $\langle c_{\nu'}^+ c_\nu \rangle$ the thermodynamic potential Φ needs to be found, which in turn is expressed via the statistical sum Z.

For estimating the statistical sum of an ideal gas of bosons or fermions can be obtained in the second quantization representation:

$$Z = \sum_{n_1, n_2 \dots n_\nu \dots} \exp\left\{-\beta \sum_\nu (\varepsilon_\nu - \zeta) n_\nu\right\}. \tag{4.97}$$

In this formula the quantities n_ν are eigenvalues of the particle number operator. They run over the values $0, 1, 2 \dots$ for the bosonic system whereas for the fermionic system take only the values 0 and 1. The right-hand side of the expression (4.97) can be rearranged to the form

$$\sum_{n_1}\sum_{n_2}\dots\sum_{n_N} e^{-\beta(\varepsilon_1 - \zeta)n_1} e^{-\beta(\varepsilon_2 - \zeta)n_2} \dots e^{-\beta(\varepsilon_N - \zeta)n_N} = \prod_\nu \sum_{n_\nu} e^{-\beta(\varepsilon_\nu - \zeta)n_\nu}. \tag{4.98}$$

Thus, performing the summation over the possible values of n_ν, we obtain the following expression:

$$Z = \begin{cases} \prod_\nu (1 + e^{-\beta(\varepsilon_\nu - \zeta)}) & \text{Fermi statistics,} \\ \prod_\nu (1 - e^{-\beta\varepsilon_\nu})^{-1} & \text{Bose statistics.} \end{cases} \tag{4.99}$$

One should note that the summation for Bose statistics is reduced to finding a sum of an infinitely decreasing geometric progression. According to the formula (4.95) the thermodynamical potential $\Phi = \ln Z/\beta$ so, taking the logarithm of (4.99), we get a simple expression for Φ:

$$\Phi = \frac{1}{\beta}\sum_{\nu}\ln\begin{cases}(1+e^{-\beta(\varepsilon_\nu-\zeta)}) & \text{Fermi statistics,}\\ (1-e^{-\beta\varepsilon_\nu})^{-1} & \text{Bose statistics.}\end{cases} \tag{4.100}$$

Using the result (4.96), we obtain

$$\langle c_{\nu'}^{+}c_\nu\rangle = -\frac{d\Phi}{d\varepsilon_\nu} = \frac{\exp\{-\beta(\varepsilon_\nu-\zeta)\}}{1+\exp\{-\beta(\varepsilon_\nu-\zeta)\}} = \frac{1}{\exp\{\beta(\varepsilon_\nu-\zeta)\}+1} \tag{4.101}$$

in the case of Fermi statistics and

$$\langle c_{\nu'}^{+}c_\nu\rangle = -\frac{d\Phi}{d\varepsilon_\nu} = \frac{\exp\{-\beta\varepsilon_\nu\}}{1-\exp\{-\beta\varepsilon_\nu\}} = \frac{1}{\exp\{\beta\varepsilon_\nu\}-1} \tag{4.102}$$

in the case of Bose statistics.

4.1.7 Phenomenon of phonon drag

We now consider a phenomenon of phonon drag. If one assumes that a phonon subsystem of a crystal forms a gas of quasi-particles (phonons), then the presence of a temperature gradient causes deviation of the gas from thermodynamic equilibrium. Consequently, there arises some phonon flux and, it is this flux that provides heat transfer through the crystal lattice. Thus, the distribution function of phonons ceases to be the equilibrium Planck function (4.85). So far as the flux of phonons is directed from the hotter face of the semiconductor to the cold one, the electrons receive a drift momentum of the phonon system at scattering. This gives rise to an additional contribution to the electron flux towards the cold conductor edge, causing the electronic component of the thermopower to increase. Such an increase is usually called *the phenomenon of phonon drag.*

Under an applied temperature gradient, another correction $\delta N_{\vec{q}\lambda}$ to the phonon distribution function is necessary to calculate the correction to the thermoelectric power associated with the electron–phonon drag effect. To find this correction we write the kinetic equation for the phonon distribution function in the relaxation time approximation. Then, in the framework of the concept of local equilibrium, we obtain

$$\left.\frac{\partial N_{\vec{q}\lambda}}{\partial t}\right|_f + \left.\frac{\partial N_{\vec{q}\lambda}}{\partial t}\right|_{\text{col}} = 0, \qquad \left.\frac{\partial N_{\vec{q}\lambda}}{\partial t}\right|_{\text{col}} = \frac{\delta N_{\vec{q}\lambda}}{\tau_{\vec{q}\lambda}}, \tag{4.103}$$

where $\tau_{\vec{q}\lambda}$ is the relaxation time of long-wavelength phonons, interacting with the electrons by thermal phonons or face of the sample.

$$\left.\frac{\partial N_{\vec{q}\lambda}}{\partial t}\right|_f = (\vec{v}_{\vec{q}\lambda}\vec{\nabla})N_{\vec{q}\lambda}(T(\vec{r})) = -(\vec{v}_{\vec{q}\lambda}\vec{\nabla}T)\frac{\partial N_{\vec{q}\lambda}}{\partial \Omega_{\vec{q}\lambda}}\frac{\Omega_{\vec{q}\lambda}}{T}. \tag{4.104}$$

In the formula (4.104) the quantity $\vec{v}_{\vec{q}\lambda}$ is a group velocity of phonons.

The results (4.103), (4.104) allow one to immediately find the correction to the phonon distribution function

$$\delta N_{\vec{q}\lambda} = \tau_{\vec{q}\lambda}\frac{\partial N_{\vec{q}\lambda}}{\partial \Omega_{\vec{q}\lambda}}\frac{\Omega_{\vec{q}\lambda}}{T}(\vec{v}_{\vec{q}\lambda}\vec{\nabla}T) \simeq -\frac{k_B\tau_{\vec{q}\lambda}}{\hbar q}\left(\frac{\vec{q}}{q}\vec{\nabla}T\right). \tag{4.105}$$

In writing the last equality, we have taken into consideration that $N_{\vec{q}\lambda} \simeq k_B T/\hbar\Omega_{\vec{q}\lambda}$, $\Omega_{\vec{q}\lambda} = sq$, $\vec{v}_{\vec{q}\lambda} = s\vec{q}/q$, where s is speed of sound in a crystal.

Let us now hark back to the kinetic equation for electrons. Taking into consideration the fact that the non-equilibrium of the phonon system leads to the appearance of an additional summand in the collision integral, we see that the integral I (4.88) falls into two summands in a linear approximation in the thermodynamic forces after substituting the quantities $N_{\vec{q}\lambda}$ for the $N_{\vec{q}\lambda} + \delta N_{\vec{q}\lambda}$:

$$I = I[f_1(\vec{k}), N_{\vec{q}\lambda}] + I[f_0(\varepsilon_{\vec{k}}), \delta N_{\vec{q}\lambda}]$$

The first summand in this expression describes the scattering of non-equilibrium electrons by phonons being in thermodynamic equilibrium. The second summand takes into account corrections responsible for the non-equilibrium of the phonon system. The electron distribution function into the second summand can be regarded as equilibrium one because the first order over thermodynamic forces is already collected. Obviously, we are interested in the second summand. Given that $N_{\vec{q}\lambda} = N_{-\vec{q}\lambda}$, and $\delta N_{\vec{q}\lambda} = -\delta N_{-\vec{q}\lambda}$, after simple transformations we arrive at an expression for that part of the collision integral that describes the correction related to the electron scattering by the non-equilibrium phonons:

$$\left.\frac{\partial f_{\vec{k}}}{\partial t}\right|_{\mathrm{dr}} = \frac{2\pi}{\hbar}\sum_{\vec{q}\lambda}|C_{\vec{q}\lambda}|^2\delta N_{\vec{q}\lambda}[f_0(\varepsilon_{\vec{k}+\vec{q}}) - f_0(\varepsilon_{\vec{k}})]$$
$$\times [\delta(\varepsilon_{\vec{k}} - \varepsilon_{\vec{k}+\vec{q}} + \hbar\Omega_{\vec{q}\lambda}) - \delta(\varepsilon_{\vec{k}} - \varepsilon_{\vec{k}+\vec{q}} - \hbar\Omega_{\vec{q}\lambda})]. \tag{4.106}$$

It should be noted that the drag effect occurs only for an inelastic scattering. The $\hbar\Omega_{\vec{q}\lambda}$ being discarded in δ-functions, the right-hand side of (4.106) immediately vanishes.

A change in the energy of the electrons in the process of the absorption (or emission) of phonons is small: $\hbar\Omega_{\vec{q}\lambda} \ll k_B T$. Therefore we can write the difference of the distribution functions in the first square bracket of (4.105) using the expansion $f_0(\varepsilon_{\vec{k}+\vec{q}})$ in the small parameter $\hbar\Omega_{\vec{q}\lambda}/k_B T$ as follows:

$$f_0(\varepsilon_{\vec{k}+\vec{q}}) - f_0(\varepsilon_{\vec{k}}) = \frac{\partial f_0(\varepsilon_{\vec{k}})}{\partial \varepsilon_{\vec{k}}}(\varepsilon_{\vec{k}+\vec{q}} - \varepsilon_{\vec{k}}).$$

Inserting this result into the formula (4.106), we obtain a simple expression appropriate to numerical evaluations:

$$\left.\frac{\partial f_{\vec{k}}}{\partial t}\right|_{\mathrm{dr}} = \frac{4\pi}{\hbar}\sum_{\vec{q}\lambda}|C_{\vec{q}\lambda}|^2\delta N_{\vec{q}\lambda}\frac{\partial f_0(\varepsilon_{\vec{k}})}{\partial\varepsilon_{\vec{k}}}\hbar\Omega_{\vec{q}\lambda}\delta(\varepsilon_{\vec{k}}-\varepsilon_{\vec{k}+\vec{q}}). \tag{4.107}$$

For further calculations it is necessary to make a replace the summation in formula (4.107) by integration and substitute both the expression for $|C_{\vec{q}\lambda}|^2$, obtained earlier (see p. 157), and the correction $\delta N_{\vec{q}\lambda}$ (4.105) into it. As a result, passing on to spherical coordinates, we get

$$\left.\frac{\partial f_{\vec{k}}}{\partial t}\right|_{dr} = -\frac{E_0^2 k_{\mathrm{B}} m}{8\pi\rho(\hbar k)^3}(\hbar\vec{k}\vec{\nabla}T)\frac{\partial f_0}{\partial\varepsilon_{\vec{k}}}\int\limits_0^{2k}\tau_{\vec{q}\lambda}q^3\,dq. \tag{4.108}$$

One should dwell on the derivation of this formula in more detail. The integration in spherical system coordinates in the formula (4.107) reduces to the integral

$$\int\limits_{q\min}^{q\max} dqq^3\tau_{\vec{q}\lambda}\int\limits_0^{\pi}\sin\theta\,d\theta\delta\left(\frac{\hbar^2 kq\cos\theta}{m}+\frac{\hbar^2q^2}{2m}\right)\times\int\limits_0^{2\pi}\left(\frac{\vec{k}}{k}\vec{\nabla}T\right)\frac{\cos\alpha}{\cos\beta}\,d\varphi. \tag{4.109}$$

where α is the angle between the vector $\vec{q}$ and the temperature gradient, vector β the angle between the vector $\vec{k}$ and the temperature gradient. The choice of the angles α, β, θ and φ is shown in Figure 4.5.

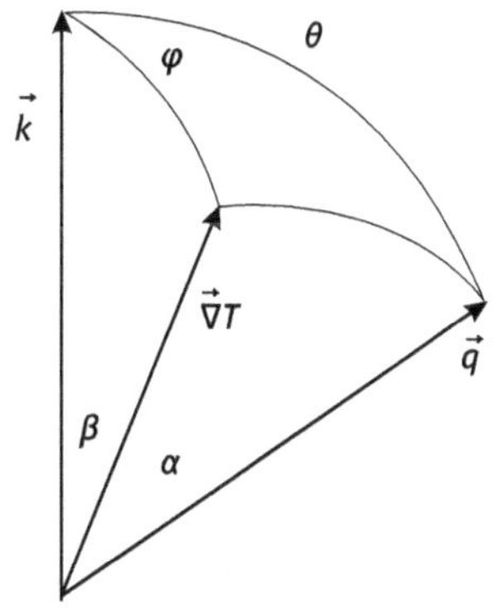

Figure 4.5: The $\vec{k}, \vec{q}, \vec{\nabla}T$ vectors and angles between them.

It can be shown [31] that there exists a simple relationship between the angles β, θ and φ:

$$\cos\alpha = \cos\theta\cos\beta + \sin\theta\sin\beta\cos\varphi.$$

As far as, as indicated by Figure 4.5, only the angle α depends on the angle φ, the integration over the angle φ yields

$$\int_0^{2\pi}\left(\frac{\vec{k}}{k}\vec{\nabla}T\right)\frac{\cos\alpha}{\cos\beta}\,d\varphi = 2\pi\left(\frac{\vec{k}}{k}\vec{\nabla}T\right)\cos\theta. \tag{4.110}$$

After replacing the variable $x = -\hbar^2 kq\cos\theta/m$ the integral (4.109) over the angle θ can be estimated as

$$\int_{-\hbar^2 kq/m}^{\hbar^2 kq/m} x\,dx\delta\left(\frac{\hbar^2 q^2}{2m} - x\right)\left(\frac{m}{\hbar^2 kq}\right)^2 = \frac{m}{2\hbar^2 k^2}; \quad q < 2k. \tag{4.111}$$

Since scattering processes comply with both laws of the momentum and energy conservation, it follows from the expression (4.111) that the electrons can interact only with the so-called long-wavelength phonons with wave vectors $q < 2k \simeq (8m\bar{\varepsilon}/\hbar)^{1/2}$ in each elementary act of scattering. Therefore, in the expression (4.111) the integral over the wave vector of phonons should be calculated between the limits 0 to $2k$.

Now, one should add a correction caused by the non-equilibrium of the phonon system to the collision integral, this makes it possible to compute the correction to the distribution function f_1 (4.14) caused by the drag effect:

$$-\frac{\partial f_0}{\partial\varepsilon_{\vec{p}}}\vec{v}\left(e\vec{\varepsilon} - \frac{\varepsilon_{\vec{p}} - \zeta}{T}\vec{\nabla}T\right) = \frac{f_1}{\tau_{\vec{p}}} + \left.\frac{\partial f_{\vec{k}}}{\partial t}\right|_{\mathrm{dr}}. \tag{4.112}$$

Inserting this expression into formula (4.108) one obtains

$$\begin{aligned} f_1 &= \tau_{\vec{p}}\left(-\frac{\partial f_0}{\partial\varepsilon_{\vec{p}}}\right)\left(\frac{\hbar\vec{k}}{m}\vec{\Phi}(\varepsilon_{\vec{k}})\right), \\ \vec{\Phi}(\varepsilon_{\vec{k}}) &= \left(e\vec{\varepsilon} - \frac{\varepsilon_{\vec{p}} - \zeta}{T}\vec{\nabla}T\right) - A_{\mathrm{dr}}(\varepsilon_{\vec{k}})\vec{\nabla}T, \\ A_{\mathrm{dr}}(\varepsilon_{\vec{k}}) &= \frac{E_0^2 k_{\mathrm{B}} m^2}{8\pi\rho(\hbar k)^3}\int_0^{2k}\tau_{\vec{q}\lambda}q^3\,dq. \end{aligned} \tag{4.113}$$

Taking into account one of the known mechanisms of relaxation of long-wavelength phonons, the integral over q needs to be also evaluated for the practical use of the result (4.113) obtained. Usually two mechanisms allow one to compute the integral: the Herring mechanism that gives the estimation

$$\tau_{\vec{q}\lambda} = \frac{\rho\hbar^2 s^3}{(k_{\mathrm{B}}T)^3 q^2}, \tag{4.114}$$

and the mechanism of Simons, giving

$$\tau_{\vec{q}\lambda} = \frac{\rho\hbar^3 s^4}{(k_{\mathrm{B}}T)^4 q}. \tag{4.115}$$

Both these mechanisms lead to a rather strong dependence of the relaxation time on a temperature ($\tau_{\vec{q}\lambda} \sim 1/T^4$ or $\tau_{\vec{q}\lambda} \sim 1/T^3$). Therefore, non-electronic mechanisms of the phonon relaxation dramatically increase a contribution at low temperatures $T \simeq 4$K. Under these conditions a component of the thermopower caused by the drag effect beats very much values of the normal diffusion component.

The numerical estimation of the contribution of the drag effect to the electron thermopower can be obtained by (4.112), (4.113). For this, it is necessary to substitute the quantity $A_{dr}(\varepsilon_{\vec{k}})$ for the quantity $(\varepsilon_{\vec{p}} - \zeta)/T$ into the formula for the integral K_1 (4.45). We shall not give these simple computations here, inviting the reader to make them as individual exercises.

4.1.8 Expressions for charge and heat fluxes in a magnetic field. Tensor structure of kinetic coefficients

We derive explicit expressions for the $\hat{\rho}$, $\hat{\alpha}$, $\hat{\Pi}$, $\hat{\bar{\kappa}}$ tensor components when an external magnetic field is not equal to zero. For this purpose, using the formulas (4.15), (4.24) and (4.27), (4.28), we find expressions for the charge and heat fluxes:

$$\vec{J} = e^2\{K_0^{\parallel}(\vec{h}\vec{\varepsilon})\vec{h} - K_0^H[\vec{h}\times\vec{\varepsilon}] - K_0^{\perp}[\vec{h}\times[\vec{h}\times\vec{\varepsilon}]]\}$$
$$- \frac{e}{T}\{K_1^{\parallel}(\vec{h}\vec{\nabla}T)\vec{h} - K_1^H[\vec{h}\times\vec{\nabla}T] - K_1^{\perp}[\vec{h}\times[\vec{h}\times\vec{\nabla}T]]\}, \tag{4.116}$$

$$\vec{J}_Q = e\{K_1^{\parallel}(\vec{h}\vec{\varepsilon})\vec{h} - K_1^H[\vec{h}\times\vec{\varepsilon}] - K_1^{\perp}[\vec{h}\times[\vec{h}\times\vec{\varepsilon}]]\}$$
$$- \frac{1}{T}\{K_2^{\parallel}(\vec{h}\vec{\nabla}T)\vec{h} - K_2^H[\vec{h}\times\vec{\nabla}T] - K_2^{\perp}[\vec{h}\times[\vec{h}\times\vec{\nabla}T]]\}. \tag{4.117}$$

Here, for convenience of the further discussion we have introduced the following notations ($l = 0, 1, 2$):

$$\left|\begin{matrix} K_l^{\parallel} \\ K_l^H \\ K_l^{\perp} \end{matrix}\right| = \frac{2\sqrt{2m}}{3\pi^2\hbar^3}\int_0^{\infty} d\varepsilon\left(-\frac{\partial f_0}{\partial\varepsilon}\right)\varepsilon^{3/2}(\varepsilon-\zeta)^l \left|\begin{matrix} \tau_{\vec{p}} \\ \frac{\omega_0\tau_{\vec{p}}^2}{1+(\omega_0\tau_{\vec{p}})^2} \\ \frac{\tau_{\vec{p}}(\varepsilon)}{1+(\omega_0\tau_{\vec{p}})^2} \end{matrix}\right|. \tag{4.118}$$

Equations (4.116), (4.117) have the same structure as the phenomenological equations (1.15), what allows the $\hat{\sigma}$, $\hat{\beta}$ and $\hat{\kappa}$ tensor components to be expressed via the introduced integrals $K_l^{\parallel}$, K_l^H, $K_l^{\perp}$. We write down the equations (4.116) and (4.117) through the components. Putting that the magnetic field H is directed along the axis Z, and then $\vec{h} \parallel \vec{e}_z$, $\vec{\varepsilon} = \varepsilon_x\vec{e}_x + \varepsilon_y\vec{e}_y + \varepsilon_z\vec{e}_z$, $\vec{\nabla}T = \nabla_x T\vec{e}_x + \nabla_y T\vec{e}_y + \nabla_z T\vec{e}_z$, where $\vec{e}_x$,

$\vec{e}_y$, $\vec{e}_z$ are unitary vectors of Cartesian coordinates, we get

$$\begin{aligned} J_x &= e^2 K_0^{\perp} \varepsilon_x + e^2 K_0^H \varepsilon_y - \frac{e}{T} K_1^{\perp} \nabla_x T - \frac{e}{T} K_1^H \nabla_y T, \\ J_y &= e^2 K_0^{\perp} \varepsilon_y - e^2 K_0^H \varepsilon_x - \frac{e}{T} K_1^{\perp} \nabla_y T + \frac{e}{T} K_1^H \nabla_x T, \\ J_z &= e^2 K_0^{\parallel} \varepsilon_z - \frac{e}{T} K_1^{\parallel} \nabla_z T, \end{aligned} \tag{4.119}$$

$$\begin{aligned} J_{Qx} &= e K_1^{\perp} \varepsilon_x + e K_1^H \varepsilon_y - \frac{1}{T} K_2^{\perp} \nabla_x T - \frac{1}{T} K_2^H \nabla_y T \\ J_{Qy} &= e K_1^{\perp} \varepsilon_y - e K_1^H \varepsilon_x - \frac{1}{T} K_2^{\perp} \nabla_y T + \frac{1}{T} K_2^H \nabla_x T, \\ J_{Qz} &= e K_1^{\parallel} \varepsilon_z - \frac{1}{T} K_2^{\parallel} \nabla_z T. \end{aligned} \tag{4.120}$$

Comparing (4.119), (4.120) with the phenomenological equation of transfer (1.15), one can find the $\hat{\sigma}$, $\hat{\beta}$ and $\hat{\kappa}$ tensor components, expressing them via the integrals $K_l^{\parallel}$, K_l^H, $K_l^{\perp}$ entered above:

$$\hat{\sigma} = \begin{pmatrix} \sigma_{\perp} & \sigma_H & 0 \\ -\sigma_H & \sigma_{\perp} & 0 \\ 0 & 0 & \sigma_{\parallel} \end{pmatrix}, \quad \hat{\beta} = \begin{pmatrix} \beta_{\perp} & \beta_H & 0 \\ -\beta_H & \beta_{\perp} & 0 \\ 0 & 0 & \beta_{\parallel} \end{pmatrix}, \tag{4.121}$$

$$\hat{\chi} = \begin{pmatrix} \chi_{\perp} & \chi_H & 0 \\ -\chi_H & \chi_{\perp} & 0 \\ 0 & 0 & \chi_{\parallel} \end{pmatrix}, \quad \hat{\kappa} = \begin{pmatrix} \kappa_{\perp} & \kappa_H & 0 \\ -\kappa_H & \kappa_{\perp} & 0 \\ 0 & 0 & \kappa_{\parallel} \end{pmatrix}, \tag{4.122}$$

where we have introduced the following notations:

$$\sigma_i = e^2 K_0^i, \quad \beta_i = \frac{e}{T} K_1^i, \quad \chi_i = e K_1^i, \quad \kappa_i = \frac{1}{T} K_2^i, \tag{4.123}$$

$i = \{\perp, \parallel, H\}$. As mentioned in Chapter 1, it is far easier to control the current $\vec{J}$ through the sample rather than the electrochemical potential gradient $\vec{\varepsilon}$. Therefore, the $\hat{\rho}, \hat{\alpha}, \hat{\Pi}, \hat{\tilde{\kappa}}$ tensor components are determined in studying the thermogalvanomagnetic phenomena. The explicit form of these components can be obtained using formulas (1.35), (1.36) and (4.121)–(4.123). Performing the necessary transformations of the tensor quantities gives

$$\hat{\rho} = \hat{\sigma}^{-1} = \begin{pmatrix} \rho_{\perp} & \rho_H & 0 \\ -\rho_H & \rho_{\perp} & 0 \\ 0 & 0 & \rho_{\parallel} \end{pmatrix}, \quad \hat{\alpha} = \hat{\rho}\hat{\beta} = \begin{pmatrix} \alpha_{\perp} & \alpha_H & 0 \\ -\alpha_H & \alpha_{\perp} & 0 \\ 0 & 0 & \alpha_{\parallel} \end{pmatrix},$$

$$\hat{\tilde{\kappa}} = \hat{\kappa} - \hat{\chi}\hat{\alpha} = \begin{pmatrix} \tilde{\kappa}_{\perp} & \tilde{\kappa}_H & 0 \\ -\tilde{\kappa}_H & \tilde{\kappa}_{\perp} & 0 \\ 0 & 0 & \tilde{\kappa}_{\parallel} \end{pmatrix}, \tag{4.124}$$

$$\rho_{\perp} = \frac{\sigma_{\perp}}{\sigma_{\perp}^2 + \sigma_H^2}, \quad \rho_H = \frac{-\sigma_H}{\sigma_{\perp}^2 + \sigma_H^2}, \quad \rho_{\parallel} = \frac{1}{\sigma_{\parallel}},$$

$$\alpha_\perp = \frac{\beta_\perp\sigma_\perp + \beta_H\sigma_H}{\sigma_\perp^2 + \sigma_H^2}, \quad \alpha_H = \frac{\beta_H\sigma_\perp - \beta_\perp\sigma_H}{\sigma_\perp^2 + \sigma_H^2}, \quad \alpha_\parallel = \frac{\beta_\parallel}{\sigma_\parallel},$$
$$\widetilde{\kappa}_\perp = \kappa_\perp - \chi_\perp\alpha_\perp + \chi_H\alpha_H, \quad \widetilde{\kappa}_H = \kappa_H - \chi_\perp\alpha_H - \chi_H\alpha_\perp,$$
$$\widetilde{\kappa}_\parallel = \kappa_\parallel - \chi_\parallel\alpha_\parallel. \tag{4.125}$$

Note the main peculiarities of the expressions obtained for the kinetic coefficients. It follows from the formulas (4.121), (4.122) and (4.124) that the structure of the tensors $\hat{\rho}, \hat{\alpha}, \hat{\Pi}, \hat{\widetilde{\kappa}}$ is characteristic for gyrotropic media, and it coincides with the structure that we suggested in Chapter 1. Furthermore, diagonal components of tensors, which characterize the phenomenon in a plane perpendicular to a magnetic field, contain only even powers of the magnetic field while longitudinal components of the tensors $\hat{\rho}, \hat{\alpha}, \hat{\Pi}, \hat{\widetilde{\kappa}}$ do not depend on the magnetic field. The non-zero off-diagonal elements that have tensor indices xy and yx are odd in the magnetic field, are equal among themselves in absolute magnitude, but have opposite signs. From what has been said above, one can claim that the expressions obtained for the kinetic coefficients satisfy the symmetry relations of Onsager:

$$\rho_{ik}(\vec{H}) = \rho_{ki}(-\vec{H}), \quad \alpha_{ik}(\vec{H}) = \alpha_{ki}(-\vec{H}),$$
$$\widetilde{\kappa}_{ik}(\vec{H}) = \widetilde{\kappa}_{ki}(-\vec{H}). \tag{4.126}$$

Another relation between the $\hat{\beta}$ and $\hat{\kappa}$ tensors follows from the formulas (4.122), (4.123): $\hat{\beta}T = \hat{\kappa}$. Hence, considering the result (1.36), we get

$$\Pi_{ik}(\vec{H}) = \alpha_{ik}(\vec{H})T. \tag{4.127}$$

4.1.9 Galvanomagnetic and thermomagnetic effects in semiconductors with a parabolic dispersion law

Consider thermogalvanomagnetic phenomena which were qualitatively discussed in Chapter 1. Using the results (4.118), (4.124), and (4.125), we calculate the kinetic coefficients concerning these effects.

It is necessary to immediately show that the outcomes obtained in such a way have limited applicability, since the simplest electronic version of the theory of kinetic transport phenomena in the relaxation time approximation when only one type of charge carriers obeying an isotropic quadratic dispersion law is considered cannot qualitatively explain the dependence of the kinetic coefficients on an amplitude and orientation of a strong magnetic field in metals. In this case, an electron moving along the Larmor orbit covers a significant portion of the Fermi surface between two scattering events and it succeeds in experiencing the real structure of this surface. Galvanomagnetic phenomena in strong magnetic fields are very sensitive to peculiarities of the energy spectrum of charge carriers and serve as a reliable way to determine the structure of the Fermi surface [28, 32].

In the case of semiconductors with a single extremum of the conduction band at the Brillouin zone center, the most serious restrictions are associated with a necessity for considering quantization, if the condition $\hbar\omega_0 \gg k_B T$ holds. Therefore, it would be proper to suppose that the magnetic field is not quantizing and the inequality $\hbar\omega_0 \ll k_B T$ is fulfilled. This allows a quasi-classical approximation to be applied to describe the motion of an electron in the magnetic field.

Both the presence of several equivalent minima (valleys) at symmetry points of the Brillouin zone and the ellipsoidal nature of the energy surfaces in some semiconductor materials (Ge, Si) may be taken into account without significant changes in main propositions of theory in question [8, 31] and, therefore will not be considered here.

The most complete review of results concerning the theory of thermomagnetic and galvanomagnetic phenomena is given in the monograph by Askerov [8], where there is also an extensive bibliography on this subject. The scope of the present book does not permit a detailed discussion of all modern outcomes in the theory thermogalvanomagnetic phenomena with rigor and completeness required. Therefore, we consider only the simplest situation, but namely, a semiconductor with a standard conduction band in the case of: (1) extremely strong degeneracy of an electron gas; (2) a nondegenerate electron gas obeying Maxwell–Boltzmann statistics.

Let us evaluate the integrals $K_l^{\parallel}$, K_l^H, $K_l^{\perp}$ defined by the expression (4.118) for the limiting cases (1) and (2) mentioned above.

Having compared the formulas (4.33) and (4.118) we see that the integrals $K_l^{\parallel}$ coincide with the integrals K_l, which have been already calculated above. Therefore, we show a way of estimating only the integrals $K_l^{\perp}$ and K_l^H.

To compute these integrals in the limit of the strongly degenerate electron gas, one should use the formula (4.35). Then in the formula (4.35) for the integrals $K_0^{\perp}$ and K_0^H it suffices to restrict oneself to a first approximation in the expansion parameter $k_B T/\zeta$ and to substitute the derivative $-\partial f_0/\partial\varepsilon$ for delta-function $\delta(\varepsilon-\zeta)$:

$$K_0^{\perp} = \frac{n}{m}\frac{\tau_{\vec{p}}(\zeta)}{1+[\omega_0\tau_{\vec{p}}(\zeta)]^2}, \quad K_0^H = \omega_0\tau_{\vec{p}}(\zeta)K_0^{\perp}. \tag{4.128}$$

To evaluate the integrals K_1^i, K_2^i where $i = \{\perp, H\}$ the quadratic expansion term in the small $k_B T/\zeta$ parameter must hold in the formula (4.33):

$$\left|\begin{array}{c} K_1^H \\ K_1^{\perp} \end{array}\right| = \frac{\pi^2}{2}\frac{(k_B T)^2}{\zeta}\frac{n}{m}\frac{\tau_{\vec{p}}(\zeta)}{1+[\omega_0\tau_{\vec{p}}(\zeta)]^2} \times \left|\begin{array}{c} \omega_0\tau_{\vec{p}}(\zeta)[1+\frac{4/3r}{1+[\omega_0\tau_{\vec{p}}(\zeta)]^2}] \\ 1+\frac{2r}{3}\frac{1-[\omega_0\tau_{\vec{p}}(\zeta)]^2}{1+[\omega_0\tau_{\vec{p}}(\zeta)]^2} \end{array}\right|, \tag{4.129}$$

$$K_2^i = \frac{\pi^2}{3}(k_B T)^2 K_0^i, \quad i = \{\perp, H\}. \tag{4.130}$$

In deriving the formula (4.129) we have assumed as before that

$$\tau_{\vec{p}}(\varepsilon_{\vec{p}}) = \tau_0(\varepsilon_{\vec{p}}/k_B T)^r,$$

and have taken into account that

$$\frac{\zeta}{\tau_{\vec{p}}(\zeta)} \frac{d\tau_{\vec{p}}(\zeta)}{d\zeta} = r.$$

In the case of a nondegenerate electron gas obeying Maxwell–Boltzmann statistics, it would be proper to discuss only the case of weak magnetic fields when the inequality $\omega_0\tau_{\vec{p}} \ll 1$ is fulfilled and to leave only the first nonvanishing term in the parameter $\omega_0\tau_{\vec{p}}$ in the integrals (4.118).

Using the definition of the average $\langle\tau_{\vec{p}}(x)x^k\rangle$ (4.49) and definition of the electron density (4.44), after simple transformations we obtain

$$\begin{vmatrix} K_l^H \\ K_l^\perp \end{vmatrix} = \tfrac{n}{m}(k_B T)^l \times \begin{vmatrix} \omega_0\langle\tau_{\vec{p}}(x)^2(x-\zeta/k_B T)^l\rangle \\ \langle\tau_{\vec{p}}(x)(x-\zeta/k_B T)^l\rangle - \omega_0^2\langle\tau_{\vec{p}}(x)^3(x-\zeta/k_B T)^l\rangle \end{vmatrix}. \tag{4.131}$$

Next, we will use the expressions obtained for the integrals K_l^i to go over to discussion of some galvanomagnetic and thermomagnetic effects.

The Hall effect

The Hall constant R based on the formula (1.57) is determined by the off-diagonal component of the conductivity tensor ρ_{xy}. Using the formulas (4.124), (4.125), we obtain an expression for the Hall constant via the $K_0^\perp$ and K_0^H integrals:

$$R = \frac{1}{H}\frac{\sigma_H}{\sigma_H^2+\sigma_\perp^2} = \frac{1}{He^2}\frac{K_0^H}{(K_0^\perp)^2+(K_0^H)^2}. \tag{4.132}$$

Inserting the results (4.128), (4.131) into this equation yields

$$R = \frac{1}{enc} \tag{4.133}$$

in the case of strong degeneracy and

$$R = \frac{\gamma}{enc}, \quad \gamma = \frac{\langle\tau_{\vec{p}}(x)^2\rangle}{\langle\tau_{\vec{p}}(x)\rangle^2} \tag{4.134}$$

for nondegenerate semiconductors. The value of the parameter γ depends on a mechanism of the charge carrier scattering and varies within $\gamma \simeq 1.18 \div 1.93$ at the value of the index scattering from $r = -1/2$ (the case of scattering by acoustic phonons) to $r = 3/2$ (the case of scattering by charged impurities).

Change in transverse resistance in a magnetic field

It should not be left unnoticed that the accuracy in evaluating the integrals $K_0^{\perp}$ and K_0^H was insufficient for metals. Plugging the results (4.128) into the formula (4.125) does not make it possible to observe a dependence of the resistance on a magnetic field. This result would be predicted beforehand, since, as noted in Chapter 1, the resistance change in the magnetic field is associated with the fact that the Hall field compensates for the magnetic component of the Lorentz force only at an average, and faster and slower electrons travel along curved trajectories, which decreases their effective free path length. Therefore, the field dependent quantity $\Delta\rho/\rho$ can be obtained by further expansion of the integrals $K_0^{\perp}, K_0^H$ in the small parameter $k_{\mathrm{B}}T/\zeta$ basing on the formula (4.35). Although these computations are reduced to elementary algebraic transformations, they do remain rather cumbersome, so we present here only the final result but detailed calculations we consider as an example:

$$\frac{\Delta\rho}{\rho} = \frac{\pi^2}{12}\left(\frac{k_{\mathrm{B}}T}{\zeta}\right)^2 \frac{[\omega_0\tau_{\vec{p}}(\zeta)]^2}{1+[\omega_0\tau_{\vec{p}}(\zeta)]^2}. \tag{4.135}$$

For nondegenerate semiconducting materials, it would be appropriate to pay attention only to the case of weak magnetic fields when the condition $\omega_0\tau_{\vec{p}} \ll 1$ is met. Use of the results (4.131) for the integrals $K_0^{\perp}$ and K_0^H gives

$$\rho_{xx}(H) = \rho_{xx}(0)\left\{1+\omega_0^2\left[\frac{\langle\tau_{\vec{p}}(x)^3\rangle}{\langle\tau_{\vec{p}}(x)\rangle} - \frac{\langle\tau_{\vec{p}}(x)^2\rangle^2}{\langle\tau_{\vec{p}}(x)\rangle^2}\right]\right\}. \tag{4.136}$$

The expression (4.136) can be written in a more convenient form, if one introduces the dimensionless parameter

$$T_r = \frac{\langle\tau_{\vec{p}}(x)^3\rangle\langle\tau_{\vec{p}}(x)\rangle - \langle\tau_{\vec{p}}(x)^2\rangle^2}{\langle\tau_{\vec{p}}(x)\rangle^4}.$$

Then one has a rather simple expression for a relative resistance change in the magnetic field:

$$\frac{\Delta\rho_{xx}}{\rho_{xx}} = \left(\omega_0\langle\tau_{\vec{p}}(x)\rangle\right)^2 T_r. \tag{4.137}$$

It follows from the formula (4.137) that the relative change resistance in the magnetic field is actually determined by the parameter $\omega_0\langle\tau_{\vec{p}}(x)\rangle$, for the dimensionless factor T_r depends weakly on the scattering coefficient r and varies in the interval from 0.38 for $r = -1/2$ to ~ 2.15 at $r = 3/2$.

The transverse Nernst–Ettingshausen effect

The transverse Nernst–Ettingshausen effect is determined by an off-diagonal tensor component of the differential thermopower α_H (4.125). Simple but rather cumbersome calculations in using formulas (4.56), (4.57), (4.123) and (4.125) being omitted, there can be given only the final result appropriate for conditions of strong degeneracy, when the $K_0^{\perp}, K_0^H$ integrals are calculated in the zero approximation in the small parameter $k_B T/\zeta$, and the $K_1^{\perp}, K_1^H$ integrals are estimated in the first nonvanishing approximation in this parameter (see the equation (4.35)):

$$Q_{NE} = \frac{\alpha_H}{H} = \frac{\beta_H \sigma_\perp - \beta_\perp \sigma_H}{H(\sigma_\perp^2 + \sigma_H^2)} = \frac{k_B}{e}\frac{\pi^2}{3}\frac{\mu_e}{c}\frac{k_B T}{\zeta} r. \quad (4.138)$$

Here $\mu_e = e\tau_{\vec{p}}(\zeta)/m$ is electron mobility ($\mu_e/c = \omega_0 \tau_{\vec{p}}(\zeta)/H$).

The expression given above implies that the effect sign is directly determined by sign of the scattering coefficient r. This fact indicates which of the electron scattering mechanisms is dominant, for example, $r = 3/2$ for scattering by neutral impurities, and $r = -1/2$ for scattering by long-wavelength acoustic vibrations.

For a nondegenerate electron gas we present the result appropriate only for the case of a weak magnetic field when the condition $\omega_0 \tau_{\vec{p}} \ll 1$ is fulfilled:

$$Q_{NE} = \frac{k_B}{e}\frac{\mu_e}{c}\left[\frac{\langle \tau_{\vec{p}}(x)^2 x\rangle}{\langle \tau_{\vec{p}}(x)\rangle^2} - \frac{\langle \tau_{\vec{p}}(x) x\rangle \langle \tau_{\vec{p}}(x)^3\rangle}{\langle \tau_{\vec{p}}(x)\rangle^3}\right]. \quad (4.139)$$

It is easy to show that the sign of the square bracket in the formula (4.139) is also determined by the sign of a magnitude of the scattering coefficient r. Thus, the inversion of the sign of the Nernst–Ettingshausen coefficient demonstrates the change in the mechanism of the electron scattering regardless of whether the electron gas is degenerate or not.

The longitudinal Nernst–Ettingshausen effect

Use of the determinations (4.125) gives

$$\alpha_{xx}(H) \equiv \alpha_\perp = \beta_\perp \rho_\perp + \beta_H \rho_H. \quad (4.140)$$

In the case of a strongly degenerate electron gas and in the limit of a weak magnetic field $\omega_0 \tau_{\vec{p}} \ll 1$ one can obtain the following expression as a result of simple calculations:

$$\alpha_{xx}(H) - \alpha_{xx}(0) = \frac{k_B}{e}\frac{\pi^2}{3}\frac{k_B T}{\zeta}\left[\omega_0 \tau_{\vec{p}}(\zeta)\right]^2 r. \quad (4.141)$$

For nondegenerate semiconducting materials in the limit of weak magnetic fields we get

$$\alpha_{xx}(H) - \alpha_{xx}(0) = -\frac{k_B}{e}\left[\omega_0 \langle \tau_{\vec{p}}(x)\rangle\right]^2 \Gamma(5/2)^2 \times \left\{\frac{\Gamma(5/2+2r)^2}{\Gamma(5/2+r)^4} - 2\frac{\Gamma(5/2+3r)}{\Gamma(5/2+r)^3}\right\} r. \quad (4.142)$$

The longitudinal Nernst–Ettingshausen effect as well as transverse effect is proportional to the scattering parameter r. However, this effect is much weaker than the transverse one because it contains square of the additional small parameter $\omega_0 \langle \tau_{\vec{p}}(x)\rangle$.

Problem 4.5. Deduce the expression (4.135) for the quantity $\Delta\rho/\rho$ under conditions of an extremely strong degeneracy and strong magnetic fields.

Solution. Note that

$$\Delta\rho_{xx}/\rho_{xx}(0) = -\Delta\sigma_{xx}/\sigma.$$

Then the following expression can be obtained on the basis of formulas (4.123), (4.125):

$$\Delta\sigma_{xx} = 1/\rho_{\perp} - \sigma = e^2\left[K_0^{\perp} + \frac{K_0^H K_0^H}{K_0^{\perp}} - K_0\right].$$

The quantity $\Delta\sigma_{xx} = 0$ in the zero approximation over the parameter $k_B T/\zeta$. Therefore, our purpose is to distinguish summands proportional to the square of the small parameter $k_B T/\zeta$ by using the expansion (4.35). After expanding, we obtain

$$\Delta\sigma_{xx} = e^2\frac{\pi^2}{6}(k_B T)^2\frac{\partial^2}{\partial\varepsilon^2}\left\{g(\varepsilon)\tau_{\vec p}(\varepsilon)\right.$$
$$\times\left.\left[\frac{1}{1+(\omega_0\tau_{\vec p}(\varepsilon))^2} + 2\frac{\omega_0^2\tau_{\vec p}(\varepsilon)\tau_{\vec p}(\zeta)}{1+(\omega_0\tau_{\vec p}(\varepsilon))^2} - \frac{\omega_0^2\tau_{\vec p}(\zeta)^2}{1+(\omega_0\tau_{\vec p}(\varepsilon))^2} - 1\right]\right\}\Bigg|_{\varepsilon=\zeta}. \tag{4.143}$$

We have introduced the notation in the formula (4.143)

$$g(\varepsilon) = \frac{2(2m)^{1/2}\varepsilon^{3/2}}{3\pi^2\hbar^3}.$$

Performing elementary algebraic transformations in the square bracket of (4.143), we are led to a form suitable for further calculations,

$$\Delta\sigma_{xx} = -e^2\frac{\pi^2}{6}(k_B T)^2\frac{\partial^2}{\partial\varepsilon^2}\left\{\frac{g(\varepsilon)\tau_{\vec p}(\varepsilon)\omega_0^2}{1+(\omega_0\tau_{\vec p}(\varepsilon))^2}\left[\tau_{\vec p}(\varepsilon) - \tau_{\vec p}(\zeta)\right]^2\right\}\Bigg|_{\varepsilon=\zeta}. \tag{4.144}$$

So far as there exists the difference $[\tau_{\vec p}(\varepsilon) - \tau_{\vec p}(\zeta)]^2$ in the expression (4.144) which vanishes at $\varepsilon = \zeta$, we obtain

$$\Delta\sigma_{xx} = -e^2\frac{\pi^2}{3}(k_B T)^2\frac{g(\zeta)\tau_{\vec p}(\zeta)\omega_0^2}{1+(\omega_0\tau_{\vec p}(\zeta))^2}\left[\tau_{\vec p}'(\zeta)\right]^2.$$

Finally, given that

$$\sigma = e^2 K_0 = e^2 g(\zeta)\tau_{\vec p}(\zeta), \quad \left[\tau_{\vec p}'(\zeta)\right]^2 = 1/4\tau_{\vec p}^2(\zeta)/\zeta^2,$$

we have

$$\frac{\Delta\sigma_{xx}}{\sigma} = -\frac{\pi^2}{12}\left(\frac{k_B T}{\zeta}\right)^2\frac{[\omega_0\tau_{\vec p}(\zeta)]^2}{1+[\omega_0\tau_{\vec p}(\zeta)]^2}. \tag{4.145}$$

This concludes the discussion of the various thermogalvanomagnetic effects. The Maggi–Riga–Leduc, Nernst–Ettingshausen, Riga–Leduc and adiabatic effects remained unconsidered. We suggest the reader to calculate kinetic coefficients characterizing these phenomena as an independent work. More detailed information on the theory thermogalvanomagnetic phenomena can be found in the monographs [8, 32].

4.2 Hydrodynamic description of a hot electrons

4.2.1 Transition to a hydrodynamic description

For a theoretical study of non-equilibrium states of an electron gas, it is not always necessary to have a solution of a complex integro-differential kinetic equation since full information contained in this solution is not often used. Indeed it is well known that there is an important class of problems in physical kinetics which are solved by means of equations of fluid dynamics [22, 23, 24, 25].

Chapter 3 discussed in detail the procedure of the transition from a kinetic description of non-equilibrium systems to hydrodynamic one through deriving the Chapman–Enskog equations. Ideas developed in this chapter can be applied for going over to the hydrodynamic description of hot electrons in semiconducting crystals. It is obvious that hydrodynamic equations are considerably simpler than the kinetic equation due to microparticle characteristics averaged over their momentum.

The transition to the description by using the averaged momentum corresponds to rougher and consequently, less complete pattern of a phenomenon under study. Nevertheless, this idea makes it possible to shorten the description of a system. It is extremely fruitful in one form or another and used to solve all kinds of problems in physical kinetics. For example, in deriving the kinetic equation in the framework of the Bogoliubov method, there appears a set of coupled equations for s-particle distribution functions. This infinite set of coupled equations is equivalent to the dynamic description. The reduction in the description becomes possible if one succeeds in expressing a two-particle distribution function via a one-particle one by using some approximations to close the system of equations. A similar approach is also used in a Green function method. The system of coupled equations of motion for all kinds of Green functions is equivalent to the complete dynamic description. As shown in Chapter 1, the state of dynamic chaos is implemented for most actual systems. In this case, the dynamic description does not make sense but a crude form (a form of averaging) of the description is important. The roughening of the description within the Green function method takes place to uncouple coupled equations of motion for the Green functions. (Assuming, for example, that the $n+1$-Green function is expressed through the lowest Green functions).

The physical reason for the possible shortening in the description is that the decay of certain correlations between dynamical variables occurs as a time scale in which dynamics of the system is investigated is growing. Moreover, chaotic dynamics of initial dynamic variables turns into regular dynamics for averaged quantities. Bogoliubov was first to formulate clearly enough the idea of reducing the description in his work.

Due to Bogoliubov one can identify four stages of evolution of a dynamical system in the case of a low-density gas. Thus, the system can be described with four different ways.

The dynamic stage of evolution corresponds to a exact mechanical description of a system. No reduction in the description happens. This stage of the evolution corresponds to the time interval

$$t < \frac{r_0}{\bar{v}},$$

where r_0 is the effective radius of the interaction in the system, $\bar{v}$ the average velocity of the particles.

As noted in Chapter 1, a dynamic description does not make sense for systems demonstrating the dynamic chaos. Such a description is appropriate only for narrow class of systems that show a regular motion of a phase point in a phase space. During the time $t_1 \approx r_0/\bar{v}$ there takes place a synchronization of distribution functions and the n-particle distribution function is expressed via the one-particle distribution function. Thus, *the kinetic stage* of evolution corresponds to a description of a system in terms of the one-particle distribution function.

Thus, the kinetic stage of evolution corresponds to a description of a system in terms of the one-particle distribution function and is characterized by the time scale that is defined by the inequalities

$$\frac{r_0}{\bar{v}} < t < \frac{l}{\bar{v}},$$

where l is the mean free path of particles.

In the period of time $l/\bar{v}$, which coincides with the relaxation time of momentum over an order of magnitude, a system is able to form the averages that have meaning of an average number of particles, the average energy, and the average momentum, i. e. meaning of hydrodynamic variables. Therefore, over periods

$$t > \frac{l}{\bar{v}}$$

there appears *the hydrodynamic stage* of evolution of a system. At this stage the system is described by specifying the hydrodynamic parameters, which are usually associated with average values of quantities that, in turn, are additive integrals of motion because, it is these quantities that are usually "long-lived" variables.

Finally, over periods

$$t > \frac{L}{\bar{v}},$$

where L is a characteristic linear dimension of a system, we have *the thermodynamic stage* of evolution of the system. At this stage non-equilibrium processes are terminated and thermodynamic equilibrium is established. To describe the system it is sufficient to use the equations of equilibrium thermodynamics.

From all that has been said, it follows that the presence of several stages of evolution of the system and the hierarchy of relaxation times is applicable to a gas of electrons interacting with phonons.

We want to get a set of hydrodynamic equations to describe the conduction electrons in the presence of a sufficiently strong electric field. Supposing that the external electric field leads not only to the emergence of the average drift momentum of electrons but also to a change in the average electron energy, we introduce several hydrodynamic parameters such as temperature T_k of kinetic degrees of freedom of the conduction electrons, the drift velocity $\vec{v}_d$ of the electrons, and the non-equilibrium chemical potential ζ.

These parameters are associated with the average kinetic energy of the electrons, the average momentum of the electrons, the average number of the conduction electrons, respectively. The profile of the dependence of the non-equilibrium distribution function on the input parameters $T_k, \vec{v}_d$, ζ is not crucial but can be just convenient. When thermodynamic equilibrium is established in the system, a non-equilibrium distribution function is represented by the equilibrium Fermi–Dirac function. Given this, we write down the non-equilibrium distribution function as

$$f_{\vec{k}} = \left\{\exp[\beta_k \varepsilon_{\vec{k}-m\vec{v}_d/\hbar} - \beta\zeta] + 1\right\}^{-1},$$
$$\varepsilon_{\vec{k}-m\vec{v}_d/\hbar} = \frac{\hbar^2(\vec{k} - m\vec{v}_d/\hbar)^2}{2m}, \quad \beta_k = \frac{1}{k_B T_k}. \tag{4.146}$$

Thus, we have introduced five effective parameters $T_k, \vec{v}_d, \zeta$, which are associated with average values of dynamic quantities that are the additive integrals of motion. This method of parameterization of non-equilibrium distribution function is also consistent with the Hilbert theorem. This theorem tells that, if a solution of the kinetic equation exists, it can be expressed through the first five moments of the distribution function. To find these parameters, five balance equations that have meaning of the energy balance equation, of the momentum and of the particle number need to be derived.

It would be proper to make a few remarks concerning applicability of the concept of the electronic subsystem temperature. Equilibrium in the electronic system is due to electron–electron collisions with the characteristic frequency ν_{ee}, and the balance in the system of electrons and phonons is established due to the electron–phonon collision frequency ν_{ep}. Then, in order to apply the effective electron temperature concept, which differs from the phonon system temperature, the condition is to be fulfilled that

$$\nu_{ee} \gg \nu_{ep}.$$

In some cases, this condition is insufficient. For example, if electrons interact with optical phonons, the electron energy in each scattering event will change by magnitude of the quantum of the optical phonon $\hbar\Omega_0$. But if there exist electron–electron collisions, the change in the energy amounts to $\simeq k_B T \ll \hbar\Omega_0$. Therefore, the electrons may not become thermalized even though the condition $\nu_{ee} \gg \nu_{ep}$ is met. In future, the applicability conditions for the description in terms of effective parameters T_k, $\vec{v}_d$, ζ are assumed to hold throughout the paper.

4.2.2 The momentum balance equation

The kinetic equation which will be used in this chapter can be symbolically written as

$$\left(\frac{\partial f}{\partial t}\right)_{\text{field}} + \left(\frac{\partial f}{\partial t}\right)_{\text{col}} = 0, \tag{4.147}$$

where the first summand on the left-hand side has a sense of the rate of change of a distribution function under the action of an external electric field:

$$\left(\frac{\partial f}{\partial t}\right)_{\text{field}} = \frac{e}{\hbar}\vec{E}\vec{\nabla}_{\vec{k}} f_{\vec{k}}, \tag{4.148}$$

The second summand presents the collision integral (4.88), which is numerically equal to the rate of change of the distribution function due to collisions.

To obtain the momentum balance equation one should multiply the equation (4.147) by $\hbar\vec{k}$ and sum up over $\vec{k}$ and σ. This results in the following expression:

$$\left(\frac{\partial}{\partial t}\langle\vec{P}\rangle\right)_{\text{field}} + \left(\frac{\partial}{\partial t}\langle\vec{P}\rangle\right)_{\text{col}} = 0. \tag{4.149}$$

The first term on the left-hand side of (4.149) is equal to the rate of change in the average momentum of an electron system due to an external field as the second term describes the change in the average momentum due to collisions with scatterers. The use of the expression (4.148) gives the first summand in the formula (4.149) in the form

$$\left(\frac{\partial}{\partial t}\langle\vec{P}^{\alpha}\rangle\right)_{\text{field}} = \sum_{\vec{k},\sigma} \hbar\vec{k}^{\alpha}\frac{e}{\hbar}E^{\beta}\nabla^{\beta}_{\vec{k}} f_{\vec{k}}. \tag{4.150}$$

To arrive at the momentum balance equation one should restrict oneself to a linear approximation over the drift velocity $\vec{v}_d$ and linear approximation in the electric field. This implies that the drift velocity in the formula (4.150) can be omitted after substituting the quantity $f^s_{\vec{k}}$ for $f_{\vec{k}}$ without the drift velocity

$$f^s_{\vec{k}} = \left\{\exp[\beta_k\varepsilon_{\vec{k}} - \beta\zeta] + 1\right\}^{-1}.$$

Passing on in the formula (4.150) to integration over the energy, we get

$$\left(\frac{\partial}{\partial t}\langle\vec{P}^{\alpha}\rangle\right)_{\text{field}} = -eE^{\alpha}\frac{(2mk_{\text{B}}T_k)^{3/2}}{2\pi^2\hbar^3}F_{1/2}\left(\frac{\zeta}{k_{\text{B}}T_k}\right) = -enE^{\alpha}. \tag{4.151}$$

Now, consider the second term on the left side of (4.149). Multiplying the expression (4.88) by $\hbar\vec{k}$ and summing up over $\vec{k}$ and σ, we have

$$\left(\frac{\partial}{\partial t}\langle\vec{P}^{\alpha}\rangle\right)_{\text{col}} = \frac{2\pi}{\hbar}\sum_{\vec{k}\sigma,\vec{q}\lambda}\hbar k^{\alpha}|C_{\vec{q}\lambda}|^2\left\{\left[(N_{\vec{q}\lambda}+1)f_{\vec{k}+\vec{q}}(1-f_{\vec{k}})\right.\right.$$
$$\left.-N_{\vec{q}\lambda}f_{\vec{k}}(1-f_{\vec{k}+\vec{q}})\right]\delta(\varepsilon_{\vec{k}}-\varepsilon_{\vec{k}+\vec{q}}+\hbar\Omega_{\vec{q}\lambda})+\left[N_{\vec{q}\lambda}f_{\vec{k}-\vec{q}}(1-f_{\vec{k}})\right.$$
$$\left.\left.-(N_{\vec{q}\lambda}+1)f_{\vec{k}}(1-f_{\vec{k}-\vec{q}})\right]\delta(\varepsilon_{\vec{k}}-\varepsilon_{\vec{k}-\vec{q}}-\hbar\Omega_{\vec{q}\lambda})\right\}. \tag{4.152}$$

So far as the summation over the wave vector $\vec{k}$ is being performed in the infinite limits, then the origin may be shifted by arbitrary vector $\vec{q}$ when the summation of terms in the second square bracket occurs, putting that $\vec{k}-\vec{q}=\vec{k}$, $\vec{k}=\vec{k}+\vec{q}$. After the replacement it is easy to see that the square brackets in the expression (4.152) are equal in magnitude but opposite in sign, and the arguments of the δ-functions coincide. As a result, the formula becomes much simpler:

$$\left(\frac{\partial}{\partial t}\langle\vec{P}^{\alpha}\rangle\right)_{\text{col}} = -\frac{2\pi}{\hbar}\sum_{\vec{k}\sigma,\vec{q}\lambda}\hbar q^{\alpha}|C_{\vec{q}\lambda}|^2[(N_{\vec{q}\lambda}+1)f_{\vec{k}+\vec{q}}(1-f_{\vec{k}})$$
$$-N_{\vec{q}\lambda}f_{\vec{k}}(1-f_{\vec{k}+\vec{q}})]\delta(\varepsilon_{\vec{k}}-\varepsilon_{\vec{k}+\vec{q}}+\hbar\Omega_{\vec{q}\lambda}). \tag{4.153}$$

On the right-hand side of the equation (4.153) we select the terms that are linear over the drift velocity. So far as the drift velocity is contained in the distribution function, contains the drift velocity, we expand it in a Taylor series, restricting ourselves to the linear terms:

$$f_{\vec{k}} = f_{\vec{k}}^{S} - \frac{\partial f_{\vec{k}}^{S}}{\partial\varepsilon_{\vec{k}}}\hbar k^{\alpha}v_d^{\alpha}. \tag{4.154}$$

Substituting this expansion for the distribution function into the expression that determines the rate of change in the average momentum of the electrons due to collisions, we can write down the right-hand side of (4.153) as

$$\frac{2\pi}{\hbar}\sum_{\vec{k}\sigma,\vec{q}\lambda}\frac{1}{3}(\hbar\vec{q})^2v_d^{\alpha}|C_{\vec{q}\lambda}|^2\left[(N_{\vec{q}\lambda}+1)f_{\vec{k}+\vec{q}}^{S\prime}(1-f_{\vec{k}}^{S})\right.$$
$$\left.+N_{\vec{q}\lambda}f_{\vec{k}}^{S}f_{\vec{k}+\vec{q}}^{S\prime}\right]\delta(\varepsilon_{\vec{k}}-\varepsilon_{\vec{k}+\vec{q}}+\hbar\Omega_{\vec{q}\lambda}),\quad f_{\vec{k}+\vec{q}}^{S\prime}=\frac{\partial f_{\vec{k}+\vec{q}}^{S}}{\partial\varepsilon_{\vec{k}+\vec{q}}}. \tag{4.155}$$

In deriving the formula (4.155) we have taken into account that the substitution of symmetric part of the distribution function $f_{\vec{k}}^{S}$ and $f_{\vec{k}+\vec{q}}^{S}$ turns the right side of (4.153) into zero which is quite clear in terms of physics. By symmetry, this can be easily proved by considering that the expression under the sum sign is odd in powers of $\vec{q}$. In accordance with (4.154) a non-zero result proportional to $\vec{k} + \vec{q}$ takes place only when the function $f_{\vec{k}+\vec{q}}$ is expanded. This yields terms proportional to q^2.

Summarizing the results (4.151), (4.155) obtained, one can now formulate the momentum balance equation. Introducing the total drift momentum of the electron system $\vec{P}_d = \langle \vec{P} \rangle = nm\vec{v}_d$, we have

$$|e|nE^{\alpha} = \frac{nm\vec{v}_d}{\tau}, \tag{4.156}$$

$$\frac{1}{\tau} = -\frac{2\pi}{\hbar}\frac{1}{3nm}\sum_{\vec{k}\sigma,\vec{q}\lambda}(\hbar\vec{q})^2|C_{\vec{q}\lambda}|^2\Big[(N_{\vec{q}\lambda}+1)f^{s\prime}_{\vec{k}+\vec{q}}(1-f^{s}_{\vec{k}}) + N_{\vec{q}\lambda}f^{s}_{\vec{k}}f^{s\prime}_{\vec{k}+\vec{q}}\Big]\delta(\varepsilon_{\vec{k}} - \varepsilon_{\vec{k}+\vec{q}} + \hbar\Omega_{\vec{q}\lambda}). \tag{4.157}$$

The momentum balance equation (4.156) makes it quite clear that the force acting on the system of electrons under the external electric field is equal in magnitude but opposite in direction to the force exerted by the lattice. The quantity τ, being defined by (4.157), is responsible for the relaxation time of the mean (total) momentum of the electron system.

In the given case, the three momentum balance equations (4.156) contain five the unknowns: three components of the drift velocity, non-equilibrium temperature T_k of kinetic degrees of freedom of the electronic system and non-equilibrium chemical potential ζ. If the external electric field is sufficiently weak and does not lead to heating of the electronic system, the temperature and chemical potential can be considered as equilibrium parameters. Then the quantity $1/\tau$ (4.157) contains no unknown parameters and the momentum balance equation (4.156) immediately allows the components of the drift velocity to be found, consequently, the expression for the conductance of the equilibrium system has the form

$$\sigma = \frac{e^2 n}{m}\tau_0,$$

where τ_0 is the electron momentum relaxation time under conditions of equilibrium.

Next, we deduce an expression for the full momentum relaxation time of the equilibrium system. For this purpose, transformation of the formula (4.157) suggests that the temperature and chemical potential correspond to equilibrium ($T_k = T, \zeta = \zeta_0$, here and elsewhere ζ_0 is an equilibrium chemical potential). To convert this formula, the properties of the equilibrium Fermi–Dirac distribution function $f^0_{\vec{k}}$ and the Planck function $N_{\vec{q}\lambda}$ should be used,

$$1 - f^0_{\vec{k}} = f^0_{\vec{k}} e^{\beta(\varepsilon_{\vec{k}} - \zeta_0)}, \quad f^{0\prime}_{\vec{k}} = -\beta f^0_{\vec{k}}(1 - f^0_{\vec{k}}),$$
$$N_{\vec{q}\lambda} + 1 = N_{\vec{q}\lambda} e^{\beta\hbar\Omega_{\vec{q}\lambda}}, \quad \beta = \frac{1}{k_B T}. \tag{4.158}$$

Given these results, we transform the square bracket of the formula (4.157):

$$(N_{\vec{q}\lambda}+1)f^{0\,\prime}_{\vec{k}+\vec{q}}(1-f^0_{\vec{k}})+N_{\vec{q}\lambda}f^0_{\vec{k}}f^{0\,\prime}_{\vec{k}+\vec{q}}$$
$$=N_{\vec{q}\lambda}f^{0\,\prime}_{\vec{k}+\vec{q}}f^0_{\vec{k}}[e^{\beta(\hbar\Omega_{\vec{q}\lambda}+\varepsilon_{\vec{k}}-\zeta_0)}+1]=-\beta N_{\vec{q}\lambda}f^0_{\vec{k}}(1-f^0_{\vec{k}+\vec{q}}). \tag{4.159}$$

In deriving the formula (4.159) we have used the energy conservation law $\hbar\Omega_{\vec{q}\lambda}+\varepsilon_{\vec{k}}=\varepsilon_{\vec{k}+\vec{q}}$.

This yields the expression for the relaxation time of the average electron momentum:

$$\frac{1}{\tau_0}=\frac{2\pi}{\hbar}\frac{\beta}{3nm}\sum_{\vec{k}\sigma,\vec{q}\lambda}(\hbar\vec{q})^2|C_{\vec{q}\lambda}|^2N_{\vec{q}\lambda}f^0_{\vec{k}}(1-f^0_{\vec{k}+\vec{q}})\delta(\varepsilon_{\vec{k}}-\varepsilon_{\vec{k}+\vec{q}}+\hbar\Omega_{\vec{q}\lambda}). \tag{4.160}$$

In conclusion, it is necessary to calculate the relaxation frequency of the average momentum $\nu_{ep}=1/\tau_0$ of the electronic system in the case of a quasi-elastic scattering of electrons by acoustic phonons. In this case, the phonon energy $\hbar\Omega_{\vec{q}\lambda}$ in the formula describing the energy conservation law can be neglected. Going over from the summation over $\vec{k}$ and $\vec{q}$ to integration over $\vec{p}$, $\vec{q}$ in the formula (4.160) and performing the summation over σ, which is simply reduced to the additional product of the result by two, we get

$$\frac{1}{\tau_0}=\frac{2\pi}{\hbar}\frac{1}{3nm}\frac{2}{(2\pi)^6\hbar^3}\int d\vec{p}\,d\vec{q}(\hbar\vec{q})^2|C_{\vec{q}}|^2\frac{k_BT}{\hbar sq}$$
$$\times\int_0^\infty d\varepsilon\left(-\frac{\partial f^0_{\vec{p}}}{\partial\varepsilon}\right)\delta(\varepsilon_{\vec{p}+\hbar\vec{q}}-\varepsilon)\delta(\varepsilon_{\vec{p}}-\varepsilon). \tag{4.161}$$

In deriving the formula (4.161) we have used the approximation (4.89) and taken into account that, in the deformation-potential method, only the interaction of electrons with longitudinal acoustic phonons makes a contribution when interactions between electrons and acoustic lattice vibrations occur. For this reason, the polarization index in (4.161) has been omitted.

Consider the integral over momenta in the expression (4.161) in more detail:

$$I(q)=\int d\vec{p}\,\delta(\varepsilon_{\vec{p}+\hbar\vec{q}}-\varepsilon)\delta(\varepsilon_{\vec{p}}-\varepsilon). \tag{4.162}$$

Passing to the integration in spherical coordinates yields

$$I(q)=2\pi\int_0^\pi\sin\theta\,d\theta\int_0^\infty p^2\,dp\,\delta\left(\frac{p^2}{2m}+\frac{\hbar^2q^2}{2m}+\frac{p\hbar q}{m}\cos\theta-\varepsilon\right)\delta\left(\frac{p^2}{2m}-\varepsilon\right). \tag{4.163}$$

After integrating in the formula (4.163) over the momentum by using the second δ-function there remains only the integral over variable $x=\cos\theta$:

$$I(q)=2\pi m\sqrt{2m\varepsilon}\int_{-1}^1 dx\,\delta\left(\frac{\hbar^2q^2}{2m}+\sqrt{2m\varepsilon}\frac{\hbar q}{m}x\right). \tag{4.164}$$

Next, having denoted

$$y = \frac{\hbar^2 q^2}{2m} + \sqrt{2m\varepsilon}\frac{\hbar q}{m}x$$

We get the integration over the new variables

$$I(q) = \frac{2\pi m^2}{\hbar q} \int_{\frac{\hbar^2 q^2}{2m} - \sqrt{2m\varepsilon}\frac{\hbar q}{m}}^{\frac{\hbar^2 q^2}{2m} + \sqrt{2m\varepsilon}\frac{\hbar q}{m}} dy\delta(y). \tag{4.165}$$

The integral (4.165) differs from zero only when the value $y = 0$ is included in the domain of the integration. This calls for imposing the restriction to the range of the wave vectors $\vec{q}$ of phonons, interacting with electrons. Considering the above, we obtain, finally, a simple expression for the integral $I(q)$:

$$I(q) = \frac{2\pi m^2}{\hbar q}, \quad 0 < q < \left(\frac{8m\varepsilon}{\hbar^2}\right)^{1/2}. \tag{4.166}$$

Further calculations based on the formula (4.161) are reduced to integration of the power function q^3. In addition, the integral over energy is replaced by the Fermi integral (4.34). We will not dwell on these simple calculations; we give just the final result:

$$\frac{1}{\tau_0} = \frac{2E_0^2(2mk_BT)^{3/2}}{3\pi^2\rho s^2\hbar^4}\frac{F_1(\zeta_0/k_BT)}{F_{1/2}(\zeta_0/k_BT)}. \tag{4.167}$$

In writing this formula we have used the determination of the equilibrium electron concentration:

$$n = \frac{(2mk_BT)^{3/2}}{2\pi^2\hbar^3}F_{1/2}(\zeta_0/k_BT). \tag{4.168}$$

Now, consider a non-equilibrium case. Let $N_{\vec{q}} \gg 1$ so $N_{\vec{q}} + 1 \simeq N_{\vec{q}}$. This condition imposes some restrictions on temperature of the system. If one takes into account the formula (4.166) and puts for the average kinetic energy of electrons $\bar{\varepsilon} \simeq k_BT$, the inequality (4.89) can be written as follows:

$$N_{\vec{q}} \simeq \frac{k_BT}{\hbar\Omega_{\vec{q}}} \simeq \frac{k_BT}{s(8mk_BT)^{1/2}} \gg 1, \tag{4.169}$$

or

$$k_BT > 8ms^2. \tag{4.170}$$

The numerical estimation indicates that the inequality (4.170) holds for semiconductors with effective mass $m \sim 0,01m_0$ at temperatures already above $1K$, where m_0 is the free electron mass.

If one accepts that the condition (4.169) is true, the expression (4.157) for the inverse relaxation time of non-equilibrium electrons has the form

$$\frac{1}{\tau} = \frac{2\pi}{\hbar}\frac{1}{3nm}\frac{2}{(2\pi)^6\hbar^3}\int d\vec{p}\,d\vec{q}(\hbar\vec{q})^2|C_{\vec{q}}|^2\frac{k_B T}{\hbar sq}$$
$$\times\int_0^\infty d\varepsilon\left(-\frac{\partial f^s_{\vec{p}}}{\partial\varepsilon}\right)\delta(\varepsilon_{\vec{p}+\hbar\vec{q}}-\varepsilon)\delta(\varepsilon_{\vec{p}}-\varepsilon). \tag{4.171}$$

Upon comparing the formulas (4.161) and (4.171) we can obtain the result:

$$\frac{1}{\tau} = \frac{2E_0^2(2mk_B)^{3/2}TT_k^{1/2}}{3\pi^2\rho s^2\hbar^4}\frac{F_1(\zeta/k_B T_k)}{F_{1/2}(\zeta/k_B T_k)}. \tag{4.172}$$

Thus, the expressions (4.156), (4.172) are three momentum balance equations, containing five the unknowns of the parameter $\vec{v}_d$, T_k, ζ. To obtain a closed set of macroscopic balance equations it is necessary to add two equations to the three equations: an energy balance equation and a particle number balance equation.

4.2.3 Balance equations of energy and particle number

Multiply the equation (4.147) by $\varepsilon_{\vec{k}}$ and then sum it up over $\vec{k}$ and σ. In contrast to the momentum balance equation, where we have restricted ourselves to a linear approximation either for intensity of an external electric field or drift velocity, quadratic terms of these parameters need to be retained in the energy balance equation.

Introducing the notation

$$\langle E_k\rangle = \sum_{\vec{k}\sigma}\varepsilon_{\vec{k}}f_{\vec{k}}$$

for the rate of change of the average kinetic energy of electrons $\langle E_k\rangle$ due to the field, we have

$$\left(\frac{\partial}{\partial t}\langle E_{\vec{k}}\rangle\right)_{\text{field}} = \sum_{\vec{k},\sigma}\varepsilon_{\vec{k}}\frac{e}{\hbar}\vec{E}\vec{\nabla}_{\vec{k}}f_{\vec{k}}. \tag{4.173}$$

Now, terms linear for the drift velocity are to be extracted from $\vec{\nabla}_{\vec{k}}f_{\vec{k}}$:

$$\nabla^\alpha_{\vec{k}}f_{\vec{k}} = \frac{\partial f_{\vec{k}}}{\partial\varepsilon_{\vec{k}}}\nabla^\alpha_{\vec{k}}\frac{1}{2m}(\hbar\vec{k}-m\vec{v}_d)^2$$
$$= \frac{\hbar}{m}(\hbar k^\alpha - mv^\alpha_d)\frac{\partial}{\partial\varepsilon_{\vec{k}}}[f^s_{\vec{k}} - f^{s\prime}_{\vec{k}}\hbar k^\beta v^\beta_d]. \tag{4.174}$$

Substituting the expression (4.174) into the formula (4.173) and considering that odd terms for $\vec{k}$ do not contribute to the sum on the right-hand side of (4.173), we get

$$\left(\frac{\partial}{\partial t}\langle E_{\vec{k}}\rangle\right)_{\text{field}} = -e\sum_{\vec{k},\sigma}\varepsilon_{\vec{k}}E^{\alpha}v_d^{\alpha}\left[f_{\vec{k}}^{s\prime} + \frac{1}{3}\frac{\hbar^2k^2}{m}f_{\vec{k}}^{s\prime\prime}\right]. \tag{4.175}$$

To calculate the sum over $\vec{k}$ on the right-hand side of (4.175) we replace, as usually, the summation by integration. Consider the contribution of the first summand in the square brackets of the formula (4.175)

$$-eE^{\alpha}v_d^{\alpha}\sum_{\vec{k},\sigma}\varepsilon_{\vec{k}}\frac{\partial f_{\vec{k}}^s}{\partial\varepsilon_{\vec{k}}} = -eE^{\alpha}v_d^{\alpha}\frac{2m^{3/2}}{\pi^2\hbar^3}\int_0^{\infty}\varepsilon_{\vec{k}}^{3/2}\frac{\partial f_{\vec{k}}^s}{\partial\varepsilon_{\vec{k}}}\,d\varepsilon_{\vec{k}}. \tag{4.176}$$

Integrating by parts the integral over energy on the right-hand side of (4.176) and making the determination of concentration n, we may write the following expression:

$$-eE^{\alpha}v_d^{\alpha}\sum_{\vec{k},\sigma}\varepsilon_{\vec{k}}\frac{\partial f_{\vec{k}}^s}{\partial\varepsilon_{\vec{k}}} = eE^{\alpha}v_d^{\alpha}\frac{3}{2}n. \tag{4.177}$$

Similarly, doing a double integration by parts, we can find also the contribution of the second term in the square brackets of the expression (4.175) to the rate of change of the average energy of the electrons at the cost of the field,

$$-\frac{1}{3}eE^{\alpha}v_d^{\alpha}\sum_{\vec{k},\sigma}\varepsilon_{\vec{k}}\frac{\hbar^2k^2}{m}\frac{\partial^2 f_{\vec{k}}^s}{\partial\varepsilon_{\vec{k}}^2} = -eE^{\alpha}v_d^{\alpha}\frac{5}{2}n. \tag{4.178}$$

Having summarized the results of (4.177), (4.178), we arrive at the final expression for the rate of change of the kinetic energy under an action of the external field:

$$\left(\frac{\partial}{\partial t}\langle E_{\vec{k}}\rangle\right)_{\text{field}} = -eE^{\alpha}v_d^{\alpha}n, \tag{4.179}$$

$$n = \frac{(2mk_{\text{B}}T_k)^{3/2}}{2\pi^2\hbar^3}F_{1/2}(\zeta/k_{\text{B}}T_k). \tag{4.180}$$

If an external electric field does not give rise to impact-ionization of electrons in donor impurities and there are no other reasons to believe that the electric field can lead to a change in the concentration of the electrons in a crystal, then the following condition must be fulfilled when the electric field is turned on:

$$\frac{(2mk_{\text{B}}T)^{3/2}}{2\pi^2\hbar^3}F_{1/2}(\zeta_0/k_{\text{B}}T) = \frac{(2mk_{\text{B}}T_k)^{3/2}}{2\pi^2\hbar^3}F_{1/2}(\zeta/k_{\text{B}}T_k). \tag{4.181}$$

Such an equation may be regarded as the particle number balance equation.

Now it would be advisable to derive an expression for determining the rate of change of the kinetic energy of electrons due to interactions with a lattice. For that, we multiply the equation (4.88) by $\varepsilon_{\vec{k}}$ and sum up over $\vec{k}$ and σ. As before, in deriving the momentum balance equation we can combine the contributions of the first and second square brackets on the right-hand side of (4.88). Summing the contribution of the second square bracket calls for shifting the origin in the $\vec{k}$-space by an arbitrary vector $\vec{q}$ by replacing $\vec{k} - \vec{q} \to \vec{k}$. Then in accordance with the conservation law

$$\varepsilon_{\vec{k}+\vec{q}} = \varepsilon_{\vec{k}} + \hbar\Omega_{\vec{q}}$$

we have

$$\left(\frac{\partial}{\partial t}\langle E_{\vec{k}}\rangle\right)_{\text{col}} = \frac{2\pi}{\hbar}\sum_{\vec{k}\sigma\vec{q}} |C_{\vec{q}}|^2 \hbar\Omega_{\vec{q}}\Big[(N_{\vec{q}}+1)f_{\vec{k}+\vec{q}}(1-f_{\vec{k}}) - N_{\vec{q}}f_{\vec{k}}(1-f_{\vec{k}+\vec{q}})\Big]\delta(\varepsilon_{\vec{k}} - \varepsilon_{\vec{k}+\vec{q}} + \hbar\Omega_{\vec{q}}). \tag{4.182}$$

Further analysis shows that the right-hand side of the expression (4.182) is proportional to the deviation of non-equilibrium temperature $\delta T_k = T_k - T$ from equilibrium one. Since the temperature deviation is due to heating the electron system by the external electric field, the quantity δT_k at least is proportional to the square of field intensity. For this reason, the distribution functions $f_{\vec{k}}$ in the square brackets of (4.182) may be replaced by their symmetric parts $f^s_{\vec{k}}$. If you keep the second term on the right side of (4.154), this would exceed the accuracy. Also, it is necessary to keep in mind that we retain the quadratic terms in the external field in the equation for balancing the kinetic energy.

Given the formulas similar to (4.158), the expression in the square brackets on the right-hand side of (4.182) can be written as

$$\begin{aligned} I(\vec{k},\vec{q}) &= (N_{\vec{q}}+1)f^s_{\vec{k}+\vec{q}}(1-f^s_{\vec{k}}) - N_{\vec{q}}f^s_{\vec{k}}(1-f^s_{\vec{k}+\vec{q}}) \\ &= f^s_{\vec{k}}(1-f^s_{\vec{k}+\vec{q}})N_{\vec{q}}\big[e^{(\beta-\beta_k)\hbar\Omega_{\vec{q}}} - 1\big]. \end{aligned} \tag{4.183}$$

The smallness of the parameter $(\beta - \beta_k)\hbar\Omega_{\vec{q}}$ being taken into account, we can expand the exponent into a series containing this parameter. Then we get for $I(\vec{k},\vec{q})$

$$I(\vec{k},\vec{q}) = f^s_{\vec{k}}(1-f^s_{\vec{k}+\vec{q}})N_{\vec{q}}\hbar\Omega_{\vec{q}}\frac{1}{k_B T}\left(1-\frac{T}{T_k}\right). \tag{4.184}$$

If one assumes that the electron–phonon scattering is quasi-elastic, then, inserting this result into the formula (4.182), the rate of change of the average energy of electrons due to collisions with a lattice is given by

$$\left(\frac{\partial}{\partial t}\langle E_{\vec{k}}\rangle\right)_{\text{col}} = \frac{2\pi}{\hbar}\beta\sum_{\vec{k}\sigma\vec{q}} |C_{\vec{q}}|^2(\hbar\Omega_{\vec{q}})^2 N_{\vec{q}}f^s_{\vec{k}}(1-f^s_{\vec{k}+\vec{q}}) \times \delta(\varepsilon_{\vec{k}} - \varepsilon_{\vec{k}+\vec{q}})\left(1-\frac{T}{T_k}\right). \tag{4.185}$$

After performing simple calculations, by analogy with calculations of the momentum relaxation time for scattering by longitudinal acoustic phonons, the rate of change of the average energy of the electrons can be determined as

$$\left(\frac{\partial}{\partial t}\langle E_{\vec{k}}\rangle\right)_{\text{col}} = \frac{8E_0^2(k_\text{B}T_k)^3m^4}{\pi^3\hbar^7\rho}F_1(\zeta/k_\text{B}T_k)\left(1-\frac{T}{T_k}\right). \tag{4.186}$$

Collecting the results (4.179) and (4.186), one is led to the energy balance equation for an electronic subsystem:

$$-eE^\alpha v_d^\alpha n = \frac{8E_0^2(k_\text{B}T_k)^3m^4}{\pi^3\hbar^7\rho}F_1(\zeta/k_\text{B}T_k)\left(1-\frac{T}{T_k}\right). \tag{4.187}$$

Equations (4.156), (4.181) and (4.187) form a closed set of five equations to determine the drift velocity components, the temperature of kinetic degrees of freedom of the conduction electrons and the chemical potential.

4.2.4 Solving a set of balance equations. Applications of hydrodynamic approach

In case of the isotropic dispersion law and isotropic scattering, the momentum balance equation is indeed a scalar equation, since drift velocity of electrons v_d is parallel to an electric-field vector. Next, we extract the quantity v_d from the equation (4.156) and substitute this result into (4.187) which yields the following expression:

$$Q_0\frac{\tau}{\tau_0} = \frac{2^{5/2}E_0^2(k_\text{B}T)^{3/2}m^{5/2}}{\pi\rho\hbar^4}\frac{F_1(\zeta/k_\text{B}T_k)}{F_{1/2}(\zeta/k_\text{B}T_k)}\left(1-\frac{T}{T_k}\right). \tag{4.188}$$

In writing the formula (4.188) we have used the quantity Q_0 which has meaning of power absorbed by the conduction electron system per an electron:

$$Q_0 = \frac{e^2\tau_0}{m}E^2. \tag{4.189}$$

In expression (4.188), the non-equilibrium chemical potential involves only as an argument of the Fermi integrals. If one restricts oneself to case of a nondegenerate electron gas and one makes use of the approximate equality

$$F_p(\zeta/k_\text{B}T_k) = \Gamma(p+1)e^{\zeta/k_\text{B}T_k}, \tag{4.190}$$

then it is easy to see that the dependence on the chemical potential ζ vanishes on the right-hand side of the equation (4.188). According to formulas (4.167), (4.172) the ratio τ/τ_0 located on the left-hand side of the equation (4.188) can be written as

$$\frac{\tau}{\tau_0} \simeq \left(\frac{T}{T_k}\right)^{1/2}, \tag{4.191}$$

since the dependence τ_0 and τ on chemical potential vanishes in the case of a non-degenerate electron gas. Thus, in the case of a nondegenerate electron gas the equation (4.188) contains only one unknown parameter—the temperature of the kinetic degrees of freedom of the electron conductivity T_k, so it can be easily solved. Substituting the results (4.189), (4.190) into the formula (4.188) and introducing the notations:

$$\frac{T_k}{T} = x, \quad \Gamma = \frac{Q_0 \hbar^4 \rho \pi \Gamma(5/2)}{2^{5/2} E_0^2 m^{5/2} (k_B T)^{3/2}}, \tag{4.192}$$

we obtain a quadratic equation for the unknown kinetic temperature,

$$x^2 - x - \Gamma = 0, \tag{4.193}$$

which has only the physically meaningful solution

$$T_k = \frac{T}{2} + T\sqrt{1/4 + \Gamma}. \tag{4.194}$$

Next, we estimate the magnitude of the deviation of the non-equilibrium temperature from equilibrium, assuming that $\Gamma \ll 1$. For that, we expand the root in the formula (4.194) into a series and restrict ourselves to only linear term over Γ. In this case, introducing the relative temperature change $\delta T_k/T = (T_k - T)/T$, we arrive at the following expression:

$$\frac{\delta T_k}{T} \simeq \Gamma. \tag{4.195}$$

For typical parameter values of semiconductor materials ($m = 0.07 m_0$, m_0 – is free electron mass, $\rho = 5.8\ \mathrm{g/cm^3}$, $s = 5 \cdot 10^5\ \mathrm{cm/s}$, $E_0 = 1.6 \cdot 10^{-18}\ \mathrm{J}$, $T = 4\ \mathrm{K}$), the estimation of Γ by using the formulas (4.192), (4.189), (4.167) gives $\delta T_k/T \simeq 1.32$.

Knowing the temperature of the kinetic degrees of freedom of the conduction electrons, we can also determine the non-equilibrium chemical potential ζ. Using the equation (4.181) for the case of the classical Maxwell–Boltzmann statistics, we have

$$\left(\frac{T_k}{T}\right)^{3/2} = \exp\left(\frac{\zeta_0}{k_B T} - \frac{\zeta}{k_B T_k}\right),$$

or taking the logarithm of this expression, we get the final result:

$$\zeta = \zeta_0 \frac{T_k}{T} - \frac{3}{2} k_B T_k \ln\left(\frac{T_k}{T}\right). \tag{4.196}$$

It follows from the formula (4.196) that the chemical potential of the non-equilibrium electrons does not depend on the drift velocity $\vec{v}_d$ in approximation in hand. In fact, such dependence, of course, takes place and it can be easily defined. For that, the terms of the second order of smallness over the parameter $\hbar k^{\alpha} v_d^{\alpha}/k_B T$ in the formula (4.154) need to be retained, but this would be a significant accuracy overestimate.

Thus, we have just considered the problem of the electron system heating by an external electric field and found expressions for the kinetic temperature (4.194), the chemical potential (4.196) and the drift velocity (4.156) involved in the nonequilibrium distribution function (4.146).

It is important to be noted that the deviation magnitude of the electron temperature T_k from the equilibrium temperature T not necessarily must be small (in fact, the magnitude of $\delta T_k/T$ is not assumed to be small in deriving the balance equations for momentum, energy and particle number).

Quite similarly, we could have considered the case of a degenerate electron gas (this problem is proposed to the reader as a self-guided work).

Let us now go over to discuss potential applications of the developed theory for solving various problems of physical kinetics.

4.2.5 Negative differential resistance

From a practical point of view, it is important to find such conditions when charge carrier heating would lead to the appearance of a segment with negative values of dJ/dE on a voltage–current characteristic. Here J is the electric current density; E is the intensity of the electric field applied to a specimen. Such a situation may arise if the current density vector $\vec{J}$ is antiparallel to the electric field $\vec{E}$ intensity or these vectors are parallel to each other but the density of the electric current inside the specimen decreases as the electric field increase.

The first case has to do with *negative resistance*, and the second one is related to the *negative differential resistance* (NDR).

Negative differential resistance may be observed experimentally either by controlling the current passing through the specimen (Figure 4.6(a)) (a large additional resistance R is series-connected with the specimen in the circuit) or by controlling the voltage measured across the specimen that is connected in parallel with a large additional load-resistance R_H (Figure 4.6(b)).

Figure 4.6(a) illustrates an S-shaped nonlinear voltage–current characteristic and Figure 4.6(b) an N-shaped one. To control the current J, flowing through the specimen with a series-connected resistance r, the condition $R \gg r$ must be met. Similarly, in the case (b) the condition $R_H \gg r$ must be fulfilled.

The simplest way required to observe NDR is to control the current passing through the specimen (Figure 4.6(a)). Then the magnitude of the electric field intensity $\vec{E}$ inside the specimen and the kinetic temperature T_k current-dependent. The power P, transmitted by the electron system to the lattice coincides with the Joule

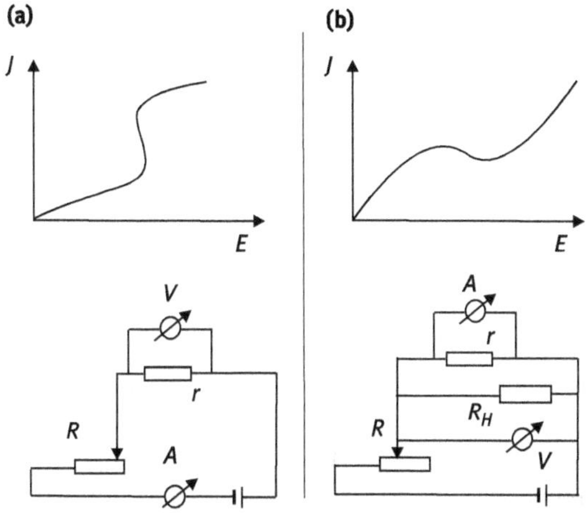

Figure 4.6: S-shaped (a) and N-shaped (b) nonlinear voltage–current characteristics and their measurement schemes under the controlled voltage and current conditions. The resistance r is a specimen.

power loss:

$$P = \sigma E^2, \quad E = \left(\frac{P}{\sigma}\right)^{1/2}. \tag{4.197}$$

On the other hand, the same power can be expressed via the current density

$$J^2 = \sigma P, \quad J = (\sigma P)^{1/2}. \tag{4.198}$$

Under conditions of heating of an electron gas, as it appears from the formula (4.172) $\sigma \sim \tau \sim T_k^m$ for the case of the Maxwell–Boltzmann statistics, where m is some exponent, and according to (4.187) the power transferred by electrons to the lattice has the following temperature dependence:

$$\left(\frac{\partial}{\partial t}\langle E_{\vec{k}}\rangle\right)_{\text{col}} \equiv P \sim (T_k - T)T_k^n. \tag{4.199}$$

Substituting the expected temperature σ and P dependence into the equations for the field strength E (4.197) and magnitude of the current J (4.198), we obtain

$$E \sim \left[(T_k - T)T_k^{n-m}\right]^{1/2}, \quad J \sim \left[(T_k - T)T_k^{n+m}\right]^{1/2}. \tag{4.200}$$

It follows from (4.200) that, if the conditions $n - m > 0$ and $n + m > 0$ are met, then E and J increase as T_k rises. Consequently, dJ/dE is positive. If, however, $n - m < 0$ and $n + m > 0$, as the current increases and, consequently, the temperature T_k, the electric field intensity will decrease. Such a situation corresponds to a segment with negative differential resistance on the voltage–current characteristic. In this case, the resulting volt–ampere curve is S-shaped. In particular, it is easy to see that the conditions for the NDR-segment to appear are satisfied if the momentum relaxation is due to the interaction of electrons with impurities ($m = 3/2$), and the relaxation energy is determined by the piezoelectric scattering by acoustic phonons ($n = -1/2$).

In fact, the NDR-phenomenon occurs along with the development of instabilities in a homogeneous semiconducting crystal. In particular, the NDR-segment of the volt–ampere curve (under conditions of the controlled current flowing through the specimen) exhibits so-called "pinching" of the current. In the plane of the longitudinal cross-section of the specimen there arise one or several channels with a lower electric resistance, which essentially shunts the specimen.

Under measured voltage conditions, an instability of a different nature appears. In other words, regions (domains) with high and low electrical resistance are formed in the specimen. The total voltage-drop occurs across the high resistance domains. These domains travel through the specimen under an action of the electric field; thus periodic electrical oscillations take place. The generator of microwave oscillations may serve as a case in point of practical application of this phenomenon (Gunn diodes).

The Gunn diode is a uniform semiconductor device which generates microwave oscillations when a direct electric field is applied to the device. The Gunn diode is based on the Gunn effect, which is to generate high-frequency oscillations of the electric current passing through the uniform semiconductor with an N-shaped volt–ampere characteristic. The Gunn effect was discovered by American physicist Gunn in 1963 in a crystal made of gallium arsenide with electrical conductivity of n-type.

As the electric field $E > 2$–3 kV/cm is applied to the homogeneous device made of gallium arsenide of n-type, spontaneous current oscillations appear. In the specimen, usually near cathode, there is a small part of a strong field ("domain") which drifts from the cathode to the anode with the velocity $v_d \simeq 10^7$ cm/s and disappears at the anode. Then a new domain is formed at the cathode and the process is periodically repeated. When the current flowing through the specimen falls down, this moment corresponds to appearance of the domain. When the current is restored to previous value, it stands for disappearance of the domain at the anode. The period of the current oscillations is approximately equal to the transit time, i. e. to the drift-time of the domain from the cathode to the anode.

From the standpoint of highly non-equilibrium thermodynamics, the occurrence of electric current pinches (domains) in a uniform semiconducting material is the typical example of self-organization and emergence of non-equilibrium structures.

It can be shown that, if a local fluctuation of the current density arises in the semiconductor with an S-shaped V–I curve within the NDR-segment, this fluctuation does not decay, as in a normal material, but grows only. This causes the current pinching. In complete analogy, if the fluctuations result in appearing of the local region with a larger value of the electric field than in neighboring regions in the semiconductor with an N-shaped voltage–current characteristic, this region does not disappear, but extends only. Consequently, a strong field domain is formed. In this regard, a semiconductor is an active medium provided that the negative differential resistivity conditions are achieved.

Further discussion of the phenomena occurring in semiconductors due to heating of conduction electron by an external electric field goes beyond the scope of the present course. More detailed information on hot electrons can be found in the monograph by Conwell [27]. References [33, 34] are devoted to the research of instability processes arising in electronic conductor plasma.

The effective parameter method developed in this chapter enables solving the quite wide range of problems of physical kinetics associated with the transfer of energy between subsystems in a crystal. Next, we can list examples of such problems: the Feher effect that is a dynamic nuclear polarization by an electric current; the effect of the change in resistance in semiconductors under a saturation of the paramagnetic resonance of impurity centers (the effect allows simple electric schemes for detecting the resonance to be used); the Overhauser effect, which is dynamic nuclear polarization under a saturation of the paramagnetic resonance by free electrons in metals or semiconductors.

A complete analysis of these problems also goes beyond the scope of the course. Nevertheless, in Chapter 6, devoted to the non-equilibrium statistical operator method, we will apply the method for generating the balance equations of momentum, energy and the number of particles for the interpretation of the Overhauser effect.

Problem 4.6. Obtain an expression for an inverse relaxation time of hot electrons, assuming that the scattering of current carriers occurs by charged centers with a screened Coulomb potential.

Solution. Using the scheme (Figure 4.4) of electron transitions between the states $\vec{k}$, $\vec{k}+\vec{q}$, we write down the rate of change of a distribution function in the state $\vec{k}$ due to interaction with the scatterers. Taking into account the Hamiltonian (4.81) as the Hamiltonian of interaction with the scatterers, we get the following expressions for the quantum-mechanical transition probability (by analogy with (4.87)):

$$1.\, W_{\vec{k}+\vec{q}\vec{k}} = \frac{2\pi t}{\hbar} \sum_{\vec{q}} |G_{\vec{q}}|^2 |\langle \vec{k}+\vec{q}|e^{i\vec{q}\vec{r}}|\vec{k}\rangle|^2 \langle \rho_q \rho_{-q}\rangle_{\text{imp}} \delta(\varepsilon_{\vec{k}+\vec{q}} - \varepsilon_{\vec{k}}),$$

$$2.\, W_{\vec{k}\vec{k}+\vec{q}} = \frac{2\pi t}{\hbar} \sum_{\vec{q}} |G_{\vec{q}}|^2 |\langle \vec{k}|e^{-i\vec{q}\vec{r}}|\vec{k}+\vec{q}\rangle|^2 \langle \rho_q \rho_{-q}\rangle_{\text{imp}} \delta(\varepsilon_{\vec{k}+\vec{q}} - \varepsilon_{\vec{k}}).$$

If one considers the determination of $\langle \rho_q \rangle$ (4.81), it is easy to see that the average over the states of the scatterers is

$$\langle \rho_q \rho_{-q} \rangle_{\text{imp}} = N_i,$$

where N_i is the number of impurity centers per unit volume.

To find the rate of change of the number of particles in state $\vec{k}$ under the influence of collisions, quantum-mechanical probability transitions must be multiplied both by occupation probability that the initial state is occupied and the probability that the final state is unoccupied. As a result, we have

$$\left.\frac{\partial f_{\vec{k}}}{\partial t}\right|_{\text{st}} = \frac{2\pi}{\hbar} N_i \sum_{\vec{q}} |G_{\vec{q}}|^2 [f_{\vec{k}+\vec{q}} - f_{\vec{k}}] \delta(\varepsilon_{\vec{k}+\vec{q}} - \varepsilon_{\vec{k}}). \tag{4.201}$$

Unlike the case of scattering by phonons, it does not make sense to consider the transitions numbers 3 and 4 in the formula (4.87) because the summation in (4.201) is being produced over all possible values of $\vec{q}$.

By analogy with (4.152), we build the balance equation of momentum:

$$\frac{\partial}{\partial t} \langle \vec{P}^\alpha \rangle|_{\text{col}} = \frac{2\pi}{\hbar} N_i \sum_{\vec{k}\sigma\vec{q}} \hbar k^\alpha |G_{\vec{q}}|^2 [f_{\vec{k}+\vec{q}} - f_{\vec{k}}] \delta(\varepsilon_{\vec{k}+\vec{q}} - \varepsilon_{\vec{k}}).$$

Replacing first the summation indices $\vec{k} + \vec{q} \to \vec{k}$, $\vec{k} \to \vec{k} - \vec{q}$ in the second summand proportional to $\vec{q} \to -\vec{q}$, we obtain

$$\frac{\partial}{\partial t} \langle \vec{P}^\alpha \rangle|_{\text{col}} = -\frac{2\pi}{\hbar} N_i \sum_{\vec{k}\sigma\vec{q}} \hbar q^\alpha |G_{\vec{q}}|^2 f_{\vec{k}+\vec{q}} \delta(\varepsilon_{\vec{k}+\vec{q}} - \varepsilon_{\vec{k}}). \tag{4.202}$$

Now, it is necessary to separate the terms linear in the drift velocity on the right side of (4.202). Since the distribution function involves the drift velocity, we expand the latter in a Taylor series retaining only linear members. Substituting this expansion (see (4.154)) in the expression that determines the rate of change of the average electron momentum due to collisions, we can rewrite the right-hand side of (4.202) as

$$\frac{\partial}{\partial t} \langle \vec{P}^\alpha \rangle|_{\text{col}} = \frac{2\pi}{\hbar} N_i \sum_{\vec{k}\sigma\vec{q}} \frac{1}{3} (\hbar q)^2 |G_{\vec{q}}|^2 f'_{\vec{k}+\vec{q}} v_d^\alpha \delta(\varepsilon_{\vec{k}+\vec{q}} - \varepsilon_{\vec{k}}). \tag{4.203}$$

Considering the momentum balance equation (4.154), the final expression for an inverse momentum relaxation time of hot electrons has the form

$$\frac{1}{\tau} = -\frac{2\pi}{\hbar} \frac{1}{3nm} N_i \sum_{\vec{k}\sigma\vec{q}} (\hbar q)^2 |G_{\vec{q}}|^2 f'_{\vec{k}+\vec{q}} \delta(\varepsilon_{\vec{k}+\vec{q}} - \varepsilon_{\vec{k}}). \tag{4.204}$$

4.3 Problems to Chapter 4

4.1. Estimate numerical values of an electric E and magnetic H fields, a temperature gradient T to describe in the semi-classical form conduction electrons with the characteristic relaxation time of momentum $\tau \approx 10^{-13}$ s in semiconductors and metals.

What value of the external magnetic field strength is responsible for a strong one, and what one for quantizing? In estimating assume that the mass of the conduction electrons is equal to the mass of a free electron; the velocity of the electrons in semiconductors is equal to the average thermal velocity of motion, but in metals to the velocity at the Fermi surface.

4.2. As far as the kinetic equation (4.13) is concerned, the correction to a distribution function is to be sought in the form

$$f_1 = -\frac{\partial f_0}{\partial \varepsilon_{\vec{p}}}(\vec{v}\vec{\chi}(\varepsilon_{\vec{p}})),$$

where $\vec{\chi}(\varepsilon_{\vec{p}})$ is an unknown only energy-dependent function.
Prove that the correction can be written as (4.24) in the presence of a gradient of electrochemical potential $-\nabla(\varphi + 1/e\zeta)$, a temperature gradient $\vec{\nabla}T$ and an external constant magnetic field $\vec{h}H$.

4.3. Using the expressions (4.112)–(4.114), estimate numerically both the contribution to the diffusion component thermopower and the electron–phonon drag effect for the Herring momentum relaxation mechanism.
Define the material parameter values and experimental conditions when the phonon drag contribution is dominant.

4.4. Using the definition of the Maggi–Righi–Leduc effect (1.73), (4.125) and the expressions (4.118)–(4.120) for charge and heat fluxes in a magnetic field, obtain a numerical estimate for the change in the thermal conductivity coefficient in the magnetic field under a strongly degenerate electron gas.

4.5. Utilizing the expression (4.139) for the transverse Nernst–Ettingshausen effect in the case of a nondegenerate electron gas, show that the effect is positive for neutral impurity scattering ($r = 3/2$) and has a negative sign for scattering by long-wavelength acoustic oscillations ($r = -1/2$), which can be applied to determine the dominant momentum relaxation mechanism in a specimen.

4.6. Derive an expression for the collision integral

$$\left.\frac{\partial f_{\vec{k}\sigma}}{\partial t}\right|_{\mathrm{col}},$$

taking into account the interaction between electrons and magnetic moments of impurity centers or nuclei. In deriving the collision integral, one should resort to the contact electron-nuclear interaction Hamiltonian (6.150), having written it down preliminarily via the cyclic components $S^z, S^+ = 1/2(S^x + iS^y), S^- = 1/2(S^x - iS^y)$ for the electron spin operators. The transition to the cyclic components should be performed for the nuclear spins as well. To simplify the derivation, consider the collision integrals for functions $f_{k\uparrow}, f_{k\downarrow}$ separately. Arrows indicate the direction of the electron spin relative to the orientation of an external magnetic field.

4.7. Using the expression (4.204), define the temperature dependence of the reverse relaxation time of the average momentum of conduction electrons in scattering the latter by a screened Coulomb potential in the cases of a nondegenerate and degenerate electron gas. Estimate numerically the inverse relaxation time.

4.8. Find the region for wave vectors of acoustic phonons that interact effectively with conduction electrons in quasi-elastic scattering processes. Answer the question: How does the size of the region depend on the degree of degeneracy of an electron gas?

4.9. With the typical values of semiconductor material parameters $m = 0.07m_0$, $\rho = 5.8\text{g/cm}^3$, $s = 5 \cdot 10^5$ cm/s, $E_0 = 1.6 \cdot 10^{-18}$ J, $T = 4$ K of energy and momentum relaxation of electrons by acoustic phonons, evaluate the electric field strength to take into account the deviation of kinetic electron temperature from equilibrium. It must be borne in mind that slight relative changes in the kinetic temperature (a few percent) can already be experimentally observed.

4.10. Define the temperature dependence of the inverse relaxation time of the average energy of electrons in scattering the latter by optical phonons. The interaction Hamiltonian can be written as (4.76)

$$H_{ep} = \sum_{\vec{q}} C_{\vec{q}}\{b_{\vec{q}}e^{i\vec{q}\vec{r}} + b_{\vec{q}}^{+}e^{-i\vec{q}\vec{r}}\}; \quad |C_{\vec{q}}|^2 = \frac{E_0^2\hbar}{2\rho\Omega_0},$$

where Ω_0 is the optical phonon frequency.

5 Theory of linear response to an external mechanical perturbation

5.1 Electrical conductivity of an electron gas. The Kubo method

5.1.1 The Liouville equation and its solution

A quantum system can exist in a pure or mixed state. If the system is in a pure state, it can be described by a wave function ψ, which obeys the Schrödinger equation:

$$i\hbar\frac{\partial\psi}{\partial t} = H\psi, \tag{5.1}$$

where H is the Hamiltonian of the system, $\hbar$ Planck's constant.

The quantum-mechanical average of the operator of some physical quantity A in the state described by the wave function ψ is given by the expression $\langle A\rangle = \langle\psi|A|\psi\rangle$. After averaging, the resulting physical quantities must be real. This leads to the fact that operators of the physical quantities are Hermitian and satisfy the condition $A^+ = A$, $A^+ = \tilde{A}^*$, where the tilde sign stands for transposition; an asterisk, as usually, means the complex conjugate of elements of a matrix. From the standpoint of quantum mechanics, a description of the system in terms of the wave functions is the most complete and in some sense corresponds to the description of particles in terms of trajectories in classical mechanics.

Let us now define the notion of a mixed state in the framework of the quantum theory. Consider a subsystem that is a part of some large system being in the pure state. Let a set of coordinates x describe the subsystem under study. A multitude of q is of the rest of coordinates of the closed system. The wave function $\psi(q,x)$ depends on the variables x and q and does not fall into the product of functions depending on the above-mentioned coordinates. For this reason, the small system of interest does not have a wave function, so quantum mechanics cannot describe such systems with maximum possible completeness.

We compute, again, the average value of the operator A, which belongs to the small system, and acts only on variables of x. Generalizing the results obtained for pure states, one is led to

$$\langle A\rangle = \int \psi^*(q,x)A\psi(q,x)\,dq\,dx. \tag{5.2}$$

For practical applications, we introduce a more convenient definition for the average (5.2). We define a complete set of eigenfunctions $\varphi_n(x)$ of some operator, such as the Hamilton operator for the selected subsystem. For the rest of the system one should define a similar set of $\theta_n(q)$. Then it is obvious that the wave function $\psi(q,x)$ can be expanded into a series

$$\psi(q,x) = \sum_{n,m} C_{nm}\varphi_n(x)\theta_m(q). \tag{5.3}$$

https://doi.org/10.1515/9783110727197-005

Substituting this result into (5.2), we get

$$\langle A \rangle = \sum_{nm;ij} C_{in}^* C_{jm} \int \theta_i^*(q)\theta_j(q)\,dq \int \varphi_n^*(x) A \varphi_m(x)\,dx. \tag{5.4}$$

Given that the eigenfunctions $\theta_i(q)$ and $\theta_j(q)$ are orthonormal, we obtain

$$\langle A \rangle = \sum_{n,m;j} C_n(j)^* C_m(j) A_{nm}. \tag{5.5}$$

To proceed, it should be noted that the $C_n(j)^*$ and $C_m(j)$ coefficients depend on the variable j, related to the large system, which gives

$$\sum_j C_n(j)^* C_m(j) = \sum_j W(j) a_n^*(j) a_m(j) = \rho_{mn}. \tag{5.6}$$

The quantity ρ_{mn} introduced above bears the name of a density matrix. To understand better the physical meaning of the density matrix one should regard the diagonal matrix elements

$$\rho_{nn} = \sum_j W(j) a_n^*(j) a_n(j), \tag{5.7}$$

which can be easily interpreted. Indeed, let the state of the small system be a mixture of pure states which are numbered by the index j. Then the quantity $W(j)$ has the meaning of the probability of the state j and the product of $a_n^*(j)a_n(j)$ stands for the probability of the n-th eigenvalue for the j-th pure state. The quantity

$$\rho_{nn} = \sum_j W(j) a_n^*(j) a_n(j)$$

has the meaning of a location probability system the n-th stationary state which can observe in any of the possible pure states of the system. Using the definition (5.6), we can write the average value of A simply enough:

$$\langle A \rangle = \sum_{n,m} \rho_{mn} A_{nm}. \tag{5.8}$$

Now, let the operator A be equal to a unit operator. The average value of such an operator is, obviously, equal to unity. Therefore, instead of the formula (5.8), we have

$$\sum_n \rho_{nn} \equiv \mathrm{Sp}\{\rho\} = 1. \tag{5.9}$$

The latter result is obvious because a diagonal matric element of the density matrix also has the meaning as noted above of the probability of finding the system in the n-th stationary state. The probability of being in one of the possible states of the complete set of states is unity.

Looking ahead, it would be appropriate to give immediately a model of a system in contact with a heat bath. The wave functions $\varphi_n(x)$ are assumed to be eigenfunctions of the Hamilton operator: $H\varphi_n = E_n\varphi_n$, where E_n is eigenvalues of the energy system. In this case, the probability of the system to be in a mixed state with the energy value E_n is determined by the Gibbs distribution:

$$\rho_{nn} = \frac{\exp(-E_n/k_B T)}{\sum_m \exp(-E_m/k_B T)}.$$

In quantum mechanics, pure and mixed states differ fundamentally. If a system was in a pure state at some instant of the time t, then, by virtue of the linearity of the Schrödinger equation, it will remain in the pure state throughout evolution. Actually, the pure states are an idealization and, apparently, cannot be implemented if the system interacts with its environment.

An interesting relationship of pure and mixed states arises in connection with the measurement problem.

Assume that we have a system that is in a pure state with the wave function $\psi = \sum_n C_n U_n$, where U_n is for the eigenfunctions, for example, an energy operator. For a pure state, a normal quantum-mechanical average

$$\langle A \rangle = \sum_{n,m} C_n^* C_m \int U_n^*(x) A U_m(x)\, dx$$

can be written by using the definition of the average (5.8). Hence there follows a simple expression for components of the density matrix of the system in a pure state:

$$\rho_{mn} = C_n^* C_m. \tag{5.10}$$

Next, we perform measurements of energy in an ensemble of identical systems. It is obvious that the probabilities P_n of being the system with values of the energy $E = E_n$ can be found by making multiple measurements. Thus, the result of such a measurement will be a mixed state, which is described by another density matrix, which does not coincide with the original one. This is clear from re-computing the average value of the operator

$$\langle A \rangle = \sum_n P_n \int U_n^*(x) A U_n(x)\, dx. \tag{5.11}$$

Upon comparing the last two outcomes, we can see that the density matrix has reduced. The matrix has lost off-diagonal elements, which lead to interference of states with different values of n in the pure state. The situation here bears a complete similarity to the case of two coherent sources of light when the electric field intensity vectors are added up at some point in space. For incoherent sources, however, the squares of the intensities are added up, and so the interference disappears.

Thus, in the course of measurements, the pure state evolves into mixed and information about the system is lost. Since the information loss means an increase of entropy, there arises a situation when the measurement process, as in classical mechanics, leads to the appearance of an irreversible behavior which increases the entropy. We have no way of dwelling on this crucial but very far from solved measurement problem in quantum mechanics; so, we refer the reader to the monograph by Prigogine [35], where one can find other sources on this subject.

Let us deduce an equation of motion for the density matrix. To do this, one should differentiate the expression (5.7) over time:

$$\frac{\partial}{\partial t}\rho_{mn}(t) = \sum_i W(i)\left[\frac{\partial a_n^*(i,t)}{\partial t}a_m(i,t) + a_n^*(i,t)\frac{\partial a_m(i,t)}{\partial t}\right]. \tag{5.12}$$

For the equations for the coefficients a_n to be obtained, one should recollect that the wave function of each pure state $\psi(i) = \sum_k a_k(i)\psi_k$ in the mixture obeys the Schrödinger equation

$$i\hbar\frac{\partial\psi(i)}{\partial t} = H\psi(i), \tag{5.13}$$

where the ψ_k are time-independent eigenfunctions of an operator. Substituting the value of the wave function $\psi(i)$ in (5.13), we arrive at the equation for the coefficients a_n:

$$i\hbar\sum_k \frac{\partial a_k(i)}{\partial t}\psi_k = H\sum_k a_k(i)\psi_k. \tag{5.14}$$

Multiplying this equation by ψ_m^* and integrating with regard to orthonormality of the eigenfunctions ψ_n, we get

$$i\hbar\frac{\partial a_m(i)}{\partial t} = \sum_k \int \psi_m^* H\psi_k \, dv a_k(i). \tag{5.15}$$

By analogy, the equation for the complex-conjugate coefficient can be written as

$$-i\hbar\frac{\partial a_n^*(i)}{\partial t} = \sum_k \int \psi_n H^* \psi_k^* \, dv a_k^*(i). \tag{5.16}$$

Substituting the expressions (5.15), (5.16) into the equation of motion of the density matrix (5.12) and taking into account the Hermitian operator of energy, we obtain

$$\frac{\partial}{\partial t}\rho_{mn}(t) = \frac{1}{i\hbar}(H_{mk}\rho_{kn} - \rho_{mk}H_{kn}). \tag{5.17}$$

Going over from the matric notations to the operator ones and using the definition of the Liouville operator iL,

$$iLA = \frac{1}{i\hbar}[A,H], \quad [A,H] = AH - HA, \tag{5.18}$$

we obtain the Liouville equation for quantum systems:

$$\frac{\partial}{\partial t}\rho(t) + iL\rho(t) = 0. \tag{5.19}$$

Equation (5.19) allows the value of $\rho(t)$ to be determined at all following instants of time, provided that the value of $\rho(t_0)$ is set in some initial time t_0.

It should be emphasized that the equation of motion for the density matrix differs by sign from the equation of motion for the operator in the Heisenberg picture:

$$\frac{d}{dt}A(t) = iLA(t), \quad A(t) = \exp(i/\hbar\, Ht)A\exp(-i/\hbar\, Ht). \tag{5.20}$$

The Liouville equation is time reversible and as in the case of classical mechanics, its solution would give the most complete description of the system. It should not be supposed, however, that the exact solution of the Liouville equation gives a possibility to describe irreversible dynamics of macroscopic systems properly. The problem is much more complicated. Chapter 1 often stressed that the irreversible behavior for classical systems exhibits weak stability of solutions responsible for the evolution of a phase point in the phase space. In the case of quantum systems, there is no such clarity but the situation seems to be similar. So, there is no point in striving to obtain the exact solution of the Liouville equation. A physically meaningful result can be reached only by a coarse-grained description. For this reason, such a kind of description plays an increasingly important role in all existing modern methods of non-equilibrium statistical mechanics. This and subsequent chapters review how to construct the description of the non-equilibrium systems by means of the best-known quantum-statistical approaches.

5.1.2 Linear response of a dynamical system to an external field

Let us consider a system described by a Hamiltonian H as regards how it reacts to the switching on of an external perturbation being defined by the correction to the Hamiltonian $H_F(t)$:

$$H_F(t) = -AF(t)e^{\epsilon t}, \quad \epsilon \to +0, \tag{5.21}$$

where A is an operator of the interaction with an external field, $F(t)$ is a C-numerical function which characterizes amplitude of the external effects. The perturbation of the type (5.21), which can be given by the correction to the Hamiltonian, usually is referred to as *mechanical* one. There is a whole class of external effects, which cannot be reduced to a mechanical force and cannot be written in the form of (5.21). Such perturbations are called *thermal*.

We suppose that the external field is turned on at the moment time $t \to -\infty$. Before switching on the external field, the system was in equilibrium and was described by the equilibrium statistical operator

$$\rho_0 = \frac{1}{Z} e^{-\beta(H-\zeta N)}, \quad Z = \mathrm{Sp}\{e^{-\beta(H-\zeta N)}\}, \tag{5.22}$$

where N is the particle number operator. After switching on the external field, the system will deviate from thermodynamic equilibrium and be described by the statistical operator $\rho(t)$. It is of interest to find how an average value of a physical quantity B is changed after switching on the external field.

To answer this question, first the explicit form of the statistical operator $\rho(t)$ is to be found. The operator must satisfy the Liouville equation,

$$\frac{\partial \rho(t)}{\partial t} + [iL + iL_F(t)]\rho(t) = 0, \tag{5.23}$$

with the boundary condition

$$\lim_{t \to -\infty} \rho(t) = \rho_0. \tag{5.24}$$

The operators iL and $iL_F(t)$ involved in (5.23) are defined by the relation

$$iLR = \frac{1}{i\hbar}[R, H], \quad iL_F(t)R = \frac{1}{i\hbar}[R, H_F(t)], \tag{5.25}$$

where R is an arbitrary operator, $[A, B] = AB - BA$.

Next, we derive a formal solution of the equation (5.23), assuming that the external field is weak. For that purpose, one should write $\rho(t)$ as

$$\rho(t) = \rho_0 + \Delta\rho(t). \tag{5.26}$$

Then we linearize the equation (5.23), holding that $H_F(t)$ and $\Delta\rho(t)$ are small values. It should be understood that the smallness of operator quantities stands for the smallness of matric elements in the representation of the diagonal operators. The linearized equation has the form

$$\frac{\partial \Delta\rho(t)}{\partial t} + iL\Delta\rho(t) + iL_F(t)\rho_0 = 0. \tag{5.27}$$

We multiply the equation (5.27) by the evolution operator e^{iLt}, defined by the relation

$$e^{iLt}R = e^{i/\hbar Ht} R e^{-i/\hbar Ht}. \tag{5.28}$$

As a result, we obtain the equation

$$\frac{\partial}{\partial t} e^{iLt} \Delta\rho(t) = -ie^{iLt} L_F(t)\rho_0. \tag{5.29}$$

Next, integrating the above equation within the limits from $-\infty$ to t and taking into account the boundary condition (5.24), we find the solution

$$e^{iLt}\triangle\rho(t) = -i\int_{-\infty}^{t} dt' e^{iLt'} L_F(t')\rho_0. \tag{5.30}$$

To arrive at the non-equilibrium statistical operator $\rho(t)$ in the explicit form, we multiply both sides of the equation (5.30) by the evolution operator e^{-iLt}. Then we make a change of variables in the integral term putting $t' - t = t_1$. The use of (5.26) yields the desired value of the statistical operator:

$$\rho(t) = \rho_0 - i\int_{-\infty}^{0} e^{iLt_1} L_F(t+t_1)\rho_0\, dt_1. \tag{5.31}$$

The substitution of the expression (5.21) for the operator $H_F(t)$ gives rise to another form of this result,

$$\rho(t) = \rho_0 + \int_{-\infty}^{0} e^{\epsilon t_1} e^{iLt_1} \frac{1}{i\hbar}[\rho_0, A]\, F(t+t_1)\, dt_1. \tag{5.32}$$

By applying the result obtained we can find a change in the average value $\triangle\langle B\rangle^t$ of the operator of a physical quantity B due to the action of an external force:

$$\triangle\langle B\rangle^t = \mathrm{Sp}\{B\rho(t)\} - \mathrm{Sp}\{B\rho_0\}. \tag{5.33}$$

For practical applications, the Fourier transform of the representation is convenient of the time t for the response function of the system to an external perturbation. For that, one should perform the Fourier transform over time t:

$$\triangle\langle B\rangle^t = \int_{-\infty}^{\infty} d\omega e^{-i\omega t}\triangle\langle B\rangle^{\omega},$$
$$F(t) = \int_{-\infty}^{\infty} d\omega e^{-i\omega t} F(\omega). \tag{5.34}$$

The Fourier-image of the response to the external impact $\triangle\langle B\rangle^{\omega}$ and the Fourier-image of the external force $F(\omega)$, obviously, are determined with the inverse Fourier transform. Using the results of (5.32), (5.33), we can write down the following expression:

$$\triangle\langle B\rangle^{\omega} = \int_{-\infty}^{0} e^{(\epsilon - i\omega)t_1}\,\mathrm{Sp}\left\{B e^{iLt_1}\frac{1}{i\hbar}[\rho_0, A]\right\} dt_1 F(\omega). \tag{5.35}$$

The formula (5.35) allows one to investigate the system's response to the external field when a mechanical perturbation does not lead to the development of thermal perturbations in the system. In this sense, an iterative procedure of solving the equation (5.23) holds only at the first step, since a correction of a second order for the $H_F(t)$ is already incorrect, if thermal disturbances induced by the mechanical force are not taken into account. Electron heating and energy transfer processes between various subsystems of a crystal may serve as a case in point of such a perturbation. Such a situation is similar to the analysis done for the energy balance equations for different subsystems of the crystal in the previous chapter.

Although the solution of (5.23) can be formally constructed as a series in powers of $H_F(t)$, as was done by Kubo in his original paper, it may make sense only for an impulsive disturbance. In this case, if the pulse duration is small enough, the system of electrons can be considered as isolated and thermal perturbations may be not taken into account, since the time required for their formation is of the order of the relaxation time for this energy in the system. The interval of time is usually one to two orders of magnitude more than the momentum relaxation time. Moreover, even if the perturbation is assumed to be weak and we can restrict ourselves to only a linear approximation of the external force, the applicability of the result (5.35) to an analysis of real physical systems apparently is not self-evident.

Indeed, before switching on the interaction at the instant of time $t \to -\infty$, the system was described by a Gibbs grand canonical distribution and, consequently, was in contact with a heat bath. After switching on the interaction, the statistical operator $\rho(t)$ satisfies the equation (5.23), which involves only a system's Hamiltonian, and the Hamiltonian of the heat bath is absent. This implies that the separation of the system from the heat bath is occurring at the very moment when the external field began acting. In practice, of course, one fails to implement such a separation, therefore, the desired solution is valid provided that the difference in thermodynamic characteristics between the isolated system and system in contact with the heat bath can be neglected.

In conclusion, we consider the important issue of the irreversible the time-evolution of the statistical operator. In the kinetic Boltzmann equation the irreversibility occurs due to the irreversible time behavior of a collision integral. Consequently, this equation is time-irreversible beforehand.

In contrast to the kinetic equation, the Liouville equation is reversible in time. The irreversibility is due to the boundary condition (5.24). It is such a method that builds a time-irreversible solution of the Liouville equation in the original work by Kubo and most presentations of this work [36]. The same result can be achieved by an equivalent but more explicit way by introducing the infinitely small source $-\epsilon(\rho(t) - \rho_0)$ into the right-hand side of the Liouville equation (5.23). The source can be interpreted as an integral of collisions between an isolated system and the environment and be the cause of the relaxation non-equilibrium statistical distribution to equilibrium (the quantity $\epsilon \to 0$ after calculating the averages). This proves to be sufficient to obtain the time-irreversible solution of the Liouville equation and to take into consideration chaos effects on the part of the heat bath. The infinitesimal source on the right side of the Liouville equation serves for removing the degeneracy with respect to time reversal.

Let us obtain the solution of the Liouville equation with the infinitesimal source on the right side:

$$\left[\frac{\partial}{\partial t} + iL + iL_F(t)\right]\rho(t) = -\epsilon(\rho(t) - \rho_0), \tag{5.36}$$

assuming now that

$$H_F(t) = -AF(t). \tag{5.37}$$

We show that one can obtain the solution (5.32) using the above definition. For this purpose, as before, we introduce the correction $\triangle\rho(t)$, using the relation (5.26) and linearize the equation (5.36) over the small parameters $H_F(t)$ and $\triangle\rho(t)$. The left side of the linearized equation will coincide with the left-hand side of the equation (5.27) and the right side will have the infinitely small sources $-\epsilon\triangle\rho(t)$. Multiplying this equation by $e^{(\epsilon+iL)t}$, we can rewrite the linearized equation (5.36) as follows:

$$\frac{\partial}{\partial t} e^{\epsilon t} e^{iLt} \triangle\rho(t) = -ie^{\epsilon t} e^{iLt} L_F(t)\rho_0. \tag{5.38}$$

Integrating this equation within the limits from $(-\infty$ to $t)$, we get the equation analogous to (5.30):

$$e^{\epsilon t} e^{iLt} \triangle\rho(t) = -i \int_{-\infty}^{t} e^{\epsilon t'} e^{iLt'} L_F(t')\, dt' \rho_0. \tag{5.39}$$

In deriving the equation (5.37) by virtue of the finiteness the parameter ϵ we have taken into account that

$$\lim_{t\to-\infty} e^{\epsilon t} e^{iLt} \triangle\rho(t) = 0.$$

The operator $iL_F(t)$ in the formulas (5.38), (5.39) is defined by the Hamiltonian (5.37). Multiplying the equation (5.39) by $e^{-(\epsilon+iL)t}$ and performing the change of variables $t' - t = t_1$ in the integral term, we indeed get the formula (5.32).

Thus, the equations (5.23) and (5.36) are essentially equivalent, however, the use of the equation (5.36) is physically more justified, since it takes into account the interaction with the heat bath, albeit in an idealized form.

5.1.3 Calculation of electrical conductivity

Consider a calculation of electrical conductivity of electrons interacting with phonons as an example of using the Kubo method. Let the system be described by the Hamiltonian

$$H(t) = H + H_F(t), \quad H = H_0 + H_{ep}, \tag{5.40}$$

where H_0 is the operator of the non-interacting electron and photon subsystems

$$H_0 = H_k + H_p, \quad H_k = \frac{P^2}{2m}, \quad P^\alpha = \sum_j^N p_j^\alpha, \tag{5.41}$$

H_{ep} is the operator of the electron–phonon interaction. An explicit form for most mechanisms of the electron–phonon interaction can be specified in the form of (4.76); p_j^α is the α-projection of the momentum operator of the j-th electron.

The summation in the formula (5.41) is being performed over all electrons in a conduction band. The Hamiltonian of the interaction of electrons with an external electric field is given by

$$H_F(t) = -eX^\alpha E^\alpha(t), \quad X^\alpha = \sum_j r_j^\alpha, \tag{5.42}$$

where r_j^α is the operator of the α-projection of a coordinate of the j-th electron, E^α the α-projection of the amplitude of the external electric field.

Recollecting the phenomenological relationship $J^\alpha(\omega) = \sigma_{\alpha\beta}E^\beta(\omega)$ between the electric current density and electric field intensity and also the result (5.35), we write down the expression for the conductivity tensor,

$$\sigma_{\alpha\beta}(\omega) = \frac{e^2}{m} \int_{-\infty}^{0} e^{(\epsilon - i\omega)t_1} \frac{1}{i\hbar} \operatorname{Sp}\{P^\alpha [\rho_0, X^\beta(t_1)]\}\, dt_1. \tag{5.43}$$

In deriving the formula we have put in use the equality $B = J^\alpha = -eP^\alpha/m$. It follows from the expression (5.43) that the electrical conductivity of an electron gas is expressed in terms of a correlation function defined for the system being in equilibrium. This physically means that processes responsible for the relaxation time of the average electron momentum, causing the electrons to drift in the electric field and for the decay time of fluctuations of the average momentum of the electrons in equilibrium are the same.

In contrast to a kinetic equation, where an expression for a correction to the distribution function virtually allows all kinetic coefficients to be immediately calculated, the use of a formal solution of the Liouville equation (5.32) only leads to the problem of computing the quantum correlation functions. Thus, instead of solving the complicated integro-differential kinetic equation, in the Kubo theory there arises the problem to "disentangle" temporal correlation functions. The correlation function on the right-hand side of the equation (5.43) may serve as a case in point of such a temporal correlation function.

To calculate the electrical conductivity $\sigma_{\alpha\beta}(\omega)$, we use the Green function method [36, 37].

One should explain why only in this method one needs to find the value of the conductivity $\sigma_{\alpha\beta}(\omega)$. The problem here is that the expression (5.43) involves the full Hamiltonian H both for the statistical operator ρ_0 and for the evolution operator that determines the time-dependence of the electron coordinate operator $X^{\beta}(t_1)$. As far as one fails to find eigenfunctions and eigenvalues of the Hamiltonian, an accurate calculation of the correlation function is impossible. The attempt to expand both the statistical operator ρ_0 and the evolution operator e^{iLt_1} into a power-series of the operator of the electron–phonon interaction H_{ep} in any finite-order in perturbation theory gives rise to a wrong result. It follows from the outcomes obtained on the basis of the kinetic equation that $\sigma \sim \tau \sim 1/H_{ep}^2$ as the expansion of the correlation function into a series in powers of H_{ep} gives some polynomial dependence on the constant of the electron–phonon interaction.

The correct result can be reached by summing up some infinite sequence of terms of a series in the perturbation theory. In addition, this infinite sequence must come to an infinite decreasing geometric progression amenable to convenient summation. It is the Green function method that improves both the diagram technique and the mass-operator method. These enable one to perform easily enough a similar summation of an infinite series in perturbation theory.

Defining the Green function $G_{\alpha\beta}(t_1)$ by the relation

$$G_{\alpha\beta}(t_1) = \theta(-t_1)e^{\varepsilon t_1}\frac{1}{i\hbar}\,\mathrm{Sp}\{P^{\alpha}e^{iLt_1}[\rho_0, X^{\beta}]\},$$
$$\theta(x) = \begin{cases} 1, & x \geq 0 \\ 0, & x < 0, \end{cases} \tag{5.44}$$

we write down the expression for electrical conductivity:

$$\sigma_{\alpha\beta}(\omega) = \frac{e^2}{m}\int_{-\infty}^{\infty} e^{-i\omega t_1}\frac{1}{i\hbar}G_{\alpha\beta}(t_1)\,dt_1 = \frac{e^2}{m}G_{\alpha\beta}(\omega), \tag{5.45}$$

where $G_{\alpha\beta}(\omega)$ is the Fourier transform of the Green function (5.44). Since in accordance with (5.45), the Fourier transform of the Green function is proportional to the electrical conductivity, a further goal is to find the quantity $G_{\alpha\beta}(\omega)$. For this purpose, it is necessary to build equations of motion for the function (5.44). Next, we differentiate (5.44) over time t_1:

$$\frac{d}{dt_1}G_{\alpha\beta}(t_1) = -\delta(t_1)\frac{1}{i\hbar}\,\mathrm{Sp}\{P^{\alpha}[\rho_0, X^{\beta}]\} + \epsilon G_{\alpha\beta}(t_1) + G_{1\alpha\beta}(t_1),$$

$$G_{1\alpha\beta}(t_1) = \theta(-t_1)e^{\varepsilon t_1}\frac{1}{i\hbar}\,\mathrm{Sp}\left\{P^\alpha e^{iLt_1}\left[\rho_0, \frac{P^\beta}{m}\right]\right\}. \tag{5.46}$$

In deriving the equation of motion for the Green function $G_{\alpha\beta}$ we have used the definition of the derivative of a theta-function

$$\frac{d}{dx}\theta(x) = \delta(x)$$

and have taken into account that

$$\begin{aligned}\frac{d}{dt_1}e^{iLt_1}[\rho_0, X^\beta] &= e^{iLt_1}iL[\rho_0, X^\beta]\\ &= e^{iLt_1}[\rho_0, \dot{X}^\beta]; \quad \dot{X}^\beta = \frac{1}{i\hbar}[X^\beta, H] = \frac{P^\beta}{m}.\end{aligned}$$

The last equality in this formula is caused by the fact that Hamilton's operators for phonons and the electron–phonon interaction commute with the operator of an electron coordinate.

It is easy to see that the equation of motion for the Green function $G_{\alpha\beta}(t_1)$ contains a new unknown quantity $G_{1\alpha\beta}(t_1)$, for which one can also set up an equation of motion. Differentiating $G_{1\alpha\beta}(t_1)$ over time t_1 in the formula (5.46), we get

$$\begin{aligned}\frac{d}{dt_1}G_{1\alpha\beta}(t_1) &= -\delta(t_1)\frac{1}{i\hbar}\,\mathrm{Sp}\left\{P^\alpha\left[\rho_0, \frac{P^\beta}{m}\right]\right\} + \epsilon G_{1\alpha\beta}(t_1) - G_{2\alpha\beta}(t_1),\\ G_{2\alpha\beta}(t_1) &= \theta(-t_1)e^{\epsilon t_1}\frac{1}{i\hbar}\,\mathrm{Sp}\left\{\dot{P}^\alpha e^{iLt_1}\left[\rho_0, \frac{P^\beta}{m}\right]\right\}.\end{aligned} \tag{5.47}$$

In deriving the formulas (5.47) we have used the cyclic permutation of the operators for a spur:

$$\mathrm{Sp}\{ABC\} = \mathrm{Sp}\{CAB\}. \tag{5.48}$$

This property is easily proved by using the definition of the spur as the sum of diagonal matric elements. In our case, the use of (5.48) gives

$$\mathrm{Sp}\left\{P^\alpha e^{iLt_1}iL\left[\rho_0, \frac{P^\beta}{m}\right]\right\} = -\,\mathrm{Sp}\left\{\dot{P}^\alpha e^{iLt_1}\left[\rho_0, \frac{P^\beta}{m}\right]\right\}.$$

The equation for the Green function $G_{1\alpha\beta}(t_1)$ also includes a new Green function $G_{2\alpha\beta}(t_1)$. Thus, virtually, there arises a chain of "coupled" equations of motion set forth for all new Green functions; consequently, it becomes impossible for us to define the function $G_{\alpha\beta}(t_1)$ exactly.

One of the possible approaches to finding the Green function approximately is to close deliberately an infinite chain of equations of motion at a certain step. Then a consequent Green function is expressed via the previous Green function. This means that an enhanced accuracy for solving the set of equations can be obtained. This procedure is to make a so-called decoupling of the Green functions; it is widely used in the theory of magnetism. Unfortunately, it is hard enough to estimate approximations by using the technique. Thus, usually, the exact treatment of the equations of motion requires the comparison process with the results of other methods.

The mass-operator method mentioned above is the more reasonable technique. An advantage of such a method is its simple and physically clear program. The Green function in frequency representation as a rule has poles in the complex plane. Therefore, it does not make sense to construct the perturbation theory for such functions. At the same time, the poles themselves can be determined by analytic functions, so a construction of the perturbation theory for them is quite possible.

Consider how this program can be fulfilled within example of calculating the electrical conductivity of an electron gas. We make the Fourier transform for the equations of motion for $G_{\alpha\beta}(t_1)$, $G_{1\alpha\beta}(t_1)$, etc. By defining the Fourier transforms

$$\begin{aligned} G_{\alpha\beta}(\omega) &= \int_{-\infty}^{\infty} dt_1 e^{-i\omega t_1} G_{\alpha\beta}(t_1), \\ G_{1\alpha\beta}(\omega) &= \int_{-\infty}^{\infty} dt_1 e^{-i\omega t_1} G_{1\alpha\beta}(t_1), \\ G_{2\alpha\beta}(\omega) &= \int_{-\infty}^{\infty} dt_1 e^{-i\omega t_1} G_{2\alpha\beta}(t_1), \end{aligned} \tag{5.49}$$

we write down the chain of equations of motion for the Green functions in the frequency representation:

$$\begin{aligned} (i\omega - \epsilon)G_{\alpha\beta}(\omega) &= -n\delta_{\alpha\beta} + G_{1\alpha\beta}(\omega), \\ (i\omega - \epsilon)G_{1\alpha\beta}(\omega) &= -G_{2\alpha\beta}(\omega), \\ &\dots. \end{aligned} \tag{5.50}$$

In the expression (5.50), the dots stand for the other equations of motion for the Green functions $G_{2\alpha\beta}(\omega)$, etc. In deriving the equations (5.50) we have used the equalities

$$\frac{1}{i\hbar}\,\mathrm{Sp}\{P^\alpha[\rho_0, X^\beta]\} = \frac{1}{i\hbar}\,\mathrm{Sp}\{[X^\beta, P^\alpha]\rho_0\} = \sum_{ij}\mathrm{Sp}\{[x_i^\beta, p_j^\alpha]\rho_0\} = -n\delta_{\alpha\beta},$$

$$\frac{1}{i\hbar}\,\mathrm{Sp}\{P^\alpha[\rho_0, P^\beta]\} = 0.$$

Next, a solution of the set of the equations (5.50) needs to be sought in the form

$$G_{\alpha\beta}(\omega) = \frac{-n\delta_{\alpha\beta}}{i\omega - \epsilon - M_{\alpha\beta}(\omega)}, \tag{5.51}$$

where $M_{\alpha\beta}(\omega)$ is *a mass-operator*. The poles of the Green function (5.51) determine the spectrum of collective excitations of the electron gas. The excitations are associated with a fluctuation of the average momentum of the electronic system. The name "mass-operator" was borrowed from the theory of elementary particles, where the energy of elementary excitations is synonymous with their mass.

Since there is reason to believe that the correction $M_{\alpha\beta}(\omega)$ to the spectrum of elementary excitations is an analytic function, one may try to find the correction, using the electron–phonon coupling as a small parameter in the perturbation theory.

The Green function $G_{1\alpha\beta}(\omega)$ is proportional to the first power, $G_{2\alpha\beta}(\omega)$ to the second power of this parameter (the proof of this important fact will be given below).

In all generality of the results, we consider a solution of the formal set of the coupled equations

$$LG = I_1 + G_1,$$

$$LG_1 = I_2 + G_2, \tag{5.52}$$

$$G = \frac{I_1}{L - M}. \tag{5.53}$$

The sense of notations in (5.52), (5.53) is quite obvious, so each term in these formulas can be easily mapped to the corresponding term in the equations (5.50), (5.51).

Substituting the Green function (5.53) into the first equation of the chain (5.52) and solving the resulting equation with respect to M, we find

$$M = \frac{G_1 L}{I_1 + G_1}. \tag{5.54}$$

We first find G_1 from the second equation (5.52) and plug them into the numerator of the expression for the mass-operator (5.54). Given that the function I_1 does not contain an interaction but G_1 is proportional to the electron–phonon interaction constant, as a result, we arrive at the expansion of the mass-operator in powers of the small parameter:

$$M = \frac{I_2}{I_1} + \frac{G_2}{I_1} - \frac{I_2}{I_1}\frac{G_1}{I_1} + \cdots. \tag{5.55}$$

In calculating the electrical conductivity, the function I_1 is equal to $n\delta_{\alpha\beta}$, as $I_2 = 0$, so the mass-operator can be written as

$$M_{\alpha\beta}(\omega) = \frac{G_{2\alpha\beta}(\omega)}{n}. \tag{5.56}$$

A more detailed form appears as

$$M_{\alpha\beta}(\omega) = \frac{1}{nm} \int_{-\infty}^{0} dt_1 e^{(\epsilon - i\omega)t_1} \, \mathrm{Sp}\left\{\dot{P}^{\alpha} e^{iLt_1} \frac{1}{i\hbar}[\rho_0, P^{\beta}]\right\}. \tag{5.57}$$

What physical meaning does the mass-operator defined by (5.51) have? For the answer, it would be advisable to recollect that, according to the classical theory of the electrical conductivity, the high-frequency conductivity can be written as

$$\sigma(\omega) = \frac{e^2}{m}\frac{n}{1/\tau - i\omega}. \tag{5.58}$$

Upon comparing the formulas (5.45), (5.51) and (5.58) it is seen that they coincide, if one assumes that the mass-operator $M_{\alpha\beta}(\omega)$ has the meaning of the momentum relaxation frequency. Thus, as already noted above, the mass-operator of the Green function (5.44) describes the spectrum of elementary excitations. Moreover, the real part of the mass-operator determines damping of the excitations, and the imaginary part (if it exists) is responsible for the frequency of its own oscillations of the average momentum of the electron system.

When calculating the mass-operator in accordance with the formula (5.57), the correlation function analysis becomes problematical again, therefore, at first glance no progress seems to have been achieved. In fact, this is far from being the case. Firstly, the mass-operator $M_{\alpha\beta}(\omega)$ has the meaning of the relaxation frequency of the average momentum and, as shown in Chapter 4 (see equation (4.160)), is proportional to the square of the electron–phonon coupling constant.

Secondly, the second order in the electron–phonon interaction constant has already been collected both for the Green function $G_{2\alpha\beta}(\omega)$ and for the mass-operator $M_{\alpha\beta}(\omega)$. Therefore, the Hamiltonian H_{ep} may be omitted both for the statistical operator ρ_0 and for the evolution operator, replacing H by H_0. Indeed, by definition, the operator $\dot{P}^\alpha$ in the right-hand side of the expression (5.57) is equal to

$$\dot{P}^\alpha = \frac{1}{i\hbar}[P^\alpha, H_0 + H_{ep}] = \frac{1}{i\hbar}[P^\alpha, H_{ep}] \equiv \dot{P}_{(l)}, \tag{5.59}$$

because the Hamiltonian H_0 commutes with the operator P^α of the total electron momentum. The commutator $[\rho_0, P^\beta]$, occurring in the spur in the formula (5.57) is also proportional to the electron–phonon coupling constant. This becomes especially evident when the Kubo identity is used:

$$[A, e^{-\beta H}] = \int_0^\beta d\lambda e^{-\lambda H}[H, A]e^{\lambda H}e^{-\beta H}. \tag{5.60}$$

To prove the Kubo identity we introduce the function $I(\lambda)$

$$I(\lambda) = [A, e^{-\lambda H}], \tag{5.61}$$

which satisfies the equation

$$\frac{d}{d\lambda}I(\lambda) = -(AHe^{-\lambda H} - He^{-\lambda H}A). \tag{5.62}$$

Now, consider the function $I(\lambda)e^{\lambda H}$ and find the derivative of its function over λ. Given the equality (5.61), we arrive at the equation

$$\frac{d}{d\lambda}(I(\lambda)e^{\lambda H}) = [H, e^{-\lambda H}Ae^{\lambda H}], \tag{5.63}$$

including the obvious initial condition $I(0) = 0$.

Integrating the equation (5.63) between the limits 0 and β and taking into account the initial condition, we obtain

$$I(\beta) = \int_0^\beta d\lambda e^{-\lambda H}[H, A]e^{\lambda H}e^{-\beta H}. \tag{5.64}$$

It can be seen that the formula (5.64) coincides with the formula (5.60). By applying the Kubo identity to the commutator $[\rho_0, P^\beta]$, we get

$$\frac{1}{i\hbar}[\rho_0, P^\beta] = \int_0^\beta d\lambda \dot{P}^\beta_{(l)}(i\hbar\lambda)\rho_0,$$
$$\dot{P}^\beta_{(l)}(i\hbar\lambda) = e^{-\lambda H}\dot{P}^\beta_{(l)}e^{\lambda H}. \tag{5.65}$$

Thus, we have shown that the second order in the interaction has already been collected for the Green function $G_{2\alpha\beta}(\omega)$. So, if one restricts oneself to the calculation of the mass-operator in the Born approximation of scattering theory, one can omit the interaction operator H_{ep} both in the statistical operator and in the evolution operator. Then, on the basis of the results (5.57), (5.59), (5.65) we have

$$\frac{1}{\tau} = \frac{1}{nm}\int_{-\infty}^0 dt_1 e^{(\epsilon - i\omega)t_1}\int_0^\beta d\lambda\langle \dot{P}^\alpha_{(l)}\dot{P}^\beta_{(l)}(t_1 + i\hbar\lambda)\rangle,$$
$$\langle\ldots\rangle = \mathrm{Sp}\left\{\ldots\frac{1}{Z}e^{-\beta H_0}\right\}. \tag{5.66}$$

In the future, we focus our attention on the case of a static external field, setting the frequency $\omega = 0$. Now, the integral over time t_1 can be extended to $+\infty$, since the integrand is an even function of the argument t_1. Indeed, consider the expression

$$\int_0^\beta d\lambda\langle \dot{P}^\alpha_{(l)}\dot{P}^\beta_{(l)}(-t_1 + i\hbar\lambda)\rangle = \int_0^\beta d\lambda\,\mathrm{Sp}\{\dot{P}^\alpha_{(l)}(t_1 - i\hbar\lambda)\dot{P}^\beta_{(l)}\rho_0\}$$
$$= \int_0^\beta d\lambda\,\mathrm{Sp}\left\{\dot{P}^\beta_{(l)}\frac{1}{Z}e^{-\beta H_0}e^{H_0\lambda}\dot{P}^\alpha_{(l)}(t_1)e^{-H_0\lambda}\right\}. \tag{5.67}$$

Introducing the new variable $\lambda' = \beta - \lambda$, we obtain

$$\int_0^\beta d\lambda \langle \dot{P}^\alpha_{(l)} \dot{P}^\beta_{(l)}(-t_1 + i\hbar\lambda)\rangle = \int_0^\beta d\lambda' \langle \dot{P}^\beta_{(l)} \dot{P}^\alpha_{(l)}(t_1 + i\hbar\lambda')\rangle = \int_0^\beta d\lambda \langle \dot{P}^\alpha_{(l)} \dot{P}^\beta_{(l)}(t_1 + i\hbar\lambda)\rangle. \quad (5.68)$$

The second equality in the formula (5.68) follows from the isotropy of the Hamiltonian H_0 with respect to rotations in a coordinate space.

Given the last result, the equation (5.66) for the mass-operator $M_{\alpha\beta}(0) = 1/\tau$ has the form

$$\frac{1}{\tau} = \frac{1}{2nm} \int_{-\infty}^{\infty} dt_1 e^{-\epsilon|t_1|} \int_0^\beta d\lambda \langle \dot{P}^\alpha_{(l)} \dot{P}^\beta_{(l)}(t_1 + i\hbar\lambda)\rangle. \quad (5.69)$$

The formula (5.69) for an inverse momentum relaxation time of conduction electrons holds for any scattering mechanisms, because the explicit form of the operator of the electron–phonon scattering by longitudinal acoustic phonons has nowhere been used in deriving this formula.

In the previous chapter, we already evaluated the momentum relaxation time $1/\tau_0$ of equilibrium electrons scattered by longitudinal acoustic phonons (see equation (4.167)). It is also of interest to compare the results obtained for the relaxation frequency of the average momentum by applying the kinetic equation and the method of linear response to an external perturbation.

Since, in this chapter, the averaging is being performed over an equilibrium ensemble, we will not label the equilibrium characteristics by the additional index "0".

We use the Hamiltonian of the electron–phonon scattering by longitudinal acoustic phonons (4.76) obtained previously, writing it in the form

$$H_{ep}(\vec{r}_j) = \sum_{\vec{q}} C_{\vec{q}} \{b_{\vec{q}} e^{i\vec{q}\vec{r}_j} + b^+_{\vec{q}} e^{-i\vec{q}\vec{r}_j}\}, \quad |C_{\vec{q}}|^2 = \frac{E_0^2 \hbar q}{2\rho s}. \quad (5.70)$$

To calculate the correlation function in the formula (5.69), it is convenient to pass to the second-quantization representation for the electron variables. The operators $\dot{P}^\alpha_{(l)}$ and $\dot{P}^\beta_{(l)}(t_1 + i\hbar\lambda)$ are represented as the sum of one-particle operators:

$$\dot{P}^\alpha_{(l)} = \frac{1}{i\hbar}\Big[\sum_j P^\alpha_j, \sum_i H_{ep}(\vec{r}_i)\Big] = \sum_j \dot{P}^\alpha_{j(l)},$$
$$\dot{P}^\alpha_{j(l)} = \frac{1}{i\hbar}[P^\alpha_j, H_{ep}(\vec{r}_j)] = -i \sum_{\vec{q}} C_{\vec{q}} q^\alpha \{b_{\vec{q}} e^{i\vec{q}\vec{r}_j} - b^+_{\vec{q}} e^{-i\vec{q}\vec{r}_j}\}. \quad (5.71)$$

Therefore, according to the transition rule to the second-quantization representation for additive operators [38], we have

$$\dot{P}^\alpha_{(l)} = -i \sum_{\vec{q},\vec{k},\vec{k}',\sigma,\sigma'} C_{\vec{q}} q^\alpha \{b_{\vec{q}} \langle \vec{k}'|e^{i\vec{q}\vec{r}}|\vec{k}\rangle - b^+_{\vec{q}} \langle \vec{k}'|e^{-i\vec{q}\vec{r}}|\vec{k}\rangle\} a^+_{\vec{k}'\sigma'} a_{\vec{k}\sigma} \delta_{\sigma'\sigma}, \quad (5.72)$$

where $a^+_{\vec{k}\sigma}$, $a_{\vec{k}\sigma}$ are the Fermi creation and annihilation operators of electrons with a wave vector $\vec{k}$ and spin projection σ on the Z-axis ($\sigma = \pm 1/2$).

The averaging in the formula (5.69) is simultaneously performed independently over electron and phonon states, since the operator H_0 contains already no interaction between electrons and phonons. Substituting the result (5.72) into (5.69) and calculating quantum-statistical averages of products of the creation and annihilation operators of the electrons and phonons, we obtain

$$
\begin{aligned}
\frac{1}{\tau} = \frac{1}{6nm} \int_{-\infty}^{\infty} dt e^{-\epsilon|t|} \int_0^{\beta} d\lambda \sum_{\vec{q},\vec{k},\vec{k}',\sigma,\sigma'} |C_{\vec{q}}|^2 q^2 \{ &|\langle \vec{k}'| e^{i\vec{q}\vec{r}} |\vec{k}\rangle|^2 \\
\times (N_{\vec{q}} + 1) f_{\vec{k}'} (1 - f_{\vec{k}}) &e^{i/\hbar(\varepsilon_{\vec{k}} - \varepsilon_{\vec{k}'} + \hbar\Omega_{\vec{q}})(t + i\hbar\lambda)} \\
+ |\langle \vec{k}'| e^{-i\vec{q}\vec{r}} |\vec{k}\rangle|^2 N_{\vec{q}} f_{\vec{k}'} (1 - f_{\vec{k}}) &e^{i/\hbar(\varepsilon_{\vec{k}} - \varepsilon_{\vec{k}'} - \hbar\Omega_{\vec{q}})(t + i\hbar\lambda)} \}.
\end{aligned}
\tag{5.73}
$$

In deriving this expression, we have used the formulas

$$
\begin{aligned}
\langle a^+_{\vec{k}'} a_{\vec{k}} \rangle &= f_{\vec{k}} \delta_{\vec{k}\vec{k}'}, & \langle a_{\vec{k}'} a^+_{\vec{k}} \rangle &= (1 - f_{\vec{k}}) \delta_{\vec{k}\vec{k}'}; \\
\langle a_{\vec{k}'} a_{\vec{k}} \rangle &= 0, & \langle a^+_{\vec{k}'} a^+_{\vec{k}} \rangle &= 0; \\
a^+_{\vec{k}}(t) &= a^+_{\vec{k}} e^{i/\hbar \varepsilon_{\vec{k}} t}, & a_{\vec{k}}(t) &= a_{\vec{k}} e^{-i/\hbar \varepsilon_{\vec{k}} t},
\end{aligned}
\tag{5.74}
$$

and the Wick–Bloch–de Dominicis statistical theorem [38], according to which an average value of an arbitrary number of creation and annihilation operators of fermions for systems with Hamiltonian represented as

$$
H_0 = \sum_{\vec{k}\sigma} \varepsilon_{\vec{k}\sigma} a^+_{\vec{k}\sigma} a_{\vec{k}\sigma}
$$

equals the sum of all possible complete pairing systems of this product. Pairing of the operators A_1 and A_2 implies the average value of these operators:

$$
\langle \overbrace{A_1 \ldots A_2} \rangle = \langle A_1 A_2 \rangle \langle \ldots \rangle = \mathrm{Sp}\{A_1 A_2 \rho_0\} \, \mathrm{Sp}\{\ldots \rho_0\}.
$$

Pairings, if there is no single unpaired operator, are called *a complete pairing system*. Then, in the case of the Fermi statistics, the sign $(-1)^P$ is ascribed to the product of average pair values of the creation or annihilation operators. Here, P is the number of permutations of the creation/annihilation operators when the latter are transferred from the original position into a given one.

According to this theorem, the product of four fermionic operators can be represented as

$$
\begin{aligned}
\langle \underbrace{\overbrace{a^+_{\vec{k}'\sigma'} a_{\vec{k}\sigma}} \; \overbrace{a^+_{\vec{\mu}'\rho'} a_{\vec{\mu}\rho}}} \rangle = {} & f_{\vec{k}\sigma} \delta_{\vec{k}\vec{k}'} \delta_{\sigma'\sigma} f_{\vec{\mu}\rho} \delta_{\vec{\mu}\vec{\mu}'} \delta_{\rho\rho'} \\
& + f_{\vec{k}'\sigma'} \delta_{\vec{k}'\vec{\mu}} \delta_{\sigma'\rho} (1 - f_{\vec{k}\sigma}) \delta_{\vec{k}\vec{\mu}'} \delta_{\sigma\rho'}.
\end{aligned}
\tag{5.75}
$$

On the right-hand side of (5.75), the first term, corresponding to the upper lines of the pairing of the creation/annihilation operators, does not contribute, because of defining the interaction operator in such a way that it should have no diagonal matrix elements in the representation of H_0. For this reason, the expression for $1/\tau$ involves only the second term on the right side of (5.75). Here, the second term corresponds to the pairing lines, depicted below.

It is not hard to calculate the frequency of the electron momentum relaxation by the formula (5.73). We first perform integration over t and λ. Using the definition of the delta-function

$$\delta(x) = \frac{1}{2\pi}\int_{-\infty}^{\infty} dt e^{-\epsilon|t|} e^{ixt} = \frac{1}{2\pi i}\left\{\frac{1}{x - i\varepsilon} - \frac{1}{x + i\varepsilon}\right\},$$

we write the first integral in the parentheses of (5.73) in the following form:

$$\begin{aligned}
&\int_{-\infty}^{\infty} dt e^{-\epsilon|t|} \int_0^{\beta} d\lambda e^{i/\hbar(\varepsilon_{\vec{k}} - \varepsilon_{\vec{k}'} + \hbar\Omega_{\vec{q}})(t + i\hbar\lambda)} \\
&= 2\pi\hbar \frac{1 - \exp\{-\beta(\varepsilon_{\vec{k}} - \varepsilon_{\vec{k}'} + \hbar\Omega_{\vec{q}})\}}{\varepsilon_{\vec{k}} - \varepsilon_{\vec{k}'} + \hbar\Omega_{\vec{q}}} \delta(\varepsilon_{\vec{k}} - \varepsilon_{\vec{k}'} + \hbar\Omega_{\vec{q}}) \\
&= 2\pi\hbar\beta\delta(\varepsilon_{\vec{k}} - \varepsilon_{\vec{k}'} + \hbar\Omega_{\vec{q}}).
\end{aligned} \tag{5.76}$$

If we transform in a similar way the second integral term of the formula (5.73) and replace the summation indices $\vec{k}' \to \vec{k}$, $\vec{k} \to \vec{k}'$, we obtain

$$\begin{aligned}
\frac{1}{\tau} = \frac{\beta}{6nm}\frac{2\pi}{\hbar} \sum_{\vec{q}\vec{k}\vec{k}'\sigma\sigma'} & |C_{\vec{q}}|^2(\hbar q)^2 |\langle\vec{k}'|e^{i\vec{q}\vec{r}}|\vec{k}\rangle|^2 \{(N_{\vec{q}} + 1) \\
& \times f_{\vec{k}'}(1 - f_{\vec{k}}) + N_{\vec{q}} f_{\vec{k}}(1 - f_{\vec{k}'})\}\delta(\varepsilon_{\vec{k}} - \varepsilon_{\vec{k}'} + \hbar\Omega_{\vec{q}}).
\end{aligned} \tag{5.77}$$

The full identity of the formulas (5.77) and (4.160) is directly proved by the quasi-momentum conservation law

$$\langle\vec{k}'|e^{i\vec{q}\vec{r}}|\vec{k}\rangle = \delta_{\vec{k}',\vec{k}+\vec{q}}$$

and by (4.158), according to which

$$(N_{\vec{q}} + 1) f_{\vec{k}'}(1 - f_{\vec{k}}) = N_{\vec{q}} f_{\vec{k}}(1 - f_{\vec{k}'}),$$

provided that the law of conservation of energy works in the system:

$$\varepsilon_{\vec{k}} - \varepsilon_{\vec{k}'} + \hbar\Omega_{\vec{q}} = 0.$$

Given the above, we may write down an expression for the inverse momentum relaxation time as follows:

$$\frac{1}{\tau} = \frac{2\pi}{\hbar}\frac{\beta}{3nm} \sum_{\vec{k}\sigma,\vec{q}} (\hbar\vec{q})^2 |C_{\vec{q}}|^2 N_{\vec{q}} f_{\vec{k}}(1 - f_{\vec{k}+\vec{q}})\delta(\varepsilon_{\vec{k}} - \varepsilon_{\vec{k}+\vec{q}} + \hbar\Omega_{\vec{q}}). \tag{5.78}$$

This expression is completely equivalent to the result (4.160) derived previously in Chapter 4 for the inverse relaxation time of the average momentum of equilibrium electrons. Some discrepancy in the notations should not lead to confusion since, as already claimed, the notation of $f_{\vec{k}}$, used in this chapter, means an equilibrium distribution function. In addition, we have omitted the phonon polarization index λ, assuming at once that the scattering occurs by the longitudinal acoustic phonons.

Thus, the example of calculating the electrical conductivity shows that there is no difference in results obtained by using the kinetic equation and the theory of linear response to an external mechanical disturbance. The Kubo method, however, is more general in the sense that the formal expression for the kinetic coefficients of the type (5.43) also retains its meaning in quantizing magnetic fields and does not contain any assumptions about the form of an electron spectrum and about the structure of the carrier–scatterer interaction Hamiltonian.

Moreover, in physical kinetics there are some problems, which are difficult to solve by using the method of kinetic equations. However, the theory of linear response to an external disturbance allows one to rather easily get results which agree closely with experiment.

The calculation of components of the dynamic paramagnetic susceptibility tensor for an electron gas and determination of a relaxation time of transverse components of spin paramagnetic susceptibility of conduction electrons may serve as a case in point of such a problem. This problem is to be discussed in the following section.

5.1.4 High-frequency magnetic susceptibility

Let a system of electrons be in an external magnetic field $\vec{H} \parallel Z$. We assume that the amplitude of the field is sufficiently small so that the quantization of the orbital motion does not occur. If a weak radiofrequency field acts on the system besides the static magnetic field, and this field is polarized in a plane perpendicular to the Z-axis, the Hamiltonian of the system can be written as

$$H(t) = H + H_F(t), \quad H = H_e + H_s + H_p + H_{ep}, \quad H_e = \frac{p^2}{2m},$$
$$H_s = -g\mu_B S^z |\vec{H}|, \quad H_F(t) = -g\mu_B \vec{S}\vec{h}(t), \quad S^\alpha = \sum_j s_j^\alpha, \tag{5.79}$$

where s_j^α is the α-projection of the spin operator of the j-th electron, g is the effective spectroscopic splitting factor for conduction electrons, μ_B is the Bohr magneton, $\vec{S}$ is the operator of the total electron spin.

The Hamiltonian H_{ep} describes an interaction of electrons with scatterers; H_p is the scatterer Hamiltonian. The explicit form of these operators we pay here no attention to; however, it is worth pointing out that the operator H_{ep}, in contrast to the operator of the electron–phonon interaction, must contain summands proportional to the components of the electron spin operator. In the case of the electron–phonon and electron–impurity interaction, the above structure of H_{ep} appears by considering the spin–orbit contribution when the electrons and scatterers collide.

Using the theory of linear response of the system to an external field, developed in Section 5.1.2, we write an expression for the projection of the average magnetic moment of electrons $\langle \vec{M} \rangle^t \equiv \vec{m}(t)$:

$$m^{\alpha}(\omega) = \frac{(g\mu_B)^2}{i\hbar} \int_{-\infty}^{0} dt_1 e^{(\epsilon - i\omega)t_1} \operatorname{Sp}\{S^{\alpha} e^{iLt_1} [\rho_0, S^{\beta}]\} h^{\beta}(\omega). \tag{5.80}$$

This result follows directly from (5.35), if one inserts into it the following equalities:

$$A = B = \vec{M} = g\mu_B \vec{S}, \quad F(\omega) = \vec{h}(\omega).$$

For further transformation of expression (5.80) one should use the Kubo identity (5.60), which now can be written as

$$\frac{1}{i\hbar}[\rho_0, S^{\beta}] = \beta \int_0^1 d\tau \dot{S}^{\beta}(i\hbar\beta\tau)\rho_0, \tag{5.81}$$

and introduce the convenient notation

$$(A, B) = \int_0^1 d\tau \operatorname{Sp}\{AB(i\hbar\beta\tau)\rho_0\}. \tag{5.82}$$

As a result, one gets

$$m^{\alpha}(\omega) = \beta(g\mu_B)^2 \int_{-\infty}^{0} dt_1 e^{(\epsilon - i\omega)t_1} (S^{\alpha}, \dot{S}^{\beta}(t_1)) h^{\beta}(\omega); \tag{5.83}$$

$$\dot{S}^{\beta} = \frac{1}{i\hbar}[S^{\beta}, H].$$

Equation (5.83) actually defines the magnetic susceptibility tensor of the electron gas. The transverse tensor components of circular variables

$$m_{\pm} = m_x \pm i m_y; \quad h_{\pm} = h_x \pm i h_y$$

have the form

$$\chi_{\pm}(\omega) = \beta \frac{(g\mu_B)^2}{2} \int_{-\infty}^{0} dt_1 e^{(\epsilon - i\omega)t_1} (S^{+}, \dot{S}^{-}(t_1)). \tag{5.84}$$

As in the case of electrical conductivity, the components of the paramagnetic susceptibility tensor $\chi_{\pm}(\omega)$ can be expressed through a correlation function. To calculate it, the method of Green functions can be used.

Now, it would be proper, considering the formula (5.84), to introduce a Green function for calculating the transverse components of the magnetic susceptibility tensor,

$$G_{+-}(t_1) = \theta(-t_1)e^{\varepsilon t_1}(S^+, S^-(t_1)). \tag{5.85}$$

Then, using the equations of motion

$$\dot{S}^{\mp} = i\omega_s S^{\mp} + \dot{S}^{\mp}_{(l)}, \quad \omega_s = \frac{g\mu_B|\vec{H}|}{\hbar}, \quad \dot{S}^{\mp}_{(l)} = \frac{1}{i\hbar}[S^{\mp}, H_{ep}], \tag{5.86}$$

we can express the transverse components of the paramagnetic susceptibility tensor via the Green functions $G_{+-}(\omega)$ and $G_{1+-}(\omega)$:

$$\chi_{+-}(\omega) = \beta\frac{(g\mu_B)^2}{2}[G_{1+-}(\omega) + i\omega_s G_{+-}(\omega)], \tag{5.87}$$

where

$$G_{1+-}(\omega) = \int_{-\infty}^{\infty} dt_1 e^{-i\omega t_1} G_{1+-}(t_1),$$

$$G_{+-}(\omega) = \int_{-\infty}^{\infty} dt_1 e^{-i\omega t_1} G_{+-}(t_1),$$

$$G_{1+-}(t_1) = \theta(-t_1)e^{\varepsilon t_1}(S^+, \dot{S}^-_{(l)}(t_1)). \tag{5.88}$$

The choice of the type of the Green function is not univocal. Given the expression (5.80), the transverse components of the magnetic susceptibility tensor for an electron gas can be expressed through the commutator Green function

$$\mathfrak{G}_{+-}(t_1) = \theta(-t_1)e^{\varepsilon t_1}\,\mathrm{Sp}\left\{S^+ e^{iLt_1}\frac{1}{i\hbar}[\rho_0, S^-]\right\}. \tag{5.89}$$

Fourier transformation of the Green function allows the transverse components of the paramagnetic susceptibility to be determined:

$$\chi_{+-}(\omega) = \frac{(g\mu_B)^2}{2}\mathfrak{G}_{+-}(\omega). \tag{5.90}$$

In virtue of the ambiguity, a question arises as to whether the final results will be the same if one takes different representations (5.85) and (5.89) for the Green function. As we shall convince ourselves later, the answer will be negative. When using the approximate methods for calculating the Green function, the expressions (5.85) and (5.89) give qualitatively different results for the magnetic susceptibility. However, the result must not depend on the type of the original Green function when performing exact computations.

We apply the mass-operator method developed in the previous section to find the components of the magnetic susceptibility $\chi_{+-}(\omega)$, taking into account the definitions (5.85) and (5.89). Also, we explain the difference in the results obtained.

We first use the definition (5.87). Setting up a chain of equations of motion for Green functions $G_{+-}(t_1)$ (5.85) and then passing on to a frequency representation, we arrive at

$$\begin{aligned} i(\omega - \omega_s + i\epsilon)G_{+-}(\omega) &= -(S^+, S^-) + G_{1+-}(\omega), \\ i(\omega - \omega_s + i\epsilon)G_{1+-}(\omega) &= -(S^+, \dot{S}^-_{(l)}) - G_{2+-}(\omega), \\ &\dots, \end{aligned} \tag{5.91}$$

$$G_{2+-}(\omega) = \int_{-\infty}^{0} dt_1 e^{(\epsilon - i\omega)t_1} (\dot{S}^+_{(l)}, \dot{S}^-_{(l)}(t_1)). \tag{5.92}$$

The set of equations (5.91) is similar to the formal system (5.52). Therefore, introducing the mass operator $M_{+-}(\omega)$ for the Green function $G_{+-}(\omega)$ by the relation

$$G_{+-}(\omega) = -\frac{(S^+, S^-)}{i(\omega - \omega_s + i\epsilon) - M_{+-}(\omega)} \tag{5.93}$$

and using the results (5.54), (5.55), one gets

$$G_{1+-}(\omega) = -\frac{(S^+, S^-)M_{+-}(\omega)}{i(\omega - \omega_s + i\epsilon) - M_{+-}(\omega)}, \tag{5.94}$$

$$M_{+-}(\omega) = \frac{(S^+, \dot{S}^-_{(l)})}{(S^+, S^-)} + \frac{G_{2+-}(\omega)}{(S^+, S^-)} + O(H^2_{ep}). \tag{5.95}$$

The last term on the right-hand side of (5.95) stands for terms, containing an electron–scatterer coupling constant of power higher than the second. Therefore, when calculating the mass operator in the Born approximation of scattering theory, we take into account only the first two terms in the formula (5.95), the last term can be dropped.

Substituting the expressions (5.93) and (5.94) into the formula (5.87) for the transverse components of the paramagnetic susceptibility tensor of conduction electrons, we obtain

$$\chi_{+-}(\omega) = \beta \frac{(g\mu_B)^2}{2} \frac{(S^+, S^-)[i\omega_s + M_{+-}(\omega)]}{i(\omega_s - \omega - i\epsilon) + M_{+-}(\omega)}. \tag{5.96}$$

To interpret the meaning of the above presentation, one should find the paramagnetic susceptibility by using phenomenological equations proposed by Bloch in 1946 to describe the motion of the magnetic moment. The set of the equations, written in a Cartesian coordinate system, is

$$\frac{d}{dt} m_x = \frac{g\mu_B}{\hbar} [\vec{m} \times \vec{\mathcal{H}}]_x - \frac{\delta m_x}{T_2}, \quad \vec{\mathcal{H}} = \vec{H} + \vec{h},$$

$$\frac{d}{dt}m_y = \frac{g\mu_B}{\hbar}[\vec{m}\times\vec{\mathcal{H}}]_y - \frac{\delta m_y}{T_2},$$
$$\frac{d}{dt}m_z = \frac{g\mu_B}{\hbar}[\vec{m}\times\vec{\mathcal{H}}]_z - \frac{\delta m_z}{T_1}, \quad \delta\vec{m} = \vec{m} - \chi_0\vec{\mathcal{H}}, \tag{5.97}$$

where χ_0 is the static magnetic susceptibility of a system. The quantities T_1 and T_2 are the relaxation times of longitudinal and transverse components of the spin magnetization, respectively; $\vec{m}$ is the total magnetic moment vector of the system. Assuming that the geometry of the external fields remains the same, we proceed to address the cyclic dynamical variables. To do so, we have to multiply the second equation of the set (5.97) by an imaginary unit. Then the first and the second equations should be added. Consequently, one gets an equation containing only one component of magnetization, m_+:

$$\frac{d}{dt}m_+ = \frac{g\mu_B}{\hbar}\{m_+H - m_zh_+\} - \frac{m_+ - \chi_0 h_+}{T_2}. \tag{5.98}$$

Indeed, the quantity m_z involved in the formula (5.98), when $h_+ \ll H$ ($H = |\vec{H}|$), can be written as

$$m_z = \chi_0 H.$$

Performing the Fourier transform of the equation (5.98), we obtain

$$-i(\omega - \omega_S)m_+(\omega) = i\omega_S\chi_0 h_+ - \frac{m_+}{T_2} + \frac{\chi_0 h_+}{T_2}. \tag{5.99}$$

Hence, using the definition of magnetic susceptibility

$$m_+(\omega) = \chi_{+-}(\omega)h_+(\omega),$$

we find

$$\chi_{+-}(\omega) = \frac{\chi_0(i\omega_s + 1/T_2)}{i(\omega_s - \omega) + 1/T_2}. \tag{5.100}$$

The expression (5.100) for transverse magnetic susceptibility, obtained from the Bloch phenomenological equations, has the same structure as the formula (5.96), if the following quantity is playing the role of static susceptibility:

$$\chi_0 = \beta\frac{(g\mu_B)^2}{2}(S^+, S^-), \tag{5.101}$$

and the real part of the mass operator the role as the frequency-dependent inverse relaxation time $\nu_2(\omega)$,

$$\frac{1}{T_2} = \nu_2(\omega) = \operatorname{Re} M_{+-}(\omega).$$

The imaginary part of the mass operator in this case describes the frequency shift of the Zeeman precession $\delta\omega_s$ due to interaction with scatterers,

$$\delta\omega_s = \operatorname{Im} M_{+-}(\omega).$$

It is interesting that the results (5.96), (5.100) hold under conditions of high and low frequencies ω. In the limit of low frequencies $\omega \to 0$, the high-frequency paramagnetic susceptibility $\chi_{+-}(\omega)$ is altered for static susceptibility χ_0. This limiting transition to the small frequency range is infringed if the commutator Green function (5.89) is used while calculating the paramagnetic susceptibility.

Indeed, making up a chain of equations of motion for the commutator Green functions in the frequency representation, one gets

$$\begin{aligned} i(\omega - \omega_s + i\epsilon)\mathfrak{G}_{+-}(\omega) &= \frac{2}{i\hbar}\langle S^z\rangle + \mathfrak{G}_{1+-}(\omega), \\ i(\omega - \omega_s + i\epsilon)\mathfrak{G}_{1+-}(\omega) &= \frac{1}{i\hbar}\langle[\dot{S}^-_{(l)}, S^+]\rangle - \mathfrak{G}_{2+-}(\omega), \end{aligned} \tag{5.102}$$

$$\begin{aligned} \mathfrak{G}_{1+-}(\omega) &= \int_{-\infty}^{0} dt_1 e^{(\epsilon - i\omega)t_1} \frac{1}{i\hbar}\langle[\dot{S}^-_{(l)}(t_1), S^+]\rangle, \\ \mathfrak{G}_{2+-}(\omega) &= \int_{-\infty}^{0} dt_1 e^{(\epsilon - i\omega)t_1} \frac{1}{i\hbar}\langle[\dot{S}^-_{(l)}(t_1), \dot{S}^+_{(l)}]\rangle. \end{aligned} \tag{5.103}$$

The solution of the set of equations (5.102) in the Born approximation of scattering theory for the mass operator $\mathfrak{M}_{+-}(\omega)$ has the form

$$\mathfrak{G}_{+-}(\omega) = \frac{1}{i\hbar}\frac{2\langle S^z\rangle}{i(\omega - \omega_s + i\epsilon) - \mathfrak{M}_{+-}(\omega)}, \tag{5.104}$$

$$\mathfrak{M}_{+-}(\omega) = \frac{\langle[\dot{S}^-_{(l)}, S^+]\rangle}{2\langle S^z\rangle} + i\hbar\frac{\mathfrak{G}_{2+-}(\omega)}{2\langle S^z\rangle}. \tag{5.105}$$

Substituting the result (5.104) into the formula (5.90) for the transverse components of the magnetic susceptibility tensor, we obtain

$$\chi_{+-}(\omega) = \frac{i}{\hbar}(g\mu_B)^2\frac{\langle S^z\rangle}{i(\omega_s - \omega - i\epsilon) + \mathfrak{M}_{+-}(\omega)}. \tag{5.106}$$

To compare the results (5.106) and (5.96) one must consider that for the zeroth-order interaction

$$\langle S^z\rangle = \frac{\beta\hbar\omega_s}{2}(S^+, S^-). \tag{5.107}$$

Thus, in the limit $\omega \to 0$, the expression $\chi_{+-}(\omega)$ which is determined by using the commutator Green function is not altered for the static susceptibility χ_0. The result (5.106) for the susceptibility $\chi_{+-}(\omega)$ can be obtained from the phenomenological system of the Bloch equations (5.97), preliminary mutilating the structure of the relaxation terms. It is easy to see that the replacement

$$\frac{m_+ - \chi_0 h_+}{T_2} \to \frac{m_+}{T_2}$$

in the equation (5.98) leads at the first onset to expression $\chi_{+-}(\omega)$. The structure of this expression is the same as well as the structure of the formula (5.106). This fact allows one to better understand the difference between the results (5.106) and (5.96).

The magnetic susceptibility, defined by (5.96), corresponds to the case when the relaxation of the magnetic moment m_+ of the system tends to an equilibrium value of the magnetic moment $\chi_0 h_+$ in an alternating magnetic field. In addition, the formula (5.106) describes the magnetic moment relaxation to zero of the transverse magnetization. For this reason, the result (5.106) is valid only for high frequencies $\omega \sim \omega_s \gg \nu_2$, when the magnetic moment lags behind the field. Then one can infer that the relaxation of the magnetic moment tends to zero.

In conclusion, we prove that the relation (5.107) holds and we can calculate the static susceptibility χ_0.

It would be proper to start with the definition of the correlation function

$$(S^+, S^-) = \int_0^1 d\tau \, \mathrm{Sp}\{S^+ \rho_0^\tau S^- \rho_0^{1-\tau}\}$$
$$= \int_0^1 d\tau \, \mathrm{Sp}\{S^+ e^{\beta\hbar\omega_s S^z \tau} S^- e^{-\beta\hbar\omega_s S^z(1-\tau)}\}. \tag{5.108}$$

Using the commutation relations for the components of the total spin operator

$$[S^\mp, S^z] = \pm S^\mp,$$

one gets the useful relation

$$S^\mp e^{\beta H_s \tau} = e^{\beta(H_s \mp \hbar\omega_s)\tau} S^\mp, \tag{5.109}$$

by means of which the expression (5.108) can be transformed as follows:

$$(S^+, S^-) = \langle S^+ S^- \rangle \frac{1 - \exp\{-\beta\hbar\omega_s\}}{\beta\hbar\omega_s} = \frac{2}{\beta\hbar\omega_s}\langle S^z \rangle. \tag{5.110}$$

In deriving this relation we have again used the formula (5.109) by setting

$$e^{-\beta\hbar\omega_s}\langle S^+ S^- \rangle = \mathrm{Sp}\left\{S^+ S^- \frac{1}{Z} e^{-\beta H_s}\right\} e^{-\beta\hbar\omega_s}$$
$$= \mathrm{Sp}\left\{S^- \frac{1}{Z} e^{-\beta H_s} S^+\right\} e^{-\beta\hbar\omega_s} = \mathrm{Sp}\left\{S^- S^+ \frac{1}{Z} e^{-\beta H_s}\right\}, \quad Z = \mathrm{Sp}\{e^{-\beta H_s}\}$$

and the commutation relations for the operators S^+, S^-

$$[S^+, S^-] = 2S^z.$$

Now, it is not hard to calculate the static paramagnetic susceptibility of an electron gas. Using the definition of χ_0, we have

$$\begin{aligned}\chi_0 &= \frac{\beta(g\mu_B)^2}{2}(S^+, S^-) = \frac{g\mu_B}{H}\langle S^z\rangle, \\ \langle S^z\rangle &= \sum_{\vec{k}\sigma} s^z_\sigma \langle a^+_{\vec{k}\sigma} a_{\vec{k}\sigma}\rangle \\ &= \frac{1}{2}\sum_{\vec{k}}(\langle a^+_{\vec{k}\uparrow} a_{\vec{k}\uparrow}\rangle - \langle a^+_{\vec{k}\downarrow} a_{\vec{k}\downarrow}\rangle), \quad s^z_\sigma = \pm\frac{1}{2},\end{aligned} \tag{5.111}$$

where the arrows ↑ and ↓ denote the orientation of the spin moment with respect to the Z-axis. Using the formulas (5.74), we express averages of electron creation/annihilation operators via occupancy functions $f_{\vec{k}\uparrow}$ and $f_{\vec{k}\downarrow}$ for electrons with the spin oriented along a magnetic field and in the opposite direction, respectively. Next, if one puts the parameter $\beta\hbar\omega_s$ small, we can expand the distribution functions $f_{\vec{k}\uparrow}$ and $f_{\vec{k}\downarrow}$ over this parameter with accuracy up to linear terms, then, going over from the summation over the wave vector $\vec{k}$ to integration over energy we arrive at the standard formula for the static paramagnetic susceptibility of the electron gas:

$$\chi_0 = \frac{2^{1/2}\mu_B^2 m^{3/2}}{\pi^2\hbar^3}\int_0^\infty \varepsilon^{1/2}\left(-\frac{\partial f(\varepsilon-\zeta)}{\partial\varepsilon}\right)d\varepsilon. \tag{5.112}$$

To calculate the transverse components of the dynamic paramagnetic susceptibility of the electron gas completely, one is required, in general, to evaluate both the mass operator defined by the formula (5.95) and the relaxation frequency of the transverse components of the spin magnetization $\nu_2(\omega)$. In the second-order interaction H_{ep}, the quantity $\nu_2(\omega)$ is determined by the real part of the function $G_{2+-}(\omega)$ and can be easily calculated. In principle, there is no significant difference between calculations of the above quantities and of the inverse relaxation time of the average momentum of electrons in the previous section. The exception is that the former are cumbersome. Therefore, we do not give them here.

5.2 Electrical conductivity in a quantizing magnetic field

5.2.1 Charge and heat fluxes in a quantizing magnetic field

In the previous chapter, the theory of thermo-galvanomagnetic phenomena was based on the Boltzmann kinetic equation. However, when external conditions such as the temperature T of a specimen and the external magnetic field H change, the approach based on the kinetic equation becomes not applicable due to possible violation of the conditions of applicability of the quasiclassical description. As shown in Section 4.1.2, if the conditions $\hbar\omega_0 \gg k_B T$ and $\omega_0 \tau_{\vec{p}} \gg 1$ are met, then quantization of the orbital motion of electrons in the magnetic field and the appearance of discrete energy levels of the electrons (Landau levels) must be taken into account in constructing the theory of transport phenomena. The presence of the discrete spectrum of the electrons in the magnetic field gives rise to a number of features exhibited by thermodynamic and kinetic phenomena. For example, in the quantizing magnetic field there may be oscillations of both thermodynamic characteristics and thermo-galvanomagnetic coefficients when changing the external magnetic field. These oscillations are associated with the passage of the next Landau level through the Fermi level. More detailed information on this issue can be seen in the next section.

Besides these rather obvious differences between the quasiclassical and quantum theories of thermo-galvanomagnetic phenomena related to restructuring the spectrum of current carriers, there are real differences in determining fluxes of charge and heat. In the last chapter, charge and heat fluxes were defined by relations (4.27), (4.28). In quantum theory, analogs of these formulas are the definitions

$$\vec{J} = \mathrm{Sp}\{\hat{\vec{J}}\rho\}, \quad \vec{J}_E = \mathrm{Sp}\{\hat{\vec{J}}_E\rho\},$$
$$\hat{\vec{J}} = \frac{e}{2}(\hat{\vec{v}}\hat{N} + \hat{N}\hat{\vec{v}}), \quad \hat{\vec{J}}_E = \frac{1}{2e}(\hat{\vec{J}}\hat{H} + \hat{H}\hat{\vec{J}}), \tag{5.113}$$

where ρ is the statistical operator, $\hat{\vec{J}}$, $\hat{\vec{J}}_E$ the electric current and energy flux density operators, $\hat{\vec{v}}$ the current-carrier velocity operator, $\hat{H}$ the energy density operator.

In a quantizing magnetic field, these definitions, however, turn out to be incorrect. As far back as 1950 s Japanese physicists Kasuya and Nakajima ascertained that the flow of charge density and heat flux $\vec{J}_Q = \vec{J}_E - \zeta/e\vec{J}$, defined in such a manner, led to violation both of the Onsager symmetry relations and of the Einstein relation. In accordance with the latter, the coefficients, standing before a gradient of the electric potential and gradient of the chemical potential (divided by the electron charge) must be equal.

The actual cause of the violations of the Einstein relations was revealed in the work by Zyryanov and Silin. They showed that the current ($c \operatorname{rot} \vec{m}$) contributed to the bulk density of the charge flux in the quantizing magnetic field in the case of spatially inhomogeneous systems. The contribution is due to the dependence of paramagnetic and diamagnetic susceptibilities of an electron gas both on chemical potential and on temperature. Therefore, the conduction current $\vec{J}_c$ involved in calculating the transport coefficients should be properly determined by excluding from the formula of the charge flux density that part, which is not directly linked to the electrotransfer:

$$\vec{J}_c = \mathrm{Sp}\{\hat{\vec{J}}\rho\} - c \operatorname{rot} \vec{m}, \quad \vec{B} - \vec{H} = 4\pi\vec{m}. \tag{5.114}$$

The heat flux in the quantizing magnetic field also calls for a new definition because even in the spatially-homogeneous case, the Poynting vector contributes to the heat flux:

$$\frac{c}{4\pi}[\vec{E}\times(\vec{H}-\vec{B})],$$

which should be subtracted from the energy flux density to obtain a correct expression for the heat flux:

$$\vec{J}_Q = \vec{J}_E - \frac{\zeta}{e}\vec{J}_c - \frac{c}{4\pi}[\vec{E}\times(\vec{H}-\vec{B})]. \tag{5.115}$$

In spatial inhomogeneity, additional summands proportional to the spatial derivatives of the magnetization current $c\,\mathrm{rot}\,\vec{m}$ appear on the right side of (5.115).

To learn more both about the problem of determining the charge and heat fluxes in a quantizing magnetic field and about the problem of calculating the thermo-galvanomagnetic coefficients, one can refer to Refs. [39] and [40]. Necessary references to original works can be found there as well.

There is no possibility to fairly completely explicate a theory of thermo-galvanomagnetic phenomena in a quantizing magnetic field in the frame of the present course. So, we would like to dwell on the problem of calculating the diagonal and off-diagonal components of the electrical conductivity tensor, basing on the Kubo linear response theory. In this case, the components of the magnetization current density $c\,\mathrm{rot}\,\vec{m}$ vanish; consequently, it is worth returning to a common definition of the conduction current (5.113). This definition will be used in the future.

5.2.2 Dynamics of electron motion in a quantizing magnetic field

Consider motion of electrons in a crystal in an external magnetic field $\vec{H}$, parallel to the Z-axis. Let the magnetic field be given by the vector potential $A = \{-Hy, 0, 0\}$, $\vec{H} = \mathrm{rot}\vec{A}$. As is well known, forces acting on a particle in a magnetic field are not potential. However, in an electromagnetic field, a generalized velocity-dependent potential function can be introduced. For a classical system, the Lagrange function L for free-moving charged particles in the electromagnetic field can be written as

$$L = \frac{mv^2}{2} - e\varphi + \frac{e}{c}\vec{A}\vec{v}. \tag{5.116}$$

Now, we introduce *a generalized (canonical)* momentum $\vec{p}$ by using the relation

$$\vec{p} = \frac{\partial L}{\partial \vec{v}} = m\vec{v} + \frac{e}{c}\vec{A}. \tag{5.117}$$

As far as the energy of an electron in a magnetic field (without spin) is $mv^2/2$, but the Hamiltonian is the energy expressed in terms of a generalized momentum, for the

Hamiltonian $\hat{H}$ for an electron in a magnetic field, one obtains

$$\hat{H}_0 = \frac{(\vec{p} - e/c\vec{A})^2}{2m}, \tag{5.118}$$

where $\vec{p} = -i\hbar\vec{\nabla}$, since it is the canonical momentum that must be replaced by the operator $-i\hbar\vec{\nabla}$ while passing on to the quantum description. If one introduces the notion of *the kinetic momentum* $\vec{\mathfrak{p}} = \vec{p} - e/c\vec{A}$, the following expressions can appear:

$$\hat{H}_0 = \frac{(\mathfrak{p})^2}{2m}, \quad \mathfrak{p}_x = -i\hbar\nabla_x + \frac{e}{c}Hy, \quad \mathfrak{p}_y = -i\hbar\nabla_y, \quad \mathfrak{p}_z = -i\hbar\nabla_z. \tag{5.119}$$

Now we present the expressions for the spectrum and eigenfunctions of the Hamiltonian operator $\hat{H}$. Details of solving this problem can be found in guidebooks on quantum mechanics:

$$\hat{H}_0\psi_{np_zp_x} = \varepsilon_{np_z}\psi_{np_zp_x}, \quad \varepsilon_{n,p_z} = \frac{\hbar^2k_z^2}{2m} + \hbar\omega_0\left(n + \frac{1}{2}\right),$$
$$\psi_{np_zp_x} = (4\pi l)^{-1/2}e^{i/\hbar(p_zz+p_xx)}\Phi_n\left(\frac{y - y_0}{l}\right),$$
$$\Phi_n(x) = N_ne^{-x^2/2}H_n(x), \quad \omega_0 = \frac{eH}{mc}, \tag{5.120}$$

where $n = 0, 1, 2, \ldots$ is the number of the Landau level. Here, we have decided to keep the traditional notation for numbering the Landau levels, albeit before the concentration of electrons was denoted by that letter. $l = (\hbar c/eH)^{1/2}$ is the magnetic length, $y_0 = -c/(eH)p_x$, N_n is a normalization factor for the eigenfunction $\Phi_n(x)$ of a harmonic oscillator. As follows from the dispersion law (5.120), the motion in the direction Z remains quasi-free. Only the motion in the plane perpendicular to the magnetic field can be quantized.

We find commutation relations for the components of the kinetic momentum operator $\vec{\mathfrak{p}}$:

$$[\mathfrak{p}_x, \mathfrak{p}_y] = \left[p_x + \frac{eH}{c}y, p_y\right] = -\frac{eH}{c}[p_y, y] = i\frac{\hbar^2}{l^2}. \tag{5.121}$$

$$[\mathfrak{p}_x, \mathfrak{p}_z] = [\mathfrak{p}_y, \mathfrak{p}_z] = [\mathfrak{p}_x, y] = [\mathfrak{p}_y, x] = [\mathfrak{p}_y, z] = [\mathfrak{p}_z, x] = 0,$$
$$[\mathfrak{p}_x, x] = [\mathfrak{p}_y, y] = [\mathfrak{p}_z, z] = -i\hbar. \tag{5.122}$$

An interesting feature of the motion of electrons in the magnetic field is that it is possible to distinguish both slowly changing variables X, Y which are coordinates of the Larmor orbit center, and coordinates of the relative motion ξ, η. The former are quasi-integrals of the motion, they commute with the Hamiltonian $\hat{H}_0$.

$$X = x - \xi, \quad Y = y - \eta, \quad \xi = -\frac{c}{eH}\mathfrak{p}_y, \quad \eta = \frac{c}{eH}\mathfrak{p}_x. \tag{5.123}$$

It is easy to verify that the newly introduced quantities satisfy the commutation relations

$$[\xi, \eta] = il^2, \quad [X, Y] = il^2. \tag{5.124}$$

The rest of the commutators are equal to zero:

$$[\xi, X] = [\eta, X] = [\xi, Y] = [\eta, Y] = 0.$$

It follows from the basic principles of quantum mechanics that if two operators do not commute, it is impossible to simultaneously measure any physical quantities corresponding to them, moreover they satisfy the uncertainty principle. It implies that

$$\triangle X \triangle Y \sim l^2,$$

i. e. a position of the Larmor orbit center is quantized, and only one center can be placed in an area of the order πl^2.

Next, let us derive the equations of motion for the X and Y operators, given that the Hamiltonian has the form $\hat{H} = \hat{H}_0 + \hat{U}$. In the future, the operator of the electron–phonon or electron–impurity interactions will be used as the operator $\hat{U}$.

We first consider the equations of motion for the $\mathfrak{p}_x$ and $\mathfrak{p}_y$ components of the momentum:

$$\dot{\mathfrak{p}}_x = \frac{1}{i\hbar}[\mathfrak{p}_x, \hat{H}_0 + \hat{U}] = \frac{i}{\hbar m}[\mathfrak{p}_y, \mathfrak{p}_x]\mathfrak{p}_y + \frac{i}{\hbar}[\hat{U}, \mathfrak{p}_x].$$

Using the previous results (5.122), we obtain

$$\dot{\mathfrak{p}}_x = \frac{eH}{c}\frac{\partial \hat{H}_0}{\partial \mathfrak{p}_y} - \frac{\partial \hat{U}}{\partial x}. \tag{5.125}$$

In perfect analogy with the above result, the equation of motion for the operator $\mathfrak{p}_y$ can be obtained:

$$\dot{\mathfrak{p}}_y = -\frac{eH}{c}\frac{\partial \hat{H}_0}{\partial \mathfrak{p}_x} - \frac{\partial \hat{U}}{\partial y}. \tag{5.126}$$

Using the definition (5.123) and the above results (5.125), (5.126), we have

$$\begin{aligned}
\dot{X} &= \frac{1}{i\hbar}[x - \xi, \hat{H}_0 + \hat{U}] = \frac{1}{i\hbar}[x, \hat{H}_0] - \frac{1}{i\hbar}[\xi, \hat{H}_0] - \frac{1}{i\hbar}[\xi, \hat{U}] \\
&= -\frac{1}{i\hbar}[\xi, \hat{U}] = \frac{1}{i\hbar}[X, \hat{U}] = -\frac{c}{eH}\frac{\partial \hat{U}}{\partial y}, \\
\dot{Y} &= \frac{1}{i\hbar}[Y, \hat{U}] = \frac{c}{eH}\frac{\partial \hat{U}}{\partial x}.
\end{aligned} \tag{5.127}$$

Thus, the coordinates of the Larmor orbit center are changed only under the action of the perturbation potential $\hat{U}$. This allows one to regard the X and Y variables as slowly varying physical quantities. Consequently, the slowly varying variables in a quantizing magnetic field modify radically the method of describing the kinetic phenomena in the quantizing magnetic field.

5.2.3 The conductivity tensor in a quantizing magnetic field

In a quantizing magnetic field, conditions for applicability of a kinetic equation are infringed. Therefore, to analyze the electrical conductivity one should use the expression (5.43), which was obtained by using the theory of linear response to weak mechanical perturbation. By applying the Kubo formula (5.60) to transform the expression (5.43) and introducing the current operators instead of momenta of P^{α}, defining them by the expression $J_{\mu} = eP^{\mu}$, one gets

$$\sigma_{\mu\nu} = \int_{-\infty}^{0} dt_1 \int_{0}^{\beta} d\lambda e^{(\epsilon - i\omega)t_1} \operatorname{Sp}\{J_{\mu} J_{\nu}(t_1 + i\hbar\lambda)\rho_0\}. \tag{5.128}$$

Making the change of variables $t_1 \to -t_1$, we finally obtain

$$\sigma_{\mu\nu} = \int_{0}^{\infty} dt_1 \int_{0}^{\beta} d\lambda e^{(i\omega - \epsilon)t_1} \operatorname{Sp}\{J_{\mu}(t_1) J_{\nu}(i\hbar\lambda)\rho_0\}. \tag{5.129}$$

Now, we write down the current components J_{μ} and J_{ν} using the definition of both coordinates of the Larmor orbit center and coordinates of the relative motion. These coordinates will be later interpreted as total quantities for an entire system of electrons:

$$J_x = e(\dot{\xi} + \dot{X}), \quad J_y = e(\dot{\eta} + \dot{Y}).$$

To simplify the notations and to reduce formulas, we introduce the so-called Kubo scalar product of the two operators $(A, B(t))$ (see also formula (5.82)). The product is an even function of time in a quantum magnetic field if the A and B operators coincide. The proof of this remarkable fact can be found in Section 5.3.1. We use

$$(A, B(t)) = \int_{0}^{\beta} d\lambda \operatorname{Sp}\{AB(t + i\hbar\lambda)\rho_0\}. \tag{5.130}$$

Then, using the expression (5.129), the following integral for $\sigma_{xx}(0)$ at zero frequency can be derived:

$$\begin{aligned} \sigma_{xx}(0) &= \int_{0}^{\infty} dt_1 e^{-\epsilon t_1} (J_x(t_1), J_x) = \int_{0}^{\infty} dt_1 e^{-\epsilon t_1} (J_x, J_x(t_1)) \\ &= e^2 \int_{0}^{\infty} dt_1 e^{-\epsilon t_1} \{(\dot{\xi}, \dot{\xi}(t_1)) + (\dot{\xi}, \dot{X}(t_1)) + (\dot{X}, \dot{\xi}(t_1)) \\ &\quad + (\dot{X}, \dot{X}(t_1))\}. \end{aligned} \tag{5.131}$$

Now, one can show that all terms except the last one are zero. To prove this, consider the correlation function

$$\int_0^\infty dte^{-\epsilon t}(\dot{\xi}, \dot{\xi}(t)) = \int_0^\infty dte^{-\epsilon t}\frac{d}{dt}(\dot{\xi}, \xi(t))$$
$$= -(\dot{\xi}, \xi(0)) + \lim_{t\to\infty} e^{-\epsilon t}(\dot{\xi}, \xi(t)) + \epsilon \int_0^\infty dte^{-\epsilon t}(\dot{\xi}, \xi(t)). \quad (5.132)$$

In obtaining the expression (5.132) we have carried out integration by parts. By making use of the correlation weakening principle, according to which a correlation between two physical quantities, taken at time t_1 and t_2, is weakened as the time interval $\triangle t = t_1 - t_2$ is increased, we have

$$\lim_{t\to\infty} e^{-\epsilon t}(\dot{\xi}, \xi(t)) \to e^{-\epsilon t}\,\mathrm{Sp}\{\dot{\xi}\rho_0\}\,\mathrm{Sp}\{\xi(t)\rho_0\} \to 0. \quad (5.133)$$

To transform the last expression in the formula (5.132) we use Abel's theorem, according to which in the thermodynamic limit, the following equality is valid:

$$\lim_{\epsilon\to+0} \epsilon \int_0^\infty dte^{-\epsilon t}(\dot{\xi}, \xi(t)) = \lim_{t\to\infty}(\dot{\xi}, \xi(t)) = \mathrm{Sp}\{\dot{\xi}\rho_0\}\,\mathrm{Sp}\{\xi(t)\rho_0\} = 0. \quad (5.134)$$

Finally, we turn to the second line of the formula (5.132) to transform the first summand. Upon making the change of the integration variable $\tau = \lambda/\beta$ in the scalar product formula (5.130), we have the following expression:

$$-(\dot{\xi}, \xi) = -\beta \int_0^1 d\tau\,\mathrm{Sp}\{\dot{\xi}\rho_0^\tau \xi \rho_0^{1-\tau}\} = -\beta \int_0^1 d\tau\,\mathrm{Sp}\{\xi\rho_0^{1-\tau}\dot{\xi}\rho_0^\tau\}$$
$$= -\beta \int_0^1 d\tau'\,\mathrm{Sp}\{\xi\rho_0^{\tau'}\dot{\xi}\rho_0^{1-\tau'}\}$$
$$= -\frac{1}{i\hbar}\mathrm{Sp}\{\xi[\rho_0, \xi]\} = -\frac{1}{i\hbar}\mathrm{Sp}\{[\xi, \xi]\rho_0\} = 0. \quad (5.135)$$

In obtaining the result (5.135), we have made the change of the variables $\tau' = 1 - \tau$ and used the Kubo formula (5.60). In perfect analogy, we can prove that

$$\int_0^\infty dte^{-\epsilon t}(\dot{\xi}, \dot{\eta}(t)) = -(\dot{\xi}, \eta) = \frac{1}{i\hbar}\mathrm{Sp}\{[\xi, \eta]\rho_0\} = \frac{nc}{eH}; \quad (5.136)$$

$$\int_0^\infty dte^{-\epsilon t}(\dot{X}, \dot{\xi}(t)) = 0; \quad \int_0^\infty dte^{-\epsilon t}(\dot{Y}, \dot{\eta}(t)) = 0. \quad (5.137)$$

The vanishing of the correlation functions in (5.137) is caused by the fact that the $[X, \xi]$ and $[Y, \eta]$ commutators are zero.

Now, it is necessary to return to (5.131) for writing the result for $\sigma_{xy}(0)$ as well as for the other components of the electrical conductivity in the quantizing magnetic field. Given the results of (5.134)–(5.137), one gets

$$\sigma_{xy} = \frac{enc}{H} + e^2 \int_0^\infty dt_1 e^{-\epsilon t_1} (\dot{X}(t_1), \dot{Y}); \tag{5.138}$$

$$\sigma_{xx} = e^2 \int_0^\infty dt_1 e^{-\epsilon t_1} (\dot{X}, \dot{X}(t_1)); \tag{5.139}$$

$$\sigma_{yy} = e^2 \int_0^\infty dt_1 e^{-\epsilon t_1} (\dot{Y}, \dot{Y}(t_1)). \tag{5.140}$$

Having analyzed the results obtained, it is easy to see that in a quantizing magnetic field, the σ_{xx} and σ_{yy} diagonal components are different from zero only due to scattering events, since they are proportional at least to the square of the electron–scatterer coupling constant, which follows from the equations of motion (5.127). The off-diagonal component σ_{xy} contains the collisionless contribution enc/H, independent of the scattering events and the correction which is quadratic in the electron–scatterer coupling constant. It is important to note that the components of the electrical conductivity tensor in a quantizing magnetic field are expressed in terms of correlation functions of the coordinates of the Larmor orbit centers. The functions mentioned above can be directly calculated in the Born approximation of scattering theory.

5.2.4 The conductivity in the case quasi-elastic scattering by phonons

Consider the calculation of the σ_{xx} and σ_{yy} components in the case of quasi-elastic scattering by phonons, assuming that the operator $\hat{U}$ in formulas (5.127) is the Hamiltonian of the electron–phonon interaction (4.76).

To better understand the previous results for the components of the electrical conductivity tensor in the quantizing magnetic field, it is useful to compare them with results obtained by means of the kinetic equation method in the limit of a strong $\omega_0 \tau_{\vec{p}} \gg 1$ (but nonquantizing) magnetic field. On the basis of formulas (4.118), (4.121) and (4.128), one gets

$$\sigma_{xy} = \frac{e^2 n}{m} \frac{\omega_0 \tau_{\vec{p}}^2}{1 + (\omega_0 \tau_{\vec{p}})^2} \simeq \frac{enc}{H}; \tag{5.141}$$

$$\sigma_{xx} = \frac{e^2 n}{m} \frac{\tau_{\vec{p}}}{1 + (\omega_0 \tau_{\vec{p}})^2} \simeq \frac{e^2 n}{m \omega_0^2 \tau_{\vec{p}}}. \tag{5.142}$$

Thus, in the limit of strong magnetic field, if there remains only the zero term after expanding the denominator of (5.141) over the small parameter $1/(\omega_0\tau_{\vec{p}})^2$, the kinetic equation for the off-diagonal component of the electrical conductivity gives the same collisionless contribution as the formula (5.138).

The expression for the diagonal component (5.142) allows one to at least in a formal way determine a momentum relaxation time in a quantizing magnetic field. Indeed, upon comparing the two expressions (5.142) and (5.139), we obtain the definition for the momentum relaxation time in the quantizing magnetic field:

$$\frac{1}{\tau_{\vec{p}}} = \frac{m\omega_0^2}{n}\int_0^\infty dt_1 e^{-\epsilon t_1}(\dot{X}, \dot{X}(t_1)). \tag{5.143}$$

The problem of calculating the σ_{xx} and σ_{xy} components of the electrical conductivity tensor in the Born approximation in scattering theory essentially reduced to quadratures. Because an interaction in the statistical operator and the operator of the evolution is neglected in this approximation, these operators have only the diagonal matrix elements in the class of eigenfunctions $|\nu\rangle$ of $\hat{H}_0$. Since the correlation function $(A, B(t))$ is an even function of argument t, the integration over the variable t_1 in the integral (5.143) can be extended to $-\infty$, which results in symmetric limits. Then the integration over time t_1 yields a δ-function.

Furthermore, the quantum-statistical average over the electronic and phonon variables can be presented in the form

$$\begin{aligned}\mathrm{Sp}\{\dot{X}\dot{X}(t)\rho_0\} = &\sum_{\nu\nu'}\langle\langle\nu'|\dot{X}|\nu\rangle\langle\nu|\dot{X}(t)|\nu'\rangle\rangle_s \\ &\times f_{\nu'}(1-f_\nu)e^{i/\hbar(\varepsilon'_\nu-\varepsilon_\nu)}.\end{aligned} \tag{5.144}$$

In the formula (5.144), angular brackets, labeled by s, mean the quantum-statistical averaging over states of scatterers. In deriving this formula we have used the Wick–Bloch–de Dominicis statistical theorem (5.75).

Finally, we can prove that (it is proposed the reader carries out the proof)

$$\int_{-\infty}^{\infty} dt e^{-\epsilon|t|}\int_0^\beta d\lambda\,\mathrm{Sp}\{\dot{X}\dot{X}(t+i\hbar\lambda)\rho_0\} = \beta\int_{-\infty}^{\infty} dt e^{-\epsilon|t|}\,\mathrm{Sp}\{\dot{X}\dot{X}(t)\rho_0\}. \tag{5.145}$$

Given the above, the expression for the component σ_{xx} can be represented in the form

$$\sigma_{xx} = \frac{e^2}{2k_BT}\int_{-\infty}^{\infty} dt e^{-\epsilon|t|}\int_{-\infty}^{\infty} dE\, f(E)\langle\delta(E-\hat{H}_0)\dot{X}(1-f(\hat{H}_0))\dot{X}(t)\rangle. \tag{5.146}$$

In this formula, the big angular brackets $\langle\ldots\rangle$ stand for a quantum-statistical average over phonon variables and quantum-mechanical average over one-particle electron states.

We write down the $\dot{X}$ and $\dot{X}(t)$ operators explicitly. Considering that the coordinate operators of the Larmor orbit center commute with the Hamiltonian $\hat{H}_0$, we obtain

$$\dot{X} = \frac{-il^2}{\hbar}\sum_q q_y\{C_{\vec{q}} b_{\vec{q}} e^{i\vec{q}\vec{r}} - C^*_{\vec{q}} b^+_{\vec{q}} e^{-i\vec{q}\vec{r}}\};$$

$$\dot{X}(t) = \frac{-il^2}{\hbar} e^{i/\hbar \hat{H}_0 t}\sum_q q_y\{C_{\vec{q}} b_{\vec{q}} e^{i\vec{q}\vec{r}} e^{-i\Omega_{\vec{q}} t} - C^*_{\vec{q}} b^+_{\vec{q}} e^{-i\vec{q}\vec{r}} e^{i\Omega_{\vec{q}} t}\} e^{-i/\hbar \hat{H}_0 t}. \tag{5.147}$$

Substituting these expressions into (5.146):

$$\begin{aligned}\sigma_{xx} = \frac{e^2\pi\hbar}{k_B T}\int_{-\infty}^{\infty} dE f(E)\sum_{\vec{q}} \frac{l^4}{\hbar^2} q_y^2 |C_{\vec{q}}|^2\{(N_{\vec{q}}+1)\\ \times \mathrm{Sp}\{\delta(E-\hat{H}_0)e^{i\vec{q}\vec{r}}(1-f(\hat{H}_0))\delta(E-\hat{H}_0-\hbar\Omega_{\vec{q}})e^{-i\vec{q}\vec{r}}\}\\ + N_{\vec{q}}\,\mathrm{Sp}\{\delta(E-\hat{H}_0)e^{-i\vec{q}\vec{r}}(1-f(\hat{H}_0))\delta(E-\hat{H}_0+\hbar\Omega_{\vec{q}})e^{i\vec{q}\vec{r}}\}\}.\end{aligned} \tag{5.148}$$

The spur in the formula (5.148) implies summation over a complete set of quantum numbers $\nu = \{n, p_x, p_z, \sigma\}$, which characterize the state of an electron in a quantizing magnetic field.

Now, to transform the expression (5.148) one should take into account the two identities

$$\begin{aligned}N_{\vec{q}}[f(E-\hbar\Omega_{\vec{q}}) - f(E)] &= f(E)[1-f(E-\hbar\Omega_{\vec{q}})];\\ -[N_{\vec{q}}+1][f(E+\hbar\Omega_{\vec{q}}) - f(E)] &= f(E)[1-f(E+\hbar\Omega_{\vec{q}})],\end{aligned} \tag{5.149}$$

which are checked by direct substitution of distribution functions. Given these relations, the expression (5.148) can be rewritten in more convenient form for further changes:

$$\begin{aligned}\sigma_{xx} = \frac{e^2 2\pi}{\hbar}\int_{-\infty}^{\infty} dE\sum_{\vec{q}} l^4 q_y^2 |C_{\vec{q}}|^2 \frac{f(E-\hbar\Omega_{\vec{q}}) - f(E)}{\hbar\Omega_{\vec{q}}}\frac{\hbar\Omega_{\vec{q}}}{k_B T}\\ \times N_{\vec{q}}[N_{\vec{q}}+1]\,\mathrm{Sp}\{\delta(E-\hat{H}_0)e^{i\vec{q}\vec{r}}\delta(E-\hat{H}_0-\hbar\Omega_{\vec{q}})e^{-i\vec{q}\vec{r}}\}.\end{aligned} \tag{5.150}$$

In deriving the formula the change of variables $E + \hbar\Omega_{\vec{q}} \to E$ haves been made to transform the second summand of the expression (5.148).

The resulting expression is valid in the case of inelastic and quasi-elastic scattering by phonons. The elastic scattering by phonons is regarded below as an example. It is worth remarking that we restrict ourselves to the simplest case of the ultra quantum limit when only electrons of the lowest Landau subband with index $n = 0$ take part in the charge transfer. In this case, the matrix elements of the exponent operators in (5.148) can be easily calculated; as a consequence, we have

$$|\langle 0, p_z + \hbar q_z, p_x + \hbar q_x | e^{i\vec{q}\vec{r}} | 0, p_z, p_x\rangle|^2 = e^{l^2 q_\perp^2/2}; \quad q_\perp = q_x^2 + q_y^2.$$

In calculating the spur over of electronic variables, it is convenient to replace the sums by integrals:

$$\sum_{\nu\sigma} \to 2\frac{L_x L_z}{(2\pi\hbar)^2}\int dp_z \int dp_x \to \frac{2V}{(2\pi l)^2\hbar}\int_{-\infty}^{\infty} dp_z. \tag{5.151}$$

Next, it is necessary to take into account the degeneracy multiplicity of electronic states in the quantum number p_x in order to get the last result. To count the quantum number, the electron wave function (5.120) calls for imposing the circularity condition along X- and Z-axes. Put it another way, we require that one and the same function should correspond to the $x + L_x$, $z + L_z$ and x, z coordinates. When account is taken of the real form of the wave function (5.120), the above requirement leads to the condition

$$p_x = \frac{2\pi\hbar}{L_x}n_x, \quad p_z = \frac{2\pi\hbar}{L_z}n_z,$$

where n_x and n_z are some integers. The circularity condition will not be now imposed along the Y-axis but the solution (5.120) must exist only when the coordinate y_0 of the Larmor orbit center falls within:

$$0 < |y_0| < L_y, \tag{5.152}$$

where L_y is a specimen size within Y-axis. It is easy to verify that y_0 is one of the coordinates of the Larmor orbit center and $y_0 = Y$. Thus, the maximum value of the coordinate of the Larmor orbit center is $|y_0|^{\max} = L_y$. Since $|y_0| = c/(eH)p_x$, the maximum quantum number of p_x can be found: $p_x = \hbar/l^2 L_y$. Then the integral over p_x in the formula (5.151) is equal to $\hbar/l^2 L_y$ and we obtain the last result in this formula.

The summation over the phonon wave vector has to be also replaced by integration in a cylindrical coordinate system:

$$\sum_{\vec{q}} \to \frac{V}{(2\pi)^3}\int d\vec{q} \to \frac{V}{(2\pi)^2}\int_0^{\infty} q_\perp dq_\perp \int_{-\infty}^{\infty} dq_z. \tag{5.153}$$

Given the remarks made above, an expression for the static conductivity σ_{xx} in the case of quasi-elastic scattering may be written as follows:

$$\sigma_{xx} = \frac{2\pi e^2 l^4}{\hbar}\frac{1}{(2\pi)^2}\int_0^{\infty} q_\perp dq_\perp \int_{-\infty}^{\infty} dq_z \int_{-\infty}^{\infty} dE q_y^2 |C_{\vec{q}}|^2\left(-\frac{\partial f(E)}{\partial E}\right)$$
$$\times\left(\frac{k_B T}{\hbar\Omega_{\vec{q}}}\right)\frac{2}{(2\pi l)^2\hbar}\int_{-\infty}^{\infty} dp_z \delta(\varepsilon_{0p_z} - E)\delta(\varepsilon_{0p_{z+\hbar q_z}} - E)e^{-l^2 q_\perp^2/2}. \tag{5.154}$$

In order to transform the expression (5.154) further, let us consider the integral:

$$I = \int_{-\infty}^{\infty} dq_z \int_{-\infty}^{\infty} dp_z q^t \delta(\varepsilon_{0p_z} - E)\delta(\varepsilon_{0p_{z+\hbar q_z}} - E). \tag{5.155}$$

The law of conservation of energy implies that

$$\frac{(p_z + \hbar q_z)^2}{2m} = \frac{(p_z)^2}{2m}.$$

Therefore, phonons participating in scattering events have a quasi-momentum $\hbar q_z \simeq p_z$ what enables one to easily evaluate the longitudinal wave-vector component for the phonons:

$$q_z \simeq \frac{\sqrt{2mk_B T}}{\hbar} \sim \lambda^{-1} \simeq 10^7 \mathrm{m}^{-1}.$$

This evaluation holds at temperatures where an actual experiment runs. The perpendicular wave-vector component $q_\perp$ for the phonons involved in scattering events is limited by the cut-off factor:

$$e^{-l^2 q_\perp^2/2},$$

so one can assume that $q_\perp \simeq 1/l$. Moreover, the perpendicular wave-vector component for the phonons, participating in scattering, coincides in order of magnitude with the inverse magnetic length. Because the magnetic length l in the quantizing field is less than the electron wavelength λ, it is supposed that the condition $\lambda \gg l$ is satisfied. By virtue of the foregoing estimates, it follows that $q_\perp \gg q_z$ and $q \simeq q_\perp$. Thus, the integrals for I in the expression (5.155) can be simply calculated:

$$I = \int_{-\infty}^{\infty} dq_z \int_{-\infty}^{\infty} dp_z q^t \delta\left(\frac{p_z^2}{2m} + \frac{\hbar\omega_0}{2} - E\right) \delta\left(\frac{p_z \hbar q_z}{m} + \frac{\hbar^2 q_z^2}{2m}\right) = q_\perp^t \frac{2m}{\hbar} \frac{1}{E - \hbar\omega_0}. \quad (5.156)$$

While performing the integration of delta-functions we have the useful well-known formula

$$\delta(\varphi(x)) = \sum_i \delta(x - x_i) \left\{ \left| \frac{d\varphi}{dx} \right|_{x=x_i} \right\}^{-1}, \quad (5.157)$$

where x_i are roots of the equation $\varphi(x) = 0$. Substituting the result (5.156) into the formula (5.154), we obtain the following expression for the diagonal static electrical conductivity:

$$\sigma_{xx} = \frac{e^2 l^2 k_B T m C'}{4\pi^3 \hbar^4 s} \int_{-\infty}^{\infty} dE \left(-\frac{\partial f}{\partial E}\right) \frac{1}{E - \hbar\omega_0/2} \int_0^{\infty} dq_\perp^2 q_\perp^{t+1} e^{-l^2 q_\perp^2/2}. \quad (5.158)$$

In obtaining this result, we have used the notation

$$|C_{\vec{q}}|^2 = C' q^t, \quad C' = \frac{E_0^2 \hbar}{2\rho s}.$$

To simplify further calculations, let us consider only the case of a highly degenerate electron gas. Then the integral over energy is in an elementary way calculated, if one uses the approximation

$$\left(-\frac{\partial f}{\partial E}\right) \simeq \delta(E - \zeta).$$

The integral over $q_\perp$, obviously, reduces to a gamma-function. Therefore, it is not hard to perform the further integration to the first onset to obtain

$$\sigma_{xx} = \frac{e^2 l^2 k_B T m C'}{4\pi^3 \hbar^4 s} \frac{1}{\zeta - \hbar\omega_0/2} \left(\frac{2}{l^2}\right)^{\frac{t+3}{2}} \Gamma\left(\frac{t+3}{2}\right). \tag{5.159}$$

A distinguishing feature of the result is the presence of a divergence when the Fermi level crosses a Landau sublevel. In a quantizing magnetic field, this peculiarity arises due to a square-root singularity of density of states for electrons at the bottom of each Landau subband in the energy space. It is especially interesting to observe the effects occurring in a two-dimensional metal in the presence of a quantizing magnetic field. A field-effect transistor may serve as a case in point. Nobel Prizes in Physics were twice awarded for work dealing with the quantum Hall effect: in 1985 for the discovery of this phenomenon and in 1998 for the discovery and interpretation of the fractional quantum Hall effect. More detailed information on this interesting theme can be found in the specialized literature [41].

Problem 5.1. Obtain an expression for density of states in energy space for the conduction electrons in a quantizing magnetic field.

Solution. The simplest way to introduce the concept of density of states in energy space is the use of the relation:

$$\sum_{np_xp_z\sigma} \rightarrow \int_0^\infty g(E)\, dE. \tag{5.160}$$

The meaning of this relation is that the total number of states for electrons may be expressed not only via the sum but through a state-density integral $g(E)$ over all possible values of the energy. With this definition, the density of states is the number of states for the electrons falling into the energy range from E to $E + dE$ in a crystal whose volume is equal to unity. To find the number of states, we use the result (5.151) obtained earlier by adding the summation over a quantum number n:

$$\sum_{np_xp_z\sigma} \rightarrow \frac{2}{(2\pi l)^2 \hbar} \sum_n \int_{-\infty}^{\infty} dp_z. \tag{5.161}$$

To find the density of states at the formula (5.161) one should pass from the integration over p_z to integration over energy using the definition of the energy spectrum of electrons in the quantum magnetic field (5.120):

$$E_{np_z} = \frac{p_z^2}{2m} + \hbar\omega_0(n + 1/2).$$

Making the required change of integration variables and comparing the expressions (5.160) and (5.161), then we get

$$g(E) = \frac{2\sqrt{2m}}{(2\pi l)^2\hbar} \sum_n \frac{1}{\sqrt{E - \hbar\omega_0(n + 1/2)}}. \tag{5.162}$$

In this formula, as always before, a unit volume of a specimen is set. The summation over n is being performed over all Landau subbands lying below the Fermi level. Equation (5.162) is valid if $E > \hbar\omega_0/2$ and the density of states is zero within the energy range $0 < E < \hbar\omega_0/2$.

The profile of the electron density of states in a magnetic field is shown in Figure 5.1 (curve (b)). The energy E subdivides a scale in units of $\hbar\omega_0/2$ along the abscissa. The quantity

$$g(E)\sqrt{\frac{\hbar\omega_0}{2}}\left\{\frac{2\sqrt{2m}}{(2\pi l)^2\hbar}\right\}^{-1}$$

generates a scale along the vertical axis. The curve (a) is responsible for the electron density of states in the absence of a magnetic field.

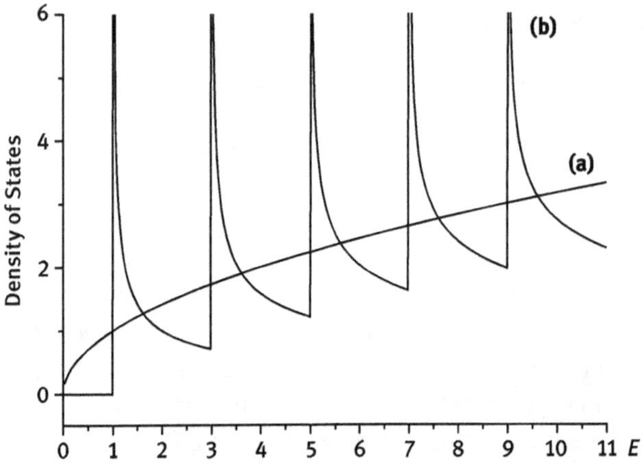

Figure 5.1: Density of states of conduction electrons in a quantizing magnetic field.

The result obtained points to the presence of the singularity in the density of states at the bottom of each Landau subband. Indeed, the density of states at the bottom of the subbands does not grow indefinitely, keeping a finite value due to a collision broadening of the Landau levels.

5.3 Symmetry properties of correlation functions

5.3.1 Additive conservation laws and selection rules for averages

Additive conservation laws lead to additional selection rules for averages. Let a system state be defined by two quantum numbers $\vec{k}$ and σ. Moreover, the total number of particles in this system remains unchangeable,

$$N = \sum_{\vec{k}\sigma} a^{+}_{\vec{k}\sigma} a_{\vec{k}\sigma}; \quad [N, H] = 0,$$

where H is a total Hamiltonian of the system. The Hamiltonian H and the statistical operator ρ_0 are invariant with respect to the conversion

$$U = e^{i\varphi N}; \quad H = U^{+} H U, \quad \rho_0 = U^{+} \rho_0 U, \tag{5.163}$$

where φ is an arbitrary real number. Applying this transformation to creation (annihilation) operators of particles, we may obtain

$$U^{+} a_{\vec{k}\sigma} U = e^{i\varphi} a_{\vec{k}\sigma}, \quad U^{+} a^{+}_{\vec{k}\sigma} U = e^{-i\varphi} a^{+}_{\vec{k}\sigma}. \tag{5.164}$$

Let us consider a quantum-statistical average of an arbitrary number of the particle creation (annihilation) operators:

$$\begin{aligned} \mathrm{Sp}\{a^{+}_{\vec{k}\sigma} \dots a_{\vec{k}'\sigma'} \dots \rho_0\} &= \mathrm{Sp}\{a^{+}_{\vec{k}\sigma} \dots a_{\vec{k}'\sigma'} \dots U^{+} \rho_0 U\} \\ &= \mathrm{Sp}\{U(a^{+}_{\vec{k}\sigma} \dots a_{\vec{k}'\sigma'} \dots) U^{+} \rho_0\} = e^{i\varphi n}\, \mathrm{Sp}\{a^{+}_{\vec{k}\sigma} \dots a_{\vec{k}'\sigma'} \dots \rho_0\}, \end{aligned} \tag{5.165}$$

where n is the difference between the numbers of the creation and annihilation operators of particles. When comparing the first and last expressions in the formula (5.165), it is not difficult to note that the condition $n = 0$ is fulfilled since the conversions are identical, i. e. the number of creation operators must match the number of the annihilation operators under the spur's sign. Otherwise, this quantum-statistical average is zero provided that the particle conservation law is satisfied.

The approach under consideration can be also applied along with other conservation laws. In particular, we consider selection rules which are imposed by the homogeneity of space on the properties of the quantum-statistical averages.

If the space is homogeneous, the momentum of the system is held constant in the absence of external forces:

$$\vec{P} = \sum_{\vec{k}\sigma} \hbar \vec{k} a^{+}_{\vec{k}\sigma} a_{\vec{k}\sigma}.$$

As consequence, the Hamiltonian of the system and the statistical operator commute with the operator of the total momentum. So, one can define a canonical transformation operator U which leaves the Hamiltonian and the statistical operator ρ_0 invariant:

$$U = e^{i\vec{\varphi}\vec{P}}; \quad H = U^{+}HU, \quad \rho_0 = U^{+}\rho_0 U, \tag{5.166}$$

where $\vec{\varphi}$ is an arbitrary vector.

Let us consider again the quantum-statistical average of the arbitrary set of the creation (annihilation) operators of particles. Repeating the foregoing steps (5.165) for the canonical transformation operator being defined by the formula (5.166), we can obtain the condition

$$e^{i\vec{\varphi}\hbar(\vec{k}+\cdots-\vec{k}'+\cdots)} = 1. \tag{5.167}$$

If this condition is not met, then

$$\mathrm{Sp}\{a^{+}_{\vec{k}\sigma}\dots a_{\vec{k}'\sigma'}\dots\rho_0\} = 0.$$

The condition has a simple physical meaning: if the total momentum of the particles is preserved in the system, the total quasi-momentum of created particles must be equal to the total quasi-momentum of annihilated particles. Similar selection rules can be obtained in the presence of other conservation laws. Each of the laws of conservation of angular momentum, spin, etc. can serve as a case in point.

The role of degeneracy of energy levels in statistical physics

As well known from quantum mechanics, the presence of degeneracy of energy levels significantly complicates a calculation of averages for operators of dynamic variables. It would seem that degenerate and non-degenerate states are regarded as completely identical in statistical mechanics. However, this is far from being the case. To see this, let us consider a problem of calculating the longitudinal component of the static magnetic susceptibility tensor χ_{zz}. Using the results (5.80) of the linear response theory, we can write an expression for longitudinal static susceptibility of an electron gas as follows:

$$\chi_{zz} = \frac{(g\mu_B)^2}{i\hbar}\int_{-\infty}^{0} dt_1 e^{\varepsilon t_1}\,\mathrm{Sp}\{S^z e^{iLt_1}[\rho_0, S^z]\}. \tag{5.168}$$

If the Hamiltonian H_0, involved in the definition of the equilibrium statistical operator ρ_0, is independent on the transverse components of the spin S^+, S^- the statistical operator commutes with the operator S^z, which results in the unreasonable solution $\chi_{zz} = 0$. It would seem that this result directly emanates from the conservation condition of the z-component of the total spin.

In fact, as far as degeneracy of energy levels in quantum-statistical mechanics is concerned, the averages of operators of dynamical quantities should be replaced by quasi-averages. *The quasi-averages* are determined as follows:

1. The Hamiltonian H_0 is replaced by the Hamiltonian $H_0 + uH'$ to remove the degeneracy property by choosing a certain addition;
2. the required quantum-statistical averages are calculated;
3. after performing the thermodynamic limiting transition, the following limiting transition $u \to 0$ needs to be fulfilled.

Thus, the correctly calculated value of the average for an arbitrary dynamic operator A is a limit:

$$\langle A \rangle = \lim_{u\to 0} \mathrm{Sp}\left\{ A \frac{1}{Z} e^{-\beta(H_0+uH')} \right\}. \tag{5.169}$$

We now go back to an analysis of the problem of calculating the longitudinal magnetic susceptibility in terms of quasi-averages. Assume that the Hamiltonian admits an infinitesimal correction, which removes the degeneracy with respect to rotations of the Z-axis in the space of the spins. In this case, it already cannot be a priori tolerated that $[\rho_0, S^z] = 0$. Therefore, to transform this commutator, we should use the Kubo formula (5.81), and write it as follows:

$$\frac{1}{i\hbar}[\rho_0, S^z] = \beta \int_0^1 d\tau \rho_0^\tau \dot{S}^z \rho_0^{1-\tau}. \tag{5.170}$$

Substituting this expression into the formula (5.168), one obtains

$$\chi_{zz} = \beta(g\mu_B)^2 \int_{-\infty}^{0} dt_1 e^{\epsilon t_1} \int_0^1 d\tau\, \mathrm{Sp}\{S^z \dot{S}^z(t_1 + i\hbar\beta\tau)\rho_0\}\,. \tag{5.171}$$

Performing the integration by parts over variable t_1 in this expression, we get

$$\begin{aligned}\chi_{zz} &= \beta(g\mu_B)^2 \int_0^1 d\tau\, \mathrm{Sp}\{S^z S^z(i\hbar\beta\tau)\rho_0\} \\ &\quad - \beta(g\mu_B)^2 \int_0^1 d\tau \epsilon \int_{-\infty}^{0} dt_1 e^{\epsilon t_1}\, \mathrm{Sp}\{S^z S^z(t_1 + i\hbar\beta\tau)\rho_0\}.\end{aligned} \tag{5.172}$$

To transform the second summand in the last formula, Abel's theorem should be used

$$\lim_{\epsilon\to 0} \epsilon \int_{-\infty}^{0} dt e^{\epsilon t} f(t) = \lim_{t\to -\infty} f(t).$$

Then we finally obtain

$$\chi_{zz} = \beta(g\mu_B)^2 \int_0^1 d\tau \,\mathrm{Sp}\{S^z S^z(i\hbar\beta\tau)\rho_0\} - \beta(g\mu_B)^2 \,\mathrm{Sp}\{S^z\rho_0\}\,\mathrm{Sp}\{S^z\rho_0\}. \tag{5.173}$$

Now, applying the standard notation

$$\mathrm{Sp}\{S^z\rho_0\} = \langle S^z\rangle,$$

the result can be written more compactly:

$$\chi_{zz} = \beta(g\mu_B)^2 \int_0^1 d\tau \langle (S^z - \langle S^z\rangle)(S^z(i\hbar\beta\tau) - \langle S^z\rangle)\rangle. \tag{5.174}$$

For a gas of noninteracting electrons $H_0 = -g\mu_B S^z H$, so $S^z(i\hbar\beta\tau) = S^z$. Then this result allows for further simplification, and we obtain a formula that coincides with the classical definition of the magnetic susceptibility :

$$\chi_{zz} = \beta(g\mu_B)^2 \langle \triangle S^z \triangle S^z\rangle, \tag{5.175}$$

where $\triangle S^z = S^z - \langle S^z\rangle$.

5.3.2 Symmetry properties of correlation functions for operations of spatial rotation, complex conjugation and time reversal

Considering the symmetry properties for correlation functions at operations of spatial rotation, complex conjugation and time reversal when determining the electrical conductivity components in a quantizing magnetic field with the aid of the correlation function, we have

$$\sigma_{\mu\nu} = \int_0^\infty dt e^{-\epsilon t} I_{\mu\nu}(t),$$

$$I_{\mu\nu}(t) = \int_0^\beta d\lambda \langle J_\mu(t) J_\nu(i\hbar\lambda)\rangle. \tag{5.176}$$

Let us first define the symmetry properties of the correlation function $I_{\mu\nu}(t)$ at the operation of the spatial rotation of a coordinate system. In this case, if a system's Hamiltonian is invariant with respect to rotation about a distinguished axis, the correlation function $I_{\mu\nu}(t)$ is presented as the product of the momentum components $P_\mu P_\nu$. In particular, if a magnetic field $H = 0$, for crystals of the cubic symmetry, one obtains

$$I_{xx}(t) = I_{yy}(t) = I_{zz}(t),$$

$$I_{xy}(t) = I_{yz}(t) = I_{xz}(t) = 0. \tag{5.177}$$

Thus, in this case, all diagonal components are equal, and the off-diagonal vanish.

In an external magnetic field, application of the same principle leads to the following result:

$$I_{xx}(t) = I_{yy}(t); \quad I_{xy}(t) = -I_{yx}(t). \tag{5.178}$$

All the other off-diagonal components are zero. It is worth saying a few words about the component $I_{zz}(t)$ in the quantizing magnetic field. In this case, since the motion along the Z-axis remains quasi-free, to calculate the longitudinal component of the electrical conductivity tensor the procedure developed in Section 5.1.3 must be used.

We find relations satisfied by the correlation function $I_{\mu\nu}(t)$ at the operation of complex conjugation. When the correlation function is a real quantity, the correlation function must also satisfy additional relations, which may come up at this operation.

We first consider when the operation of complex conjugation is applied to the spur of two operators:

$$\begin{aligned}(\mathrm{Sp}\{AB\})^* &= \Big(\sum_{nm} \langle n|A|m\rangle \langle m|B|n\rangle\Big)^* \\ &= \sum_{nm} \int \psi_n A^* \psi_m^* d\tau_1 \int \psi_m B^* \psi_n^* d\tau_2 = \mathrm{Sp}\{B^+ A^+\}.\end{aligned} \tag{5.179}$$

In the formula (5.179) B^+ denotes the Hermitian conjugate operator.

Applying the ratio found for the correlation function $I_{\mu\nu}(t)$, we have

$$I_{\mu\nu}(t)^* = \int_0^\beta d\lambda \,\mathrm{Sp}\{(J_\nu(i\hbar\lambda)\rho_0)^+ J_\mu(t)^+\} = \int_0^\beta d\lambda \,\mathrm{Sp}\{\rho_0 J_\nu(-i\hbar\lambda) J_\mu(t)\}. \tag{5.180}$$

In deriving this relation we have taken into account that the statistical operator ρ_0 is self-adjoint and $\rho_0^+ = \rho_0$. The current operator is also self-adjoint one and therefore it satisfies the relation

$$J_\alpha(t + i\hbar\lambda)^+ = J_\alpha(t - i\hbar\lambda).$$

Making the change of the integration variables $\lambda - \beta = \lambda'$ in the last integral of (5.180) we get

$$I_{\mu\nu}(t)^* = I_{\mu\nu}(t). \tag{5.181}$$

Thus, we have shown that the correlation function $I_{\mu\nu}(t)$ has a real value.

Now, consider the symmetry properties of correlation functions with respect to the operation of time reversal.

In quantum mechanics, symmetry of motion in relation to change of the sign on time makes itself evident in the fact that if the function ψ is a wave function of some stationary state, the time-reversed wave function, being designated here by ψ_-, also describes some possible state with the same energy.

Consider the Schrödinger equation,

$$i\hbar\frac{\partial\psi}{\partial t} = H\psi. \tag{5.182}$$

If the Hamiltonian is invariant with respect to the time-reversal operation, we can obtain another equation, reversing time,

$$-i\hbar\frac{\partial\psi_-}{\partial t} = H\psi_-. \tag{5.183}$$

This equation is like the complex conjugate of (5.182):

$$-i\hbar\frac{\partial\psi^*}{\partial t} = H^*\psi^*. \tag{5.184}$$

Comparing the equations (5.183) and (5.184), we try to determine an operator, which would play the role of the time-reversal operator. Let an operator O be unitary, and satisfy the conditions

$$OH^* = HO, \quad O^{-1}O = 1, \quad O^{-1} = O^+.$$

We apply this operator to the equation (5.184), and the following expression appears:

$$-i\hbar\frac{\partial O\psi^*}{\partial t} = HO\psi^*. \tag{5.185}$$

Comparing this equation with (5.183), we find that

$$O\psi^* \equiv K\psi = \psi_-.$$

The entity $K = OK_0$, where K_0 is the operator of complex conjugation, can be called the time-reversal operator.

The explicit form of the operator O depends on the particularly chosen Hamiltonian. If the Hamiltonian has the form

$$\frac{1}{2m}\left(\vec{P} - \frac{e}{c}A\right)^2 - g\mu_B\vec{\sigma}\,\mathrm{rot}\,\vec{A} + V(\vec{r}), \tag{5.186}$$

the operator O can be chosen as

$$O = i\sigma_y O_A,$$

where the operator O_A changes the sign of the vector potential and the direction of a magnetic field is reversed.

It is easy to verify that the property $OH^* = HO$ for the Hamiltonian of (5.186) is fulfilled for the operator $O = i\sigma_y O_A$. It also holds true for the first and third summands in (5.186). In order to ensure that this property works, it is sufficient to call attention to the second summand of Hamiltonian (5.186):

$$i\sigma_y \vec{\sigma} = -\vec{\sigma} i\sigma_y.$$

That is because the Pauli matrices,

$$\sigma_x = \frac{1}{2}\begin{pmatrix} 0 & 1 \\ 1 & 0 \end{pmatrix}, \quad \sigma_y = \frac{1}{2}\begin{pmatrix} 0 & -i \\ i & 0 \end{pmatrix}, \quad \sigma_z = \frac{1}{2}\begin{pmatrix} 1 & 0 \\ 0 & -1 \end{pmatrix},$$

anticommute:

$$\sigma_x\sigma_y + \sigma_y\sigma_x = 0, \quad \sigma_y\sigma_z + \sigma_z\sigma_y = 0,$$

and the obvious relation $i\sigma_y\sigma_y^* = -\sigma_y i\sigma_y$ is met.

Thus, the time-reversal operator K has the form

$$K = OK_0 = i\sigma_y O_A K_0; \quad K_0^{-1} = K_0, \ O^{-1} = O^+, \tag{5.187}$$

where the operator K_0 performs an operation of complex conjugation and the operator O_A changes the sign of $A \to -A$ or $H \to -H$.

Now, we establish the symmetry properties for the correlation functions, arising due to the invariance of the Hamiltonian with respect to the time-reversal operation. We first consider the matrix element of the operator

$$\begin{aligned} \langle\psi_n|K^{-1}AK|\psi_m\rangle &= \langle\psi_n|(O^+AO)^*|\psi_m\rangle \\ &= \langle\psi_n^*|O^+AO|\psi_m^*\rangle^* = \langle O\psi_n^*|A|O\psi_m^*\rangle^* = \langle K\psi_n|A|K\psi_m\rangle^*. \end{aligned} \tag{5.188}$$

Thus, we have proved that

$$\langle\psi_n|K^{-1}AK|\psi_m\rangle = \langle K\psi_n|A|K\psi_m\rangle^*.$$

After generalizing this result, the relation for the spur of two operators can be written as

$$(\mathrm{Sp}\{AB\})_H^* = (\mathrm{Sp}\{A^\times B^\times\})_{-H}. \tag{5.189}$$

In deriving the formula (5.189) we have taken into account the fact that the numerical value of the spur is calculated regardless of which a complete system of eigenfunctions is applied: ψ or ψ_-, and we have used the notation $K^{-1}AK = A^\times$.

The subscript H or $-H$ in the correlation functions only serves as a reminder (the operation of the sign change in the magnetic field is included in the time-reversal operator).

Using the relations (5.179) and (5.189) in successive order, we obtain another useful relation:

$$(\mathrm{Sp}\{AB\})_H = (\mathrm{Sp}\{B^{+\times}A^{+\times}\})_{-H}. \tag{5.190}$$

Applying the above result (5.190) to the correlation function $I_{\mu\nu}(t)$, we can write the integral

$$I_{\mu\nu}(t)_H = \int_0^\beta d\lambda\, \mathrm{Sp}\{(J_\nu(-t+i\hbar\lambda)\rho_0)^{+\times} J_\mu^{+\times}\}_{-H},$$
$$\rho_0^{+\times} = \rho_0, \quad (J_\nu(-t+i\hbar\lambda))^{+\times} = -J_\nu(t-i\hbar\lambda), \quad J_\mu^{+\times} = -J_\mu. \tag{5.191}$$

Then the expression for the correlation function takes the form

$$\begin{aligned} I_{\mu\nu}(t)_H &= \int_0^\beta d\lambda (\mathrm{Sp}\{\rho_0 J_\nu(t-i\hbar\lambda)J_\mu\})_{-H} \\ &= \int_0^\beta d\lambda (\mathrm{Sp}\{J_\nu J_\mu(-t+i\hbar\lambda)\rho_0\})_{-H} = I_{\nu\mu}(t)_{-H}. \end{aligned} \tag{5.192}$$

If one recalls the definition (5.176), the relation (5.192) allows one to write the Onsager symmetry relation for the components of the electrical conductivity tensor in a magnetic field,

$$\sigma_{\mu\nu}(H) = \sigma_{\nu\mu}(-H). \tag{5.193}$$

After generalizing this result, the components of the generalized susceptibility χ_{AB} are given by

$$\chi_{AB}(H) = \varepsilon_A \varepsilon_B \chi_{BA}(-H), \tag{5.194}$$

where the ε_A and ε_B quantities are equal to ±1, depending on parity of the A and B operators for the time-reversal operation.

The relations (5.192), (5.193) imply that the diagonal components of the electrical conductivity tensor can contain only even powers of the magnetic field.

Using the symmetry properties of the current operators relating to the time-reversal operation, we can write another useful relation:

$$I_{\mu\nu}(t)_H = I_{\mu\nu}(-t)_{-H}. \tag{5.195}$$

To prove this relation, we consider the correlation function $I_{\mu\nu}(t)_H$:

$$\begin{aligned}
I_{\mu\nu}(t)_H &= \int_0^\beta d\lambda(\mathrm{Sp}\{J_\mu J_\nu(-t + i\hbar\lambda)\rho_0\})_H \\
&= \int_0^\beta d\lambda(\mathrm{Sp}\{(J_\mu J_\nu(t + i\hbar\lambda)\rho_0)^\times\})_{-H} \\
&= \int_0^\beta d\lambda\Big(\sum_m \langle\psi_m^*|O^+ J_\mu J_\nu(t + i\hbar\lambda)\rho_0 O|\psi_m^*\rangle\Big)^*_{-H} \\
&= \int_0^\beta d\lambda\Big(\sum_m \langle K\psi_m|J_\mu J_\nu(t + i\hbar\lambda)\rho_0|K\psi_m\rangle\Big)^*_{-H} \\
&= \int_0^\beta d\lambda(\mathrm{Sp}\{J_\mu J_\nu(t + i\hbar\lambda)\rho_0\})^*_{-H} = I_{\mu\nu}(-t)^*_{-H}.
\end{aligned} \tag{5.196}$$

By virtue of the equality (5.181), this yields the result (5.195). As far as the diagonal components of the correlation function $I_{\mu\nu}(t)$ are even in the magnetic field, it follows that

$$I_{xx}(t) = I_{xx}(-t), \quad I_{yy}(t) = I_{yy}(-t), \quad I_{zz}(t) = I_{zz}(-t). \tag{5.197}$$

5.4 Problems to Chapter 5

5.1. Argue that the following identities hold for arbitrary operators:

1. $[AC, B] = A[C, B] + [A, B]C$;
2. $\mathrm{Sp}\{B[A(t_1), \rho_0]\} = \mathrm{Sp}\{[B, A(t_1)]\rho_0\}; \quad A(t_i) = \exp\{iLt_i\}A$;
3. $\mathrm{Sp}\{B[A(t_1), [A(t_2), \rho_0]]\} = \mathrm{Sp}\{[[B, A(t_1)], A(t_2)]\rho_0\}$.

5.2. The Heaviside theta-function is given by

$$\theta(x) = \begin{cases} 1, & x \geq 0, \\ 0, & x < 0. \end{cases}$$

Prove that its derivative coincides with the Dirac delta-function

$$\frac{d}{dx}\theta(x) = \delta(x).$$

5.3. Formulate physical applicability conditions for the theory of a system's response to an external mechanical perturbation. What is the meaning of the relaxation time in the expression (5.78)?

5.4. By means of the formula (5.57), show that one can arrive at the expression (4.204) again in the case of electron scattering by a screened Coulomb potential for the reverse relaxation time of the average momentum of the electrons.

5.5. Prove that the Kubo identity can be written as

$$\frac{1}{i\hbar}[\rho_0, S^\beta] = \beta \int_0^1 d\tau \dot{S}^\beta(i\hbar\beta\tau)\rho_0,$$

where

$$H = H_e + H_s; \quad H_e = \frac{p^2}{2m}; \quad H_s = -g\mu_B S^z |\vec{H}|; \quad \dot{S}^\beta = \frac{1}{i\hbar}[S^\beta, H];$$

$$\rho_0 = \frac{1}{Z}e^{-\beta(H-\zeta N)}, \quad Z = \mathrm{Sp}\{e^{-\beta(H-\zeta N)}\}; \quad A(i\hbar\beta\tau) = \rho_0^\tau A \rho_0^{-\tau}.$$

5.6. It is well known that the theory of transport phenomena includes the so-called sum rule: the integral over frequency of generalized susceptibility as well as kinetic coefficients can be expressed through a static correlation function (in some cases, the correlation function can be easily calculated).
Prove that the sum rule holds for the electrical conductivity

$$\int_{-\infty}^{\infty} d\omega \sigma_{\mu\nu}(\omega) = \pi \frac{e^2 n}{m},$$

where $\sigma_{\mu\nu}$ is given by (5.43).

5.7. The system of electrons is in an external magnetic field H, oriented along the Z-axis and in a variable (periodic) field with amplitude h and frequency ω that is oriented in a plane perpendicular to the Z-axis. The interaction Hamiltonian between the electrons and the external field can be written as

$$H_{sf} = -\frac{\hbar\omega_{1s}}{2}(S^+ e^{i\omega t} - S^- e^{i\omega t}),$$

where S^α is the α-component of the total electron spin.
Prove that, in this case, the power Q absorbed by the spin system can be represented as

$$Q = \frac{1}{2}\omega h^2 \chi''_{+-}(\omega),$$

where $\chi''_{+-}(\omega)$ is the imaginary part of the high-frequency transverse magnetic susceptibility.

5.8. To analyze experimental data of how kinetic coefficients behave in a quantizing magnetic field, it is very important to know the behavior of the Fermi level. Prove that the position of the Fermi level in the magnetic field is weakly dependent on its amplitude. Calculate the value of the Fermi level for cases lying three, two or one Landau levels below the Fermi level. Compare the results obtained with the value of the Fermi level in the absence of the magnetic field.

5.9. Argue that the relation

$$\langle A_1(t_1)A_2(t_2)\rangle_H = \epsilon_{A_1}\epsilon_{A_2}\langle A_2(t_2)A_1(t_1)\rangle_{-H}$$

is valid for the correlation function $\langle A_1(t_1)A_2(t_2)\rangle = \mathrm{Sp}\{A_1(t_1)A_2(t_2)\rho_0\}$ where $\epsilon_{A_1}\epsilon_{A_2}$ are equal to ± 1 depending on the parity of the operators A_1, A_2 in the time-reversal operation.

5.10. The quantization of electron orbital motion is most obvious in a two-dimensional metal of MOS transistors. According to (5.138), (4.132), the Hall constant remains intact and is equal to $1/enc$ in the zeroth-order interaction. In practice, it is more convenient to analyze a different quantity, namely the Hall resistance $R_H = R{\cdot}H$. Then the quantity $R_H = H/enc$ must increase linearly with increasing a magnetic field. However, an experiment demonstrates a completely different situation on changing the magnetic field, the Hall resistance obeys the law

$$R_H = \frac{2\pi\hbar}{ie^2},$$

where i is the number of Landau levels below the Fermi level. Explain the above laws of the quantum Hall effect in a two-dimensional metal.
Hint. Prove that the multiplicity of degeneration of the Landau levels is the same for all levels and is $1/(\pi l^2)$.

6 Non-equilibrium statistical operator method

6.1 Non-equilibrium and quasi-equilibrium statistical operators

6.1.1 Quasi-equilibrium distribution

This chapter reviews the method of non-equilibrium statistical operator (NSO), which is conceptually related to the projection operator method developed by Mori. D. N. Zubarev and V. P. Kalashnikov also made great strides in advancing the NSO method. As to this method, the books [36, 37] contain a sufficiently complete picture of early work by these authors. To get acquainted with the method, the monograph G. Repke [42] is recommended, but, unfortunately, it gives too few examples of applying the NSO method for various applications. An overview of more recent papers, containing the modern development of this fairly promising method, can be found in [43].

The authors of the present book do not pretend to give a fairly complete overview of recent research on the use of the NSO method for solving problems of physical kinetics. The goal of the authors is to draw the reader's attention, first of all of students, to the NSO method that is simple and modern. The method is comparable, all in all, with the kinetic equation approach. Nevertheless, it has still not found a proper practice.

The time evolution of a non-equilibrium state of a macroscopic system can be set forth by means of the non-equilibrium statistical operator $\rho(t,0)$, which satisfies the Liouville equation (5.19):

$$\left(\frac{\partial}{\partial t}+iL\right)\rho(t,0)=0,\quad iLA=\frac{1}{i\hbar}[A,H]\equiv\dot{A}. \tag{6.1}$$

In the equation (6.1), the quantity of $\rho(t,0)$ has two time arguments. The first describes the dependence of the statistical operator on time t, i. e. some parameters are connected with time explicitly. For example, this may be the dependence of temperature or drift velocity on time. The second time argument t describes the time dependence of the operator on time in the Heisenberg notation. In addition, since the quantity $\rho(t)$ is an integral of motion, the following expression must be fulfilled:

$$\rho(t,t)=\exp\{iLt\}\rho(t,0)=\rho(0,0). \tag{6.2}$$

The Liouville equation within these notations can be written also in the form

$$\frac{d\rho(t,t)}{dt}=0. \tag{6.3}$$

If the statistical operator is known and equal to $\rho(t_0,0)$ at an initial moment of time t_0, the solution of the Cauchy problem for the NSO is given by

$$\rho(t,0)=\exp\{-iL(t-t_0)\}\rho(t_0,0). \tag{6.4}$$

https://doi.org/10.1515/9783110727197-006

At the same time, a time dependence of averages for the operator of some physical quantity A appears:

$$\langle A\rangle^t = \mathrm{Sp}\{A\rho(t,0)\} = \mathrm{Sp}\{\rho(t_0,0)\exp\{iL(t-t_0)\}A\}. \tag{6.5}$$

In deriving the last relation we have used both cyclic commutativity of operators under the spur sign and expression (5.20) for the Heisenberg evolution operator. It should be noted that the above relations are treated as a particular case for systems whose Hamiltonian is not time dependent.

The formulas (6.2)–(6.5) correspond to a precise dynamical system description, which, as it follows from results of the previous chapters, is unobservable for systems with weak stability. Suppose that at some point of time τ, which is more time mixing in the system, measurable quantities are the average values $\langle P_n\rangle^t$ of some set of operators P_n. Following this line of reasoning, the system's memory of the initial distribution $\rho(t_0,0)$ is expected to fade away over time τ, and, as a consequence, the evolution of the system will be determined by dint of general statistical properties.

So we do not take into account the correlations that decay in time $t \simeq \tau$, when considering the fairly distant asymptotic $t \gg \tau$. This idea, due to Bogoliubov, is at the heart of the NSO method. If one accepts it, the true initial condition for the Liouville equation (which is in any case unknown)

$$\lim_{t\to t_0}\rho(t) = \rho(t_0)$$

can be replaced without prejudice by an idealized condition, consisting in the fact that the NSO at an initial time is a functional only of the same variables of $\langle P_n\rangle^t$, which prove to be long-living or measurable over periods of time $t \gg \tau$. Therefore, as follows from the solution of the Liouville equation (6.4), $\rho(t,0)$ will be also a functional of $\langle P_n\rangle^t$ at all subsequent time points.

Let us now discuss another important level of the method under consideration. Suppose we have a system whose state at a given stage of evolution is described by a set of average (measurable) quantities of $\langle P_n\rangle^t$. Along with the non-equilibrium statistical operator $\rho(t,0)$, we introduce a quasi-equilibrium statistical operator $\overline{\rho}(t,0)$ which is equivalent to NSO in the sense that the average values of operators P_n are equal among themselves at all time points for non-equilibrium and quasi-equilibrium distributions:

$$\langle P_n\rangle^t = \mathrm{Sp}\{P_n\rho(t,0)\} = \mathrm{Sp}\{P_n\overline{\rho}(t,0)\}. \tag{6.6}$$

The condition (6.6) is a new assumption and not a consequence of the program of constructing a theory of irreversible phenomena, which was discussed in the previous chapter. So one should postpone an explanation of the physical meaning of this condition before deriving an explicit expression for the quasi-equilibrium distribution. Now it is worth pointing out but that the condition (6.6) allows thermodynamics of a non-equilibrium system to be constructed.

The treatment of the quasi-equilibrium distribution will be accounted for as the reader advances through the text. Emanating from the fact that such a distribution may be entered and it is to be some functional of the average values of observable quantities of $\langle P_n\rangle^t$, we think that the distribution $\overline{\rho}(t,0)$ is a functional of the observed averages of $\langle P_n\rangle^t$, taken at one and the same time t. Then, if $\overline{\rho}(t,0)$ is time dependent on the time-dependent averages of $\langle P_n\rangle^t$, the differential of $\overline{\rho}(t,0)$ is given by

$$\frac{\partial\overline{\rho}(t,0)}{\partial t}=\sum_n\frac{\partial\overline{\rho}(t,0)}{\partial\langle P_n\rangle^t}\frac{\partial}{\partial t}\langle P_n\rangle^t. \tag{6.7}$$

Equation (6.7) allows one to yield another interpretation of the operators P_n. These operators are basic operators on a Hilbert space. Therefore, the time evolution of any operator can be expressed through the evolution of an aggregate of the basic operators. From the equation (6.7) it follows that the quasi-equilibrium distribution does not satisfy the Liouville equation. The expression for the time derivative for the quantities of $\langle P_n\rangle^t$ can be obtained by using equation (6.6). Considering the Liouville equation (5.19) and differentiating the equation (6.6) over time, we get

$$\frac{\partial\langle P_n\rangle^t}{\partial t}=\langle\dot{P}_n\rangle^t. \tag{6.8}$$

In deriving the last expression we have used the definition of the Liouville operator (5.18) and taken into account that

$$\langle\dot{P}_n\rangle^t=-\mathrm{Sp}\{P_n iL\rho(t,0)\}=\mathrm{Sp}\{\dot{P}_n\rho(t,0)\}. \tag{6.9}$$

Equation (6.8) can be regarded as a generalized kinetic equation. In particular, this equation can have the meaning of an equation for a one-particle distribution function, if the quantity $P_k=a^+_{\vec{k}}a_{\vec{k}}$ where $a^+_{\vec{k}}$, $a_{\vec{k}}$ are creation and annihilation operators of a particle such as an electron in a state $\vec{k}$.

To understand the sense of the introduced quasi-equilibrium distribution, it is necessary to calculate a system's entropy, suggesting that a quasi-equilibrium ensemble of systems is already prepared. Let the entropy of the quasi-equilibrium system be defined by the expression

$$S(t)=-\mathrm{Sp}\{\overline{\rho}(t,0)\ln\overline{\rho}(t,0)\}. \tag{6.10}$$

In addition, the quantity

$$\hat{S}(t)=-\ln\overline{\rho}(t,0) \tag{6.11}$$

will be called the entropy operator.

Let us find the entropy production in a system. The term "entropy production" was borrowed from phenomenological thermodynamics of irreversible processes [5] and stands for the time derivative of an average value of the system entropy. For equilibrium systems, entropy production is zero and for non-equilibrium ones it is positive. Differentiating the equation (6.10) over time, we arrive at

$$\dot{S}(t) = -\frac{d}{dt}\mathrm{Sp}\{\rho(t,0)\ln\overline{\rho}(t,0)\} = \mathrm{Sp}\{\dot{\hat{S}}(t,0)\rho(t,0)\}, \tag{6.12}$$

$$\dot{\hat{S}}(t,0) = \left(\frac{\partial}{\partial t} + iL\right)\hat{S}(t,0). \tag{6.13}$$

In deriving the formula (6.12) we have taken into account the fact that $\ln\overline{\rho}(t,0)$ is linear in operators P_n (this will be shown in the following section), and therefore, the following expression is valid:

$$\mathrm{Sp}\{\overline{\rho}(t,0)\ln\overline{\rho}(t,0)\} = \mathrm{Sp}\{\rho(t,0)\ln\overline{\rho}(t,0)\}.$$

The quantity $\dot{\hat{S}}(t,0)$ is called the entropy production operator.

Because $S(t)$ is also a functional of $\langle P_n\rangle^t$, using the expression (6.8), we have

$$\frac{\partial S(t)}{\partial t} = \sum_n \frac{\delta S(t)}{\delta\langle P_n\rangle^t}\langle\dot{P}_n\rangle^t . \tag{6.14}$$

Introducing the notation

$$\frac{\delta S(t)}{\delta\langle P_n\rangle^t} \equiv F_n(t) \tag{6.15}$$

for the entropy production we obtain the simple equation

$$\frac{\partial S(t)}{\partial t} = \sum_n F_n(t)\langle\dot{P}_n\rangle^t, \tag{6.16}$$

which coincides in form with the equation of entropy production in phenomenological non-equilibrium thermodynamics of Onsager [5]. The sign of δ in the formula (6.15) means a functional derivative. According to Onsager, entropy production in a system is equal to the sum of products of a generalized thermodynamic force by a conjugate thermodynamic flux. The expression (6.15) has the desired structure and allows one to interpret the quantity $F_n(t)$ as a generalized thermodynamic force, and $\langle\dot{P}_n\rangle^t$ as a generalized thermodynamic flux.

6.1.2 Extremal properties of a quasi-equilibrium distribution. Thermodynamics of a quasi-equilibrium ensemble

It is interesting to see what the explicit form of the quasi-equilibrium distribution should be. It is clear that the definition of $\overline{\rho}(t)$ can be ambiguous due to only one requirement to this distribution, i. e., it must be a functional of $\langle P_n\rangle^t$. The expression (6.10), defining the relationship between the quasi-equilibrium distribution and entropy, allows one to unambiguously determine the $\overline{\rho}(t)$. So, we require that $\overline{\rho}(t)$ should satisfy the maximum information entropy

$$S(t) = -\mathrm{Sp}\{\overline{\rho}(t,0)\ln\overline{\rho}(t,0)\}$$

under the additional conditions.

1. No matter how the distribution is varied, the observed average values of basic operators must remain unchanged:

$$\mathrm{Sp}\{P_n\overline{\rho}(t,0)\} = \langle P_n\rangle^t; \tag{6.17}$$

2. In varying the distribution, the normalization condition must be preserved,

$$\mathrm{Sp}\{\overline{\rho}(t,0)\} = 1. \tag{6.18}$$

The extremality conditions (6.10) along with the restrictions (6.17) and (6.18) imposed on all possible variations put the problem on a conditional extremum of the functional $S(t)$. It is well known that the problem on the conditional extremum of the functional $S(t)$ can be reduced to a problem of an unconditional extremum of another functional $£(\overline{\rho}(t))$ by introducing Lagrange multipliers:

$$£ = -\mathrm{Sp}\{\overline{\rho}\ln\overline{\rho}\} - \sum_n F_n(t)\mathrm{Sp}\{\overline{\rho}P_n\} - [\phi(t)-1]\mathrm{Sp}\{\overline{\rho}\}. \tag{6.19}$$

Here, $F_n(t)$ and $[\phi(t)-1]$ are Lagrange multipliers. Calculating the variation of both sides of (6.19) over $\overline{\rho}$, we obtain

$$\delta £ = -\mathrm{Sp}\left\{\left[\ln\overline{\rho} + \sum_n F_n(t)P_n + \phi(t)\right]\delta\overline{\rho}\right\}. \tag{6.20}$$

From the condition of extremality it follows that $\delta £ = 0$. Therefore, considering that the quantity $\delta\overline{\rho}$ is arbitrary, and the spur on the right-hand side of (6.20) must still be equal to zero, we have

$$\ln\overline{\rho} + \sum_n F_n(t)P_n + \phi(t) = 0. \tag{6.21}$$

From (6.21), it is easy to obtain an explicit form of the quasi-equilibrium statistical operator:

$$\overline{\rho}(t) = \exp\left\{-\left[\phi(t) + \sum_n F_n(t)P_n\right]\right\}. \tag{6.22}$$

The Lagrange multipliers are not yet determined in the formula (6.22). So, it is necessary to use the equations (6.17) and (6.18) to find them. To better understand the meaning of the parameters involved in the definition (6.22), let us compare it with the Gibbs' canonical distribution

$$\rho_0 = \frac{1}{Z}\exp\{-\beta(H - \zeta N)\}. \tag{6.23}$$

Here, Z is the statistical sum, ζ the chemical potential of a system, H the Hamiltonian, N the number particle operator and β the inverse temperature in energy units.

Upon comparing the formulas (6.22) and (6.23), it is seen that the equilibrium distribution is the distribution with a certain value of energy and number of particles, at the same time the quasi-equilibrium distribution is a distribution with a certain value of averages of $\langle P_n\rangle^t$. The quantity $\phi(t)$ in the expression (6.22) is known as the Massieu–Planck function, which, along with the statistical sum Z, is also defined by the normalization condition:

$$\phi(t) = \ln \mathrm{Sp}\left\{\exp\left\{-\sum_n P_n F_n(t)\right\}\right\}. \tag{6.24}$$

The choice of parameters P_n and functions $F_n(t)$ depends on the particular problem. In particular, in the case of the hydrodynamic regime when the energy of a system, the drift momentum and the number of particles are measurable, see the table presented below, we find the way to select a set of operators P_n and thermodynamic functions $F_n(t)$ conjugate to them:

Operators	H	$\vec{P}$	N
Thermodynamic functions	$\beta(t)$	$\beta(t)m\vec{V}(t)$	$\beta(t)\zeta(t)$

Here, $\vec{P}$ is the total momentum operator of particles in a system, $\vec{V}$ the drift velocity of the particles, m the mass.

Let us address the thermodynamics of a quasi-equilibrium distribution.

Using the definitions (6.10) and (6.22), we write down the expression for a system's entropy:

$$S(t) = \phi(t) + \sum_n \langle P_n\rangle^t F_n(t). \tag{6.25}$$

This equation can be regarded as the Legendre transformation, namely as the transition from one thermodynamic potential to another (from $\phi(t)$ to $S(t)$) for the non-equilibrium system. This becomes obvious if one performs the variation of the Massieu–Planck function (6.24):

$$\begin{aligned}\delta\phi(t) &= \delta \ln \mathrm{Sp}\left\{\exp\left\{-\sum_n P_n F_n(t)\right\}\right\}\\ &= -\left[\mathrm{Sp}\left\{\exp\left\{-\sum_n P_n F_n(t)\right\}\right\}\right]^{-1} \sum_m \mathrm{Sp}\left\{P_m \delta F_m(t)\right.\\ &\quad\times \left.\exp\left\{-\sum_n P_n F_n(t)\right\}\right\} = -\sum_m \langle P_m\rangle^t \delta F_m(t).\end{aligned} \tag{6.26}$$

The last expression on the right-hand side of (6.26) is written in conformity with the relations (6.6), (6.22) and (6.24).

On the other hand, using the definition of entropy (6.25) and considering the explicit form of the quasi-equilibrium distribution (6.22), we get

$$\delta S(t) = \delta\phi(t) + \sum_n (\delta\langle P_n\rangle^t F_n(t) + \langle P_n\rangle^t \delta F_n(t)). \tag{6.27}$$

Substituting the value of $\delta\phi(t)$, defined by the expression (6.26), into the formula above, we obtain

$$\delta S(t) = \sum_n F_n(t)\delta\langle P_n\rangle^t. \tag{6.28}$$

The relations (6.26), (6.28) can be interpreted as follows: in writing the entropy, the role of independent variables are playing quantities of $\langle P_n\rangle^t$, but in writing the Massieu–Planck function we have the quantities $F_n(t)$.

The results obtained allow one to generalize the Gibbs–Helmholtz relations in the case of non-equilibrium thermodynamics. Calculating the derivative of the Massieu–Planck function and using the equation (6.26), we have

$$\langle P_m\rangle^t = -\frac{\delta\phi(t)}{\delta F_m(t)}. \tag{6.29}$$

Substituting this result into the expression for entropy, we obtain a generalization of the Gibbs–Helmholtz relations in the case of non-equilibrium thermodynamics:

$$S(t) = \phi(t) - \sum_m \frac{\delta\phi(t)}{\delta F_m(t)} F_m(t). \tag{6.30}$$

This formula expresses the system's entropy through the Massieu–Planck functional. Also, it is easy to get the inverse ratio. Indeed, from the expression for the variation of entropy, $F_n(t)$ appears:

$$F_n(t) = \frac{\delta S(t)}{\delta\langle P_n\rangle^t}. \tag{6.31}$$

Then the formula for entropy yields again

$$\phi(t) = S(t) - \sum_m \frac{\delta S(t)}{\delta\langle P_n\rangle^t}\langle P_n\rangle^t. \tag{6.32}$$

The difference between these relations and their equilibrium analogs reduces to only a substitution of functional derivatives for partial ones.

To get the meaning of the quasi-equilibrium distribution $\bar{\rho}(t)$, it is very important to find whether one can use this distribution for description of non-equilibrium processes.

Now, we calculate entropy production of a quasi-equilibrium state. Averaging the entropy production operator (6.13) over the quasi-equilibrium distribution, we obtain

$$\langle \dot{S}(t)\rangle_q^t = \mathrm{Sp}\left\{\overline{\rho}(t)\left[\dot{\phi}(t) + \sum_n \dot{P}_n F_n(t) + \sum_n P_n \dot{F}_n(t)\right]\right\}. \tag{6.33}$$

Applying the relation (6.26), we arrive at

$$\dot{\phi}(t) = -\sum_m \langle P_m\rangle^t \dot{F}_m(t).$$

Substituting this result into (6.33), we find that

$$\begin{aligned}\langle \dot{\hat{S}}(t)\rangle_q^t &= \mathrm{Sp}\left\{\overline{\rho}(t)\sum_n [(P_n - \langle P_n\rangle^t)\dot{F}_n(t) + \dot{P}_n F_n(t)]\right\}\\ &= \sum_n (\mathrm{Sp}\{\overline{\rho}(t)P_n\} - \langle P_n\rangle^t)\dot{F}_n(t) + \mathrm{Sp}\{\overline{\rho}(t) iL\hat{S}(t)\} = 0.\end{aligned} \tag{6.34}$$

In deriving the last relation we have taken into account that $\overline{\rho}(t)$ and the entropy operator $\hat{S}(t)$ commute among themselves and therefore

$$\mathrm{Sp}\{\overline{\rho}(t) iL\hat{S}(t)\} = 0.$$

Thus, entropy production in a quasi-equilibrium state is equal to zero. This means that there are no fluxes in the quasi-equilibrium state, and so such a distribution cannot describe the non-equilibrium state of the system. Given the above, we can point out that the quasi-equilibrium distribution characterizes an ensemble whose thermodynamic forces as if are compensated by some causes. Therefore, these thermodynamic fluxes do not develop.

However, there is another point of view. The quasi-equilibrium distribution describes a newly formed non-equilibrium ensemble of particles, the evolution of which just begins, so the thermodynamic fluxes have not yet been in progress. Obviously, the quasi-equilibrium distribution can be used as an initial condition for the true non-equilibrium distribution, which we expect to make further.

At the end of the section, we find the relationship between the second functional derivatives of the $S(t)$ and $\phi(t)$ potentials and correlation functions in the quasi-equilibrium state:

$$\frac{\delta\langle P_m\rangle^t}{\delta F_n(t)} = -\frac{\delta^2\phi(t)}{\delta F_n(t)\delta F_m(t)} = \frac{\delta}{\delta F_n(t)}\mathrm{Sp}\left\{P_m \exp\left\{-\left[\phi(t) + \sum_k P_k F_k(t)\right]\right\}\right\}. \tag{6.35}$$

Let us make a small mathematical digression and calculate the derivative over the parameter of the operator exponent. We first consider the simpler question of expanding the exponent

$$\exp\{(A+B)t\}$$

into a power series. Here, A and B are operators noncommuting among themselves, and t is some parameter. We introduce the following notation:

$$\exp\{(A+B)t\} \equiv D(t) \equiv G(t)\exp(At).$$

Let us make up an equation of motion for the function $D(t)$:

$$\frac{dD(t)}{dt} = (A+B)D(t) = \frac{dG(t)}{dt}\exp(At) + G(t)A\exp(At). \tag{6.36}$$

The operator A commutes with the operator exponent $\exp(At)$. Therefore, the second equality in the expression (6.36) can be written as

$$(A+B)D(t) = \frac{dG(t)}{dt}\exp(At) + D(t)A.$$

$A + B$ also commutes with $D(t)$, so

$$(A+B)D(t) = D(t)(A+B).$$

Substituting this result into the formula (6.36) and reducing the identical terms, one is led to

$$D(t)B = \frac{dG(t)}{dt}\exp(At),$$

or

$$\frac{dG(t)}{dt} = \exp\{(A+B)t\}B\exp\{-At\}. \tag{6.37}$$

Given that $\exp\{(A+B)t\} = G(t)\exp(At)$ and using the last equation, we get

$$\frac{d\ln G(t_1)}{dt_1} = \exp(At_1)B\exp(-At_1)\,dt_1.$$

Integrating this differential equation under the boundary conditions $G(0) = 1$, $\ln G(0) = 0$, we have

$$G(t) = \exp\left\{\int_0^t \exp(A\lambda)B\exp(-A\lambda)\,d\lambda\right\},$$

$$\exp\{(A+B)t\} = \exp\left\{\int_0^t \exp(A\lambda)B\exp(-A\lambda)d\lambda\right\}\exp(At). \tag{6.38}$$

If the operator B is small (smallness of the operator is understood as smallness of corresponding matrix elements) and one can envisage only the first terms of expansion, the following expression is arrived at instead of (6.38):

$$G(t) = 1 + \int_0^t \exp(A\lambda)B\exp(-A\lambda)\,d\lambda;$$

$$\exp\{(A+B)t\} = \exp(At) + \int_0^t \exp(A\lambda)B\exp(-A\lambda)\,d\lambda\exp(At). \tag{6.39}$$

Now, on account of this formula, one can derive a rule of differentiation of the operator exponent over a parameter. Using the definition of the derivative, we have

$$\frac{d}{d\lambda_2}\exp(P_1\lambda_1 + P_2\lambda_2) = \lim_{\Delta\lambda_2\to 0}\frac{1}{\Delta\lambda_2}[\exp(P_1\lambda_1 + P_2\lambda_2 + P_2\Delta\lambda_2) - \exp(P_1\lambda_1 + P_2\lambda_2)]. \tag{6.40}$$

Assuming that the $P_2\Delta\lambda_2$ is a small operator and $t = 1$, based on the formula (6.39) we obtain

$$\exp(P_1\lambda_1 + P_2\lambda_2 + P_2\Delta\lambda_2) = \exp(P_1\lambda_1 + P_2\lambda_2) + \int_0^1 \exp[(P_1\lambda_1 + P_2\lambda_2)\lambda]P_2\Delta\lambda_2\exp[-(P_1\lambda_1 + P_2\lambda_2)\lambda]\,d\lambda. \tag{6.41}$$

Given the last result, in the long run, we have

$$\frac{d}{d\lambda_2}\exp(P_1\lambda_1 + P_2\lambda_2) = \int_0^1 \exp[(P_1\lambda_1 + P_2\lambda_2)\lambda]P_2\exp[-(P_1\lambda_1 + P_2\lambda_2)(\lambda - 1)]\,d\lambda. \tag{6.42}$$

Let us go back again to the formula (6.35) and find the functional derivative by means of (6.42),

$$\frac{\delta}{\delta F_n(t)}\exp\Big\{-\sum_k P_kF_k(t)\Big\} = -\int_0^1 \exp\Big[-\sum_k P_kF_k(t)\tau\Big]P_n\exp\Big[\sum_k P_kF_k(t)(\tau - 1)\Big]\,d\tau. \tag{6.43}$$

Similarly, one can deduce the formula

$$\exp(-\phi(t)) = \Big[\mathrm{Sp}\Big\{\exp\Big(-\sum_k P_kF_k(t)\Big)\Big\}\Big]^{-1},$$

considering that

$$\frac{\delta}{\delta F_n(t)}\exp(-\phi(t)) = \frac{\mathrm{Sp}\{P_n\exp(-\sum_k P_kF_k(t))\}}{[\mathrm{Sp}\{\exp(-\sum_k P_kF_k(t))\}]^2}.$$

Now, an expression for the functional derivative of an average value of the basic operator can be found:

$$\frac{\delta}{\delta F_n(t)}\mathrm{Sp}\{P_m\bar{\rho}(t)\} = \langle P_n\rangle^t\langle P_m\rangle^t - \int_0^1 d\tau \mathrm{Sp}\{P_m\bar{\rho}(t)^\tau P_n\bar{\rho}(t)^{1-\tau}\}. \qquad (6.44)$$

Finally, determining the scalar of the two operators by dint of the relation

$$(P_m, P_n)_q^t = \int_0^1 d\tau \mathrm{Sp}\{(P_m - \langle P_m\rangle^t)\bar{\rho}(t)^\tau (P_n - \langle P_n\rangle^t)\bar{\rho}(t)^{1-\tau}\},$$

we obtain
$$\frac{\delta\langle P_m\rangle_q^t}{\delta F_n(t)} = -(P_m, P_n)_q^t. \qquad (6.45)$$

Let us sum up. Emanating from the extremality principle of informational entropy we have constructed the expression for the quasi-equilibrium statistical operator (6.22). The meaning of this distribution is that it describes the just prepared ensemble of non-equilibrium systems where evolution has not yet occurred and any fluxes are absent.

The key to understanding the NSO method is the relation (6.6), establishing the equality of average values of the basic operators P_n, which in turn we have calculated with the use of both non-equilibrium and quasi-equilibrium distributions. This relation can be interpreted as follows. By the time when the quasi-equilibrium ensemble had formed, the set of variables of P_n was the only set of quantities measurable in a non-equilibrium system. In the future, evolution of the system occurs so that new slowly changing dynamic variables do not appear, and the average values of operators $\langle P_n\rangle^t$ slowly evolve due to time-dependent conjugate thermodynamic forces $F_n(t)$.

As for the thermodynamic forces $F_n(t)$, they form in the course of the system's real evolution and will depend on non-equilibrium processes in this system. Finding the thermodynamic forces $F_n(t)$ is the theme of the section devoted to linear relaxation equations in the NSO method.

The results obtained allow thermodynamics of a non-equilibrium system to be also constructed. However, the explicit form of the quasi-equilibrium distribution is still unknown. Therefore, we will formulate an equation of motion for NSO in the next section. This enables one to restore the explicit form of the quasi-equilibrium distribution and to develop the thermodynamics of the non-equilibrium system.

6.1.3 Boundary conditions and the Liouville equation for the NSO

Consider a non-equilibrium system whose state over sufficiently large periods is described by a set of macroscopic variables of $\langle P_n\rangle^t$. As has been repeatedly noted, this means that only these quantities are measurable in the given system, and the assumption made above does not infringe on the reasoning's generality. Usually, the set of the quantities P_n is a set of hydrodynamic quasi-integrals of motion such as energy, the drift momentum, the number of particles, etc. However, we may also assign more small structural variables as the quantities P_n. The occupation number of quantum states may serve as a case in point.

Let us assume that a quasi-equilibrium ensemble of systems described by the quasi-equilibrium distribution $\overline{\rho}(t)$ is already prepared at a moment of time t_0. For convenience, the time t_0 should be referred to negative infinity, which implies "physical infinity". The physical infinity means periods much greater than some characteristic mixing time for the given system during which insignificant correlations for further evolution die out.

We first formulate an initial condition for a non-equilibrium statistical operator $\rho(t)$. Let non-equilibrium statistical and quasi-equilibrium operators coincide at the moment of time t_0.

Next, we should define a condition allowing writing down the non-equilibrium statistical operator in the form of a functional of the quasi-equilibrium distribution. As already noted, the quasi-equilibrium statistical operator $\overline{\rho}(t)$ does not satisfy the Liouville equation. Consequently, it will be transformed under the action of the evolution operator in contrast to the non-equilibrium distribution $\rho(t)$, which is an integral of motion.

We think that if one prepares the quasi-equilibrium distribution, the system to develop, then the transformation of the quasi-equilibrium distribution $\overline{\rho}(t)$ to the non-equilibrium distribution $\rho(t)$ after a certain time of order of the mixing time will take place.

In terms of mathematics, considering the definitions (6.2)–(6.4), the last condition and the boundary condition for NSO formulated above can be written as

$$\lim_{t_1 \to -\infty} \exp(it_1 L)\overline{\rho}(t + t_1, 0) = \lim_{t_1 \to -\infty} \exp(it_1 L)\rho(t + t_1, 0). \tag{6.46}$$

Equation (6.46) enables one not only to express the non-equilibrium statistical operator $\rho(t)$ via a quasi-equilibrium distribution $\overline{\rho}(t)$, but also brings irreversibility in behavior of the quantity $\rho(t)$. Indeed, in this equation, it suffices to send $t_1 \to +\infty$ to describe a decrease in entropy of the system rather than an increase of it. The reason of this is understandable. In the equation (6.46), the quasi-equilibrium distribution, formed at the time $t_0 = -\infty$, in the course of evolution is transformed into a non-equilibrium distribution at $t > t_0$.

In other words, a direction of the spontaneous process gets assigned, and a more ordered state corresponds to the lower temporal value. If one puts $t_0 = +\infty$, the system as time goes by will transit from a less ordered to more ordered state, which corresponds to the decrease in the entropy over time. The application Abel's theorem yields

$$\lim_{t \to -\infty} f(t) = \lim_{\epsilon \to 0} \epsilon \int_{-\infty}^{0} \exp(\epsilon t) f(t)\, dt. \tag{6.47}$$

If this limit exists, the equation (6.46) can rewrite in the following form:

$$\lim_{\epsilon\to 0} \epsilon \int_{-\infty}^{0} \exp(\epsilon t_1)\overline{\rho}(t+t_1, t_1)\, dt_1 = \lim_{\epsilon\to 0} \epsilon \int_{-\infty}^{0} \exp(\epsilon t_1)\rho(t+t_1, t_1)\, dt_1. \tag{6.48}$$

Equation (6.48) gives an interesting interpretation. Essentially, the formula (6.48) claims that the $\rho(t+t_1, t_1)$ and $\overline{\rho}(t+t_1, t_1)$ statistical operators smoothed (averaged) over a sufficiently large time interval are equal among themselves. Often, smoothing defined by the formula (6.48), is referred to as taking the invariant part. Obviously, $\rho(t+t_1, t_1) = \rho(t)$ and therefore:

$$\lim_{\epsilon\to 0} \epsilon \int_{-\infty}^{0} \exp(\epsilon t_1)\overline{\rho}(t+t_1, t_1)\, dt_1 = \rho(t). \tag{6.49}$$

From equations (6.48), (6.49) it follows that, in the course of evolution, a quasi-equilibrium distribution is transformed into a non-equilibrium one. This, in fact, is the physical meaning of (6.48). The result (6.49) can be obtained otherwise. Integrating the right-hand side of the equation (6.48) by parts, we get

$$\begin{aligned} &\epsilon \int_{-\infty}^{0} dt_1 \exp(\epsilon t_1) \exp(iLt_1)\overline{\rho}(t+t_1, 0) \\ &\quad = \rho(t, 0) - \lim_{t_1\to -\infty} \exp(\epsilon t_1)\rho(t+t_1, t_1) \\ &\qquad - \int_{-\infty}^{0} dt_1 \exp(\epsilon t_1) \exp(iLt_1)\left(\frac{\partial}{\partial t_1} + iL\right)\rho(t+t_1, 0). \end{aligned} \tag{6.50}$$

Let us require that the last integral in (6.50) should vanish. This requirement is satisfied automatically provided that $\rho(t, 0)$ is an exact integral of motion. Strictly speaking, as it will be shown later, $\rho(t, 0)$ is not an integral of the Liouville equation. But that expression for $\rho(t, 0)$ that we will obtain further also leads to the vanishing of the integral

$$\int_{-\infty}^{0} dt_1 \exp(\epsilon t_1) \exp(iLt_1)\left(\frac{\partial}{\partial t_1} + iL\right)\rho(t+t_1, 0). \tag{6.51}$$

Next,

$$\lim_{t_1\to -\infty} \exp(\epsilon t_1)\rho(t+t_1, t_1) = 0,$$

since the quantity ϵ in this formula is finite and must tend to zero after performing the thermodynamic limit and calculating averages. Therefore, the expression (6.50) is essentially the definition of the non-equilibrium statistical operator:

$$\rho(t,0) = \epsilon \int_{-\infty}^{0} dt_1 \exp(\epsilon t_1) \exp(iLt_1)\overline{\rho}(t + t_1, 0). \tag{6.52}$$

We now find an equation of motion, which is satisfied by NSO (6.52). To do this, the equation (6.52) needs to be differentiated over time t:

$$\begin{aligned}\frac{\partial\rho(t)}{\partial t} &= \epsilon \int_{-\infty}^{0} dt_1 \exp(\epsilon t_1) \exp(iLt_1)\frac{d}{dt}\overline{\rho}(t + t_1, 0) \\ &= \epsilon \exp(\epsilon t_1) \exp(iLt_1)\overline{\rho}(t + t_1, 0)|_{-\infty}^{0} - \epsilon\rho(t,0) - iL\rho(t).\end{aligned} \tag{6.53}$$

Given that the exp $\epsilon t_1 \to 0$ at the $t_1 \to -\infty$ one is led to the Liouville equation containing an infinitesimal source on the right-hand side:

$$\frac{\partial\rho(t,0)}{\partial t} + iL\rho(t,0) = -\epsilon\big(\rho(t,0) - \overline{\rho}(t,0)\big). \tag{6.54}$$

It is necessary to note that the vanishing of (6.51) is performed, as is easily seen if one recalls the formula (6.48).

One should say a few words about the meaning of the infinitesimal sources in the right-hand side of the equation of motion for NSO (6.54). As is known the Liouville equation (5.19) is time-reversible. However, in real systems, there is a spontaneous infringement of dynamical equations symmetry with respect to the operation of time reversal. Thus, in reliance on the fact that the second law of thermodynamics removes degeneracy of states associated with the symmetry with respect to operation of time reversal, the dynamical equations becomes corrected.

The main basis for more consistent interpretation of the emergence of the sources in the right-hand side of equation (6.54) is ideology of Bogoliubov's quasiaverages (see Section 5.2.4). Obviously, that under this aspect all averages, which are calculated by using the NSO method are quasiaverages; moreover, the term $-\epsilon(\rho(t,0) - \overline{\rho}(t,0))$ removes the degeneracy of the Liouville equation with respect to the operation of time reversal. Therefore, if the system is by itself and in contact with a heat bath, this term in some idealized form takes into account the contact, which leads to relaxation of the non-equilibrium distribution. Then the quantity ϵ can be interpreted as an inverse relaxation time of the non-equilibrium distribution to the quasi-equilibrium one.

6.1.4 Linear relaxation equations in the NSO-method

It would be proper to start solving problems by means of NSO method with the simplest case when a weakly non-equilibrium state of a system within a hydrodynamic approach can be described by a set of average values of thermodynamic coordinates of $\langle P_n \rangle^t$ or set of their conjugate thermodynamic forces $F_n(t)$ (6.31).

Consider a problem of determining the spectrum of hydrodynamic excitations for such a system. In other words, one has to put the problem of determining the decay times of associated fluctuations of the averages:

$$\delta\langle P_n \rangle^t = \langle P_n \rangle^t - \langle P_n \rangle_0^t,$$

where

$$\langle P_n \rangle_0^t = \mathrm{Sp}\{P_n \rho_0\},$$

and ρ_0 is the equilibrium Gibbs distribution. Because the non-equilibrium is weak, it is natural to assume that the set of equations describing the associated relaxation of the deviations $\delta\langle P_n \rangle^t$ should be linear.

To construct the linear relaxation equations with respect to the quantities $\delta\langle P_n \rangle^t$, linear expansion of the $\overline{\rho}(t,0)$, and $\rho(t,0)$ statistical operators need to be obtained.

We first expand the quasi-equilibrium statistical operator $\overline{\rho}(t,0)$. For simplicity, we adopt the following convention: we think of the quantities P, $\langle P \rangle^t$, $F(t)$ as column-vectors with components P_n, $\langle P_n \rangle^t$, $F_n(t)$, respectively. Then the quasi-equilibrium distribution (6.22) can be written as follows:

$$\overline{\rho}(t) = \exp(-\hat{S}(t,0)), \quad \hat{S}(t,0) = \phi(t) + P^{+}F(t). \tag{6.55}$$

Performing the expansion of $\hat{S}(t,0)$, we arrive at

$$\begin{aligned} \hat{S}(t,0) &= \hat{S}_0 + \delta\hat{S}(t,0), \quad \delta\hat{S}(t,0) = \delta\phi(t) + P^{+}\delta F(t), \\ \delta\phi(t) &= \ln \mathrm{Sp}\{\exp[-P^{+}(F_0 + \delta F(t))]\} - \ln \mathrm{Sp}\{\exp[-P^{+}F_0]\}. \end{aligned} \tag{6.56}$$

The quantities marked with a subscript 0 pertain to an equilibrium system.

To find the increment $\delta\phi(t)$ of a functional, it is necessary to expand the operator exponent in the last expression of (6.56) in the small parameter $P^{+}\delta F(t)$.

Using the formula (6.39) for the expansion of the operator exponent and considering the fact that the operator exponents under the spur sign can be cyclically traded, we get

$$\delta\phi(t) = -\mathrm{Sp}\{P^{+}\rho_0\}\delta F(t), \quad \rho_0 = \exp(-S_0). \tag{6.57}$$

Substituting the result (6.57) into the second equality of the expression (6.56), one is led to

$$\delta\hat{S}(t) = -\Delta P^{+}\delta F(t), \quad \Delta P^{+} = P^{+} - \mathrm{Sp}\{P^{+}\rho_0\}. \tag{6.58}$$

Using this representation, the expression (6.55) for the quasi-equilibrium distribution can be written as

$$\overline{\rho}(t) = \exp[-\hat{S}_0 - \delta\hat{S}(t, 0)]. \tag{6.59}$$

Performing again the expansion of the operator exponent (6.59) by means of the formula (6.39), we have

$$\overline{\rho}(t) = \rho_0 - \int_0^1 d\tau \rho_0^\tau \Delta P^+ \rho_0^{1-\tau} \delta F(t). \tag{6.60}$$

We perform a similar expansion of the non-equilibrium statistical operator $\rho(t, 0)$. Integrating this equation by parts yields

$$\rho(t) = \overline{\rho}(t) - \int_{-\infty}^0 dt_1 \exp(\epsilon t_1) \exp(iLt_1) \left(\frac{\partial}{\partial t} + iL \right) \overline{\rho}(t + t_1, 0). \tag{6.61}$$

We substitute the result as previously obtained into equation (6.61) for the expansion of the quasi-equilibrium distribution (6.60). In the long run, performing simple transformations, we get

$$\begin{aligned} \rho(t) = \rho_0 &- \int_0^1 d\tau \rho_0^\tau \Delta P^+ \rho_0^{1-\tau} \delta F(t) \\ &+ \int_{-\infty}^0 dt_1 \exp(\epsilon t_1) \exp(iLt_1) \\ &\times \int_0^1 d\tau \rho_0^\tau \{\Delta \dot{P}^+ \delta F(t + t_1) + \Delta P^+ \delta \dot{F}(t + t_1)\} \rho_0^{1-\tau}. \end{aligned} \tag{6.62}$$

The expression (6.62) allows one to solve the posed problem and to obtain the set of linear relaxation equations for fluctuations of the thermodynamic parameters $\delta\langle P_n \rangle^t$. For this it is only necessary to use the condition (6.6):

$$\langle P_n \rangle^t = \mathrm{Sp}\{P_n \rho(t, 0)\} = \mathrm{Sp}\{P_n \overline{\rho}(t, 0)\}.$$

However, there is a more convenient and elegant form of these equations by passing to the Fourier representation. Let us define Fourier transforms of the quantities $\delta\langle P_n \rangle^t$, $\delta\rho(t) = \rho(t) - \rho_0$, $\delta\overline{\rho}(t) = \overline{\rho}(t) - \rho_0$, $\delta F(t)$ by the following relations:

$$\delta\langle P \rangle^t = \int_{-\infty}^{\infty} \frac{1}{2\pi} d\omega \exp(-i\omega t) \delta\langle P \rangle^\omega,$$

$$\delta F(t) = \int_{-\infty}^{\infty} \frac{1}{2\pi} d\omega \exp(-i\omega t) \delta\langle F(\omega) \rangle,$$

$$\delta\rho(t) = \int_{-\infty}^{\infty} \frac{1}{2\pi}\, d\omega \exp(-i\omega t)\delta\rho(\omega),$$
$$\delta\overline{\rho}(t) = \int_{-\infty}^{\infty} \frac{1}{2\pi}\, d\omega \exp(-i\omega t)\delta\overline{\rho}(\omega). \tag{6.63}$$

Then, using the obvious relation

$$\delta\langle P\rangle^{\omega} = \mathrm{Sp}\{P\delta\overline{\rho}(\omega)\} = \mathrm{Sp}\{P\delta\rho(\omega)\}, \tag{6.64}$$

from the first side of the equality (6.64) one obtains the important result (see (6.60))

$$\delta\langle P\rangle^{\omega} = -(P,P^{+})\delta F(\omega); \quad (P,P^{+}) = \int_{0}^{1} d\tau \langle \Delta P \Delta P^{+}(i\hbar\beta\tau)\rangle. \tag{6.65}$$

Next, using the definition of $\delta\rho(t)$ and $\delta\overline{\rho}(t)$, one arrives at

$$\delta\rho(\omega) = \delta\overline{\rho}(\omega) + \int_{-\infty}^{0} \exp[(\epsilon - i\omega)t_1]\exp(iLt_1)\, dt_1$$
$$\times \int_{0}^{1} d\tau \rho_0^{\tau}\Delta(\dot{P}^{+} - i\omega P^{+})\rho^{1-\tau}\delta F(\omega). \tag{6.66}$$

If one integrates this expression over t_1, instead of (6.66) one gets a simple expression for $\delta\rho(\omega)$, which is convenient for practical applications:

$$\delta\rho(\omega) = \delta\overline{\rho}(\omega) + \int_{0}^{1} d\tau \frac{1}{\epsilon - i\omega + iL}\rho_0^{\tau}\Delta(\dot{P}^{+} - i\omega P^{+})\rho^{1-\tau}\delta F(\omega). \tag{6.67}$$

In the above expression, the operator resolvent is understood as a certain infinite series.

Now, we can construct the linear relaxation equations. Through logical reasoning, it becomes clear that such equations in temporal representation must be of the form

$$\frac{\partial}{\partial t}\delta\langle P\rangle^{t} = \int_{-\infty}^{t} T(t-t_1)\delta\langle P\rangle^{t_1}\, dt_1, \tag{6.68}$$

where $T(t-t_1)$ is a some core. Similar equations can be written for the deviations $\delta F(t)$.

An equation of the form (6.68) is easily obtained from the condition

$$\mathrm{Sp}\{P\delta\overline{\rho}(\omega)\} - \mathrm{Sp}\{P\delta\rho(\omega)\} = 0.$$

Substituting the results for $\delta\rho(\omega)$ and $\delta\overline{\rho}(\omega)$ as previously obtained, we have

$$\int_0^1 d\tau \mathrm{Sp}\left\{P\frac{1}{\epsilon - i\omega + iL}\rho_0^\tau(\Delta\dot{P}^+ - i\omega P^+)\rho_0^{1-\tau}\right\}\delta F(\omega) = 0. \tag{6.69}$$

Introducing the correlation functions for brevity,

$$\begin{aligned}(A,B)^\omega &= \int_0^1 \mathrm{Sp}\left\{\Delta A\frac{1}{\epsilon - i\omega + iL}\rho_0^\tau \Delta B \rho_0^{1-\tau}\right\} d\tau \\ &= \int_{-\infty}^0 dt_1 \exp[(\epsilon - i\omega)t_1](A, B(t_1)),\end{aligned} \tag{6.70}$$

gives an equation for the deviations $\delta F(\omega)$ of the thermodynamic forces

$$i\omega(P,P^+)^\omega \delta F(\omega) - (P,\dot{P}^+)^\omega \delta F(\omega) = 0. \tag{6.71}$$

It should be recalled that the equation (6.71) is of matrix form and the quantity $\delta F(\omega)$ is a column-vector.

It is convenient for further analysis to introduce the so-called transport matrix,

$$\overline{T}(\omega) = \frac{1}{(P,P^+)^\omega}(P,\dot{P}^+)^\omega. \tag{6.72}$$

Then the set of the linear relaxation equations takes the simple form

$$[i\omega - \overline{T}(\omega)]\delta F(\omega) = 0. \tag{6.73}$$

A completely analogous equation can be obtained for the quantities $\delta\langle P\rangle^\omega$. To do this, using the equation

$$\delta\langle P\rangle^\omega = -(P,P^+)\delta F(\omega)$$

(see formula (6.65)), it is necessary to express $\delta F(\omega)$ via $\delta\langle P\rangle^\omega$ and to substitute this result into the equation (6.72). Then a dispersion equation for $\delta\langle P\rangle^\omega$ appears:

$$\begin{aligned}&[i\omega - T(\omega)]\delta\langle P\rangle^\omega = 0, \\ &\overline{T}(\omega) = \frac{1}{(P,P^+)}T(\omega)(P,P^+).\end{aligned} \tag{6.74}$$

Equations (6.73) and (6.74) allow one to solve the problem about the related relaxation of hydrodynamic excitations in the weakly non-equilibrium systems. As far as the sets of equations (6.73) or (6.74) are homogeneous, the spectrum of elementary excitations to be sought is found by equating the system's determinant to zero:

$$\det|T(\omega) - i\omega| = 0. \tag{6.75}$$

Naturally, a transition to normal coordinates is a more accurate approach in solving such a problem. The normal coordinates are introduced so that the transport matrix of new variables would be diagonal.

The monograph by Forster [44] contains examples of collective hydrodynamic excitations in multiparticle systems. For this reason we will not discuss model systems and restrict ourselves to envisaging principal questions to develop a technique for calculating components of the transport matrix.

We define the matrix Green function of the relaxation equations (6.73) and (6.74) by the relations

$$\begin{aligned} \{T(\omega) - i\omega + \epsilon\}G(\omega) &= 1, \\ \{\overline{T}(\omega) - i\omega + \epsilon\}\overline{G}(\omega) &= 1. \end{aligned} \tag{6.76}$$

An explicit definition of the Green function $\overline{G}(\omega)$ can be easily obtained via the correlation functions $(P, P^+)^\omega$ and (P, P^+) by using the definition for $\overline{T}(\omega)$ (6.72). Integrating the numerator of (6.72) by parts brings about

$$\overline{T}(\omega) = \frac{1}{(P, P^+)^\omega}\{(P, P^+) + i(\omega + i\epsilon)(P, P^+)^\omega\}. \tag{6.77}$$

Substituting this result into the expression (6.76), one is led to

$$\overline{G}(\omega) = \frac{1}{(P, P^+)}(P, P^+)^\omega. \tag{6.78}$$

Similarly, the function can be determined:

$$G(\omega) = (P, P^+)^\omega \frac{1}{(P, P^+)}. \tag{6.79}$$

It follows from the definition (6.76) that the introduced Green functions (6.78), (6.79) are indeed Green functions of the relaxation equations, and their poles coincide with the spectrum of normal modes of the system.

We next sum up and outline the next steps of solving the posted Problem in determining the spectrum of the hydrodynamic excitations in a system whose state is determined by a set of dynamic parameters P_n.

Likely, the above results determine a formal problem, since the explicit calculation of the poles of the Green functions (6.78), (6.79) is a rather daunting problem. Typically, to determine the poles of a Green function either the mass operator method or a method based on a diagram technique is used. It should be emphasized that the use of the diagram technique to estimate the Green functions involved in kinetic coefficients leads, in the authors' opinion, to an unreasonably intricate theory. In addition, it is clear that all the results, which we can achieve by using the diagram technique, can be obtained by means of the mass operator method. However, the reverse assertion is incorrect.

Here, we will demonstrate another method to analyze the Green function. It is known as the Mori method [56]. The Mori projection operator method derives equations of motion both for correlation functions and for operators of dynamical quantities. The method mentioned above jointly with NSO method intended for constructing both the statistical operator and the generalized relaxation equations enables one to speak about a creation of a new method for solving the problems of physical kinetics. This new method is based on a successive use of ideology of the projection operator.

6.2 The projection operators method in non-equilibrium statistical mechanics

6.2.1 Why is it necessary to introduce projection operators?

To begin constructing a theory of irreversible phenomena, it is natural that the dynamic Liouville equation (5.19) to be taken as basic. But in this case, the question immediately arises: how to develop a theory which results in irreversible behavior of a system?

It is well known due to Boltzmann's classic work that a nondecreasing function can be found for a non-equilibrium system:

$$H_B = -\int d\vec{p} f(\vec{p}, t) \ln(f(\vec{p}, t)), \tag{6.80}$$

where $f(\vec{p}, t)$ is the one-particle distribution function, $\vec{p}$ the momentum of a particle.

The function H_B coincides with statistical entropy of a system with accuracy up to a multiplier that determines the dimension. The quantity $f(\vec{p}, t)$ satisfies the Boltzmann equation, which is not dynamical and more similar to a phenomenological diffusion equation in a phase space. One can try to generalize the definition (6.80), using the functional

$$S = -\int dp\, dq \rho(t) \ln(\rho(t)), \tag{6.81}$$

where ρ is a statistical operator and the integration is being performed over the entire surface of constant energy (classical case). We determine yet the more general form of the functional

$$S' = \int dp\, dq \rho(t) M(p, q), \tag{6.82}$$

where $M(p,q)$ is some function of sufficiently general form. If the quantity S' is a non-decreasing function (a Lyapunov function), the derivative $dS'/dt \geq 0$. To calculate this derivative a formal solution of the equation (5.19) needs to be written as

$$\rho(t) = \exp(-iLt)\rho(0), \tag{6.83}$$

where $\rho(0)$ is the statistical operator (at the initial instant of time, i. e. immediately after preparation of an ensemble). The definitions (5.19), (5.20) also imply that $d\rho(t)/dt = 0$. Emanating from this and differentiating (6.82) over time, we have

$$\frac{dS'}{dt} = \int dp\, dq\rho(t)iLM(p,q) \geq 0. \tag{6.84}$$

To write down this equation, we have used the definition

$$d/dtM(p,q) = iLM(p,q) \equiv \{M(p,q), H(p,q)\},$$

which holds true for classical mechanics (the role of the operator iL is playing the Poisson bracket $\{A,B\}$). Let us introduce the notation

$$iLM(p,q) = D(p,q),$$

where the quantity $D(p,q)$ can be simply both function and operator acting on the variables p, q. It can be shown that if $D(p,q)$ is simply a function of the variables p, q the Lyapunov function cannot be defined by the relation (6.82).

Indeed, consider the special case of an equilibrium system. Then $\rho(0) = \text{const}$ because one assumes that the system is ergodic. If $D(p,q)$ is a function of the variables p,q then $dS'/dt = 0$ for the thermodynamic equilibrium state and it follows from (6.82) that:

$$\frac{dS'}{dt} = \int dp\, dqD(p,q) = 0, \tag{6.85}$$

which by the arbitrariness of the system, immediately leads to the conclusion that $D(p,q) = 0$. Consequently, the functional (6.82) does not exist, if $D(p,q) = iLM(p,q)$ is a common function of the variables p, q.

An important conclusion follows from the result (6.85) obtained by Prigogine [35]. If we want to construct a Lyapunov function from first principles in the classical theory, we have to assume that the quantity $M(p,q)$, involved in the equation (6.82) should be an operator. As far as $D(p,q) = iLM(p,q) \neq 0$, then, according to the ideology developed in quantum theory, it follows that a system's energy and the quantity $M(p,q)$ cannot be measured simultaneously.

This fact can be interpreted as follows: the irreversible behavior of the system cannot be obtained staying within the concept of single particle trajectories, so the Lyapunov function cannot be built. We can reject the concept of the trajectories by introducing, as in quantum mechanics, a new operator quantity. It is worth recalling that, in quantum mechanics, a momentum operator is such a quantity, in the theory of irreversible phenomena; the operator $M(p,q)$ is closely related to the operator of microscopic entropy.

In the quantum case, when the quantities $H, \rho(t)$ are operators themselves, a Lyapunov function can be introduced by generalizing the relation (6.81), (6.82):

$$S' = -k\mathrm{Sp}\{\rho(t)\ln\rho(t)\}, \tag{6.86}$$

or in the more general form

$$S' = \mathrm{Sp}\{\rho(t)M\rho(t)\}. \tag{6.87}$$

It is quite clear, however, that the former cannot be a Lyapunov function due to the fact that $d\rho/dt = 0$; consequently $dS'/dt = 0$.

The expression (6.87) can play the role of the Lyapunov function only if the quantity M is a super operator, i. e. an operator acting not to functions, but operators. In addition, the operator M must not commute with Hamiltonian and, perhaps most importantly, the operator M must be a nonfactorizable operator. In other words, in quantum mechanics, it must not preserve differences between pure and mixed states.

Recall that all other quantum mechanical operators, acting on a wave function of a system in a pure state, leave it in its pure state. The nonfactorizability condition is less obvious and requires some explanations. It is clear that a description of the system in terms of wave functions is the most complete in the quantum theory, where irreversible behavior does not occur. In systems which are characterized by irreversible behavior, the difference between pure and mixed states is lost. Note that this does not mean that the Schrödinger equation is no longer valid. In these systems, differences between pure and mixed states become unobservable. Such a point of view belongs to Prigogine [35] who together with coworkers was developing it intensively.

The above analysis allows one to conclude that the irreversible behavior can be introduced neither in classical nor in quantum mechanics without significant additional assumptions that go beyond the standard classical or quantum theory. Hence, in particular, it follows that we will not succeed in creating the irreversible behavior of a system directly from dynamic equations without new physical ideas. The reason is that the dynamic description is insufficiently developed. There is to date a possibility of describing only integrable systems in classical mechanics or systems that are in a pure state in quantum mechanics.

Such reasoning is not new. In either case, Boltzmann himself and all founders of the theory of transport phenomena recognized this and extended their own methods of dynamics generalization to case of nonintegrable systems.

For example, Boltzmann used the collision hypothesis (Stosszahlansatz), according to which states of particles are assumed to be uncorrelated before they collide. Moreover, these states are described by one-particle distribution functions. Somewhat different arguments were suggested by Bogoliubov in deriving the Boltzmann kinetic equation of the system for s-particle distribution functions (see [17, 20]). The main idea of Bogoliubov is to distinguish several characteristic time scales where the system should be described by means of different approaches in a crucial respect.

If one adopts that the particle has a characteristic dimension R_0 and a characteristic velocity v, the system can be described only dynamically over the periods $t \approx \tau_{col} = R_0/v$.

The next temporal scale has a lot to do with mean free time of a particle. If one denotes an average distance between particles by letter $l \gg R_0$, the mean free time is equal to $\tau = l/v \gg \tau_{col}$. The kinetic stage of evolution begins when $\tau \leq t \gg \tau_{col}$. At these time intervals, according to Bogoliubov, a two-particle and each successive distribution function is some functional of a one-particle distribution function. It is this idea that allows a chain of Bogoliubov equations to be closed and one to obtain an equation for a one-particle distribution function. It is clear that the Bogoliubov approach is based on the assumption that the exact dynamics of a system taking into account all correlations becomes insignificant, starting from some point of time. The same idea underlies Boltzmann's hypothesis of collisions. Put differently, the above techniques are an attempt to make allowance for peculiarity of dynamics of nonintegrable systems demonstrating instability.

The method of projection operators dating back to the work of Zwanzig [57] is widely used to obtain irreversible dynamics. This method allows a statistical operator to fall into two orthogonal, in some sense, parts. It is worth pointing out that the next paragraph discusses properties of projection operators in detail, but here, we restrict ourselves to only certain remarks. To construct a statistical operator $\mathcal{P}\rho(t)$ Zwanzig managed to derive a time-irreversible equation of motion, which is usually referred to as *master equation*. Part of the statistical operator $\mathcal{P}\rho(t)$ is sometimes called the relevant part. The quantity $(1 - \mathcal{P})\rho(t)$ oscillates rather rapidly and it is not usually taken into account when calculating averages. Chapter 9 presents the method of constructing the description of non-equilibrium systems.

Another approach, based on the application of projection operators was used by Mori [56]. He advanced the method of constructing equations of motion for operators of physical quantities. This method was intended for determining dynamics of an arbitrary operator with the aid of dynamics of a set of the basic operators. In this case, as for the projection of the operator $\mathcal{P}A(t)$ we can deduce the time-irreversible equation of motion, which reminds one of the Langevin equation for a Brownian particle.

It is worth noting the distinct advantages to construct the theory of irreversible phenomena by using the projection operator technique, leaving aside details concerning their determination and practical utilization. First of all, the projection operator method enables one to set up new dynamical equations to describe irreversible and non-Hamiltonian evolution of dynamical quantities. Secondly, Zwanzig noted that the derivation of the master equations was a fairly simple and compact procedure.

In order for the irreversibility to occur, it is necessary to find an appropriate mechanism when invariance of a simple dynamic description with respect to time reversal would be infringed. Symmetry breaking that is of interest to us must be internal, i. e. not be related to new interactions. At the same time, this mechanism must be universal. Put another way, it must take place both in classical and in quantum systems.

Such a general and internal cause of the symmetry breaking may take place subject to realizing not all possible states but some set of states, which possesses by asymmetry of a required type. In substance, this idea is a new theoretic postulate, which is equivalent to status of the second law of thermodynamics. (See the monograph [35].)

It is interesting to note that as far back as 1909 Ritz came up with such a statement of the second law of thermodynamics. He deemed that the second law of thermodynamics permitted some solutions of dynamic equations to be eliminated when appropriate.

The consistent construction of the theory of irreversible processes as being dynamic is best carried out with the projection operator method. The above theory is suitable to describe either weak stability or spontaneous symmetry breaking in systems. The projection operator approach was specially developed for selecting essential states of evolution.

There is, however, a bolder idea. Developing the projection operator method, we take a step towards a creation of new dynamics where the second law of thermodynamics becomes brought to the level of a dynamic principle as picking up only physically realizable solutions.

6.2.2 The Mori projection operator method

It follows from all the foregoing discussions that the investigation of the dynamics of hydrodynamic fluctuations leads to the problem of computing correlation functions of the basic operators i. e. correlation functions dynamical variables as measurable quantities, on the one hand, and as sufficient ones to describe physical phenomena under consideration, on the other hand. The calculation of these correlation functions is a complex independent problem. In essence, we have moved forward here only in the sense that we succeeded in reducing the problem of relaxation in a weakly non-equilibrium system to investigation of the correlation functions, which are ascertained for an equilibrium state.

It is well known that there is a crucial possibility of such a reduction, or otherwise, a possibility of expressing kinetic coefficients of a weakly non-equilibrium system via equilibrium correlation functions. The above statement is the fluctuation–dissipation theorem (FDT), which in turn is a powerful tool in statistical physics for predicting the behavior of non-equilibrium thermodynamical systems.

The fluctuation–dissipation theorem relies on the assumption that the response of a system in thermodynamic equilibrium to a small applied force is the same as its response to a spontaneous fluctuation. Put another way, micro processes, giving rise to the relaxation in the non-equilibrium system are the same as the causes of the dissipation of the fluctuations in the equilibrium system.

Now, we should take a step further and develop a procedure for calculating the equilibrium correlation functions of the basic operators. In substance, this is a little different set up of the problem discussed in Section 6.2.1 where we analyzed the reasons that proved to be convenient for introducing the projection operators.

There are a lot of various definitions of the projection operators, which are used for building equations of motion for the dynamical variables. To gain insight into the technique of the projection operators, it would be proper to start with a projection operator approach proposed by Mori (see [56]).

The Mori projection operator method is based on the simple idea that any dynamic operator $A(t)$ can be divided into two components: one of them is expressed through basic operators and c-number functions, and another part is the rest:

$$A(t) = \mathcal{P}A(t) + QA(t), \quad Q = (1 - \mathcal{P}),$$
$$\mathcal{P}A(t) = (A(t), P^{+})(P, P^{+})^{-1}P, \quad \mathcal{P}^2 = \mathcal{P}. \tag{6.88}$$

The scalar product of two operators is defined as before (see equation (6.65)):

$$(A, B) = \int_0^1 d\tau \mathrm{Sp}\{\Delta A \rho_0^{\tau} \Delta B \rho_0^{1-\tau}\}. \tag{6.89}$$

It is quite clear that such a division is accurate and it can be always done. The whole point of the mathematical operation consists of that the operators $\mathcal{P}A(t)$ and $QA(t)$ pertain to completely different types of time dependence. The operators P and P^{+} are quasi-integrals of motion, i. e. almost preserved quantities. They change in time due to only relatively weak perturbations of a basic Hamiltonian.

The quantity $QA(t)$, on the contrary, oscillates rapidly with a characteristic time of atomic scales. It is the fact that enables one to separate slow evolution of an operator and fast oscillations as determining only the processes with a characteristic atomic frequency scale.

We should say a few words about the meaning of the concepts “slow evolution” and “rapid evolution” of the operators. The fact is that the equation of motion for the correlation function is obtained from the operator equation by multiplying all its terms on the right-hand side by some time-independent operator and then by computing the average over an equilibrium state. Therefore, the behavior of the operator is comparable with the behavior of the correlation functions.

It is easy to understand the meaning of the projection operator if one uses the geometric analogy shown in Figure 6.1 for the case when there is the only one operator in the set of P.

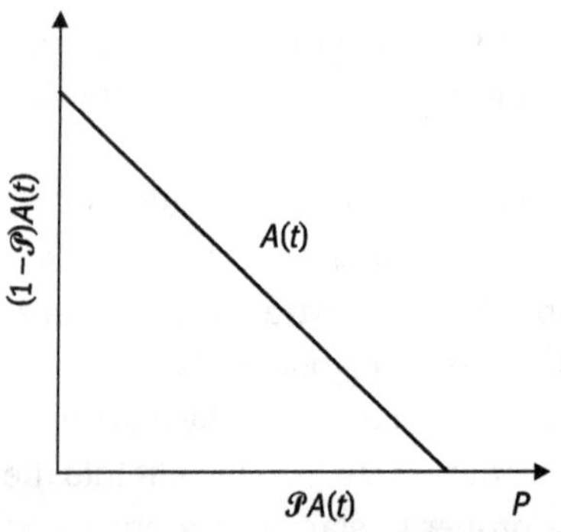

Figure 6.1: The geometric meaning of the projection operator.

Using the definition of the projection operator (6.88), it is easy to prove that the most important condition to project a vector on the axis of the orthogonal basis is fulfilled: operators $\mathcal{P}A(t)$ and $(1 - \mathcal{P}A(t))$ are orthogonal in the sense of the scalar product of (6.89):

$$(\mathcal{P}A(t), (1 - \mathcal{P})A^{+}(t)) = 0. \tag{6.90}$$

To prove the relation (6.90) we consider the effect of the projection operator upon the adjoint operator $A^{+}(t)$. Applying the definition of the Mori projection operator (6.88), we get

$$\begin{aligned} \mathcal{P}A^{+}(t) &= (\mathcal{P}A(t))^{+} = \left((A(t), P^{+})\frac{1}{(P, P^{+})}P\right)^{+} \\ &= P^{+}\frac{1}{(P, P^{+})}(A(t), P^{+})^{+} = P^{+}\frac{1}{(P, P^{+})}(P, A^{+}(t)). \end{aligned} \tag{6.91}$$

The last equality in the formula (6.91) is immediately obtained, if one recalls that $\mathrm{Sp}\{AB\}^{+} = \mathrm{Sp}\{B^{+}A^{+}\}$. Now, a proof of (6.90) is simply to take into account the formula (6.91) and performing the algebraic manipulations in the expression (6.90).

It must be emphasized that the operator Q is also idempotent projection operator and the condition $Q^2 = Q$ satisfies it.

Consider the equation of motion for the operator P, belonging to a set of basic operators:

$$\frac{d}{dt}P(t) = iLP(t). \tag{6.92}$$

We apply the operator $Q = (1 - \mathcal{P})$ to this equation. Since the operator $(1 - \mathcal{P})$ is time-independent, it can be permuted with the differentiation operator with respect to time. Introducing the notation $QP(t) = (1 - \mathcal{P})P(t) = P'(t)$, one is led to

$$\frac{d}{dt}P'(t) = QiL(1 - \mathcal{P})P(t) + QiL\mathcal{P}P(t). \tag{6.93}$$

Next, it would be convenient to introduce the notation to simplify the formulas:

$$\mathcal{P}P(t) = (P(t), P^{+})(P, P^{+})^{-1}P = \Theta(t)P,$$
$$\Theta(t) = (P(t), P^{+})(P, P^{+})^{-1}. \tag{6.94}$$

Given these definitions, the equation (6.93) can be rewritten in the form

$$\frac{d}{dt}P'(t) - (1 - \mathcal{P})iLP'(t) = \Theta(t)(1 - \mathcal{P})\dot{P}. \tag{6.95}$$

Equation (6.95) can be easily integrated. For this, the left-hand side should be multiplied by the operator exponent:

$$\exp\{-(1 - \mathcal{P})iLt\}.$$

Then the first two terms in equation (6.95) can be merged as one, and integration in the range from 0 to t yields

$$P'(t) = \int_0^t dt_1 \Theta(t_1) \exp\{(1 - \mathcal{P})iL(t - t_1)\}(1 - \mathcal{P})\dot{P}(t_1). \tag{6.96}$$

This result is intermediate and will be used later.

Consider an equation of motion for the correlation function $\Theta(t)$ (6.94). Using again the relation

$$iLP(t_1) = iL\mathcal{P}P(t_1) + iL(1 - \mathcal{P})P(t_1), \tag{6.97}$$

we get

$$\frac{d}{dt_1}\Theta(t_1) = \left(\frac{d}{dt_1}P(t_1), P^{+}\right)(P, P^{+})^{-1} = (\dot{P}, P^{+}(-t_1))(P, P^{+})^{-1}$$
$$= (\mathcal{P}\dot{P}, P^{+}(-t_1))(P, P^{+})^{-1} + ((1 - \mathcal{P})\dot{P}, P^{+}(-t_1))(P, P^{+})^{-1}, \tag{6.98}$$

or

$$\frac{d}{dt_1}\Theta(t_1) = i\Omega\Theta(t_1) + ((1 - \mathcal{P})\dot{P}, P^{+}(-t_1))(P, P^{+})^{-1}, \tag{6.99}$$

where $i\Omega$ is so-called frequency matrix, $i\Omega = (\dot{P}, P^{+})(P, P^{+})^{-1}$.

Consider the scalar product $((1 - \mathcal{P})\dot{P}, P^{+}(-t_1))$. As far as the equality

$$((1 - \mathcal{P})C, \mathcal{P}B^{+}) = 0$$

holds true for arbitrary operators C and B, the scalar product can be written as

$$((1 - \mathcal{P})\dot{P}, (1 - \mathcal{P})P^{+}(-t_1)).$$

Taking into account the result (6.96) the equation of motion for a correlation function can be represented as

$$\frac{d}{dt}\Theta(t) = i\Omega\Theta(t) + \int_{0}^{-t} dt_1((1-\mathcal{P})\dot{P}, (\exp\{-i(1-\mathcal{P})L(t+t_1)\}(1-\mathcal{P})\dot{P})^{+})\Theta(t_1)^{+}(P,P^{+})^{-1}. \tag{6.100}$$

Consider the correlation function

$$\Theta(t_1)^{+} = \frac{1}{(P,P^{+})^{+}}\int_{0}^{1} \mathrm{Sp}\{P(t_1)\rho_0^{\tau}P^{+}\rho_0^{1-\tau}\}^{+}\, d\tau.$$

Given the above expression and the symmetry properties of correlation functions under a Hermitian conjugation operation, we obtain

$$\Theta(t_1)^{+} = \frac{1}{(P,P^{+})}(P(-t_1),P^{+}).$$

Finally, we make a change of the variables in the integral by introducing the new variable $s = t_1 + t$, and define a magnitude of the random force f by the relation $f = (1-\mathcal{P})\dot{P}$. Taking into account all observations previously made, instead of (6.100) we obtain the following expression:

$$\frac{d}{dt}\Theta(t) = i\Omega\Theta(t) - \int_{0}^{t} ds(f,f^{+}(-s))\frac{1}{(P,P^{+})}\Theta(t-s). \tag{6.101}$$

If one has recourse to the fact that $\Theta(t) = (P(t),P^{+})(P,P^{+})^{-1}$, an equation of motion for the dynamical variable $P(t)$ can be also obtained:

$$\frac{d}{dt}P(t) = i\Omega P(t) - \int_{0}^{t} ds\Sigma(s)P(t-s), \tag{6.102}$$

where $\Sigma(s)$ is the so-called memory function, which takes into account the pre-history of a system at time $0 < s < t$

$$\Sigma(s) = (f,f^{+}(-s))(P,P^{+})^{-1}. \tag{6.103}$$

Now, it makes sense to summarize and discuss the physical meaning of the results obtained. The form of the equations (6.101), (6.102) resembles in appearance a Langevin equation for a Brownian particle. Consequently, they describe non-Markovian dynamics of the quantities P_n in question. It is important to emphasize that the temporal evolution of the memory function

$$\Sigma(s) \sim (f,f^{+}(-s)) = \int_{0}^{1} d\tau\mathrm{Sp}\{(1-\mathcal{P})\dot{P}\rho_0^{\tau}[\exp\{-(1-\mathcal{P})iLs\}(1-\mathcal{P})\dot{P}]^{+}\rho_0^{1-\tau}\}$$

is non-Hamiltonian and is determined by only one part of the Hamiltonian in which the terms responsible for the slow evolution of dynamical variables are excluded by using the projection operator Q.

Note that the above separation of the rapidly changing kernel of the integral equations (6.101), (6.102) has been produced exactly. In addition, we have not yet made any assumptions about weakness of interactions in the system.

Finally, we discuss the meaning of the "identical" transformations, which we have done in deriving the equations (6.101), (6.102). It seems necessary to do this right now because the reader is likely to wonder: does it make any sense to perform the identical transformations of the dynamic equations, entering the projection operators if anything new cannot be obtained?

Actually, this is a quite complex question. For an answer to be arrived at, it is necessary again to turn to the problem of describing the systems, showing irreversible behavior (see Chapter 1).

It seems reasonable to simplify the problem having considered the situation in the Markov limit when a correlator of random forces (6.103) is supposed to have a δ-shaped time dependence. In the case of electrical conductivity, such a situation may arise provided that the duration of interaction of colliding particles is much less than the time between the collisions. We would recall for the reader that the Boltzmann kinetic equation for the case of a low-density gas is also a Markov equation.

Substituting the value $\Sigma(s) = \Gamma\delta(s)$ into the expression (6.102), an equation of motion for the operator in the Markov limit can be given as

$$\frac{d}{dt}P(t) = i\omega P(t) - \gamma P(t). \tag{6.102a}$$

In writing this expression we have distinguished the real and imaginary parts:

$$\omega = \mathrm{Re}\,\Omega + \mathrm{Im}\,\Gamma; \quad \gamma = \mathrm{Re}\,\Gamma + \mathrm{Im}\,\Omega.$$

The meaning of equation (6.102a) is obvious. If $\Gamma = 0$, the dynamic quantity $P(t)$ oscillates with a characteristic frequency ω. But if the value of $\Gamma \neq 0$, the precession damps and the quantity γ makes sense of reverse decay time.

It is this division of the dynamic equation into the two summands that constitutes basic sense of using the projection operators. Here, the first and second terms describe precession and damping, respectively. Moreover, it should be noted that the temporary evolution of random forces within the memory function is not Hamiltonian because it is determined only by part of Hamilton's function as orthogonal in some sense to a set of the basic operators.

As far as, usually, hydrodynamic quasi-integrals of motion should be chosen as basic operators, the partition of the dynamic equation of motion for the physical quantity $P(t)$ on a regular and dissipative component is responsible for distinguishing two different time scales of evolution. It is this idea which was proposed by Bogoliubov to derive a kinetic equation for a one-particle distribution function.

The emergence of damping in the equation of motion for the dynamic variable in terms of quantum mechanics can be interpreted slightly differently. If a spectrum of elementary excitations is characterized by an actual value of energy or frequency, the elementary excitation is well defined and becomes permanent. Such a system is not dissipative. But if the elementary excitation fails to be well-defined and there is an imaginary part in the spectrum, there arises an analogue of the uncertainty relation. It should, however, be recognized that now the uncertainty relation is due to failure of isolating a subsystem from surroundings. Therefore, the subsystem is considered to be in a mixed-state and is a part of some other system. That is why a phase surface of constant energy turns into a layer with width ΔE and an exact value of the energy cannot be specified. Consequently, this stands for information loss about the system or its irreversible behavior.

Let us go back again to the further analysis of the equations of motion which were obtained by means of the Mori projection operator method.

Equations (6.101), (6.102) appear to be simpler, if we write them for Laplace-images of the $\Theta(t)$ and $P(t)$ functions, having performed the Laplace transforms. Specialized literature (see [58]) contains more detailed information on this matter, therefore, basic relations required to perform Laplace transforms of the equations (6.101), (6.102) are presented below.

The direct and inverse Laplace transforms of a function $f(x)$ are given by the expressions

$$f(s) = \int_0^{\infty} f(x)e^{-sx}dx,$$

$$f(x) = \frac{1}{2\pi i}\int_{C-i\infty}^{C+i\infty} f(s)e^{sx}ds. \tag{6.104}$$

In the second formula of (6.104), the integration is being performed along a line s in the complex plane, for which $\mathrm{Re}\, s = C$.

To transform the equations (6.101), (6.102) other formulas of Laplace-transforms are required both for the derivative $f'(x)$ and for the convolution of two functions:

$$g(x) = \int_0^{x} dt f_1(t)f_2(x-t).$$

We present these formulas without proof [58]:

$$\int_0^{\infty} dxe^{-sx}f'(x) = sf(s) - f(0),$$

$$g(s) = f_1(s)f_2(s). \tag{6.105}$$

Now one can write down the result, which is obtained by applying the relations (6.104), (6.105) and after performing the Laplace-transforms of (6.101), (6.102):

$$\Theta(z) = \frac{\Theta(0)}{z - i\Omega + \Sigma(z)}, \tag{6.106}$$

$$P(z) = \frac{P(0)}{z - i\Omega + \Sigma(z)}, \tag{6.107}$$

$$\Sigma(z) = \int_0^\infty dt e^{-zt}(f, [f(-t)]^+)(P, P^+)^{-1}. \tag{6.108}$$

In essence, the expressions obtained speak for themselves. Indeed, the structure of the expression (6.106) is reminiscent of the Fourier transform of an autocorrelation function when writing in the standard procedure the equations of motion for Green functions with subsequent use of the mass-operator approach. Meanwhile the quantities Ω and Σ correspond to real and imaginary parts of the mass operator.

By analogy with the mass-operator method, the correlation function can be expanded into a continued fraction. For that purpose, it suffices to make the transformations for the function $\Sigma(z)$ by taking the successive steps from the formula (6.94) to the formula (6.106). Thus, we descend to a level below. This down-path as a matter of fact means that more subtle correlations in a system should be taken into account. Of course, it can be extended further. Indeed, using such an approach, we can write an infinite chain of coupled equations as the continued fraction expansion.

What is the practical advantage of using the technique based on the Mori projection operator? The aim of this method under a proper choice of dynamical variables is to obtain immediately an expression for the memory function $\Sigma(z)$, containing the interaction of at least a second power.

Due to this, the interaction with scatterers (phonons, impurities, etc.) for the statistical operator and operators of evolution can be immediately ignored when calculating the kinetic coefficients in the Born approximation of scattering theory. Then the quantity $\Sigma(z)$ can be found at the first onset.

In the next section, we will demonstrate the application of the Mori projection operator method and NSO method in the simplest cases to estimate both electrical conductivity and magnetic susceptibility for a free-electron system in conducting crystals.

Quite similarly, one can find, in principle, the poles of the Green functions (6.78) and (6.79) as defining a spectrum of hydrodynamic excitations in a system. However, it is worth pointing out that the transition to normal coordinates is a necessary preliminary condition in order for a Green function matrix to be diagonal.

6.2.3 Using the Mori projection operators to calculate conductivity

The formal expression for the electrical conductivity, known as the Kubo formula [36], can be obtained in two ways. First, the electrical conductivity can be defined as a response of a system to an external high-frequency electric field. Another method implies some connection between fluctuations of the drift momentum of the electron system and fluctuations of the internal electric field. In the given case, the two approaches lead to identical results, which we can easily demonstrate by using the results of the present chapter.

Let us first find an expression for the electrical conductivity in the form of the system response to the external electric field. The formula in question can be derived quite simply without involving the NSO method. For that we restrict ourselves to the theory of Kubo's linear response to an external mechanical disturbance (see Section 5.1.3).

However, it would be reasonable to utilize the NSO method to solve the above task, bearing in mind a more complicated case in further consideration, but namely a linear response of a non-equilibrium system to a weak-measurable field.

Consider a non-equilibrium system described by the Hamiltonian H. We assume that the disturbance is given by the Hamiltonian H_v and acts on this system. The explicit form of this Hamiltonian will be determined later. In particular, we are interested in the case when the external electric or magnetic field induces the disturbance.

The Liouville equation (6.54) for a non-equilibrium statistical operator (NSO) can be rewritten now in the form

$$\frac{\partial \rho(t,0)}{\partial t} + (iL + iL_v)\rho(t,0) = -\epsilon(\rho(t,0) - \overline{\rho}(t,0)),\tag{6.109}$$

where L_v is a Liouville operator corresponding to part of the Hamilton operator H_v.

Next, we transform the equation (6.109) to an equivalent integral equation. Subtracting from the left-hand and right-hand sides of the equation (6.109) the expression

$$\left(\frac{\partial}{\partial t} + iL\right)\overline{\rho}(t,0)$$

can be brought into the form

$$\left(\frac{\partial}{\partial t} + iL + \epsilon\right)\delta\rho(t,0) = -\left(\frac{\partial}{\partial t} + iL\right)\overline{\rho}(t,0) - iL_v\rho(t,0),$$

$$\delta\rho(t,0) = \rho(t,0) - \overline{\rho}(t,0).\tag{6.110}$$

Introducing the evolution operator $\exp(iLt)$ with the Hamiltonian H and multiplying the first equation of (6.110) by a factor

$$\exp(\epsilon t)\exp(iLt),$$

we have

$$\frac{d}{dt}\exp(\epsilon t)\exp(iLt)\delta\rho(t,0) = -\exp(\epsilon t)\exp(iLt)\left[\left(\frac{\partial}{\partial t} + iL\right)\overline{\rho}(t,0) + iL_v\rho(t,0)\right].\tag{6.111}$$

Assuming that

$$\lim_{t \to -\infty} \exp(\epsilon t) \exp(iLt) \delta\rho(t, 0) = 0,$$

we integrate the equation (6.111) over time between the limits $-\infty$ and t:

$$\rho(t, 0) = \bar{\rho}(t, 0) - \int_{-\infty}^{0} dt_1 \exp(\epsilon t_1) \exp(iLt_1)$$

$$\times \left\{ \frac{\partial}{\partial t_1} \bar{\rho}(t + t_1) + iL\bar{\rho}(t + t_1) + iL_v\rho(t + t_1) \right\}. \tag{6.112}$$

To derive this formula, the result of integrating the equation (6.111) must be multiplied on the left by $\exp(-\epsilon t) \exp(-iLt)$. After that, it is necessary to make a change of variables in the integral, putting $t_1 - t \to t_1$.

In fact, this is the desired integral equation. If the interaction operator H_v does not come into the picture explicitly in the basic operators P_n (which is expected in the future), the equation (6.112) admits a simple interpretation. Since the first two terms in the integral in the formula (6.112) depend on the H_v only implicitly via the parameters of $F_n(t)$, they describe so-called thermal perturbations, whereas the third term, containing explicitly the interaction H_v, describes a mechanical perturbation.

The last statement is obvious if one considers the case where the quantities F_n are equal to their equilibrium values, and the operators P_n commute with the Hamiltonian. Then the expression (6.112) coincides with the result, which gives Kubo's linear response theory.

Equation (6.112) can have another form, which will be used in below. For that purpose, it should be noted that

$$\bar{\rho}(t, 0) - \int_{-\infty}^{0} dt_1 \exp(\epsilon t_1) \exp(iLt_1) \left\{ \frac{\partial}{\partial t_1} \bar{\rho}(t + t_1) + iL\bar{\rho}(t + t_1) \right\}$$

$$= \epsilon \int_{-\infty}^{0} dt_1 \exp(\epsilon t_1) \exp(iLt_1) \bar{\rho}(t + t_1). \tag{6.113}$$

This result is obtained by simple integration by parts in the left-hand side of (6.113), because

$$\exp(iLt_1) \left\{ \frac{\partial}{\partial t_1} \bar{\rho}(t + t_1) + iL\bar{\rho}(t + t_1) \right\} = \frac{d}{dt_1} \exp(iLt_1) \bar{\rho}(t + t_1).$$

Introducing the notation

$$\rho^0(t, 0) = \epsilon \int_{-\infty}^{0} dt_1 \exp(\epsilon t_1) \exp(iLt_1) \bar{\rho}(t + t_1), \tag{6.114}$$

we write down an integral equation for the NSO in the final form:

$$\rho(t,0) = \rho^0(t,0) - \int_{-\infty}^{0} dt_1 \exp(\epsilon t_1) \exp(iLt_1) iL_v \bar{\rho}(t + t_1). \tag{6.115}$$

The quasi-equilibrium distribution $\bar{\rho}(t,0)$ yields the distribution $\rho^0(t,0)$ as a consequence of an evolutionary perturbation-free system with the Hamiltonian H whereas the distribution $\rho(t)$ is a result of the evolution with the total Hamiltonian $H + H_v$. It should be noted that the distributions as a matter of fact are not independent, since $\rho^0(t,0)$ depends on accurate values of the functions $F_n(t)$. These values should be determined by the generalized kinetic equations (6.8).

Now one can return to the problem of calculating the electrical conductivity. Let $\rho^0(t,0)$ be equal ρ_0 and the system be at equilibrium before switching on the electric field. Here, ρ_0 is Gibb's equilibrium distribution. In addition, one must confine oneself to a linear approximation of the electric field when calculating the system response and make the substitution of ρ_0 for $\bar{\rho}(t,0)$ in the integral (6.115). Furthermore, the operator of interacting electrons with a uniform external electric field $E(t)$ needs to be taken as the operator H_v:

$$H_v(t) = -e \sum_j X_j^\alpha E^\alpha(t). \tag{6.116}$$

The summation with respect to j is being performed over coordinates X_j of all electrons. The index α denotes a projection on the axis of a Cartesian system. We find the average value of electric current $J^\alpha(t)$ of the system by calculating the average:

$$J^\alpha(t) = \mathrm{Sp}\left\{ e\frac{P^\alpha}{m}\rho(t,0) \right\} = \frac{e^2}{m} \int_{-\infty}^{0} dt_1 \exp(\epsilon t_1)$$
$$\times \mathrm{Sp}\left\{ P^\alpha \frac{1}{i\hbar}[\rho_0, X^\beta(t_1)] \right\} E^\beta(t + t_1), \quad P^\alpha = \sum_j p_j^\alpha, \tag{6.117}$$

where p_j is the momentum of the jth electron. Performing the Fourier transformation of the equation (6.117) and considering the phenomenological definition of the electrical conductivity tensor:

$$J^\alpha(\omega) = \sigma_{\alpha\beta}(\omega) E^\beta(\omega),$$

one can obtain the well-known expression for electrical conductivity:

$$\sigma_{\alpha\beta}(\omega) = \frac{e^2}{m} \int_{-\infty}^{0} dt_1 \exp[(\epsilon - i\omega)t_1] \mathrm{Sp}\left\{ P^\alpha \frac{1}{i\hbar}[\rho_0, X^\beta(t_1)] \right\}. \tag{6.118}$$

The direct calculation of the electrical conductivity at the finite order of perturbation theory is not possible, since, in the given case, it gives rise to physically unreasonable result. Indeed, conductivity of a system at zero frequency should be inversely proportional to the effective constant of electron–scatterer interactions. But it occurs only if one sums up an infinite series, for example, an infinitely decreasing geometric progression. That is why the mass-operator method is usually used to calculate the electrical conductivity according to the formula (6.118).

Now, one should show that exactly the same result is obtained when applying the Mori projection operator method. We first prepare the expression (6.118), using the Kubo formula (5.60):

$$\frac{1}{i\hbar}[\rho_0, X^\alpha] = \frac{\beta}{m}\int_0^1 d\tau \rho_0^\tau P^\alpha \rho_0^{1-\tau}. \tag{6.119}$$

Here, β is the inverse temperature in energy units. Substituting the result (6.119) into (6.118), the expression for conductivity can be written through the Mori scalar product:

$$\sigma_{\alpha\beta}(\omega) = \frac{e^2\beta}{m^2}\int_{-\infty}^0 \exp\{(\epsilon - i\omega)t_1\}(P^\alpha, P^\beta(t_1))\,dt_1. \tag{6.120}$$

In order to utilize the results (6.106) and (6.107) in calculating the components of the electrical conductivity tensor, the Cartesian components of the total electron momentum operator P^α should be taken as basic operators P_n involved in the formula (6.106). In addition, the complex variable z needs to be introduced by the relation $\epsilon - i\omega = z$ instead of frequency. This results in modifying the expression (6.120):

$$\begin{aligned}\sigma_{\alpha\beta}(z) &= \frac{e^2\beta}{m^2}\int_{-\infty}^0 \exp\{zt_1\}(P^\alpha, P^\beta(t_1))\,dt_1 \\ &= \frac{e^2\beta}{m^2}\int_0^\infty \exp\{-zt_1\}(P^\alpha(t_1), P^\beta)\,dt_1 = \frac{e^2\beta}{m^2}\Theta(z)(P^\alpha, P^\beta).\end{aligned} \tag{6.121}$$

Here, attention should be given to the fact that the substitution of $t_1 \to -t_1$ has been made in order for the second equation in the formula (6.121) should be obtained.

Finally, using the expression (6.106) for the correlation function $\Theta(z)$, we represent the formula to calculate the electrical conductivity:

$$\sigma_{\alpha\beta}(z) = \frac{e^2\beta}{m^2}\frac{\Theta(0)(P^\alpha, P^\beta)}{z - i\Omega + \Sigma(z)}. \tag{6.122}$$

In order to compare the result (6.122) with the expression obtained through the mass-operator method (5.51), it should be noted that $\Theta(0) = 1$, $\Omega = 0$. The first follows from the definition (6.94) of the correlation function $\Theta(t)$. To prove the second equality we first consider the correlation function (P^α, P^β) in the numerator of the formula (6.122)

$$(P^\alpha, P^\beta) = \beta \int_0^1 d\tau \mathrm{Sp}\{P^\alpha \rho_0^\tau P^\beta \rho_0^{1-\tau}\}$$
$$= m\mathrm{Sp}\left\{P^\alpha \frac{1}{i\hbar}[\rho_0, X^\beta]\right\} = m\mathrm{Sp}\left\{\frac{1}{i\hbar}[X^\beta, P^\alpha]\rho_0\right\} = mn\delta_{\alpha\beta}, \tag{6.123}$$

where n is the electron concentration. Repeating the same operations for the frequency matrix $i\Omega$ according to its definition (see formula (6.99)), we get

$$i\Omega \sim \mathrm{Sp}\left\{\frac{1}{i\hbar}[P^\beta, P^\alpha]\rho_0\right\} = 0.$$

Finally, the following expression appears as the memory function, which, in this case, is the inverse relaxation time of a total momentum for the electronic system:

$$\Sigma(\omega) = \frac{1}{nm} \int_{-\infty}^{0} dt_1 \exp\{(\epsilon - i\omega)t_1\}$$
$$\times ((1 - \mathcal{P})\dot{P}^\alpha, \exp\{(1 - \mathcal{P})iLt_1\}(1 - \mathcal{P})\dot{P}^\beta). \tag{6.124}$$

For comparison, the expression for the inverse relaxation time, obtained by using the Green function (see Section 5.1.3), may be represented by

$$\frac{1}{\tau(\omega)} = \frac{1}{nm} \int_{-\infty}^{0} dt_1 \exp\{(\epsilon - i\omega)t_1\}(\dot{P}^\alpha, \exp\{iLt_1\}\dot{P}^\beta). \tag{6.125}$$

It is easy to see that the difference between the formulas (6.124) and (6.125) is the absence of projection operators in the last expression. The natural question arises: which expression of the two is correct? The question is very relevant, since such a type of formulas as (6.125) for the relaxation time is widely enough used in the literature. Moreover, it is well known that these formulas often give results which are in good agreement with experimental data.

It can be argued that the expression (6.125) for the total momentum relaxation time of an electronic system is correct only in the Born approximation. This can be easily verified. First, if the operator $\dot{P}^\alpha$ is proportional to an interaction, the formulas (6.124) and (6.125) just coincide in the Born approximation. Indeed, in this case, the projection operators in (6.124) can be omitted. Otherwise this would lead to the retention of the fourth- and higher-order interaction terms. The proof is to be accomplished by the reader itself.

It can be shown that an exact value of the inverse relaxation time, determined by the expression (6.125) is exactly equal to zero at $\omega = 0$ in the constant electric field. Therefore, strictly speaking, this formula is incorrect.

Consider the diagonal components of the electrical conductivity tensor at $\omega = 0$:

$$\sigma_{\alpha\alpha} = \frac{e^2\beta}{m^2} \int_{-\infty}^{0} dt_1 \exp\{\epsilon t_1\}(P^\alpha, P^\alpha(t_1)). \tag{6.126}$$

On the other hand

$$\sigma_{\alpha\alpha} = \frac{e^2 n\tau}{m}; \quad \frac{1}{\tau} = \frac{1}{nm} \int_{-\infty}^{0} dt_1 \exp\{\epsilon t_1\}(\dot{P}^\alpha, \dot{P}^\alpha(t_1)). \tag{6.127}$$

We integrate the above integral in (6.127) by parts twice. Performing the integration for the first time, we obtain

$$\begin{aligned} &\frac{1}{nm} \int_{-\infty}^{0} dt_1 \exp\{\epsilon t_1\} \frac{d}{dt_1}(\dot{P}^\alpha, P^\alpha(t_1)) \\ &= \frac{1}{nm}(\dot{P}^\alpha, P^\alpha) - \frac{\epsilon}{nm} \int_{-\infty}^{0} dt_1 \exp\{\epsilon t_1\}(\dot{P}^\alpha, P^\alpha(t_1)). \end{aligned} \tag{6.128}$$

Since the correlation function $(\dot{P}^\alpha, P^\alpha) = 0$, applying integration by parts for the second time, we get

$$\begin{aligned} &\frac{1}{nm} \int_{-\infty}^{0} dt_1 \exp\{\epsilon t_1\}(\dot{P}^\alpha, \dot{P}^\alpha(t_1)) \\ &= \frac{\epsilon}{nm}(P^\alpha, P^\alpha) - \frac{\epsilon^2}{nm} \int_{-\infty}^{0} dt_1 \exp\{\epsilon t_1\}(P^\alpha, P^\alpha(t_1)). \end{aligned} \tag{6.129}$$

Since all the correlation functions in the right-hand side of (6.129) are finite and they are multiplied by the parameters ϵ or ϵ^2 which, after performing the thermodynamic limiting transition,

$$n \to \infty, \quad V \to \infty, \quad \frac{n}{V} \to \text{const}$$

(n is the number of particles in a system, V the volume), should tend to zero, the formula (6.129) implies the vanishing of the inverse relaxation time.

The physical cause for this result is quite obvious. It follows from Section 6.2.1 that the irreversible behavior does not appear by itself due to some mathematical subterfuges. It should be reinforced that the emergence of the irreversibility is associated with realizing only a limited set of states, leading to the time-irreversible behavior rather than with all possible states admitted by the dynamic equations.

Certainly, there arise several questions. First of all, why does the use of the projection operators rather than the standard method of Green functions yield the correct result? This question becomes even more relevant in the case of carrying out only the identity transformations in the derivation of the equation of motion for the correlation function $\Theta(t)$ (6.99).

Secondly, why is it that the NSO method to satisfy the time-irreversible equation is not a sufficient condition? Must not the correct expressions for the kinetic coefficients be immediately obtained from this?

It is easier to answer the second question. The time-irreversible equation for the NSO provides but the correct structure of the kinetic coefficients or generalized kinetic equations. Moreover, the correct calculation of the transport coefficients is associated with the problem of finding equilibrium or non-equilibrium correlation functions. Even though this is a completely different problem, there are certain essential similarities.

As to the first question, it is dominant: how to enter those dynamic variables by means of which can describe the irreversible behavior?

In the equation of motion for the total momentum operator, the projection operator method allows one to separate out the terms, describing both precession and damping Γ (see equation (6.102a)). This criterion turns out to be a sufficient condition to obtain the correct result. One can rigorously prove that the term Γ, describing the damping, is taken into account twice with different signs in calculating the inverse relaxation time in the form (6.125). Therefore, this term can be exactly compensated.

A brief scheme to prove this curious fact will be represented at the end of the next section.

6.2.4 Relationship between a linear variant of the NSO-method and Mori's method

Let us now consider the question of how to further develop the usage-based approach of the transport matrix $T(\omega)$ and the Green functions $G(\omega)$ (6.78), (6.79) introduced earlier. Our objective is to obtain equations of motion in the Mori form (6.102) instead of the generalized equations of motion for the averages (6.68) within the NSO method.

Upon comparing the expressions (6.68) and (6.102), one can see that they are structurally identical if one manages to write the transport matrix $T(\omega)$ in the form

$$T(\omega) = i\Omega + \Sigma(\omega).$$

The difference between the values in the lower limit of the integral is not significant due to the choice of an initial moment of time.

To prove the possibility of such a representation, we should perform a number of identical transformations, which result in the derivation of the equation (6.102) by the Mori method.

To simplify the form of the equality, the notation should be introduced of

$$P^{+}(E) = \frac{1}{iL - iE}P^{+}, \quad iE = i\omega - \epsilon.$$

Also, one has to consider the identity

$$i(L - E)P^+(E) = P^+. \tag{6.130}$$

Let $\mathcal{P}$ and $(1-\mathcal{P})$ be projection operators acting step-by-step on both sides of this identity. Then, acting by the operator $\mathcal{P}$ and taking into account the identity

$$P^+(E) = \mathcal{P}P^+(E) + (1 - \mathcal{P})P^+(E),$$

we have

$$(-iE + \mathcal{P}iL)\mathcal{P}P^+(E) + \mathcal{P}iL(1 - \mathcal{P})P^+(E) = P^+. \tag{6.131}$$

In deriving this equality it has been taken into account that $\mathcal{P}P^+ = P^+$. After that, acting by the operator $(1 - \mathcal{P})$, we find

$$(-iE + (1 - \mathcal{P})iL)(1 - \mathcal{P})P^+(E) = -(1 - \mathcal{P})iL\mathcal{P}P^+(E). \tag{6.132}$$

Now, the quantity $(1-\mathcal{P})P^+(E)$ should be found from equation (6.132). Multiplying the left-side equation (6.132) by the quantity $(-iE + (1 - \mathcal{P})iL)^{-1}$, we get

$$(1 - \mathcal{P})P^+(E) = -\frac{1}{-iE + (1 - \mathcal{P})iL}(1 - \mathcal{P})iL\mathcal{P}P^+(E).$$

Substituting this result into the equation (6.131), we obtain

$$(-iE + \mathcal{P}iL)\mathcal{P}P^+(E) - \mathcal{P}iL\frac{1}{-iE + (1 - \mathcal{P})iL}(1 - \mathcal{P})iL\mathcal{P}P^+(E) = P^+. \tag{6.133}$$

Consider now an action of the projection operator $\mathcal{P}$ on the quantity $P^+(E)$. Emanating from the definition of the Mori projection operator (6.88) and (6.91), we have

$$\mathcal{P}P^+(E) = P^+\frac{1}{(P, P^+)}(P, P^+(E)) = P^+\overline{G}(E),$$

$$(P, P^+(E)) = \int_{-\infty}^{0} dt_1 \exp\{(\epsilon - i\omega)t_1\}\int_0^1 d\tau \mathrm{Sp}\{P, \rho_0^\tau \exp(iLt_1)P^+\rho_0^{1-\tau}\}. \tag{6.134}$$

In deriving the equality (6.134) the definition of the Green function (6.78) has been used. Next,

$$\mathcal{P}iLP^+ = P^+i\overline{\Omega}; \quad i\overline{\Omega} = (P, P^+)^{-1}(P, \dot{P}^+).$$

Scalar-multiplying the equation (6.133) on the left (in terms of Mori's scalar product) by P, we obtain

$$(P, P^+)(-iE + i\overline{\Omega})\overline{G}(E) - \left(P, \mathcal{P}iL\frac{1}{-iE + (1 - \mathcal{P})iL}(1 - \mathcal{P})iLP^+\right)\overline{G}(E) = (P, P^+). \tag{6.135}$$

Using the definition of a projection operator, we transform the second term on the left-hand side of the equation to obtain the following form:

$$(P,P^+)\frac{1}{(P,P^+)}\left(P, iL\frac{1}{-iE+(1-\mathcal{P})iL}(1-\mathcal{P})iLP^+\right)\overline{G}(E).$$

By reducing the same factor (P,P^+) on both sides, one is led to

$$\begin{aligned}&[-iE+i\overline{\Omega}+\overline{\Sigma}(E)]\overline{G}(E)=1;\\ &\overline{\Sigma}(E)=\frac{1}{(P,P^+)}\left(\dot{P},\frac{1}{-iE+(1-\mathcal{P})iL}(1-\mathcal{P})iLP^+\right).\end{aligned}\tag{6.136}$$

Now, if one considers that by virtue of the definition of a projection operator the following expression for any operators A and B appears:

$$(\mathcal{P}A,(1-\mathcal{P})B)=0$$

the expression for a memory function can be written in a form identical to Mori's definition (6.108):

$$\overline{\Sigma}(E)=\frac{1}{(P,P^+)}\left(f,\frac{1}{-iE+(1-\mathcal{P})iL}f^+\right),\quad f=(1-\mathcal{P})\dot{P}.\tag{6.137}$$

The difference between the definitions for $\overline{\Sigma}(E)$ (6.137) and (6.103) is not significant; it is connected only with a difference between designations.

From the expression (6.136) and equations (6.77) it follows that the transport matrix can indeed be expressed as $\overline{T}(\omega)=i\overline{\Omega}+\overline{\Sigma}(\omega)$.

Now, one may back to the issue of how to write the relaxation frequencies properly. Here, we should show that the expressions of the type (6.125) for the inverse relaxation time are, in the concept of a rigorous analysis, incorrect. However, these expressions are widely used in the literature to calculate the relaxation frequencies in the Born case. We define a new projection operator $\mathcal{P}(E)$ as follows:

$$\begin{aligned}\mathcal{P}(E)A&=(A,P^+)^E\frac{1}{(P,P^+)^E}P,\\ \mathcal{P}(E)A^+&=P^+\frac{1}{(P,P^+)^E}(P,A^+)^E,\end{aligned}\tag{6.138}$$

$$(A,B^+)^E=\int_{-\infty}^{0}dt_1\exp\{(\epsilon-i\omega)t_1\}\int_0^1 d\tau\,\mathrm{Sp}\{A,\rho_0^{\tau}\exp(iLt_1)B^+\rho_0^{1-\tau}\}.\tag{6.139}$$

Considering the definitions (6.138), (6.139), (6.72), and (6.76) it is easy to verify that the following equations are valid:

$$\mathcal{P}(E)\dot{P}=-T(E)P,\quad \mathcal{P}(E)\dot{P}^+=P^+\overline{T}(E),$$

and to prove that the expression for $\overline{\Sigma}(E)$ can be written in the form

$$\overline{\Sigma}(E) = \frac{1}{(P,P^+)}([1-\mathcal{P}(E)]\dot{P}, [1-\mathcal{P}(E)]\dot{P}^+)^E. \tag{6.140}$$

The reader is recommended to perform the proof of (6.140) as an exercise.

The results obtained allow the expression for the inverse relaxation time (6.125) at zero frequency to be written in another way. Using the definitions (6.139), (6.123), we get

$$\begin{aligned}\frac{1}{\tau} &= \Sigma'(\epsilon) = (\dot{P},\dot{P}^+)^\epsilon \frac{1}{(P,P^+)} \\ &= (\mathcal{P}(\epsilon)\dot{P}, \mathcal{P}(\epsilon)\dot{P}^+)^\epsilon \frac{1}{(P,P^+)} + ([1-\mathcal{P}(\epsilon)]\dot{P}, [1-\mathcal{P}(\epsilon)]\dot{P}^+)^\epsilon \frac{1}{(P,P^+)}.\end{aligned} \tag{6.141}$$

Here, $\mathcal{P}(\epsilon)$ is the projection operator $\mathcal{P}(E)$ at $\omega = 0$.

It is not hard to notice that the expression (6.141) is related with the inverse relaxation time in an obvious manner. It is sufficient to replace the P, P^+ by components P^α of the total electron momentum operator.

We first show that

$$(\mathcal{P}(\epsilon)\dot{P}, \mathcal{P}(\epsilon)\dot{P}^+)^\epsilon \frac{1}{(P,P^+)} = -(i\Omega + \Gamma).$$

For that, we use the relations

$$\mathcal{P}(\epsilon)\dot{P} = -T(\epsilon)P, \quad \mathcal{P}(\epsilon)\dot{P}^+ = P^+\overline{T}(\epsilon),$$

which can be easily proved by applying the definition (6.138) and formulas (6.72), (6.76). Then

$$\begin{aligned}(\mathcal{P}(\epsilon)\dot{P}, \mathcal{P}(\epsilon)\dot{P}^+)^\epsilon \frac{1}{(P,P^+)} &= -T(\epsilon)(P,P^+)^\epsilon \overline{T}(\epsilon)\frac{1}{(P,P^+)} \\ &= -T(\epsilon)(P,P^+)^\epsilon \frac{1}{(P,P^+)}(P,P^+)\overline{T}(\epsilon)\frac{1}{(P,P^+)} = -T(\epsilon)G(\epsilon)T(\epsilon).\end{aligned}$$

In deriving the above relation it has been taken into account that the $\overline{T}(\epsilon)$ and $T(\epsilon)$ matrices have a link through the relation (6.74). As far as, by analogy with (6.136), the expression can be written as

$$[\epsilon + i\Omega + \Sigma'(\epsilon)]G(\epsilon) = 1,$$

it follows that

$$\lim_{\epsilon\to 0} T(\epsilon)G(\epsilon) = 1,$$

and we obtain the desired relation

$$(\mathcal{P}(\epsilon)\dot{P}, \mathcal{P}(\epsilon)\dot{P}^+)^\epsilon (P,P^+)^{-1} = -(i\Omega + \Sigma'(\epsilon)).$$

On the other hand, as it follows from the relation (6.140), the last summand on the right-hand side of (6.141) is just $\Sigma'(\epsilon)$. Summing up the two results, we find

$$\Sigma'(\epsilon) = (\dot{P}, \dot{P}^{+})^{\epsilon} \frac{1}{(P, P^{+})} = -i\Omega.$$

In other words, the formulas of such a type do contain no damping at all. In the previous paragraph, the same result was obtained through direct integration for the special case when the basic operators are the components of the total momentum of an electronic system.

6.2.5 High-frequency susceptibility

Consider another example of applying the projection operator method to obtain an expression for transverse components of the magnetic susceptibility tensor of an electronic system.

We assume that an external perturbation with the Hamiltonian $H_F(t)$ begin acting on the system with the Hamiltonian

$$H = H_e + H_s + H_{ep}, \quad H_e = P^2/2m, \quad H_s = -g\mu_B S^z H^z$$

at some initial moment of time. Here: H_e, H_s are the Hamiltonian of kinetic and Zeeman degrees of freedom of conduction electrons, respectively; H_{ep} is the electron–scatterer interaction Hamiltonian, g the Zeeman spectroscopic splitting factor μ_B the Bohr magneton.

$$S^\alpha = \sum_{i=1}^{n} s_i^\alpha,$$

where n is the number of the conduction electrons.

The interaction Hamiltonian of the system with a variable magnetic field $H_F(t)$ can be written as

$$H_F(t) = -g\mu_B S^\alpha h^\alpha(t),$$

where $h^\alpha(t)$ is an induction vector of the high-frequency magnetic field.

Let us find the magnetic moment m^α induced by the high-frequency field $h^\alpha(t)$ in the electronic system.

Using, as in Section 6.2.3, the integral equation (6.115) for the non-equilibrium statistical operator (NSO), and assuming that $\rho^0(t,0) = \rho_0$, the following equation can be obtained for the Fourier transforms of the high-frequency magnetic moment:

$$m^\alpha(\omega) = \frac{(g\mu_B)^2}{i\hbar} \int_{-\infty}^{0} dt_1 \exp\{(\epsilon - i\omega)t_1\} \mathrm{Sp}\{S^\alpha \exp(iLt_1)[\rho_0, S^\beta]\} h^\beta(\omega). \tag{6.142}$$

Again using the Kubo formula (6.119) (S^β is now playing the role of the operator X^β) and introducing the circular components by the relations $m_\pm = m^x \pm im^y$, $h_\pm = h^x \pm ih^y$, the following can be obtained:

$$\chi_{+-}(\omega) = \frac{\beta(g\mu_B)^2}{2} \int_{-\infty}^{0} dt_1 \exp\{(\epsilon - i\omega)t_1\}(S^+, \dot{S}^-(t_1)),$$

$$\dot{S}^- = i\omega_S S^- + \dot{S}^-_{(l)}, \quad \dot{S}^-_{(l)} = \frac{1}{i\hbar}[S^-, H_{ep}], \tag{6.143}$$

where ω_S is frequency of the Zeeman electron spin precession. It is obvious that the relation (6.143) can be rewritten as

$$\chi_{+-}(\omega) = \frac{\beta(g\mu_B)^2}{2} \int_{-\infty}^{0} dt_1 \exp\{(\epsilon - i\omega)t_1\} \frac{d}{dt_1}(S^+, S^-(t_1)). \tag{6.144}$$

Similarly, as in the case of electrical conductivity, we introduce the notation $\epsilon - i\omega = z$ and make the change of variables under the sign of integral $t_1 \to -t_1$. Then, according to the notation, the expression (6.144) can be written as

$$\chi_{+-}(z) = -\frac{\beta(g\mu_B)^2}{2} \int_{0}^{\infty} dt_1 \exp\{-zt_1\} \frac{d}{dt_1}\Theta(t_1)(S^+, S^-). \tag{6.145}$$

The function $\Theta(t_1)$, appearing in this expression, is defined in accordance with the formula (6.94) and, in the given case, has the form

$$\Theta(t_1) = (S^+(t_1), S^-)\frac{1}{(S^+, S^-)}.$$

Now, applying a generalized Langevin equation to the correlation function $\Theta(t_1)$ (6.101), we have

$$\chi_{+-}(z) = -\frac{\beta(g\mu_B)^2}{2} \int_{0}^{\infty} dt_1 \exp\{-zt_1\} \Bigg[i\Omega\Theta(t_1)$$

$$- \int_{0}^{t_1} ds(f, f^+(-s))\frac{1}{(S, S^+)}\Theta(t_1 - s)\Bigg](S^+, S^-),$$

$$f = (1 - \mathcal{P})\dot{S}^+, \quad i\Omega = (\mathcal{P}\dot{S}^+, S^-)\frac{1}{(S^+, S^-)}. \tag{6.146}$$

Performing Laplace transforms in the equation (6.146) and considering the definitions (6.105), we obtain

$$\chi_{+-}(z) = \frac{\beta(g\mu_B)^2}{2} \frac{(S^+, S^-)[\Sigma(z) - i\Omega]}{z - i\Omega + \Sigma(z)}; \tag{6.147}$$

$$\Sigma(z) = \int_0^\infty dt_1 \exp(-zt_1)((1-\mathcal{P})\dot{S}^+, \exp\{-(1-\mathcal{P})iLt_1\}(1-\mathcal{P})\dot{S}^-)\frac{1}{(S^+, S^-)}. \qquad (6.148)$$

Equation (6.147) for the transverse components of the magnetic susceptibility tensor coincides completely with the result obtained by means of the Bloch equations (5.100), if one takes into consideration that in the given case $i\Omega = -i\omega_s$ as the memory function determines the inverse relaxation time of the transverse electron spin components.

6.2.6 Determination of non-equilibrium parameters by the NSO-method

The analysis of the NSO-method is incomplete without discussing the main question of how to calculate the non-equilibrium parameters $F_n(t)$, which define a quasi-equilibrium and non-equilibrium distribution.

Certainly, the problem of finding the non-equilibrium parameters can be treated in a general way without specifying the form of the system. However, due to limitations of this book, it would be appropriate to regard such a physical effect as, on the one hand, being typical enough, but, on the other hand, being not complex enough to carry out the analysis to the end.

So we should dwell on the Overhauser effect that consists in amplifying the signal of nuclear magnetic resonance by saturation of the free-electrons paramagnetic resonance in metals or semiconductors.

The Overhauser effect is a typical effect and can be explained simply enough when the effective temperature of Zeeman subsystem of conduction electrons and nuclei is used as the parameter of $F_n(t)$. From a physical point of view, the nature of this effect is quite understandable. As far as the magnetic subsystem of the conduction electrons and nuclei interact mainly with each other, their total magnetic moment is preserved. The magnetic moment of the electronic system decreases under conditions of paramagnetic resonance saturation by conduction electrons; consequently, the magnetic moment of the nuclear system must increase. The increase in the magnetic moment of the nuclear system is exhibited as a drop in effective temperature of nuclei, which leads to an enhancement of a signal of nuclear magnetic resonance.

The Feher effect is very similar in essence; it is the phenomenon of nuclei polarization by a direct electric current in semiconductors. The nature of this effect is the same, i. e. energy is "pumped" into the kinetic degrees of freedom of electrons, and then is transferred to a thermostat in scattering with an accompanying spin-flip of the electrons.

There is another distinguishing feature of the Feher and Overhauser effects. In essence, both are an example of realization of an ordinary refrigerator. If the temperature T_s of the spin system is greater than the temperature of the kinetic degrees of freedom of the conduction electrons T_k, the energy transfers from the electron Zeeman system into the kinetic degrees of freedom. Then in every elementary scattering event involving electron spin, nuclei and kinetic degrees of freedom, the nuclear Zeeman system also loses a certain part of the energy. In the case of the Feher effect, the energy transfers from the subsystem of the kinetic degrees of freedom into the thermostat. Moreover, according to the energy and momentum conservation laws, when considered regarding the probability of changing in orientation in every scattering event, the nuclei's spins, oriented along an applied field have it slightly greater than others. This as matter of fact leads to the phenomenon of dynamic nuclear polarization.

There are some other effects, which can be interpreted in the framework of the effective temperature method. They are exhibited, for example, as a change in electrical resistance in the vicinity of the resonance under conditions of magnetic resonance saturation that occurs by conduction electrons or by donor impurities, or ferromagnetic resonance in magnetic semiconductors. Moreover, the resulting curve of the change in resistance accurately coincides with an absorption resonance curve of the high-frequency energy in a specimen. In spite of the smallness of the effect (of order 30 % for ferromagnetic semiconductors), it allows one to detect the resonance through the change in electrical resistance.

It should be emphasized that this refers to only the kinetic degrees of freedom of conduction electrons rather than an increase in temperature of the whole specimen.

Now, after this brief, qualitative overview of the effects, there is every reason for going over to a detailed description of the Overhauser effect.

Consider the simplest case, when heterogeneity of an electromagnetic microwave field in specimen volume can be neglected. Furthermore, a subsystem of long-wave-length phonons interacting with electrons is supposed to be in a state of thermodynamic equilibrium.

To describe the non-equilibrium system by the NSO method, it is required that the system's Hamiltonian and a set of non-equilibrium parameters characterizing the system should be chosen. Let us represent the Hamiltonian in the form

$$\begin{aligned} H(t) &= H + H_F(t), \\ H &= H_e + H_s + H_p + H_n + H_{ep} + H_{en}. \end{aligned} \tag{6.149}$$

The H_e, H_s Hamiltonians were determined earlier,

$$H_n = -\hbar\omega_n I^z, \quad I^\alpha = \sum_j I_j^\alpha,$$

where ω_n is the Zeeman precession frequency of nuclear spins in a static magnetic field $\vec{H}$; I^α the component of a total spin of the nuclear system, the summation is being performed over all nuclei with the spin; H_{en} the electron–nuclear contact interaction Hamiltonian, which in second-quantization over electronic variables can be written as

$$H_{en} = \sum_{\nu'\sigma',\nu\sigma} U^\alpha_{en\nu'\nu} S^\alpha_{\sigma'\sigma} a^+_{\nu'\sigma'} a_{\nu\sigma},$$

$$U^{\alpha}_{env'v} = \sum_{\vec{q}} J_{\vec{q}} \langle v' | e^{i\vec{q}\vec{x}} | v \rangle I^{\alpha}_{-\vec{q}},$$
$$I^{\alpha}_{-\vec{q}} = \sum_{j} I^{\alpha}_{j} e^{i\vec{q}\vec{x}_j}, \tag{6.150}$$

$J_{\vec{q}}$ is the Fourier transform of a contact interaction between conduction electrons and magnetic nucleus; $\vec{x}_j$ the coordinate of a nucleus with the spin I^{α}_j; H_{ep} the electron–phonon interaction Hamiltonian. The explicit form is presented in Chapter 4.

Let the alternating electromagnetic field with frequency ω and amplitude $h(t)$ be polarized in the plane perpendicular to direction of the static field $\vec{H}$. In this case, the interaction Hamiltonian of Zeeman degrees of freedom of conduction electrons with the external field coincides with the Hamiltonian $H_F(t)$, used in the previous paragraph. After introducing circular components, the Hamiltonian $H_F(t)$ appears as

$$H_F(t) = -\frac{\hbar\omega_s}{2}(S^+ e^{i\omega t} + S^- e^{i\omega t}). \tag{6.151}$$

Here ω_s is the Zeeman frequency precession of the electron spin in the alternating magnetic field.

Consider the behavior of the conduction electrons, phonons and nuclear spins over periods longer than the time required to establish equilibrium inside each subsystem. Then a description of the subsystems in terms of effective non-equilibrium temperatures is valid.

We write down the entropy operator (6.11) for the system under study in the form

$$S(t,0) = \phi(t) + \beta_k(t)(H_e - \zeta(t)N) + \beta_s(t)(H_s + H_F(t))$$
$$+ \beta_n(t)(H_n + H_{en}) + \beta(H_p + H_{ep}). \tag{6.152}$$

Here $\beta_k(t)$, $\beta_s(t)$, $\beta_n(t)$ are the inverse temperatures of kinetic and spin degrees of freedom of conduction electrons and nuclear spins, respectively; β is the inverse equilibrium temperature, $\zeta(t)$ the non-equilibrium chemical potential.

For the interacting subsystems of a crystal, the scheme at hand is shown in Figure 6.2. Here, the crystal is divided into the subsystems which are marked by the rectangles: S is the subsystem of spin degrees of freedom, k is the subsystem of kinetic degrees of freedom of the conduction electrons, n is the subsystem of the nuclear spins, and T (thermostat) is for all other degrees of freedom of the crystal. The straight arrows indicate channels of the energy transfer between the subsystems but a curly arrow depicts a radio-frequency energy pumping (Rf) into the subsystem S.

The above scheme shows that there may exist a direct channel to transfer the energy from the subsystem n into the thermostat, but we leave this process aside. Similarly, the subsystem S can transfer its energy into the thermostat not only directly but as a result of electron–phonon interaction with both spin-flip and participation of the kinetic degrees of freedom. The phonon system, which is supposed to be in a state of equilibrium, may serve as a case in point of such a thermostat.

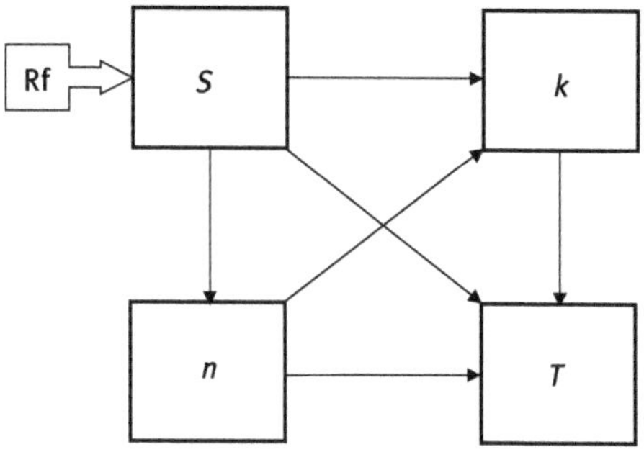

Figure 6.2: The Overhauser effect: scheme for interacting subsystems of a crystal.

The direct channel of the energy transfer is not worth taking into account due to its extreme inefficiency despite its probability (the corresponding arrow is shown in Figure 6.2). In this sense, the arrow connecting the S and T subsystems demonstrates only that the system of the electron spins can drop its energy to the thermostat even without participating of the subsystem k.

To construct a set of energy balance equations for the S, k, n subsystems, which in this case are playing a role of the generalized kinetic equations (6.8), it is necessary to write down an expression for the NSO.

We use the integral equation obtained in Section 6.2.3 for the NSO (6.115). It is natural for the amplitude of the radio-frequency field h to be small. So we restrict ourselves to terms no higher than the second order over the parameter in these equations. Then the new integral equation for the NSO can be written as

$$\begin{aligned}
\rho(t,0) &= \rho^0(t,0) - i\int_{-\infty}^{0} dt_1 \exp\{\epsilon t_1\}\exp\{iLt_1\}L_F(t+t_1)\rho_0, \\
\rho^0(t,0) &= \epsilon\int_{-\infty}^{0} dt_1 \exp\{\epsilon t_1\}\exp\{iLt_1\}\overline{\rho}(t+t_1,0), \qquad (6.153) \\
\overline{\rho}(t,0) &= \exp\{-S(t,0)\}.
\end{aligned}$$

Here, we have replaced the non-equilibrium distribution on the right-hand side of the first equation (6.153) by the equilibrium one. The reason is that a deviation of the non-equilibrium parameters from their equilibrium is of second-order smallness in interacting with an external electromagnetic field. Since this term itself already contains the first order of the smallness in the field, the deviation of the thermodynamic parameters from equilibrium can be neglected.

As to the second equation in (6.153) we can transform it by using the result (6.62). As follows from the expression for entropy (6.152), the thermodynamic coordinates P_n and thermodynamic forces F_n can be chosen as follows:

P_n	H_k	$H_s + H_F(t)$	$H_n + H_{en}$	N
$F_n(t)$	$\beta_k(t)$	$\beta_s(t)$	$\beta_n(t)$	$\beta_k(t)\zeta(t)$

Let us write the equations of motion for basic operators with using of the Hamiltonian H. Denoting as before $\dot{A} = \frac{1}{i\hbar}[A, H]$, we have

$$\begin{aligned} \dot{N} &= 0, \quad \dot{H}_k = \dot{H}_{k(p)} + \dot{H}_{k(n)}, \\ \dot{H}_{i(m)} &= \frac{1}{i\hbar}[H_i, H_{em}], \quad i = k, s, \quad m = p, n, \\ \dot{H}_s &= \dot{H}_{s(p)} + \dot{H}_{s(n)}, \\ \dot{H}_n + \dot{H}_{en} &= -\dot{H}_{k(n)} - \dot{H}_{s(n)}. \end{aligned} \tag{6.154}$$

We substitute the results obtained into the formula (6.62). Considering again the fact that the deviations $\delta\beta_k$, $\delta\beta_s$, $\delta\beta_n$ and $\delta\zeta$ of the inverse temperatures and chemical potential, respectively, from equilibrium values are proportional to the second-order interaction, and assuming that the steady state is realized when the thermodynamic parameters δF_n are time-independent and therefore $\delta\dot{F}_n(T) = 0$, we rewrite (6.62) in the form

$$\rho^0(t,0) = \rho_0 - \int_0^1 d\tau \rho_0^\tau \delta S(t,0)\rho_0^{1-\tau} + \int_{-\infty}^0 dt_1 \exp\{\epsilon t_1\} \int_0^1 d\tau \exp\{iLt_1\}\rho_0^\tau \frac{d}{dt_1}\delta S(t+t_1,0)\rho_0^{1-\tau}, \tag{6.155}$$

$$\begin{aligned} \delta S(t,0) &= \Delta\{\beta H_F(t) + \delta\beta_k(H_k - \zeta_0 N) - \delta\zeta\beta N + \delta\beta_s H_s + \delta\beta_n(H_n + H_{en})\}, \\ \frac{d}{dt_1}\delta S(t+t_1,0) &= \Delta\Big\{\beta\frac{\partial}{\partial t_1}H_F(t+t_1) + \delta\beta_k(\dot{H}_{k(p)} \\ &\quad + \dot{H}_{k(n)}) + \delta\beta_s(\dot{H}_{s(p)} + \dot{H}_{s(n)}) - \delta\beta_n(\dot{H}_{k(n)} + \dot{H}_{s(n)})\Big\}. \end{aligned} \tag{6.156}$$

We now construct a set of macroscopic energy balance equations for the subsystems by means of the NSO (6.155) for further using to find values of the non-equilibrium temperatures of the subsystems of the crystal.

The equations of motion for the energy operators of the S, k, n subsystems should be sought with respect to the total Hamiltonian $H(t) = H + H_F(t)$. It is clear that the equations of motion for the k and n subsystems will coincide with the equations (6.154). The equation of motion for the subsystem S should be written in the following form:

$$\begin{aligned} \frac{d}{dt}(H_s + H_F(t)) &= \frac{1}{i\hbar}[H_s + H_F(t), H(t)] + \frac{\partial}{\partial t}H_F(t) \\ &= \dot{H}_{s(p)} + \dot{H}_{s(n)} - \frac{i\hbar\omega_s\omega}{2}(S^+\exp\{i\omega t\} - S^-\exp\{-i\omega t\}). \end{aligned}$$

As has been said above, the average energy values do not depend on time, consequently, the partial derivative over time on the left-hand side of (6.8) is equal to zero. Further, by averaging the operator equations of motion of the energy of the S, k, n subsystems, one is led to

$$\delta\beta_k \mathrm{L}_{kk(p)} + \delta\beta_s \mathrm{L}_{ks(p)} = 0; \tag{6.157}$$

$$\delta\beta_k \mathrm{L}_{sk(p)} + \delta\beta_s \mathrm{L}_{ss(p)} + Q_s = 0; \tag{6.158}$$

$$-\delta\beta_k \mathrm{L}_{ek(n)} - \delta\beta_s \mathrm{L}_{es(n)} + \delta\beta_n \mathrm{L}_{ee(n)} = 0. \tag{6.159}$$

Correlation functions, which have appeared in the balance equations, have the form

$$\mathrm{L}_{ij(m)} = \int_{-\infty}^{0} dt_1 \exp\{\epsilon t_1\}(\dot{H}_{i(m)}, \dot{H}_{j(m)}(t_1)), \tag{6.160}$$

$$i, j = k, s, e, \quad m = p, n,$$

$H_e = H_k + H_s$. The quantity Q_s in the equation (6.158) makes sense as the Rf-power absorbed by the electron spin system and is expressed via the transverse components of high-frequency magnetic susceptibility $\chi_{+-}(\omega)$ (6.143). We have

$$Q_s = \omega\,\mathrm{Im}\,\chi_{+-}(\omega)|h|^2. \tag{6.161}$$

The previous paragraph discussed the calculation of these components. Note that, in deriving the set of the coupled balance equations (6.157), (6.158), the weak electron–nuclear interaction has been neglected because it is not significant when it comes to the kinetics of an electronic system.

The solution of the set of equations (6.157)–(6.159) allows one to express the corrections to the temperatures of non-equilibrium subsystems both through the correlation functions $\mathrm{L}_{ij(m)}$ and through the absorbed power Q_s.

There is no necessity to calculate the correlation functions $\mathrm{L}_{ij(m)}$ since it would require a more in-depth discussion of mechanisms of electron scattering in conducting crystals, which goes beyond the scope of this textbook. Therefore, a solution of set of (6.157), (6.158) can be written in the general form

$$\delta\beta_k = Q_s \frac{\mathrm{L}_{ks(p)}}{\mathrm{L}_{kk(p)}\mathrm{L}_{ks(p)} - \mathrm{L}^2_{ks(p)}},$$

$$\delta\beta_s = -Q_s \frac{\mathrm{L}_{kk(p)}}{\mathrm{L}_{kk(p)}\mathrm{L}_{ks(p)} - \mathrm{L}^2_{ks(p)}},$$

$$\delta\beta_n = -\delta\beta_k \frac{\mathrm{L}_{ek(n)}}{\mathrm{L}_{ee(n)}} - \delta\beta_s \frac{\mathrm{L}_{es(n)}}{\mathrm{L}_{ee(n)}}. \tag{6.162}$$

It is seen from the solution of (6.162) that the Overhauser effect is exhibited as a change in temperature of the nuclear spins when the Rf-energy pumping into the subsystem

S occurs. It is recommended for the reader to run a complete analysis of the solution obtained as independent work to consider all possible modes of the Overhauser effect implementation and to estimate numerical values of an effective temperature deviation from the equilibrium state.

The set of equations (6.162) permits the temperature values of the non-equilibrium subsystems S, k and n to be found. In addition, the constancy condition for the number of electrons yields a non-equilibrium chemical potential:

$$\mathrm{Sp}\{N\overline{\rho}\} = \mathrm{Sp}\{N\rho_{0},\},$$

where N is the particle number operator.

Thus, the example of the Overhauser effect demonstrates the possibility of constructing generalized kinetic equations. Moreover, any parameters defining the quasi-equilibrium and non-equilibrium distributions can be ascertained.

6.3 Hydrodynamic modes and singularity of dynamic correlation functions

6.3.1 Spin diffusion

Spin diffusion describes a situation where relaxation time of longitudinal and transverse spin components of conduction electrons in conductive crystals often proves to be several orders of magnitude longer than the momentum relaxation time. So, the spin relaxation time in a metal $T_s \simeq 10^{-9}$ s, whereas the momentum relaxation time $\tau_{\vec{p}} \simeq 10^{-12}$ s. This leads to conservation of the spin orientation within many electron scattering events. Therefore, if there arises a deviation of the magnetization of conduction electrons from an equilibrium state in any point in space, then motion of the spin magnetization emerges. This phenomenon is naturally called *spin diffusion.*

If one assumes that the behavior of a system changes over periods longer than $\tau_{\vec{p}}$, but less than T_s, the spin orientation is supposed to be preserved. Therefore, it would be expedient to consider only motion of particles carrying a magnetic moment. Then, if one introduces the concept of magnetic moment density,

$$M^{\alpha}(\vec{r},t) = g\mu_B \sum_i S_i^{\alpha}\delta(\vec{r}-\vec{r}_i(t)),$$

then the macroscopic continuity equation can be written for this quantity:

$$\frac{\partial}{\partial t}\langle \vec{M}^{\alpha}(\vec{r},t)\rangle + \mathrm{div}\langle \vec{J}_{M^{\alpha}}(\vec{r},t)\rangle = 0,$$
$$\vec{J}_{M^{\alpha}} = g\mu_B \sum_i S_i^{\alpha}\left\{\frac{\vec{p}_i(t)}{m}, \delta(\vec{r}-\vec{r}_i(t))\right\},$$

$$\{A, B\} = \frac{1}{2}(AB + BA). \tag{6.163}$$

Obviously, the continuity equation (6.163) does not give a full description of the dynamics of the system's magnetic moment. In order to determine the temporal behavior of the $M^\alpha(\vec{r}, t)$, another equation required for linking the $\vec{J}_{M^\alpha}$ and $M^\alpha(\vec{r}, t)$ among themselves must be added. As far as there is a tendency to equalization of the magnetic moments, such a connection may be found by using the phenomenological Fick law:

$$\langle \vec{J}_{M^\alpha}(\vec{r}, t)\rangle = -D\vec{\nabla}\langle M^\alpha(\vec{r}, t)\rangle. \tag{6.164}$$

The averages in the expression (6.164) are being calculated with respect to a non-equilibrium distribution. Upon substituting this result into the continuity equation (6.163), one is led to a closed expression for the components of the magnetic moment density of the system:

$$\frac{\partial}{\partial t}\langle \vec{M}^\alpha(\vec{r}, t)\rangle - D\nabla^2\langle M^\alpha(\vec{r}, t)\rangle = 0. \tag{6.165}$$

This expression allows one to find the value of the components of mean magnetization at any given time, if the initial density of the magnetization is known.

Assuming that a medium is unbounded, one can perform the Fourier transform of (6.165) over the variable $\vec{r}$ and Laplace transform over t:

$$\langle \vec{M}^\alpha(\vec{k}, t)\rangle = \int d\vec{r}\langle \vec{M}^\alpha(\vec{r}, t)\rangle e^{-i\vec{k}\vec{r}},$$

$$\langle \vec{M}^\alpha(\vec{k}, z)\rangle = \int_0^\infty dt\langle \vec{M}^\alpha(\vec{k}, t)\rangle e^{izt}. \tag{6.166}$$

After that, we obtain the simple equation

$$\frac{\partial}{\partial t}\langle \vec{M}^\alpha(\vec{k}, t)\rangle + Dk^2\langle M^\alpha(\vec{k}, t)\rangle = 0, \tag{6.167}$$

whose solution appears as

$$\langle \vec{M}^\alpha(\vec{k}, t)\rangle = \langle \vec{M}^\alpha(\vec{k}, 0)\rangle e^{-Dk^2 t}, \tag{6.168}$$

where $\langle \vec{M}^\alpha(\vec{k}, 0)\rangle$ is the Fourier-image of the magnetization density at the initial time $t = 0$. After substituting the last result in the definition of $\langle \vec{M}^\alpha(\vec{k}, z)\rangle$ (6.166), we have

$$\langle \vec{M}^\alpha(\vec{k}, z)\rangle = \int_0^\infty dt\langle \vec{M}^\alpha(\vec{k}, 0)\rangle e^{-Dk^2 t} e^{izt}.$$

Integration over the time argument yields

$$\langle \vec{M}^\alpha(\vec{k}, z)\rangle = -\frac{\langle \vec{M}^\alpha(\vec{k}, 0)\rangle}{iz - Dk^2} = i\frac{\langle \vec{M}^\alpha(\vec{k}, 0)\rangle}{z + iDk^2}. \tag{6.169}$$

The result found can be interpreted as follows: the diffusion process leads to the appearance of the pole of the function $\langle \vec{M}^{\alpha}(\vec{k}, z)\rangle$ on the negative imaginary axis,

$$z = -iDk^2.$$

Such a peculiarity may be treated as a consequence of emergence of the system's collective excitations called hydrodynamic modes.

The hydrodynamic mode is commonly referred to as a sinusoidal collective fluctuation, slowly decaying near $k \to 0$ with the characteristic temporal scale

$$\tau = \frac{1}{Dk^2}.$$

In contrast to a propagating mode with real and imaginary parts of a spectrum of the collective excitations, the hydrodynamic mode can have only an imaginary component, but the lifetime of the excitation tends to infinity as $k \to 0$.

Let us find a link between the spin diffusion coefficient D and the spin correlation function being introduced as

$$S_{\alpha\beta}(\vec{r}, t) = \mathrm{Sp}\{M^{\alpha}(\vec{r}, t)M^{\beta}(0, 0)\rho_0\}. \tag{6.170}$$

In the formula (6.170), the averages are the quantities, calculated by using the equilibrium distribution ρ_0, and therefore

$$\mathrm{Sp}\{M^{\alpha}(\vec{r}, t)\rho_0\} = 0.$$

Thus, the function $S_{\alpha\beta}(\vec{r}, t)$ describes the fluctuations. Assuming in accordance with the correlation attenuation principle that the function $S_{\alpha\beta}(\vec{r}, t)$ decreases rapidly as $\vec{r}$ and t increase, one can use the Fourier transform:

$$S_{\alpha\beta}(\vec{k}, \omega) = \int_{-\infty}^{\infty} dt \int d\vec{r} S_{\alpha\beta}(\vec{r}, t) e^{i(\omega t - \vec{k}\vec{r})}. \tag{6.171}$$

The quantity $S_{\alpha\beta}(\vec{k}, \omega)$ has the meaning of the spectral density of spin magnetization fluctuation and is a real positive quantity. Furthermore, it would be appropriate to regard the system's Hamiltonian as invariant with respect to the operations of spatial rotation and time reversal. In this case, the function $S_{\alpha\beta}(\vec{r}, t)$ is an even function of $\vec{r}$ and t and a diagonal one with respect to the indices α and β.

We define the function $\widetilde{S}(\vec{k}, \omega)$ by the Laplace transformation

$$\widetilde{S}(\vec{k}, \omega) = \int_{0}^{\infty} dt S(\vec{k}, t) e^{i\omega t} \tag{6.172}$$

and find a relationship between functions $\widetilde{S}(\vec{k}, \omega)$ and $S(\vec{k}, \omega)$. For that, the complex conjugate function should be taken into account:

$$\{\widetilde{S}(\vec{k}, \omega)\}^* = \int_0^\infty dt S(\vec{k}, t) e^{-i\omega t}.$$

Making the change of the variables $t \to -t$ in the last integral and considering the parity of the function $S(\vec{k}, t)$, one is led to

$$\{\widetilde{S}(\vec{k}, \omega)\}^* = \int_{-\infty}^0 dt S(\vec{k}, t) e^{i\omega t}. \tag{6.173}$$

Hence, it follows that for real ω

$$\widetilde{S}(\vec{k}, \omega) + \{\widetilde{S}(\vec{k}, \omega)\}^* = S(\vec{k}, \omega),$$

or

$$2\mathrm{Re}\widetilde{S}(\vec{k}, \omega) = S(\vec{k}, \omega). \tag{6.174}$$

Now, emanating from the general principles of a hydrodynamic description of the system we define the function $\widetilde{S}(\vec{k}, \omega)$ under the assumption that the diffusion equation (6.165), created for the averages, also holds true at the operator level:

$$\frac{\partial}{\partial t} M^\alpha(\vec{r}, t) - D\nabla^2 M^\alpha(\vec{r}, t) = 0. \tag{6.175}$$

Upon multiplying this equation by operator $M^\beta(0, 0)$ on the right-hand side and averaging it over the equilibrium distribution, we get an equation for the function $S(\vec{r}, t)$, which, as noted above, is diagonal with respect to the indices α, β:

$$\frac{\partial}{\partial t} S(\vec{r}, t) - D\nabla^2 S(\vec{r}, t) = 0. \tag{6.176}$$

The assumption made above means that the spontaneous equilibrium fluctuations, being described by the function $S(\vec{r}, t)$, and the non-equilibrium quantities $\langle \vec{M}(\vec{r}, t) \rangle$ are relaxing in accordance with the same diffusion equations. As early as 1931 Onsager put forth this hypothesis, but one has still not found disagreeing empirical facts.

Equation (6.176) is solved in exactly the same manner as equation (6.165) for the non-equilibrium averages $\langle \vec{M}(\vec{r}, t) \rangle$, and therefore one can immediately arrive at

$$\widetilde{S}(\vec{k}, z) = i \frac{\widetilde{S}(\vec{k}, 0)}{z + iDk^2}. \tag{6.177}$$

It should be kept in mind that for this formula $\widetilde{S}(\vec{k}, 0)$ is $\widetilde{S}(\vec{k}, t = 0)$. Below, we show that

$$\lim_{\vec{k} \to 0} \widetilde{S}(\vec{k}, 0) = \frac{1}{\beta} \chi,$$

where χ is the static magnetic susceptibility of the system. Consequently, in the limit of the small $\vec{k}$, the following representation is valid:

$$\widetilde{S}(\vec{k},z) = i\frac{\beta^{-1}\chi}{z + iDk^2}. \tag{6.178}$$

Using the previously obtained result (6.174): $2\,\mathrm{Re}\,\widetilde{S}(\vec{k},\omega) = S(\vec{k},\omega)$, we can find the representation for the function $S(\vec{k},\omega)$ in a long-wave approximation:

$$S(\vec{k},\omega) = 2\mathrm{Re}\widetilde{S}(\vec{k},\omega) = \frac{2}{\beta}\chi\frac{Dk^2}{\omega^2 + (Dk^2)^2}. \tag{6.179}$$

This result is quite important and can be easily verified experimentally, since the quantity $S(\vec{k},\omega)$ is closely associated with a structural factor, determining the particle scattering by fluctuations of a magnetic moment. Also, it enables the spin diffusion coefficient D, expressed via the correlation function of the magnetic moment operators in the equilibrium states, to be found. Indeed, it is not hard to notice that, performing the limiting transitions in the correct sequence, but namely $k \to 0$ and then $\omega \to 0$, we obtain

$$\begin{aligned} &\beta \lim_{\omega\to 0}\lim_{k\to 0}\omega^2 k^{-2} S(\vec{k},\omega) = 2D\chi, \\ &D = \beta\frac{1}{2\chi}\lim_{\omega\to 0}\lim_{k\to 0}\omega^2 k^{-2} S(\vec{k},\omega). \end{aligned} \tag{6.180}$$

Thus, we have succeeded in expressing the ratio between a spin diffusion coefficient and a correlation function of spin fluctuations in an equilibrium state. The result obtained can be considered as yet another confirmation of the fluctuation-dissipation theorem, described by Kubo.

6.3.2 The fluctuation–dissipation theorem

The fluctuation–dissipation theorem (FDT) relates either correlation functions of operators of physical quantities or corresponding spectral functions with an imaginary part of generalized susceptibility as a characteristic of dissipative processes to describe the response of a system to an external perturbation. In other words, according to the fluctuation–dissipation theorem, relaxation mechanisms of fluctuations of dynamical variables in an equilibrium state obey the same physical laws as those mechanisms that are responsible for the system's relaxation behavior in the presence of the external perturbations.

There are a lot of ways to formulate the fluctuation–dissipation theorem. The fluctuation–dissipation theorem was originally formulated by Nyquist in 1928, and later proven by Callen and Welton. To date, the formulations of Kubo and Callen–Welton are the most popular. In essence, Kubo's formulation gives the exact mathematical expression for the kinetic characteristics such as electrical conductivity and magnetic susceptibility, in terms of the correlation functions of operators of dynamic variables being in an equilibrium state. The formulas (2.11), (5.43), (5.84) and (6.180) may serve as examples of the implementation of the fluctuation–dissipation theorem.

As far back as 1951 Callen and Welton generalized the Nyquist theorem on noise in electrical circuits.

To understand the formulation of the fluctuation–dissipation theorem by Callen – Welton, it is expedient to cite an example of the magnetic susceptibility tensor $\chi_{\alpha\beta}$, being defined by the relation

$$m^{\alpha}(\omega) = \chi_{\alpha\beta}(\omega)h^{\beta}(\omega).$$

The explicit form of this relation can easily be obtained from the formula (5.80):

$$\chi_{\alpha\beta}(\omega) = \frac{i}{\hbar}\int_{-\infty}^{0} dt_1 e^{(\epsilon-i\omega)t_1}\langle[M^{\alpha}, M^{\beta}(t_1)]\rangle_0 . \qquad (6.181)$$

Here, $\langle AB\rangle_0 = \mathrm{Sp}\{AB\rho_0\}$, $M^{\alpha} = g\mu_B S^{\alpha}$.

Let us determine the spectral intensity $f_{\alpha\beta}(\omega)$ and its classical analogue $g_{\alpha\beta}(\omega)$ along with the Fourier transform of the magnetic susceptibility tensor. It is worth drawing attention to the fact that the quantity $g_{\alpha\beta}(\omega)$ preserves its meaning when passing to the classical case:

$$f_{\alpha\beta}(\omega) = \frac{i}{\hbar}\int_{-\infty}^{\infty} dt_1 e^{-i\omega t_1}\langle[M^{\alpha}, M^{\beta}(t_1)]\rangle_0, \qquad (6.182)$$

$$g_{\alpha\beta}(\omega) = \int_{-\infty}^{\infty} dt_1 e^{-i\omega t_1}\langle\{M^{\alpha}M^{\beta}(t_1)\}\rangle_0, \qquad (6.183)$$

where

$$\{AB\} = \frac{1}{2}(AB + BA).$$

Now, we find a relationship between the functions $\chi_{\alpha\beta}(\omega)$, $f_{\alpha\beta}(\omega)$ and $g_{\alpha\beta}(\omega)$. For that, we first consider the expression

$$(\chi_{\beta\alpha}(\omega))^* = \frac{-i}{\hbar}\int_{-\infty}^{0} dt_1 e^{(\epsilon+i\omega)t_1}(\langle[M^{\beta}, M^{\alpha}(t_1)]\rangle_0)^* . \qquad (6.184)$$

Given the hermiticity of operators of physical quantities (see the formula (5.179)), it is easy to show that

$$(\langle[M^{\beta}, M^{\alpha}(t_1)]\rangle_0)^* = \langle[M^{\alpha}, M^{\beta}(-t_1)]\rangle_0 .$$

Substituting this result into (6.184) and making the change of variables $t_1 \to -t_1$, we obtain

$$(\chi_{\beta\alpha}(\omega))^* = \frac{-i}{\hbar}\int_0^\infty dt_1 e^{-(\epsilon+i\omega)t_1}\langle[M^\alpha, M^\beta(t_1)]\rangle_0. \tag{6.185}$$

Upon comparing the results of (6.181), (6.182) and (6.185) the relationship between the spectral intensity function $f_{\alpha\beta}(\omega)$ and the components of the magnetic susceptibility tensor $\chi_{\alpha\beta}(\omega)$ is given by

$$f_{\alpha\beta}(\omega) = \chi_{\alpha\beta}(\omega) - (\chi_{\beta\alpha}(\omega))^*. \tag{6.186}$$

We now find a link between the functions $f_{\alpha\beta}(\omega)$ and $g_{\alpha\beta}(\omega)$. For this, it is necessary to find how the expressions differ from each other:

$$\int_{-\infty}^\infty dt_1 e^{-i\omega t_1}\langle M^\alpha M^\beta(t_1)\rangle_0; \quad \int_{-\infty}^\infty dt_1 e^{-i\omega t_1}\langle M^\beta(t_1) M^\alpha\rangle_0.$$

Let us consider the integral

$$\begin{aligned}\int_{-\infty}^\infty & dt_1 e^{-i\omega t_1}\langle M^\alpha M^\beta(t_1)\rangle_0 \\ &= \int_{-\infty}^\infty dt_1 e^{-i\omega t_1}\,\mathrm{Sp}\{M^\alpha e^{i/\hbar H t_1} M^\beta e^{-i/\hbar H t_1}\rho_0\} \\ &= \int_{-\infty}^\infty dt_1 e^{-i\omega t_1}\,\mathrm{Sp}\left\{M^\alpha \frac{1}{Z} e^{-\beta H} e^{i/\hbar H(t_1 - i\hbar\beta)} M^\beta e^{-i/\hbar H(t_1 - i\hbar\beta)}\right\}.\end{aligned}$$

Making the change of variables $t_1 - i\hbar\beta \to t_1$ in the last integral and taking into account that $e^{-i\omega t_1} \to e^{-i\omega t_1} \cdot e^{\beta\hbar\omega}$, we obtain

$$\int_{-\infty}^\infty dt_1 e^{-i\omega t_1}\langle M^\alpha M^\beta(t_1)\rangle_0 = e^{\beta\hbar\omega}\int_{-\infty+i\hbar\beta}^{\infty+i\hbar\beta} dt_1 e^{-i\omega t_1}\langle M^\beta(t_1) M^\alpha\rangle_0. \tag{6.187}$$

The poles of the integrand in (6.187) are on the real axis, so one can shift down the path of integrations by the magnitude $i\hbar\beta$. Then instead of (6.187) we have

$$\int_{-\infty}^\infty dt_1 e^{-i\omega t_1}\langle M^\alpha M^\beta(t_1)\rangle_0 = e^{\beta\hbar\omega}\int_{-\infty}^\infty dt_1 e^{-i\omega t_1}\langle M^\beta(t_1) M^\alpha\rangle_0. \tag{6.188}$$

The equality (6.188) can be easily verified to hold true, and consequently, the possibility to shift down the integration contour can become feasible provided that eigenfunctions of the total Hamiltonian H are known. In the given case, there arise delta functions when integrating both sides of (6.188) over the time t_1, therefore, the equality (6.188) becomes clear. Because the value of the spur of an arbitrary set of operators does not depend on the total system of eigenfunctions to evaluate matrix elements, this relation is taken to be proved.

Inserting the result (6.188) in the definition of the functions $f_{\alpha\beta}(\omega)$ and $g_{\alpha\beta}(\omega)$, one gets

$$f_{\alpha\beta}(\omega) = \frac{i}{\hbar}(1 - e^{-\beta\hbar\omega}) \int_{-\infty}^{\infty} dt_1 e^{-i\omega t_1} \langle M^\alpha M^\beta(t_1) \rangle_0; \tag{6.189}$$

$$g_{\alpha\beta}(\omega) = \frac{1}{2}(1 + e^{-\beta\hbar\omega}) \int_{-\infty}^{\infty} dt_1 e^{-i\omega t_1} \langle M^\alpha M^\beta(t_1) \rangle_0. \tag{6.190}$$

We combine these two results to obtain the desired relationship between the functions $f_{\alpha\beta}(\omega)$ and $g_{\alpha\beta}(\omega)$:

$$f_{\alpha\beta}(\omega) = 2\frac{i}{\hbar}\frac{1 - e^{-\beta\hbar\omega}}{1 + e^{-\beta\hbar\omega}} g_{\alpha\beta}(\omega). \tag{6.191}$$

It should be kept in mind that the relationship between the function $f_{\alpha\beta}(\omega)$ and the imaginary part of magnetic susceptibility (6.186) exists. Then the spectral intensity of the symmetrized correlation function $g_{\alpha\beta}(\omega)$ can be expressed through the imaginary part of the magnetic susceptibility tensor:

$$g_{\alpha\beta}(\omega) = \frac{\hbar}{2i}\frac{1 + e^{-\beta\hbar\omega}}{1 - e^{-\beta\hbar\omega}} \cdot (\chi_{\alpha\beta}(\omega) - (\chi_{\beta\alpha}(\omega))^*). \tag{6.192}$$

If one multiplies the numerator and denominator of the last expression by $e^{\beta\hbar\omega/2}$ and introduces the notation

$$\frac{1}{2i}(\chi_{\alpha\beta}(\omega) - (\chi_{\beta\alpha}(\omega))^*) = \mathrm{Im}\chi^s_{\alpha\beta},$$

which has meaning being the imaginary part of a symmetrical component of the magnetic susceptibility tensor, the expression (6.192) can be represented in compact form:

$$g_{\alpha\beta}(\omega) = \hbar \cdot \mathrm{cth}\left(\frac{\beta\hbar\omega}{2}\right) \cdot \mathrm{Im}\chi^s_{\alpha\beta}(\omega). \tag{6.193}$$

It is the expression (6.193) that is a formulation of fluctuation–dissipation theory by Callen and Welton. The results (6.193) and (6.186) hold true in the spatially inhomogeneous case, when the functions $f_{\alpha\beta}(\vec{k},\omega)$, $g_{\alpha\beta}(\vec{k},\omega)$, $\chi_{\alpha\beta}(\vec{k},\omega)$ depend on the wave vector $\vec{k}$ and frequency ω.

In the previous paragraph, another function $S_{\alpha\beta}(\vec{k},\omega)$ was introduced. Now, recollecting this fact, one should establish its connection with the functions $f_{\alpha\beta}(\vec{k},\omega)$ and

$\chi_{\alpha\beta}(\vec{k},\omega)$. For simplicity, we first consider the spatially homogeneous case. Then, using the definitions (6.171) and (6.170) and setting $\vec{k}=0$, we obtain

$$S_{\alpha\beta}(\omega) = \int_{-\infty}^{\infty} dt_1 e^{i\omega t_1} \langle M^\alpha(t_1) M^\beta \rangle_0$$
$$= \int_{-\infty}^{\infty} dt_1 e^{i\omega t_1} \langle M^\alpha M^\beta(-t_1) \rangle_0 = \int_{-\infty}^{\infty} dt_1 e^{-i\omega t_1} \langle M^\alpha M^\beta(t_1) \rangle_0 .$$

In writing the last equality in this formula the change of variables $t_1 \to -t_1$ has been made. Given this result, the expressions (6.189), (6.190) can be rewritten as follows:

$$f_{\alpha\beta}(\omega) = \frac{i}{\hbar}(1 - e^{-\beta\hbar\omega}) S_{\alpha\beta}(\omega) ; \tag{6.194}$$

$$g_{\alpha\beta}(\omega) = \frac{1}{2}(1 + e^{-\beta\hbar\omega}) S_{\alpha\beta}(\omega). \tag{6.195}$$

As far as the relations (6.194), (6.195) are directly generalized for a spatially inhomogeneous case, and considering (6.193), we have

$$g_{\alpha\beta}(\vec{k},\omega) = \frac{1}{2}(1 + e^{-\beta\hbar\omega}) S_{\alpha\beta}(\vec{k},\omega) = \hbar \cdot \mathrm{cth}\left(\frac{\beta\hbar\omega}{2}\right) \cdot \mathrm{Im}\chi^s_{\alpha\beta}(\vec{k},\omega).$$

Hence there follows the simple equality in the limit of low frequencies $\beta\hbar\omega \ll 1$

$$\mathrm{Im}\,\chi^s_{\alpha\beta}(\vec{k},\omega) = \frac{\beta\omega}{2} S_{\alpha\beta}(\vec{k},\omega). \tag{6.196}$$

Finally, substituting the value of the function $S_{\alpha\beta}(\vec{k},\omega)$ of (6.179) into the expression (6.196) which is valid in the limit of small $\vec{k}$, we arrive at the representation of an imaginary part of magnetic susceptibility

$$\mathrm{Im}\,\chi^s_{\alpha\beta}(\vec{k},\omega) = \chi \frac{Dk^2\omega}{\omega^2 + (Dk^2)^2}. \tag{6.197}$$

This expression holds true also in the long-wave approximation $\vec{k} \to 0$. In connection with the result obtained, it is important to clarify that the structure of the imaginary part of the magnetic susceptibility of (6.197) is "dictated" by conservation laws and symmetry properties of the system under consideration. Moreover, it is independent on the specific form of the system's Hamiltonian.

Problem 6.1. Prove that

$$\lim_{\vec{k}\to 0} S(\vec{k}, t=0) = 1/\beta \cdot \chi,$$

where χ is the static susceptibility in a spatially homogeneous case.

Solution. We start with the definition of

$$\lim_{\vec{k}\to 0} S(\vec{k}, t = 0) = \lim_{\vec{k}\to 0} \int d\vec{r}\, \mathrm{Sp}\{M(\vec{r})M(0)\rho_0\} e^{-i\vec{k}\vec{r}},$$

where M is magnetization of the system along an external magnetic field with an amplitude h. Furthermore, $M(0) = \sum_i M_i$;

$$\lim_{\vec{k}\to 0} \int d\vec{r} M(\vec{r}) e^{-i\vec{k}\vec{r}} = \lim_{\vec{k}\to 0} \int d\vec{r} \sum_i M_i \delta(\vec{r} - \vec{r}_i) e^{-i\vec{k}\vec{r}} = \sum_i M_i.$$

Thus,

$$\lim_{\vec{k}\to 0} S(\vec{k}, t = 0) = \mathrm{Sp}\{M_T M_T \rho_0\},$$

where M_T is the total magnetic moment of the specimen.

The magnetic susceptibility of the specimen in the framework of phenomenological thermodynamics is given by

$$\chi = \lim_{h\to 0} \frac{\partial}{\partial h} \langle M\rangle^h.$$

If one assumes that the system's Hamiltonian H in an external field h can be represented as $H = H_0 - hM_T$, then the average magnetization can be calculated by averaging over an equilibrium ensemble,

$$\langle M\rangle^h = \frac{\mathrm{Sp}\{M_T \exp[-\beta(H_0 - hM_T)]\}}{\mathrm{Sp}\{\exp[-\beta(H_0 - hM_T)]\}}.$$

Now, we compute the derivative over h. In this case this can be easily done due to preserving the total magnetic moment. Consequently, it commutes with the Hamiltonian H_0. As a result, we get

$$\lim_{h\to 0} \frac{\partial}{\partial h} \langle M\rangle^h = \beta \cdot [\langle M_T^2\rangle_0 - \langle M_T\rangle_0^2] = \beta \cdot \mathrm{Sp}\{M_T M_T \rho_0\}.$$

The last equality implies that $\langle M_T\rangle_0 = 0$ in the absence of spontaneous magnetization. Thus, we have shown that indeed

$$\lim_{\vec{k}\to 0} S(\vec{k}, t = 0) = 1/\beta \cdot \chi.$$

6.3.3 Long-range correlations and slow modes

In Section 6.3.1 of the present chapter in the context of spin diffusion, we discussed the conditions of the emergence of hydrodynamic modes weakly damped in the limit

$k \to 0$ for the collective excitations. We showed, in particular, that if a dynamical variable satisfies some conservation law and is a quasi-integral of motion, the corresponding autocorrelation function will have a hydrodynamic pole in a complex z-plane. Now, it is expedient to generalize these results by using the Mori projection operator method, developed in Sections 6.2.2–6.2.5.

Let us define an autocorrelation function of the operators $A(\vec{k}, t)$ and $A(\vec{k}', t')$ by the relation

$$C_{AA}(\vec{k}, t) = (A(\vec{k}, t), A^{+}(\vec{k}, 0)) = \int_0^1 d\tau \mathrm{Sp}\{\Delta A(\vec{k}, t)\rho_0^{\tau}\Delta A^{+}(\vec{k}, 0)\rho_0^{1-\tau}\}. \tag{6.198}$$

By the homogeneity of space, it follows from the formula (5.167) in Section 5.2.4 that the nonzero averages are only those that satisfy the condition $\vec{k}' = -\vec{k}$. Let the operator A be self-conjugate, and then $A^{+}(\vec{k}) = A(-\vec{k})$. Next, we define the correlation function $C_{AA}(\vec{k}, z)$ by performing the Laplace transform of the correlation function $C_{AA}(\vec{k}, t)$ with respect to the variable t:

$$C_{AA}(\vec{k}, z) = \int_0^{\infty} dt C_{AA}(\vec{k}, t) e^{-zt}. \tag{6.199}$$

It is worth drawing attention to the fact that the function $C_{AA}(\vec{k}, z)$ differs from the function $\Theta(z)$ (6.106) only by the factor of the type $(A, A^{+})^{-1}$ and an additional dependence on $\vec{k}$. So by repeating all the foregoing calculations leading us to (6.106) from (6.94), we can seek the representation

$$C_{AA}(\vec{k}, z) = \frac{\beta^{-1}\chi_{AA}(\vec{k})}{z + \mathfrak{S}_A(\vec{k}, z) \cdot [\chi_{AA}(\vec{k})]^{-1}}, \tag{6.200}$$

where

$$\mathfrak{S}_A(\vec{k}, z) = \beta\Big((1 - \mathcal{P})\dot{A}(\vec{k}), \frac{1}{z - (1 - \mathcal{P})iL}(1 - \mathcal{P})\dot{A}^{+}(\vec{k})\Big). \tag{6.201}$$

In the formulas (6.200), (6.201), the following definition has been used:

$$C_{AA}(\vec{k}, t = 0) = \langle A(\vec{k})A^{+}(\vec{k})\rangle = \beta^{-1}\chi_{AA}(\vec{k}).$$

In addition, for simplicity, it would be proper to assume that

$$i\Omega = (\dot{A}(\vec{k}), A^{+}(\vec{k}))(A(\vec{k}), A^{+}(\vec{k}))^{-1} = 0.$$

Had the quantity $A(\vec{k}, t)$ been a preserving quantity, the function $\mathfrak{S}_A(\vec{k}, z)$, as it was shown in Section 6.3.1, would be proportional to k^2, and a hydrodynamic pole would appear as well. This is easily seen, if one assumes that $A(\vec{k})$ is the only basis operator satisfying the equation of motion:

$$\dot{A}(\vec{k}) = a \cdot A(\vec{k}) + i\vec{k}\vec{J}_A(\vec{k}),$$

where a is some numerical coefficient, $\vec{J}_A(\vec{k})$ the flow vector, associated with the physical quantity A. Then, since the correlator of the random forces $\mathfrak{S}_A(\vec{k}, z)$ contains the construction $(1 - \mathcal{P})\dot{A}(\vec{k})$, the component $a \cdot A(\vec{k})$ does not contribute. Therefore, $(1 - \mathcal{P})\dot{A}(\vec{k}) \sim k$ and

$$\mathfrak{S}_A(\vec{k}, z) \sim k^2.$$

If

$$\lim_{\vec{k}\to 0} \mathfrak{S}_A(\vec{k}, z) \cdot [\chi_{AA}(\vec{k})]^{-1} \neq 0$$

remains a finite quantity at $\vec{k} \to 0$, this means that the correlation function $C_{AA}(0, t)$ satisfies the equation

$$\frac{\partial}{\partial t} C_{AA}(0, t) = -\frac{C_{AA}(0, t)}{\tau_A},$$
$$\tau_A^{-1} = \lim_{\vec{k}\to 0} \mathfrak{S}_A(\vec{k})[\chi_{AA}(\vec{k})]^{-1}, \tag{6.202}$$

showing the usual relaxation behavior. Indeed, if one converts the equation (6.202) by means of a Laplace-transform and uses the relation (6.105), one is led to

$$-C_{AA}(0, 0) + zC_{AA}(0, z) = \frac{-1}{\tau_A} C_{AA}(0, z),$$

or

$$C_{AA}(0, z) = \frac{C_{AA}(0, 0)}{z + \tau_A^{-1}}.$$

Assume now that a nonconserved quantity has static correlations of an infinite radius and

$$\lim_{\vec{k}\to 0} \chi_{AA}(\vec{k}) \sim \frac{M_0}{R_A k^2},$$

where M_0 and R_A are constants. Obviously, then

$$\mathfrak{S}_A(\vec{k}, z) \cdot [\chi_{AA}(\vec{k})]^{-1} \sim k^2$$

and we shall again have a hydrodynamic pole.

It is of interest to know in what systems the $1/k^2$ singularities of the static susceptibility are possible. First of all, let us consider an isotropic ferromagnetic material. It is well known that there arises spontaneous ordering of its magnetic moments resulting in their lining up along a certain direction. The Z-axis can be chosen as the direction. In fact, nothing marks this direction, only a very weak anisotropy in the specimen may occur. The anisotropy trends to orient the spontaneous magnetization only along this direction.

If we now apply an external field $h_x(\vec{r})$ along the X-axis, there arises nonzero magnetization $M_x(\vec{r})$ under the action of this field. After performing the Fourier transform of a constitutive equation, we obtain

$$\langle M_x(\vec{k})\rangle = \chi_{xx}(\vec{k})h_x(\vec{k}).$$

Consider the transition $k \to 0$ in this equation. Obviously, in the given limiting case

$$\lim_{\vec{k}\to 0}\langle M_x(\vec{k})\rangle = M_0 = \chi_{xx}h_x,$$

where M_0 is an equilibrium magnetic moment of the specimen. Since a turn of the spontaneous magnetization vector is not associated with any work, such a rotation will occur in an infinitesimal field as well. According to this fact one can conclude that $\lim_{\vec{k}\to 0}\chi_{xx}(\vec{k}) \to \infty$.

To arrive at the same conclusions, another argument can be used. As to the rotation of the spontaneous magnetization vector in a local region of the specimen, only a very small amount of energy is required. Thus, to create a sinusoidal magnetization fluctuation with a large wavelength in the space, one must spend an infinitesimal energy. This is due to fact, that the energy interaction between domains with different orientations of the magnetic moments is too small. If fluctuations exist, they damp out too slowly because the spin orientation in each domain is in equilibrium. The interaction between the domains with different orientations of magnetization is the only interaction causing the relaxation processes.

Thus, the divergence of $\chi_{xx}(\vec{k})$ is assumed to be related to static long-range correlations, which, in turn, are caused by spontaneous symmetry breaking in the ground state. This question calls for more attention when it comes to symmetry breaking in the ground state. Consider a system of spins whose interaction is described by Heisenberg's Hamiltonian:

$$H = -\frac{1}{2}\sum_{i\neq j} J(|\vec{r}_i - \vec{r}_j|)\vec{S}_i\vec{S}_j. \tag{6.203}$$

It is well known that the total spin is preserved in this system, so

$$\left[\sum_i \vec{S}, H\right] = 0.$$

Assuming that the net magnetic moment is directed along the Z-axis, we calculate the average value of the z-component of the total spin:

$$\left\langle \sum_i S_i^z \right\rangle_0 = \frac{1}{i\hbar} \operatorname{Sp}\left\{\left[\sum_i S_i^x, \sum_j S_j^y\right]\rho_0\right\}$$
$$= \frac{1}{i\hbar} \operatorname{Sp}\left\{\left[\sum_i S_i^x, \rho_0\right]\sum_j S_j^y\right\} \neq 0. \tag{6.204}$$

From this result it follows that the equilibrium statistical operator

$$\rho_0 \neq \frac{1}{Z} e^{-\beta H}.$$

However, it is more correctly to assume that

$$\rho_0 = \frac{1}{Z} \mathcal{P}_Z e^{-\beta H},$$

i. e. only those states are selected out of all possible states with a given energy, for which the total spin momentum is oriented along the Z-axis. In particular, one should think that in order to distinguish the Z-direction, the projection operation consists in:

$$\rho_0 \to \frac{1}{Z} \exp\left\{-\beta\left(H - \sum_i S_i^z h\right)\right\},$$

where h is an infinitesimal parameter. In this case, the z-component the total spin commutes with ρ_0, but the x-and y-components are not commute. In spite of the extreme smallness of the external field h, it gives somewhat greater statistical weight for the states in which the total spin along the Z-axis is not zero. In the ferromagnetic state, it is sufficient to line up the spins to be parallel to the Z-axis.

Singularity of $\chi_{xx}(\vec{k})$ is associated with symmetry breaking whereas the magnetic moments are aligned along the Z-axis, while the Hamiltonian of the system is rotationally invariant. That is why the rotation of the resultant magnetic moment occurs in an infinitely weak field h, applied along the X-axis, which leads to a singularity of $\chi_{xx}(\vec{k})$. Moreover, as far as $\chi_{xx}(\vec{k})$ is an even function of k ($\chi_{xx}(\vec{r})$ depends only on $|\vec{r}|$), this peculiarity has the form $1/k^2$.

It is interesting also to note that $\chi_{zz}(\vec{k})$ does not have any unusual behavior because an increase of the magnetic moment along the Z-axis by applying an infinitesimal field along this axis is also infinitesimal.

The result of a quadratic singularity of static susceptibility of systems with spontaneously broken symmetry is called the Bogoliubov $1/k^2$ theorem. The next section reviews the most basic ideas of the proof of the Bogoliubov theorem on the example of the static magnetic susceptibility. More detailed information on this theme can be found in the monographs [44, 59].

6.3.4 Bogoliubov inequality and $1/k^2$ divergence theorem

It would be proper to start the present section with a simplified derivation of Bogoliubov's inequality and the theorem of the singularity of the static components of generalized susceptibility in systems with broken symmetry in the ground state. In this case, the static magnetic susceptibility may serve as an example. We first derive the Kramers–Kronig dispersion relations. Then we establish the sum rule that relates real and imaginary parts of the generalized susceptibility.

The generalized susceptibility is defined as a response of an operator value of a physical quantity B to an external action, determined by the perturbation $-A^+F$:

$$\delta\langle B(\vec{k},\omega)\rangle = \chi_{BA}(\vec{k},\omega)F(\vec{k},\omega),$$
$$\chi_{BA}(\vec{k},\omega) = \chi'_{BA}(\vec{k},\omega) + i\chi''_{BA}(\vec{k},\omega), \tag{6.205}$$

where $\chi'_{BA}(\vec{k},\omega)$ and $\chi''_{BA}(\vec{k},\omega)$ are real and imaginary parts of a generalized susceptibility tensor. Suppose now that

$$\lim_{\omega\to\pm\infty} \chi_{BA}(\vec{k},\omega) \to 0. \tag{6.206}$$

If this is not so, the renormalization of the generalized susceptibility should be made, for example, by subtracting the value of $\chi_{BA}(\vec{k},\infty)$ from $\chi_{BA}(\vec{k},\omega)$, the limiting transition (6.206) to be true for this case as well. Then there is every reason to assume that $\chi_{BA}(\vec{k},\omega)$ is an analytic function in a complex plane z. Consequently, by Cauchy's theorem for analytic functions

$$\int_C dz \frac{\chi_{BA}(\vec{k},z)}{z-\omega} = 0, \tag{6.207}$$

if the integration contour is chosen so that a pole of the resolvent can be bypassed along the segment of the circle centered at ω. The point ω lies on the real axis in a way as shown in Figure 6.3.

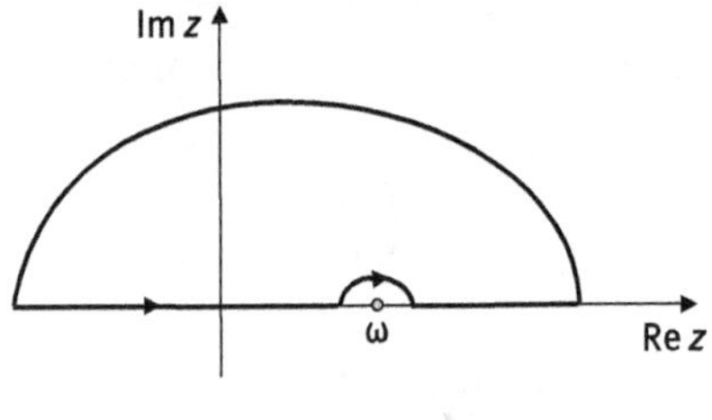

Figure 6.3: Contour around the pole of the resolvent in equation (6.207).

Let ρ be the radius of the circle along which the pole can be bypassed. Then the contour integral can be written as the sum of the three integrals

$$\lim_{\rho\to 0}\left\{\int_{-\infty}^{\omega-\rho} dz\frac{\chi_{BA}(\vec{k},z)}{z-\omega}+\int_{\omega+\rho}^{\infty} dz\frac{\chi_{BA}(\vec{k},z)}{z-\omega}+\chi_{BA}(\vec{k},\omega)\int_{-\pi}^{0}\frac{i\rho e^{i\varphi}\,d\varphi}{\rho e^{i\varphi}}\right\},$$

$$z-\omega=\rho e^{i\varphi},\quad dz=i\rho e^{i\varphi}\,d\varphi. \tag{6.208}$$

Performing the limiting transition, we find the ratio

$$i\pi\chi_{BA}(\vec{k},\omega)=P\int_{-\infty}^{\infty} dz\frac{\chi_{BA}(\vec{k},z)}{z-\omega}. \tag{6.209}$$

The real and imaginary parts of the above expression having been separated, the Kramers–Kronig relations can be easily obtained:

$$\chi'_{BA}(\vec{k},\omega)=\frac{1}{\pi}P\int_{-\infty}^{\infty} dz\frac{\chi''_{BA}(\vec{k},z)}{z-\omega}; \tag{6.210}$$

$$\chi''_{BA}(\vec{k},\omega)=-\frac{1}{\pi}P\int_{-\infty}^{\infty} dz\frac{\chi'_{BA}(\vec{k},z)}{z-\omega}. \tag{6.211}$$

In the formulas (6.209)–(6.211) the symbol P is used for the principal value of the integral.

The Kramers–Kronig relations allow for formulating the sum rule for tensor components of generalized susceptibility. Putting $\omega=0$ and considering that

$$\mathrm{Re}\chi_{BA}(\vec{k},0)=\chi_{BA}(\vec{k}),$$

we get

$$\chi_{BA}(\vec{k})=\frac{1}{\pi}P\int_{-\infty}^{\infty} dz\frac{\chi''_{BA}(\vec{k},z)}{z}. \tag{6.212}$$

Thus, the static generalized susceptibility can be found by integrating the imaginary part over the entire frequency interval.

We now show that the generalized static susceptibility $\chi_{BA}(\vec{k})$ can be determined by the Kubo (Mori) "scalar" product of the two operators A and B, previously defined (5.82), (6.89):

$$\chi_{BA}(\vec{k})=\beta(B(\vec{k}),A^+(\vec{k}))=\beta\int_0^1 d\tau\,\mathrm{Sp}\{B(\vec{k})\rho_0^\tau A^+(\vec{k})\rho_0^{1-\tau}\}. \tag{6.213}$$

The proof of this expression comes from the linear-response formula (5.35). For that purpose, one has to put $\omega = 0$ and perform a series of identical transformations (see also (5.43), (5.80), (6.118)):

$$\chi_{BA}(\vec{k}) = \frac{i}{\hbar}\int_0^\infty dt e^{-\epsilon t}\,\mathrm{Sp}\{[\rho_0, B(\vec{k},t)]A^+(\vec{k})\}. \tag{6.214}$$

We use the Kubo identity (5.81)

$$\frac{i}{\hbar}[\rho_0, B(\vec{k},t)] = -\beta\frac{d}{dt}\int_0^1 d\tau B(\vec{k}, t + i\hbar\beta\tau)\rho_0$$

to modify the right-hand side of (6.214). In the formula, integrating by parts the integral over the variable t, we arrive at

$$\begin{aligned}\chi_{BA}(\vec{k}) = &-\beta\int_0^1 d\tau\,\mathrm{Sp}\{B(\vec{k}, t + i\hbar\beta\tau)\rho_0 A^+(\vec{k})\}\big|_{t=0}^{t=\infty}\\ &-\epsilon\int_0^\infty dt e^{-\epsilon t}\beta\int_0^1 d\tau\,\mathrm{Sp}\{B(\vec{k}, t + i\hbar\beta\tau)\rho_0 A^+(\vec{k})\}.\end{aligned} \tag{6.215}$$

Applying Abel's theorem

$$\lim_{\epsilon\to 0}\epsilon\int_0^\infty dt e^{-\epsilon t} f(t) = \lim_{t\to\infty} f(t)$$

and the correlation weakening principle, it is easy to show that the second summand on the right-hand side of (6.215) is zero, assuming that the B and A operators are defined so that their non-equilibrium average is equal to zero. By virtue of the correlation weakening principle, substitution of the upper limit $t = \infty$ in the first summand on the right side of (6.215) also yields a zero result. The desired result can be obtained when substituting the lower limit $t = 0$:

$$\chi_{BA}(\vec{k}) = \beta\int_0^1 d\tau\,\mathrm{Sp}\{B(\vec{k}, i\hbar\beta\tau)\rho_0 A^+(\vec{k})\}. \tag{6.216}$$

The right-hand side of (6.216) coincides with the right part of the formula (6.213), as is easily seen, making the change of variables $\tau - 1 \to \tau'$ in the formula (6.216).

Furthermore, since the correlation function

$$(A, B) = \int_0^1 d\tau\mathrm{Sp}\{A\rho_0^\tau B\rho_0^{1-\tau}\}$$

possesses all the properties of the scalar product, it provides the Schwarz inequality for the A and B vectors in a Hilbert space:

$$(A,A)\cdot(B,B) \geq |(A,B)|^2. \tag{6.217}$$

Next, the generalized static susceptibility according to (6.213) and (6.212) being written as the Kubo scalar product of the A and B operators, the Schwarz inequality (6.217) can be written as follows:

$$\frac{1}{\pi}P\int_{-\infty}^{\infty} d\omega \frac{\chi''_{AA}(\vec{k},\omega)}{\omega} \cdot \frac{1}{\pi}P\int_{-\infty}^{\infty} d\omega \frac{\chi''_{BB}(\vec{k},\omega)}{\omega} \geq \left|\frac{1}{\pi}P\int_{-\infty}^{\infty} d\omega \frac{\chi''_{AB}(\vec{k},\omega)}{\omega}\right|^2. \tag{6.218}$$

Equation (6.218) is another way of writing the Bogoliubov inequality for correlation functions. To show that

$$\lim_{\vec{k}\to 0} \chi_{AA}(\vec{k}) \sim \frac{1}{k^2},$$

the expression (6.218) needs to be used, whereas the $A(\vec{k})$ and $B(\vec{k})$ operators can be arbitrary. Now, one should take an operator $M_x(\vec{k})$ as the operator $A(\vec{k})$, but the operator $\dot{M}_y(\vec{k})$ as the operator $B(\vec{k})$. Therefore, the Schwarz inequality for this particular case can be expressed as

$$\frac{1}{\pi}P\int_{-\infty}^{\infty} d\omega \frac{\chi''_{M_x M_x}(\vec{k},\omega)}{\omega}$$
$$\geq \left|\frac{1}{\pi}P\int_{-\infty}^{\infty} d\omega \frac{\chi''_{\dot{M}_y M_x}(\vec{k},\omega)}{\omega}\right|^2 \cdot \left[\frac{1}{\pi}\int_{-\infty}^{\infty} d\omega\,\omega\,\chi''_{M_y M_y}(\vec{k},\omega)\right]^{-1}. \tag{6.219}$$

When writing the second co-factor on the right-hand side of (6.219) we have used the representation (6.186) for the imaginary part of the generalized susceptibility and performed two-fold integration by parts with respect to time argument t. Furthermore, one can show that the below expression is true with regard to the correlation weakening principle:

$$\chi''_{\dot{M}_y \dot{M}_y}(\vec{k},\omega) = \omega^2 \chi''_{M_y M_y}(\vec{k},\omega).$$

We convert the first co-factor on the right-hand side of the inequality (6.219). Using the sum rule (6.212) and formula (6.186), we have

$$\chi_{\dot{M}_y M_x}(\vec{k}) = \frac{1}{\pi}P\int_{-\infty}^{\infty} d\omega \frac{\chi''_{\dot{M}_y M_x}(\vec{k},\omega)}{\omega}$$
$$= \frac{1}{2\pi}\int_{-\infty}^{\infty} d\omega \int_{-\infty}^{\infty} dt e^{-i\omega t}\frac{i\omega}{\hbar\omega}\langle[M_y(\vec{k}), M_x(-\vec{k},t)]\rangle_0 = \frac{g\mu_B}{\hbar}\langle M_z\rangle_0, \tag{6.220}$$

where $\langle M_z \rangle_0 \equiv M_0$ is a vector of the magnetic moment in an equilibrium state. In deriving this expression we have used the definition of the delta-function.

$$\frac{1}{2\pi}\int_{-\infty}^{\infty} d\omega e^{-i\omega t} = \delta(t),$$

and the representation (6.186) for the imaginary part of generalized susceptibility. Then integration by parts over the time argument t in accordance with the correlation weakening principle yields

$$\chi''_{\dot{M}_y M_x}(\vec{k},\omega) = i\omega\chi''_{M_y M_x}(\vec{k},\omega).$$

In addition, we have taken into account the commutation rules of the Fourier spin components

$$\begin{aligned}[S^\alpha(\vec{k}), S^\beta(\vec{k}')] &= \sum_{i,j}[S_i^\alpha e^{i\vec{k}\vec{r}_i}, S_j^\beta e^{\vec{k}'\vec{r}_j}] \\ &= \sum_i [S_i^\alpha, S_i^\beta] e^{(\vec{k}+\vec{k}')\vec{r}_i} = i\epsilon_{\alpha\beta\gamma} S^\gamma(\vec{k}+\vec{k}').\end{aligned} \tag{6.221}$$

The relation (6.220) can be treated as a variant of writing the sum rule for the components of the magnetic susceptibility tensor. Here $\epsilon_{\alpha\beta\gamma}$ is a unit third-rank anti-symmetric tensor.

Thus, the first co-factor on the right side of the inequality (6.219) is independent on k and proportional to square of the equilibrium magnetization M_0.

We now transform the second factor on the right side of the inequality (6.219). In temporal representation, the components of a total magnetization vector satisfy the continuity equation:

$$\frac{\partial}{\partial t} M_y(\vec{r},t) + \operatorname{div}\vec{J}_{M_y}(\vec{r},t) = 0.$$

Performing the Fourier transform of this equation, we have

$$i\omega M_y(\vec{k},\omega) + i\vec{k}\vec{J}_{M_y}(\vec{k},\omega) = 0.$$

Hence, it follows that

$$\omega^2\chi''_{M_y M_y}(\vec{k},\omega) = k_i k_j \chi''_{J^i_{M_y} J^j_{M_y}}(\vec{k},\omega).$$

In the long run, performing the limiting transition $k \to 0$ in the inequality (6.219) and considering the sum rule (6.212), we obtain the desired result,

$$\lim_{k\to 0}\chi_{M_x M_x}(\vec{k},\omega) \geq \frac{M_0^2}{k^2 \cdot \text{Const}}. \tag{6.222}$$

In this formula, the constant is determined by a correlation function of magnetization currents.

Thus, the brief introduction to the proof of the $1/k^2$ singularity theorem is completed. It is worth emphasizing that the different behavior of the static susceptibility components $\chi_{M_y M_x}$ and $\chi_{M_x M_x}$ within the limit $k \to 0$ constitutes the base of the broken symmetry concept in deriving the formula (6.222). The former remains finite as the latter diverges as $1/k^2$.

Having summarized the results obtained, it can be argued that there are two mechanisms of emergence of hydrodynamic modes. The first is associated with the presence of conserved physical quantities (quasi-integrals of motion). The phenomenon of spin diffusion may serve as an example. The second mechanism reflects spontaneous symmetry breaking in the ground state. In this case, there also arise the long-wave hydrodynamic modes, long-lived, as $k \to 0$, but the nature of their appearance is somewhat different.

If an original symmetry group is continuous (for example, there is invariance with respect to translations or rotations), there may appear a branch of excitations as a result of the phase transition, which spontaneously breaks the original symmetry. In the long-wavelength limit, the branch is characterized by vanishing excitation energy. Goldstone, a British-born theoretical physicist, was first to discover this phenomenon. The statement put forward by him is often called Goldstone's theorem. According to this theorem, massless particles must exist in a relativistic system with broken symmetry and a corresponding degenerate vacuum state.

In the nonrelativistic case, this theorem should read as follows: there must exist a branch of elementary excitations without an energy gap in systems with broken symmetry. It later became clear that such a formulation had a lot of contradictions, consequently it was quite pregnable. Therefore, Lange in his work "Nonrelativistic Theorem Analogous to Goldstone Theorem" postulated some restrictions concerning the nature of the interaction between particles in such systems. In particular, it turns out that the long-range Coulomb interaction can suppress the emergence of the hydrodynamic modes.

Magnons in ferromagnetic (antiferromagnetic) materials, superfluid helium and three branches of acoustic phonons may serve as an example of Goldstone's modes in solids. In the first case, the spontaneous orientation of the magnetic moment infringes spherical symmetry of an original Hamiltonian. The second case is characterized by broken gauge invariance [60]. The latter reflects the breakdown of invariance with respect to infinitesimal translations of atoms in three mutually perpendicular directions as a result of their ordering in a crystal lattice.

Goldstone's theorem provides only the general statement that a long-wave branch without energy gap at $k \to 0$ can be expected when continuous symmetry breaking occurs. However, such an assertion does not exclude that there are other reasons for such modes to appear. As far as a hydrodynamic description of excitations is applicable in the long-wavelength limit, the Goldstone-modes are nothing but the hydrodynamic modes. As mentioned above, the hydrodynamic behavior is based on the existence of conserved physical quantities. The conserved quantities as generators of transformations of broken symmetry at the phase transition can also be linked with the presence of the Goldstone modes. For example, a rotation operator of spin moments in a coordinate space through an infinitesimal angle can be treated as the generator of the transformations. Since a system's Hamiltonian in the Heisenberg model is invariant so far as infinitesimal rotations in the coordinate space are concerned, this operator is a conserved quantity.

More information on the issues raised in Section 6.3.3 and 6.3.4 is contained in the paper by Forster [44].

6.4 Problems to Chapter 6

6.1. To study an evolving hydrodynamically system, it is necessary to define the quasi-integrals of motion such as kinetic and spin degrees of freedom of conduction electrons, the energy of long-wavelength acoustic phonons, the number and average drift momentum of the electrons.
Write down quasi-equilibrium statistical distribution for this system $\overline{\rho}_q$ by introducing a set of suitable macroscopic parameters P_n and appropriate Lagrange multipliers $F_n(t)$.

6.2. Prove that if the quasi-equilibrium state of a system is characterized by the vector of electric polarization $P^\alpha = -e\sum_i x_i^\alpha$, where x_i^α is the coordinate of ith electron, we have the well-known formula

$$T\frac{dS}{dt} = \langle J^\alpha \rangle E^\alpha(t),$$

where $\langle \vec{J} \rangle$ is the electric current density, $\vec{E}(t)$ is the external electric field strength, ensuing from the definition of the entropy (6.10).

6.3. Using the expansion of the NSO (6.62) linear in thermodynamic forces, prove that the correction to the mean value of any physical quantity assigned by the operator A is given by

$$\delta\langle A\rangle^\omega = \sum_n \{-(A, P_n^+) + (A, \dot{P}_n^+)^\omega - i\omega(A, P_n^+)^\omega\}\delta F_n(\omega),$$

where

$$(A, B)^\omega = \int_{-\infty}^{0} dt_1 \exp\{(\epsilon - i\omega)t_1\} \int_0^1 d\tau\, \mathrm{Sp}\{A\rho_0^\tau B(t_1)\rho_0^{1-\tau}\}.$$

6.4. Prove that the idempotence property of the Mori projection operators $\mathcal{P}$ and $Q = 1 - \mathcal{P}$ holds for any set of basic operators P_n. The projection operator is given by

$$\mathcal{P}A(t) = \frac{1}{(P, P^+)}(A(t), P^+)P.$$

6.5. Argue that the identity

$$((1 - \mathcal{P})C, \mathcal{P}B^{+}) = 0$$

holds for arbitrary operators C and B.

6.6. Derive formulas for Laplace transforms to calculate the derivative of a function $f'(x)$ and convolution of two arbitrary functions

$$g(x) = \int_0^x dt f_1(t) f_2(x - t).$$

6.7. Using the expression (6.124) for the memory function, compute the average momentum relaxation time of conduction electrons in scattering by charged impurity centers with the electron–impurity interaction Hamiltonian (4.82) in the static case ($\omega = 0$).

6.8. Consider a system of conduction electrons. The system interacts with long-wavelength acoustic phonons and is in an external static electric field. The electrons gain energy due to this interaction. In the process of quasi-elastic scattering, the electrons transfer this energy to the long-wavelength acoustic phonons; those in turn transport it to thermal phonons. The subsystem of the thermal phonons can be regarded as a thermostat. As a result, a stationary non-equilibrium state is established with non-equilibrium electron and phonon temperatures T_k and T_l, and non-equilibrium chemical potential ζ. Write down a set of energy balance equations for the electrons, long-wavelength phonons and the number of particles. Deduce expressions for the corrections to the above temperatures and chemical potential provided that the energy relaxation of the long-wavelength phonons by the thermal phonons is characterized by the relaxation time

$$\frac{1}{\tau_{ph}} = \frac{qT^4}{4\pi\rho s^4 \hbar^3}.$$

Hint. To solve the problem, the plan should include the following:

1. Write down the system's Hamiltonian $H = H_k + H_p + H_{kp} + H_T + H_{Tp} + H_{kf}$, where H_k, H_p are the Hamiltonians of the electron and phonon subsystems, H_{kp} is the electron–phonon interaction Hamiltonian, H_T is the Hamiltonian of the thermal phonons, H_{Tp} is the interaction Hamiltonian between the thermal phonons and long-wavelength phonons (it is obvious that there is no need to add the two latter), H_{kf} is the interaction Hamiltonian between the electrons and external field.
2. Write down operator equations of motion for the number of particles and subsystem energies:

$$\dot{n} = 0; \quad \dot{H}_k = \frac{1}{i\hbar}[H_k, H_{kp}] + \frac{e}{m} p^{\alpha} E^{\alpha};$$

$$\dot{H}_p + \dot{H}_{kp} = -\frac{1}{i\hbar}[H_k, H_{kp}] + \frac{1}{i\hbar}[H_p, H_{Tp}].$$

Write down the non-equilibrium statistical operator, using expressions (6.153), (6.155) and assuming that

$$S(t,0) = \phi + \beta_k(H_k - \zeta n) + \beta_l(H_p + H_{kp}) + \beta(H_T + H_{TP});$$
$$\delta S(t,0) = \Delta\{\delta\beta_k(H_k - \zeta_0 n) + \delta\beta_l(H_p + H_{kp}) - \delta\zeta\beta n\};$$
$$\dot{S}(t,0) = (\beta_k - \beta_l)(H_k - \zeta_0 n) + \delta\beta_l \frac{1}{i\hbar}[H_p + H_{kp}, H_{Tp}].$$

3. Write down energy balance equations for the subsystems and the number of particles,

$$n = n^0; \quad (\beta_k - \beta_l)L_{kk} = \sigma_{\alpha\beta}E^\alpha E^\beta$$
$$(\beta_k - \beta_l)L_{kk} = \frac{d}{dt}\langle H_p + H_{kp}\rangle|_T;$$
$$\frac{d}{dt}\langle H_p + H_{kp}\rangle|_T = \sum_{q\lambda} \hbar\omega_{q\lambda}\frac{N_{q\lambda} - N^0}{\tau_{ph}};$$
$$L_{kk} = \frac{2\pi}{\hbar}\sum_{\vec{k}\vec{q}\lambda\sigma} |C_{q\lambda}|^2(\hbar\omega_q)^2 N^0_{q\lambda} f^0_{\vec{k}+\vec{q}}(1 - f^0_{\vec{k}})\delta(\varepsilon_{\vec{k}+\vec{q}} - \varepsilon_{\vec{k}} - \hbar\omega_{q\lambda}).$$

4. Calculate non-equilibrium parameters from the equations obtained previously.

6.9. Using the definition of the generalized kinetic coefficient

$$L_{AB}(\omega) = \int_{-\infty}^{0} e^{(\varepsilon - i\omega)t_1}\,\mathrm{Sp}\left\{Be^{iLt_1}\frac{1}{i\hbar}[\rho_0, A]\right\} dt_1,$$

relate frequency-dependent real and imaginary parts of the electrical conductivity.

6.10. Formulate the fluctuation–dissipation theorem in the Callen–Welton form for the high-frequency conductivity tensor components.

7 Physical principles of spintronics

7.1 Spin current

7.1.1 Nature of spin current emergence

The standard analysis of kinetic phenomena described in the previous chapters agrees with the fact that electrons (holes) carry an electric charge and does not account for their spin degrees of freedom. The few publications devoted to nuclear polarization induced by hot electrons (Feher effect), the polarization of spin nuclei upon saturation of paramagnetic resonance of conduction electrons (Overhauser effect), and detection of saturation of paramagnetic resonance at conduction electrons by a change in electrical resistance in semiconductors have sparked no interest concerning the problem of spin transport. This could be partly explained by the circumstance that the characteristic length l_s to retain spin orientation turns out to be of the order of several tens of nanometers. For electronic devices designed in 60–70 years of the last century this was a too small dimension.

The development of microelectronics dramatically changed the situation. It became possible to apply structures with a layer of several tens of angstroms thick. Under these conditions, the spin polarization created in any way holds during a charge-carrier motion in the layer. Then there can be spoken of spin-dependent transport.

Naturally, exposing external magnetic fields or optical pumping is completely unacceptable for microelectronics to generate spin polarization. In fact, it turned out that there is a simple, compact, and energy-efficient way to create a flow of spin-polarized electrons. The main distinguishing features of spintronics are the creation, use, and control of spin currents. Therefore, it is worth delving into the physical principles of the generation of spin currents in modern microelectronics.

The simplest device for generating a spin current is a ferromagnetic 3d-metal thin film deposited onto a substrate of a sufficiently pure paramagnetic metal. To understand how an electric current can arise in such a system to carry a spin magnetic moment, let dwell on the structural features of the electronic spectrum of ferromagnetic 3d-metals, typical representatives of which are Fe, Co, Ni, their alloys and compounds. These substances are usually called band magnetics.

The energy spectrum of transition metals of the 3d-group is a wide sp-band as resulting from the broadening of the energy levels of 4s- and 4p-electrons. The sp-band covers five narrow mutually overlapping 3d-bands combined through the broadening of the levels of 3d-electrons. Against the typical conduction bands of s- and p-electrons, 3d-bands have a smaller width. However, the density of energy states in the latter turns out to be much higher than those of s- and p-electrons in the same energy range. Figure 7.1 illustrates a schematic arrangement of the sp- and 3d-bands in transition metals of the 3d-group.

https://doi.org/10.1515/9783110727197-007

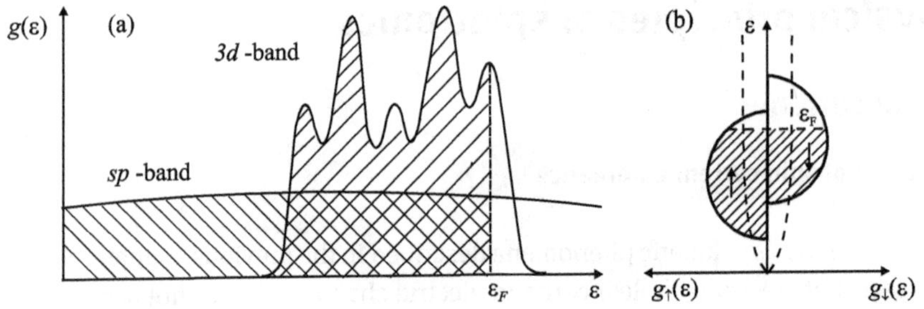

Figure 7.1: (a) Schematic arrangement of sp- and d-bands in transition metals of the 3d-group. The filled states are shaded. $g(\varepsilon)$ is the density of states; (b) shift of 3d-electron subbands with different spin orientations within the Stoner model. The dotted line indicates the energy dependence of the density of states of sp-electron. Arrows point the direction of the spin of d-electrons.

Although the charge density distribution of d-electrons is close to the atomic one, the overlap of the atomic orbitals leads to partial collectivization of d-electrons, and they can participate in electric transport. However, their mobility remains much less than that of sp-electrons.

Another characteristic feature of d-electrons is the presence of a strong interaction between them. The largest value is the energy of the exchange interaction $E_{\text{ex}} \simeq 10$ eV of electrons with opposite directions of the spin projection, located near the same site of the crystal lattice. The d-band being rather narrow (its width is estimated as $W \simeq 1$ eV), the d-electron subband splits into two subbands with different populations. It is this difference in the population of the spin d-subbands that provokes the so-called band magnetism (see Figure 7.1(b)). It should be noted that as far back as 1938 Stoner, having taken a shot to explain the magnetic properties of ferromagnets, put forward the hypothesis that the Coulomb interaction between electrons gives rise only to a separation of subbands of electrons with different spin projections, with the dispersion law and the density of states remaining unchanged (the Stoner model). Having compiled an equation for the magnetization, Stoner arrived at the condition for the occurrence of spontaneous magnetization

$$E_{\text{ex}} g(\varepsilon_F) > 1,$$

where $g(\varepsilon_F)$ is the density of states at the Fermi level. This is the Stoner criterion that, as it later was found out, is quite well satisfied for ferromagnets of the 3d-group.

If one applies an electric field to a sample of a ferromagnetic conductor, there arises an electric current defined as the total contribution of electrons with both spin-up and spin-down. The resulting electrical conductivity is equal to the sum of contributions:

$$\sigma = \sigma^{\uparrow} + \sigma^{\downarrow} = \frac{e^2}{m}(n^{\uparrow}\tau^{\uparrow} + n^{\downarrow}\tau^{\downarrow}), \tag{7.1}$$

where e, m are the charge and mass of electrons, $n^{\uparrow}$, $n^{\downarrow}$ – are the concentrations of electrons with spin orientations ↑, ↓ and $\tau^{\uparrow}$, $\tau^{\downarrow}$ – are relaxation times of a momentum with different spin orientations. So far as a highly degenerate electron gas is concerned, these relaxation times, according to formula (4.37), coincide with the relaxation times of electrons with different spin orientations at the Fermi level ($\varepsilon_{\vec{p}} = \varepsilon_F$). 3d-electrons are strongly polarized but have rather low mobility. Therefore, formula (7.1) does not make an allowance for their contribution. s-electrons are collectivized, represent a quasi-free electron gas in a metal, have high mobility but are very weakly polarized. Nevertheless, the spin current (flux of the intrinsic mechanical momentum of electrons) can be realized due to the fact that the relaxation time of the momentum of the mobile s-electrons depends on the orientation of their spins.

Since the main mechanism of scattering of mobile electrons is scattering on the vibrations of the crystal lattice without a spin flip (transitions with a spin flip occur only if a sufficiently weak spin–orbit interaction is taken into account), as a result of scattering s-electron with energy $\varepsilon_{\vec{p}\sigma}$ can pass into a state $\varepsilon_{\vec{p}'\sigma}$ that can belong to either electrons of the s-band or electrons of the d-band.

If we assume that the quantum mechanical amplitude of the probability of transition from the state $\vec{p}$, σ to the state $\vec{p}'$, σ does not depend on what band (s or d) belongs to the end state, then the inverse relaxation time of the mobile electrons will be determine by the density of energy levels in the final state. Then the inverse relaxation time of mobile electrons is controlled by the energy level density in the end state. This result can be easily obtained if, in the expression (4.92), we go from summing over the final states $\vec{p}'$ to integrating over the energy.

The number of d-states per atom is ten; the number of states per atom in the s-band is two. Therefore, the conductivity of transition metals is chiefly governed by the scattering of mobile charge carriers from the s-band to the d-states with retention of spin orientation.

After the above remarks, let us turn again to Figure 7.1. The subbands with different spin orientations are shifted by E_{ex} due to the polarization of d-electrons. Therefore, the density of energy states at the Fermi level in these subbands is significantly different. This also implies a dramatic difference in the relaxation times $\tau^{\uparrow}$ and $\tau^{\downarrow}$, appearing in the formula (7.1). This, in turn, means that the current that emerges in a ferromagnetic 3d metal is spin polarized despite the lack of spin polarization of s-electrons. In other words, the electric current is accompanied by the transfer of mechanical and magnetic moments.

It is interesting to note that the above considerations about the dependence of the relaxation times of s-electrons on the spin orientation gained fame in the papers by Mott as far back as 1936. He used them for evaluating the electrical resistance of 3d-transition metals within the so-called two-channel model. The latter offered that the conductivity in ferromagnetic 3d-metals consists of the conductivities of each of the channels with its own characteristic impulse relaxation time. However, the seemingly obvious conclusion that the electric current, in this case, is accompanied by transport mechanical and magnetic momenta was made only more than 50 years after and is associated with the observation of the giant magnetoresistance effect (up to 100 %) discovered by Fert and Grunberg in 1988.

A spin current can be also observed in a non-ferromagnetic compound. The injection of a current from a ferromagnetic material into a non-magnetic metal inside a heterostructure as a thin ferromagnet film deposited onto a paramagnetic metal substrate can exemplify this situation. When passing an electrical current through such a structure, spin-polarized electrons injected into the paramagnet retain non-equilibrium spin polarization for $t \simeq \tau_s$, where τ_s- is the relaxation time of spin magnetization. In 1988, Johnson and Silsbee experimentally confirmed the existence of such a current. In the same year, the obtained results of their research were published.

There are numerous other experimental proofs of the existence of a spin current. Among them are giant magnetoresistance already mentioned above, spin accumulation, the spin Hall effect, the spin Seebeck effect, and as well as effects that exhibit the transfer of mechanical and magnetic momentum by spin current (spin-rotational effect). These results will be discussed in the following sections of this chapter.

7.1.2 Kinetic equation in the relaxation time approximation for the two-channel Mott model

Consider the problem of calculating the electrical resistance for a ferromagnetic metal. As noted above, to describe kinetic phenomena in such a material, it is necessary to take into account the difference in the relaxation times of different spin-orientation mobile electrons (along and opposite the direction of spontaneous magnetization). It is these speculations that brought about the formula (7.1). However, a more rigorous treatment requires accounting for possible transitions between different spin-orientation states, which change the electron distribution function.

Let us utilize the results of Chapter 4, namely formulas (4.6)–(4.13). Suppose that there is no magnetic field, the temperature and chemical potential are homogeneous. Then we can generalize the equation (4.13) for the case of two charge-transfer channels, each of which is characterized by its own distribution function $(f_\uparrow, f_\downarrow)$:

$$\begin{aligned} e(\vec{E}\cdot\vec{v})\frac{\partial f_0}{\partial \varepsilon_{\vec{p}}} &= -\frac{f_\uparrow - f_0}{\tau_\uparrow} - \frac{f_\uparrow - f_\downarrow}{\tau_{\uparrow\downarrow}}; \\ e(\vec{E}\cdot\vec{v})\frac{\partial f_0}{\partial \varepsilon_{\vec{p}}} &= -\frac{f_\downarrow - f_0}{\tau_\downarrow} - \frac{f_\downarrow - f_\uparrow}{\tau_{\uparrow\downarrow}}. \end{aligned} \tag{7.2}$$

In writing the set of equations (7.2), the relaxation of the non-equilibrium distributions $(f_\uparrow, f_\downarrow)$ to the equilibrium value f_0 in each charge-transfer channel is assumed to meet its time $(\tau_\uparrow, \tau_\downarrow)$. The last summand in each equation of (7.2) describes the process of changing the distribution function in the channel due to the spin-flip scattering with a characteristic frequency $\tau_{\uparrow\downarrow}^{-1}$.

For seeking the solution of this set of equations, we introduce deviations of the non-equilibrium distribution function ($\delta f_\uparrow$, $\delta f_\downarrow$) in the channels from the equilibrium value f_0:

$$\delta f_\uparrow = f_\uparrow - f_0, \quad \delta f_\downarrow = f_\downarrow - f_0$$

and rewrite the set of equations in terms of these notations:

$$\begin{aligned} e(\vec{E}\cdot\vec{v})\frac{\partial f_0}{\partial \varepsilon_{\vec{p}}}\tau_{\uparrow\downarrow} &= -\delta f_\uparrow \frac{\tau_{\uparrow\downarrow}}{\tau_\uparrow} - (\delta f_\uparrow - \delta f_\downarrow), \\ e(\vec{E}\cdot\vec{v})\frac{\partial f_0}{\partial \varepsilon_{\vec{p}}}\tau_{\uparrow\downarrow} &= -\delta f_\downarrow \frac{\tau_{\uparrow\downarrow}}{\tau_\downarrow} + (\delta f_\uparrow - \delta f_\downarrow). \end{aligned} \tag{7.3}$$

The solution of this set of equations

$$\begin{aligned} \delta f_\uparrow &= -e(\vec{E}\cdot\vec{v})\frac{\partial f_0}{\partial \varepsilon_{\vec{p}}}\frac{(\tau_{\uparrow\downarrow}/\tau_\downarrow + 2)\tau_\uparrow\tau_\downarrow}{\tau_{\uparrow\downarrow} + \tau_\uparrow + \tau_\downarrow}, \\ \delta f_\downarrow &= -e(\vec{E}\cdot\vec{v})\frac{\partial f_0}{\partial \varepsilon_{\vec{p}}}\frac{(\tau_{\uparrow\downarrow}/\tau_\uparrow + 2)\tau_\uparrow\tau_\downarrow}{\tau_{\uparrow\downarrow} + \tau_\uparrow + \tau_\downarrow} \end{aligned} \tag{7.4}$$

allows one to find the electrical current density and electrical resistance of the channels, $\rho_\uparrow, \rho_\downarrow$, expressed them through integrals $K_{0\uparrow}, K_{0\downarrow}$. The latter are defined by analogy with the relation (4.31) for the case of a strongly degenerate gas:

$$\begin{aligned} \frac{1}{\rho_\uparrow} &= e^2 K_{0\uparrow}; \quad K_{0\uparrow} = \frac{n}{2m}\frac{(\tau_{\uparrow\downarrow}/\tau_\downarrow + 2)\tau_\uparrow\tau_\downarrow}{\tau_{\uparrow\downarrow} + \tau_\uparrow + \tau_\downarrow}, \\ \frac{1}{\rho_\downarrow} &= e^2 K_{0\downarrow}; \quad K_{0\downarrow} = \frac{n}{2m}\frac{(\tau_{\uparrow\downarrow}/\tau_\uparrow + 2)\tau_\uparrow\tau_\downarrow}{\tau_{\uparrow\downarrow} + \tau_\uparrow + \tau_\downarrow}. \end{aligned} \tag{7.5}$$

The reverse resistance of the sample being equal to the sum of the reverse resistances of the channels, it is easy to arrive at a rather simple formula for calculating the electrical resistance of a ferromagnetic metal having a strong difference in the relaxation times of different-spin-orientation conduction electrons

$$\rho = \frac{\rho_\uparrow\rho_\downarrow + \rho_{\uparrow\downarrow}(\rho_\uparrow + \rho_\downarrow)}{\rho_\uparrow + \rho_\downarrow + 4\rho_{\uparrow\downarrow}}, \tag{7.6}$$

where

$$\rho_\uparrow = \frac{m}{e^2 n\tau_\uparrow}, \quad \rho_\downarrow = \frac{m}{e^2 n\tau_\downarrow}, \quad \rho_{\uparrow\downarrow} = \frac{m}{e^2 n\tau_{\uparrow\downarrow}}.$$

Problem 7.1. Obtain an expression for the differential thermopower coefficient α for the two-channel Mott model, assuming that the relaxation times $\tau_\uparrow, \tau_\downarrow, \tau_{\uparrow\downarrow}$ are energy-independent.

Solution. We write down the set of equations (7.2) for two thermodynamic forces: electrochemical potential gradient $\vec{\varepsilon}$ and temperature gradient $\vec{\nabla} T$.

$$\frac{\partial f_0}{\partial \varepsilon_{\vec{p}}} \vec{v} \left(e\vec{\varepsilon} - \frac{\varepsilon_{\vec{p}} - \zeta}{T} \vec{\nabla} T \right) = -\frac{f_\uparrow - f_0}{\tau_\uparrow} - \frac{f_\uparrow - f_\downarrow}{\tau_{\uparrow\downarrow}},$$

$$\frac{\partial f_0}{\partial \varepsilon_{\vec{p}}} \vec{v} \left(e\vec{\varepsilon} - \frac{\varepsilon_{\vec{p}} - \zeta}{T} \vec{\nabla} T \right) = -\frac{f_\downarrow - f_0}{\tau_\downarrow} - \frac{f_\downarrow - f_\uparrow}{\tau_{\uparrow\downarrow}}. \tag{7.7}$$

Having solved the set of equations (7.7), we obtain corrections to the electron distribution functions in the channels $\delta f_\uparrow = f_\uparrow - f_0$, $\delta f_\downarrow = f_\downarrow - f_0$;

$$\delta f_\uparrow = -\frac{\partial f_0}{\partial \varepsilon_{\vec{p}}} \vec{v} \left(e\vec{\varepsilon} - \frac{\varepsilon_{\vec{p}} - \zeta}{T} \vec{\nabla} T \right) \frac{(\tau_{\uparrow\downarrow}/\tau_\downarrow + 2)\tau_\uparrow \tau_\downarrow}{\tau_{\uparrow\downarrow} + \tau_\uparrow + \tau_\downarrow},$$

$$\delta f_\downarrow = -\frac{\partial f_0}{\partial \varepsilon_{\vec{p}}} \vec{v} \left(e\vec{\varepsilon} - \frac{\varepsilon_{\vec{p}} - \zeta}{T} \vec{\nabla} T \right) \frac{(\tau_{\uparrow\downarrow}/\tau_\uparrow + 2)\tau_\uparrow \tau_\downarrow}{\tau_{\uparrow\downarrow} + \tau_\uparrow + \tau_\downarrow}. \tag{7.8}$$

For finding the differential thermopower coefficient, we use the definition for the current density (4.29) upon the electrochemical potential gradient $\vec{\varepsilon}$ and the temperature gradient $\vec{\nabla} T$

$$\vec{j} = e^2 (K_{0\uparrow} + K_{0\downarrow})\vec{\varepsilon} - \frac{e}{T}(K_{1\uparrow} + K_{1\downarrow})\vec{\nabla} T. \tag{7.9}$$

An explicit form of the integrals $K_{0\uparrow}$, $K_{0\downarrow}$ appearing in the formula (7.9) was given above. The integrals $K_{1\uparrow}$, $K_{1\downarrow}$ evaluated for the case of a strongly degenerate electron gas are defined by the formula (4.36), where the first summand in square brackets on the right side of the equation vanishes if the relaxation times $\tau_\uparrow$, $\tau_\downarrow$, $\tau_{\uparrow\downarrow}$, as posted in the problem statement, are energy-independent. Taking this fact into account, we can immediately produce the result of computing the integrals $K_{1\uparrow}$, $K_{1\downarrow}$ in the following form:

$$K_{1\uparrow} = \frac{\pi^2}{4} \frac{n}{m} \frac{(k_B T)^2}{\zeta} \frac{(\tau_{\uparrow\downarrow}/\tau_\downarrow + 2)\tau_\uparrow \tau_\downarrow}{\tau_{\uparrow\downarrow} + \tau_\uparrow + \tau_\downarrow};$$

$$K_{1\downarrow} = \frac{\pi^2}{4} \frac{n}{m} \frac{(k_B T)^2}{\zeta} \frac{(\tau_{\uparrow\downarrow}/\tau_\uparrow + 2)\tau_\uparrow \tau_\downarrow}{\tau_{\uparrow\downarrow} + \tau_\uparrow + \tau_\downarrow}. \tag{7.10}$$

Using the definition (4.32) for the differential thermopower coefficient, we write down

$$\alpha = \alpha_\uparrow + \alpha_\downarrow = \frac{K_{1\uparrow}}{eT(K_{0\uparrow} + K_{0\downarrow})} + \frac{K_{1\downarrow}}{eT(K_{0\uparrow} + K_{0\downarrow})}. \tag{7.11}$$

Plugging the values found for the integrals $K_{0\uparrow}$, $K_{0\downarrow}$, $K_{1\uparrow}$, $K_{1\downarrow}$, it is easy to deduce the simple expression

$$\alpha = \frac{\pi^2}{2} \frac{k_B}{e} \frac{k_B T}{\zeta}, \tag{7.12}$$

that completely coincides with the expression for thermopower of a degenerate electron gas in the event of the energy-independent electron-momentum relaxation time.

Indeed, it is necessary to understand that the assumption made about the energy-independence of the momentum relaxation time is rather rough because it is well known that some typical electron conductors have a thermopower with a positive sign. This fact is possible due to the decay of the momentum relaxation time with the growth of energy. If we reject the assumption made, then the thermopower in the two-channel model is determined by a rather cumbersome expression and therefore is not given here.

Experimental confirmation of the validity of the two-channel model has been obtained in early work on the measurement of low-temperature electrical resistivity in dilute nickel-based alloys [45]. Other experimental results confirming the fact that the electrons with spin $\uparrow$ and with spin $\downarrow$ in ferromagnetic conductors transfer not only charge, but also magnetization, which will be discussed in the following sections of this chapter.

7.2 Magnetoresistance and spin accumulation in layered structures

7.2.1 Giant magnetoresistance

Magnetoresistance is the relative change of electrical resistance of a conductor in an externally-applied magnetic field. The influence of a magnetic field on kinetic phenomena in non-magnetic metals and semiconductors was already discussed in the fourth and fifth chapters. There it was shown (see (4.135)) that the magnitude of the effect of transverse magnetoresistance in classical magnetic fields for a typical metal is of the order of

$$\frac{\Delta\rho}{\rho} \simeq \left(\frac{k_{\mathrm{B}}T}{\zeta}\right)^2$$

and amounts to no more than one percent. In this formula, as before, k_B is the Boltzmann constant, T is the absolute temperature, and ζ is the chemical potential (the Fermi level for a degenerate electron gas in metals).

A completely different type of phenomenon called *giant magnetoresistance* (GMR) is implemented in multilayered structures of the $[F/N]_n$ type, where F is slim, order a few angstroms, the layer of ferromagnetic metal (such as iron) and N is the layer of non-magnetic metal (e. g., chromium), n is the number repetition of such a bilayer.

GMR in multilayers can exhibit in two geometries. In the first case, it is observed when the current and magnetic field are oriented in the plane of the layers (*current in-plane*, CIP). Another case implies the direction of current transverse to the layers (*current is perpendicular to the plane*, CPP). Before 1993, experiments on exploring the influence of a magnetic field on the resistance of layered structures were carried out only in CIP-geometry. In particular, in 1988, Fert and Grunberg discovered a significant change in the electrical resistance of layered structures Fe/Cr (up to 100 %). This phenomenon gained the name of giant magnetoresistance. The approximate shape of typical dependencies of the resistance of various layered structures $[Fe/Cr]_n$ on the value of the magnetic field induction H at liquid helium temperature are shown in Figure 7.2.

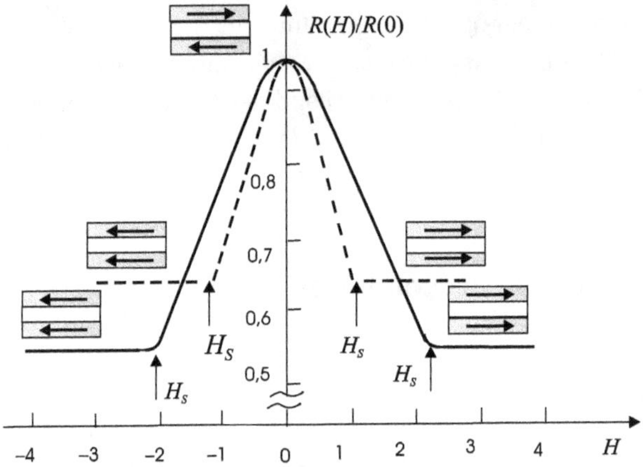

Figure 7.2: Dependence of the resistance of $[Fe/Cr]_n$ structures on the magnetic field induction H (T): H_s is the field that completely re-orients the magnetization in the layer; Fe- and Cr-layer thicknesses amount to 30 Å and 12 Å, respectively; the dotted and solid curves correspond to structures with $n = 35$ and $n = 80$, respectively.

The discovery of the giant magnetoresistance effect has served as the starting point for the development of a new trend in transfer phenomenon research, which is now commonly referred as to spintronics. The early history of the development of spintronics and its prospects are presented in the Nobel lectures by Fert [46] and Grunberg [47].

The GMR effect is observed when using the CIP and the CPP geometry, and when the electric current passes perpendicular to the plane of the layers, the effect value is greater, all other things being equal.

Let us look qualitatively at the nature of the GMR effect discovered by Fert and Grunberg. An important feature of layered structures such as $[Fe/Cr/Fe]_n$ is the fact that, due to indirect exchange interactions through the electrons of the intermediate layer (chromium layer), the magnetization of adjacent iron layers changes from parallel to almost antiparallel with a chromium layer thickness of about 12 Å. The curious specifics of the $[Fe/Cr/Fe]_n$ structure allows for controlling the orientation of magnetization in neighboring iron layers through an externally-applied magnetic field. In other words, a structure consisting, for example, of 30 Å thick iron layers and 12 Å thick chromium layers alters the original antiparallel ordering magnetization in the iron layers under a sufficiently strong magnetic field with induction $H \simeq 2\,T$: the magnetizations of the iron layers become parallel to each other and parallel to the direction of the external magnetic field.

Although the CIP and CPP geometries share similar dependencies of the electrical resistance of the $[Fe/Cr]_n$ structures on the magnitude of the magnetic field the interpretation of the GMR effect for this cases is quite different. The fact is that on a qualitative level, the effect of GMR for CPP configuration can be explained on the basis of two-channel model conductivity discussed above, while the application of this model to CIP geometry gives a null result, if we assume a local concept conduction and that in the process of transport of the electrons from one layers do not fall into the adjacent layers (this fact is proved below).

The GMR effect in CPP-geometry

To interpret the GMR effect in the CPP-geometry, we address Figure 7.3. Consider the resistance of a structure consisting of two magnetic layers separated by a non-magnetic layer. Let the resistance of non-magnetic layers be denoted by the symbol R_N. Suppose that this resistance is spin-orientation independent. Designate the resistance for spin-up ↑ electrons as it passes through the layers with the magnetizations ↑ and ↓ as $R_{\uparrow\uparrow}$ and $R_{\uparrow\downarrow}$, respectively.

If one denies the spin-flip scattering and assumes a local dependence between the current and the electrical field, there are all grounds to believe that the two spin channels behave as two parallel conduction channels with their own resistance, as depicted in Figures 7.3(c) and 7.3(d). The above statement underlies a very simple mechanism to explain the GMR effect in the CPP-geometry.

Let R^P and R^{AP} be the resistances of a three-layered structure for parallel and antiparallel layer magnetizations, respectively. Then, using the circuits 7.3(c) and 7.3(d) for the layer resistances, we can write down

$$\begin{aligned} \frac{1}{R^P} &= \frac{1}{R_{\uparrow\uparrow} + R_N + R_{\uparrow\uparrow}} + \frac{1}{R_{\downarrow\uparrow} + R_N + R_{\downarrow\uparrow}}; \\ \frac{1}{R^{AP}} &= \frac{1}{R_{\uparrow\uparrow} + R_N + R_{\uparrow\downarrow}} + \frac{1}{R_{\downarrow\uparrow} + R_N + R_{\downarrow\downarrow}}. \end{aligned} \tag{7.13}$$

We denote the structure resistance for the spin-up ↑ electrons by the symbol $R^{\uparrow}$. From the relation (7.13) it follows that $R^{\uparrow} = R_{\uparrow\uparrow} + R_N + R_{\uparrow\uparrow}$. Analogously, for the spin ↓, we have $R^{\downarrow} = R_{\downarrow\uparrow} + R_N + R_{\downarrow\uparrow}$. Then the resistance of the structure with a parallel

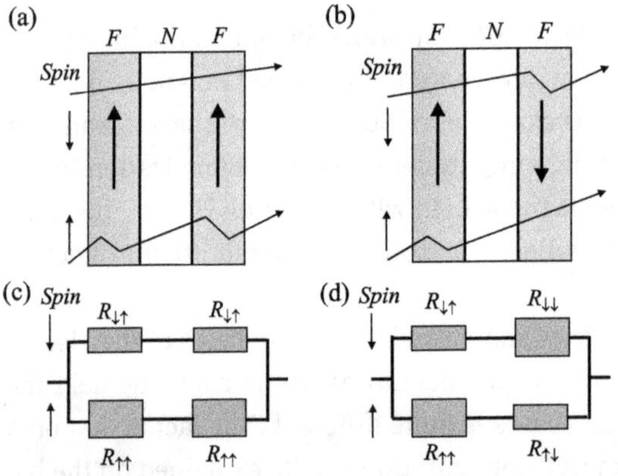

Figure 7.3: Application of the two-channel conduction model to explain the GMR effect in a three-layered structure (ferromagnet/normal metal/ferromagnet in CPP-geometry): (a) schematic representation of passage the structure with the ferromagnetic ordering of the layers by electrons with different spin-orientation; (b) the same upon antiparallel layer ordering of a ferromagnet; (c) and (d) a resistor model for the structure. The resistance of a normal metal is not shown.

orientation of magnetization can be represented as

$$\frac{1}{R^P} = \frac{1}{R^\uparrow} + \frac{1}{R^\downarrow}; \quad R^P = \frac{R^\uparrow R^\downarrow}{R^\uparrow + R^\downarrow}. \tag{7.14}$$

Next, we focus on the resistance of the structure with an antiparallel orientation of layer magnetization. Given the $R^\uparrow$ and $R^\downarrow$ notations previously entered, it is easy to notice that the channel resistances satisfy the relation

$$R^\uparrow + R^\downarrow = 2(R_{\uparrow\uparrow} + R_N + R_{\uparrow\downarrow}). \tag{7.15}$$

If one assumes that the physical properties of the magnet layers are identical, we can accept that $R_{\uparrow\uparrow} = R_{\downarrow\downarrow}$, $R_{\uparrow\downarrow} = R_{\downarrow\uparrow}$.

Accounting for the made assumptions and simple transformations yields

$$\frac{1}{R^{AP}} = \frac{2}{R_{\uparrow\uparrow} + R_N + R_{\uparrow\downarrow}} = \frac{4}{R^\uparrow + R^\downarrow}; \quad R^{AP} = \frac{1}{4}(R^\uparrow + R^\downarrow). \tag{7.16}$$

Plugging the found values of R^{AP}, R^P into the definition for the structure magnetoresistance

$$\frac{\Delta R}{R} = \frac{R^{AP}}{R^P} - 1, \tag{7.17}$$

we end up with

$$\frac{\Delta R}{R} = \frac{(R^\uparrow - R^\downarrow)^2}{4R^\uparrow R^\downarrow}. \tag{7.18}$$

It is granted that the electrical resistance of the layer of the normal metal can be neglected in formula (7.13) and the $R_{\uparrow\uparrow}$ and $R_{\uparrow\downarrow}$ resistances can be replaced by appropriate specific resistances $\rho_\uparrow$ and $\rho_\downarrow$, where $\rho_\uparrow$ is the specific resistance in the channel, in which the directions of spin of an conduction electron and magnetization coincide (*majority electrons*), $\rho_\downarrow$ is the specific resistance of the channel, in which the directions of the spin and magnetization are opposite (*minority electrons*). Consequently, the change in multilayered structure resistance can be expressed through the specific resistances of the channels,

$$\frac{\Delta R}{R} = \frac{(\rho_\uparrow - \rho_\downarrow)^2}{4\rho_\uparrow\rho_\downarrow}. \tag{7.19}$$

The formula given above allows purposefully searching for materials to expect a large magnitude of magnetoresistance. For example, according to the data given in [48], in the [Co/Cu/Co] structure the specific resistance for a cobalt film $\rho_\uparrow = 32\,\mu\Omega\,\text{cm}$, $\rho_\downarrow = 141\,\mu\Omega\,\text{cm}$, and the specific resistance of copper in the layer is only $4.6\,\mu\Omega\,\text{cm}$. Thus, formula (7.19) predicts the effect magnitude for samples of such a structure as

$$\frac{\Delta R}{R} = \frac{(\rho_\uparrow - \rho_\downarrow)^2}{4\rho_\uparrow\rho_\downarrow} \simeq 66\,\%.$$

Despite the offered model with a series connection of resistors giving a simple picture to estimate the GMR effect magnitude, it needs improvement in two ways. For this purpose, it is mandatory to bear in mind both spin-flip scattering to get a realistic interpretation of the CPP GMR and the effects of chaotic scattering of electrons in a layer and their finite mean path length to explain the CIP GMR.

The GMR effect in CIP-geometry

As already underscored above, the two-channel model that corresponds with local conductivity is adequate for explaining GMR in CPP-geometry and gives plausible numerical estimates of the effect. However, this model absolutely fails to provide a satisfactory description of the GMR effect in the CIP-geometry since it is easy enough to prove that the change in resistance in the CIP geometry is exactly zero if the conductivity is completely local even in the presence of spin-dependent conductivity.

Let us explore again a structure $[F/N/F]_n$, where F and N are layers of ferromagnetic and nonmagnetic materials, respectively. When an electric current flows through the layer planes, the total current is the sum of currents in each layer. In the case of completely local conductivity, the current inside each layer does not depend on the orientation of the magnetization of other layers. So, if the orientation of the magnetization of two ferromagnetic layers is parallel, the total current I^P flowing in the three-layered structure consists of the currents in each layer:

$$I^P = I_{1\uparrow} + I_{1\downarrow} + I_N + I_{2\uparrow} + I_{2\downarrow}, \tag{7.20}$$

where I_{1s}, I_{2s} ($s = \uparrow, \downarrow$) are the currents in spin channels of the first and second ferromagnetic layers, I_N is the current in the non-magnetic metal layer. If the magnetization of the layers is antiparallel, the total current I^{AP} flowing in the three-layered structure is equal to the sum of the currents in the spin channels:

$$I_{\uparrow}^{AP} = I_{1\uparrow} + \frac{1}{2}I_N + I_{2\downarrow};$$
$$I_{\downarrow}^{AP} = I_{1\downarrow} + \frac{1}{2}I_N + I_{2\uparrow};$$
$$I^{AP} = I_{1\uparrow} + I_{1\downarrow} + I_N + I_{2\uparrow} + I_{2\downarrow}. \tag{7.21}$$

Comparing the expressions (7.20) and (7.21), it is easy to see that they are fully identical, and the magnetization reorientation in one of the layers leads to no change in the current. Consequently, the resistance of the three-layered structure does not vary too.

Despite the result obtained above, it is possible, having made a number of serious assumptions, to apply the two-channel model to the interpretation of the GMR in the CIP geometry.

Suppose that the isotropic scattering of electrons in the layers occurs at randomly distributed impurity centers with a sufficiently high concentration of n_i, $n_i l_s^3 \gg 1$, and the thickness of the ferromagnetic layers a and the thickness of the nonmagnetic layer b are significantly less than the characteristic length l_s, at which the orientation of the electron spin moment is preserved. Under these conditions, we can assume that when the electrons pass through the layered structure, the residence time of the electron in each of the layers is proportional to its thickness, and introduce the average resistivity for each of the channels, ρ^s:

$$\rho^s = \frac{2a\rho_s + b\rho_N}{2a + b}, \quad s = \uparrow\downarrow, \tag{7.22}$$

where $\rho_{\uparrow}$ and $\rho_{\downarrow}$ are the specific resistances for electrons whose spin coincides with the direction of magnetization in a layer and opposite to it, respectively, ρ_N is the specific resistance of a non-magnetic metal.

Hence, the inverse resistivity of the three-layered structure for the parallel ($1/\rho^P$) and antiparallel ($1/\rho^{AP}$) orientations of the layers can be written as

$$\frac{1}{\rho^P} = \left(\frac{2a + b}{2a\rho_{\uparrow} + b\rho_N} + \frac{2a + b}{2a\rho_{\downarrow} + b\rho_N}\right);$$
$$\frac{1}{\rho^{AP}} = \left(\frac{2(2a + b)}{a\rho_{\uparrow} + a\rho_{\downarrow} + b\rho_N}\right). \tag{7.23}$$

Accounting for the relationship between the resistivity and the resistance for the parallel R^P and antiparallel R^{AP} orientations of magnetization of the layers leaves us with

$$R^P = \frac{L}{(2a + b)S}\left(\frac{(2a\rho_{\uparrow} + b\rho_N)(2a\rho_{\downarrow} + b\rho_N)}{2a(\rho_{\uparrow} + \rho_{\downarrow}) + 2b\rho_N}\right); \tag{7.24}$$

$$R^{AP} = \frac{L}{2(2a+b)S}\left[a(\rho_\uparrow + \rho_\downarrow) + b\rho_N\right]. \tag{7.25}$$

Here L is the sample length, S is the layer cross section area.

Using the results obtained (7.24), (7.25) and the definition of relative change in resistance (7.17) in result of simple transformations, we obtain the estimate of the GMR effect in the CIP geometry, which almost coincides with the result of (7.19), previously obtained for the CPP geometry:

$$\frac{\Delta R}{R} = \frac{(\rho_\uparrow - \rho_\downarrow)^2}{(2\rho_\uparrow + \xi\rho_N)(2\rho_\downarrow + \xi\rho_N)}, \tag{7.26}$$

where $\xi = b/a$.

GMR in CPP-geometry considering spin-flip processes

The above-examined GMR model in CPP geometry is assumed to keep the spin of electrons traveling through the structure layers unchanged. Obviously, this assumption does not reflect the actual situation despite fulfilling the condition $\tau_{sf} \gg \tau_{ps}$ for the samples used, where τ_{sf} is the time of retaining the electron spin, τ_{ps} is the momentum relaxation time for electrons with spin s. Spin-flip-electron processes can be allowed for within either the kinetic Boltzmann equation or an approach based on the usage of local equilibrium macroscopic balance equations [48]. The second technique is simpler and compact, and we will focus on it.

Let the x- and y-axes of the Cartesian coordinate system lie in the layer planes and the z-axis be directed perpendicular to them. The system is spatially homogeneous along the x- and y-axes. However, since we intend to take into account spin-flip processes, the density of the electric current $\vec{J}_s$ and the chemical potential ζ_s ($s = \uparrow, \downarrow$) are z-coordinate dependent.

Bearing in mind the first of equations of (1.15), we write an expression for the density of the current $\vec{J}_s$ in the spin channel s conditional upon a gradient of the electric field potential φ and a gradient of the chemical potential ζ_s are imposed

$$\vec{J}_s = -\frac{\sigma_s}{e}\vec{\nabla}(\zeta_s + e\varphi). \tag{7.27}$$

In what follows in the text, e is the module of the electron charge, σ_s is the conductivity of the spin channel s.

When an electric current flows through the system, the electroneutrality conditions are not violated and $\operatorname{div}\vec{J} = 0$. However, spin-flip processes break this condition for electrons with a fixed spin orientation. In this case, their concentration n_s is no longer stationary:

$$\operatorname{div}\vec{J}_s = -e\frac{dn_s}{dt}. \tag{7.28}$$

The electron density n_s depends on the coordinate z and on the time t only due to the fact that the local macroparameter ζ_s depends on these values. Given this circumstance, we can express the quantity n_s through the locally equilibrium distribution function f_0

$$n_s(z,t) = \frac{1}{V}\sum_{\vec{k}} f_0(\varepsilon_{\vec{k}s} - \zeta_s(z,t)), \tag{7.29}$$

and the rate of changing in $n_s(t)$ can be written via density of states, $g_s(\zeta_s)$, at the Fermi quasi-level ζ_s

$$\frac{dn_s}{dt} = -\frac{1}{V}\sum_{\vec{k}} \frac{df_0(\varepsilon_{\vec{k}s} - \zeta_s)}{d\varepsilon_{\vec{k}s}}\frac{d\zeta_s}{dt} = g_s(\zeta_s)\frac{d\zeta_s}{dt}; \tag{7.30}$$

$$g_s(\zeta_s) = -\frac{1}{V}\sum_{\vec{k}} \frac{df_0(\varepsilon_{\vec{k}s} - \zeta_s)}{d\varepsilon_{\vec{k}s}}, \tag{7.31}$$

where V is the system's volume.

Under the conditions of the problem, the only reason for the change in the chemical potential of ζ_s is the spin-flip processes that change the concentration of n_s. Therefore, we can write

$$\frac{d\zeta_s}{dt} = -\frac{\zeta_s - \zeta_{-s}}{\tau_{sf}}, \tag{7.32}$$

where the quantity $1/\tau_{sf}$ governs the rate of the spin-flip processes.

Let us transform the equation (7.28), using the relations (7.30) and (7.32),

$$\operatorname{div}\vec{J}_s = eg(\zeta_s)\frac{\zeta_s - \zeta_{-s}}{\tau_{sf}}. \tag{7.33}$$

The density $g(\zeta_s)$ of states can be expressed in terms of electrical conductivity in the channel with spin s and velocity v_F at the Fermi level. For this purpose, we adduce formulas for electrical conductivity, the density of the number of states at the Fermi level, and the electron concentration in the channel with spin s:

$$\sigma_s = \frac{e^2 n_s \tau_{ps}}{m}; \quad g_s = \frac{m^{3/2}\zeta_s^{1/2}}{2^{1/2}\pi^2\hbar^3}; \quad n_s = \frac{2^{1/2}(m\zeta_s)^{3/2}}{3\pi^2\hbar^3}. \tag{7.34}$$

Derivation of the last two formulas will be suggested to the reader at the end of the section as a challenge.

By combining the formulas in the last expression, it is easy to see that

$$eg(\zeta_s) = \frac{3\sigma_s}{ev_F^2\tau_{ps}}.$$

Let us define the diffusion length for the spin s by the relation $l_s^{sf} = v_F\sqrt{\tau_s\tau_{sf}/3}$, which allows us to write the continuity equation in the spin channel in a convenient form

$$\operatorname{div}\vec{J}_s = \frac{\sigma_s}{e}\frac{\bar{\zeta}_s - \bar{\zeta}_{-s}}{(l_s^{sf})^2}. \tag{7.35}$$

In (7.35), we have replaced the chemical potential ζ_s of the electrons in the spin channel by the electrochemical potential $\bar{\zeta}_s$, defining it as $\bar{\zeta}_s = \zeta_s + e\varphi$.

Equations (7.27) and (7.35) enable deriving a closed differential equation for the difference of electrochemical potentials $\Delta\bar{\zeta} = \bar{\zeta}_\uparrow - \bar{\zeta}_\downarrow$. Indeed, if we substitute the expression for the current in the spin channel s (7.27) into the left side of the equation (7.35) and neglect the dependence of σ_s on the coordinates, we will get two diffusion equations,

$$\nabla^2\bar{\zeta}_\uparrow = \frac{\bar{\zeta}_\uparrow - \bar{\zeta}_\downarrow}{(l_\uparrow^{sf})^2}; \quad \nabla^2\bar{\zeta}_\downarrow = \frac{\bar{\zeta}_\downarrow - \bar{\zeta}_\uparrow}{(l_\downarrow^{sf})^2}, \tag{7.36}$$

and subtracting from the first equation the second one, we have the diffusion equation for $\Delta\bar{\zeta}$,

$$\frac{d^2\Delta\bar{\zeta}}{dz^2} = \frac{\Delta\bar{\zeta}}{l_{sf}^2}; \quad \frac{1}{l_{sf}^2} = \frac{1}{(l_\uparrow^{sf})^2} + \frac{1}{(l_\downarrow^{sf})^2}. \tag{7.37}$$

If we neglect the dependence of the diffusion length l_{sf} on the coordinate z, then the resulting equation is a linear homogeneous second-order differential equation with constant coefficients and its general solution is the sum of partial solutions that are easily obtained by using the Euler substitution

$$\Delta\bar{\zeta} = A\exp(z/l_{sf}) + B\exp(-z/l_{sf}), \tag{7.38}$$

where A and B are unknown constant coefficients. Since the total current density satisfies the equation continuity $\operatorname{div}(\vec{J}_\uparrow + \vec{J}_\downarrow) = 0$ (current flow does not lead to the accumulation of charge in an arbitrary cross-section of the sample), one more condition must be met:

$$\frac{d^2J}{dz^2} = \frac{d^2}{dz^2}(\sigma_\uparrow\bar{\zeta}_\uparrow + \sigma_\downarrow\bar{\zeta}_\downarrow) = 0. \tag{7.39}$$

Obviously, the solution to (7.39) can be written as

$$J = \sigma_\uparrow\bar{\zeta}_\uparrow + \sigma_\downarrow\bar{\zeta}_\downarrow = Cz + D, \tag{7.40}$$

where C, D are unknown constants.

For a multi-layer structure, the joint solution of the set equations (7.38), (7.40), with taking into account the boundary conditions of equality of the current density and electrochemical potentials for each of the spin directions at the boundary of each layer (if there are no additional spin reversal mechanisms in the boundary region), is a very complex, but solvable problem. As a result, the $\overline{\zeta}_\uparrow$- and $\overline{\zeta}_\downarrow$-dependence as a function of the z-coordinate and total current can be found. In this case, the density of the electric current J_s can be determined through the standard formula (7.27).

Here we will not deal with the implementation of this program, referring the reader to the original publication [49]. Nevertheless, in the section on spin accumulation, based on the equations (7.38), (7.40), we shall find the coordinate dependence of the electrochemical potentials $\overline{\zeta}_\uparrow$, $\overline{\zeta}_\downarrow$ for the F/N structure.

Problem 7.2. Derive formulas for calculating the density of states and concentration of electrons in the spin channel within an electron gas model with a parabolic dispersion law.

Solution. We proceed from the definition (7.31) for the density states of an electron gas, assuming that the dispersion law is parabolic. Next, in this formula, we go over from the summation over a wave vector $\vec{k}$ to integration over k in a spherical coordinate system. In doing so, the sum is replaced by the integral

$$\sum_{\vec{k}} \to \int_0^\infty \frac{V}{(2\pi)^3} k^2\, dk \sin\theta\, d\theta\, d\varphi,$$

where $V/(2\pi)^3$ is the density of states in $\vec{k}$-space. The integration over angular variables gives 4π. Relying on these results, we have

$$g(\varepsilon_{\vec{k}s}) = -\frac{1}{2\pi^2} \int_0^\infty \frac{df_0(\varepsilon_{\vec{k}s} - \zeta_s)}{d\varepsilon_{\vec{k}s}} k^2\, dk. \tag{7.41}$$

In the formula (7.41), let us pass from integration over the wave vector to integration over energy, having made the substitutions

$$kdk = \frac{m}{\hbar^2} d\varepsilon_{\vec{k}s}; \quad k = \frac{(2m)^{1/2}}{\hbar} \sqrt{\varepsilon_{\vec{k}s}}.$$

Entering these results into the formula (7.41) and using the fact that the derivative of the distribution function $f_0(\varepsilon_{\vec{k}s} - \zeta_s)$ for a highly degenerate electron gas is approximated by the δ-function, we get an expression for the density of states in the spin channel:

$$g_s = \frac{m^{3/2} \zeta_s^{1/2}}{2^{1/2} \pi^2 \hbar^3}.$$

It is easy to see that the formula deduced is identical to (7.34).

Furthermore, the concentration of electrons in the spin channel can be computed through the relation

$$n_s = \frac{1}{V}\sum_{\vec{k}} f_0(\varepsilon_{\vec{k}s} - \zeta_s). \tag{7.42}$$

If in the formula (7.42) we go from summation by $\vec{k}$ to integration by k, and then to integration by energy, then the expression for the concentration in the spin channel can be written as an integral over energy:

$$n_s = \frac{m^{3/2}}{2^{1/2}\pi^2\hbar^3}\int_0^{\infty} \varepsilon_{\vec{k}s}^{1/2} f_0(\varepsilon_{\vec{k}s} - \zeta_s)\,d\varepsilon_{\vec{k}s}. \tag{7.43}$$

Performing partial integration in the formula (7.43) and using the δ-shaped behavior of the derivative of the distribution function, we obtain the desired expression for the electron concentration in the spin channel:

$$n_s = \frac{2^{1/2} m^{3/2} \zeta_s^{3/2}}{3\pi^2\hbar^3}.$$

7.2.2 Spin accumulation

In virtue of the difference in the momentum relaxation times, as previously noted (see Section 7.1), an electric current in a ferromagnet is spin-polarized and accompanied by a transfer of magnetic momentum of the electrons. To observe a spin-polarized current turned out to be possible as a result of a simple, at first glance, experiment carried out by Johnson and Silsbee in 1988 [50].

Figure 7.4 schematically illustrates this experiment. Their experiment includes the following steps. A foil made of ultrapure monocrystalline aluminum is placed between ferromagnetic permalloy films (NiFe). For Al-foil designated as N in the figure, a mean free path of electrons amounts to $l_p \simeq 17\,\mu m$ and their spin length diffusion is $l_s \simeq 500\,\mu m$. The magnetization direction in F_1 and F_2 films (see Figure 7.4) can vary under an applied external field.

An electric current was passed through this system (the direction of movement of the positive charges is indicated by an arrow). In this configuration, the left ferromagnetic film served as an injector, creating a nonequilibrium concentration of electrons n_s, ($s = \uparrow\downarrow$) in the aluminum foil, and the right a detector that allowed detecting the presence of a non-equilibrium spin concentration of electrons.

Let us examine the interface between the left ferromagnetic film and the paramagnetic metal. The electric current passing through this system injects electrons into the paramagnetic metal. A current in a ferromagnetic metal being carried mainly majority

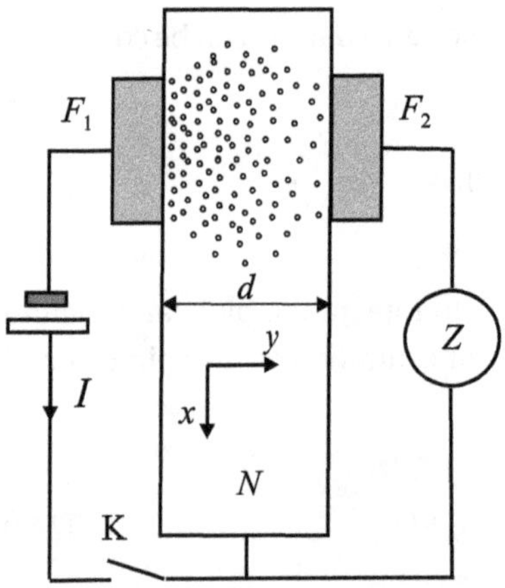

Figure 7.4: Principal scheme of the experiment by Johnson and Silsbee.

electrons, the injected electrons have different concentration n_s for ↑ and ↓ spins. In other words, the injected electrons are non-equilibrium in the paramagnetic metal.

The thermodynamic equilibrium in the electronic system of a paramagnetic metal will be restored as a result of two processes. First, the non-equilibrium concentration will decrease due to the spin-flip processes with a characteristic time of τ_{sf}, and secondly, the non-equilibrium will fade out at distances of the order of l_s due to the spin diffusion. Both of these processes are shown schematically in Figure 7.4. The density of points in the figure reflects the non-equilibrium electron concentration that declines along the x- and y-axes.

If the distance d between the injector-film and detector-film to choose sufficiently small ($d \simeq l_s$), and the size of the paramagnetic along x axis to be large enough, then the non-equilibrium spin state of the injected electrons can change the population of the spin sublevels in the ferromagnetic detector-film, but the electrons in the lower contact of aluminum foil will not have the spin polarization.

A measuring device with an impedance of Z connected to an electrical circuit, as is shown in the figure, depending on the value of the impedance of the device, can measure either the potential difference if Z is large (a change in the contact potential difference due to a shift of the Fermi level for electrons in the detector), or the current to be caused by a change in the potential difference at $Z \to 0$.

Why a change in the contact potential difference arises is easy to understand from Figure 1.1 of the first chapter. Since the work function of electrons in a vacuum for various metals is different, the chemical potentials of the aluminum foil and ferromagnetic film are different too.

Suppose the level of the chemical potential of electrons in the foil is higher than that in the ferromagnet. Then there will be an overflow of electrons in the boundary layer due to the difference in chemical potentials, and this overflow will stop when the

Coulomb potential difference appears. This difference of the Coulomb potentials is a cause for preventing further redistribution of electrons. As a result, the ferromagnetic will have an excess negative charge, and the foil will have a positive charge

Figure 7.5 presents schematic diagrams of energy zones in an idealized Stoner model. For simplicity, d-electrons of the lower spin subband are assumed to fail to participate in the electrotransfer. In reality, both spin subbands make a contribution to the charge transfer and magnetization. Figure 7.5(a) reflects a scenario of the lack of an electrical current in the circuit and thermodynamic equilibrium of the system. The horizontal dotted line depicts the level of chemical potential (for simplicity, it is assumed that the contact potential difference in the absence of injection is zero). All electronic states below the Fermi level under conditions of a strongly degenerate electron gas are also assumed to be occupied. For a normal metal, this situation is displayed as a solid fill of the spin subbands.

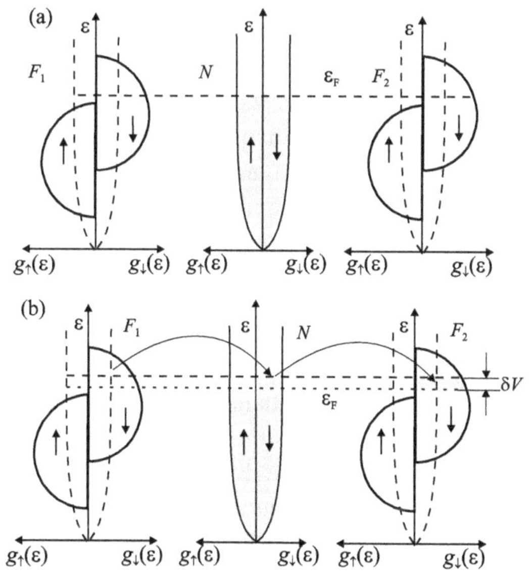

Figure 7.5: A scheme of diagrams of the density of states and occupancy of spin subbands of the injector F_1, paramagnetic metal N, and detector F_2, which explains the possibility of detecting spin accumulation.

As seen in Figure 7.5(b), the spin current induced by the electrical current in the circuit turns out to be polarized, which causes a misbalance of the spin subband population in the normal metal. Namely, one of the spin subbands has a larger number of electrons. Besides, in the event of preserving the imbalance at the normal metal–detector boundary, the quasi-chemical potential for $\downarrow$ spin electrons in the spin subband of the detector is proven to be higher than the equilibrium chemical potential (the upper dotted line in Figure 7.5(b)). With the selected sample geometry (see Figure 7.4), the electron gas has time to thermalize in the area of the lower contact of the aluminum foil with the conductor, the spin system will be in a state of thermodynamic equilibrium, and the level of the chemical potential in this area will remain in equilibrium (the lower dotted line in Figure 7.5(b)). The difference in the chemical potentials of the aluminum foil and ferromagnetic metal–detector amounts to δV, if the impedance of the measuring device tends to infinity $Z \to \infty$.

7.2.3 Spin-induced voltage detection

Let us estimate the value of the spin-induced voltage. A spin-polarized current injected into a normal metal induces a magnetic moment with a rate J_m per unit contact area

$$J_m = \eta \frac{\mu_B J}{e}, \tag{7.44}$$

where J is the electric current density, μ_B is the Bohr magneton, η is a dimensionless parameter reflecting the degree of spin polarization of the electric current, the degree of depolarization of the electron flow during the passage of interface contact region, and a number of other factors that can lead to depolarization of the electron flow.

Let the non-equilibrium spins in the region of a normal metal relax in a steady-state regime with a rate of $1/T_2$. Then the steady-state non-equilibrium magnetization M per unit volume in a paramagnetic metal can be evaluated as

$$M = \frac{J_m T_2}{d} = \frac{\eta \mu_B J T_2}{ed}, \tag{7.45}$$

where d is the distance between the injector and the detector. For implementation of the diffusion regime, the distance d must satisfy two inequalities: $d \gg v_F \tau$, where τ is the momentum relaxation time, and $d \simeq l_s$. Here l_s is the spin diffusion length determined from the expression $l_s \simeq \sqrt{DT_2}$, where D is the diffusion coefficient for electrons, and T_2 is the relaxation time of the non-equilibrium magnetization.

It is obvious that the concentration n_s of the non-equilibrium electrons and magnetic moment are connected by the relation

$$n_s = \frac{M}{\mu_B} = g(\zeta) e \delta V. \tag{7.46}$$

In this formula, ζ is the equilibrium chemical potential a degenerate electron gas. The magnitude δV of the shift of the quasi-Fermi level is assumed to be small, which allows one to regard the density of states as a constant equal to the density of states at the Fermi level.

According to (7.34), the density of states at the Fermi level and the equilibrium concentration of electrons are related to each other by a simple ratio

$$g(\zeta) = \frac{3n}{2\zeta}, \tag{7.47}$$

where n is the bulk density of electrons in a normal metal. Keeping in mind the relations (7.45)–(7.47), we write down an expression for the potential difference in the detector circuit

$$\delta V = \frac{\eta 2\zeta T_2 I}{3e^2 nSd}, \tag{7.48}$$

where I is the current flowing in the injector circuit, S is the injector area adjacent to the normal metal. Having expressed ζ and performed the simplest transformations, we get a formula convenient for estimation:

$$\delta V = \frac{\eta I \pi^{4/3} \hbar^2 T_2}{(3n)^{1/3} e^2 mSd}. \tag{7.49}$$

If the parameters close to the real values of the experiment [50] are selected for the numerical estimation of δV: $d = 10^{-4}$ m, $T_2 = 10^{-8}$ c, $S = 10^{-5}$ m^2, $n = 18 \cdot 10^{28}$ m^{-3}, $\eta \simeq 0, 1$, $I = 20$ mA, then the formula (7.49) yields the value of 10^{-11} V for δV. After dividing this voltage by the current $I = 20$ mA, the effective resistance value R for the detector circuit amounts to $\simeq 5$ nΩ. The value of $R \simeq 2$ nΩ found in the experiment of [50] indicates the operability of this simple spin accumulation model.

7.2.4 Using the Hanle effect to detect spin accumulation

As the above estimate shows, the voltage at the detector induced by the spin current is very small, and, therefore, the author in the work [50] uses the method of registering the voltage δV based on the use of the Hanle effect, discovered in 1924. The Hanle effect, applied to the problems of magneto-optics of semiconductors, consists in the fact that when a magnetic field H is applied perpendicularly to the direction of propagation of circularly polarized light incident on the surface of the semiconductor, the degree of photoluminescence polarization decreases.

The ideas that allow for controlling the polarization of photoluminescence using a magnetic field can also be applied to controlling the magnetization of the nonequilibrium electrons injected into a normal metal using a weak magnetic field with an induction of the order of hundredths of a T.

Let a constant magnetic field H be applied along the Z-axis and perpendicularly to the xy-plane, as in Figure 7.4. Then the magnetic moment of the injected nonequilibrium electrons precesses around the Z-axis. To make sure of this, it is enough to refer to the Bloch equations (5.97). If the relaxation terms are not taken into account in these equations, then for the components of our interest of the magnetic moment of the electrons m_x, m_y we get

$$\frac{dm_x}{dt} = \frac{g\mu_B}{\hbar} m_y H;$$

$$\frac{dm_y}{dt} = -\frac{g\mu_B}{\hbar} m_x H. \tag{7.50}$$

Having differentiated the first equation over time and transformed its right side taking the second equation of (7.50) into account, we arrive at a homogeneous second-order differential equation for m_x. Analogously, we can deduce an identical equation for m_y. We have

$$\frac{d^2 m_x}{dt^2} = -\left(\frac{g\mu_B H}{\hbar}\right)^2 m_x;$$
$$\frac{d^2 m_y}{dt^2} = -\left(\frac{g\mu_B H}{\hbar}\right)^2 m_y. \tag{7.51}$$

Here $\omega = g\mu_B H/\hbar$ is the cyclic frequency of spin precession in a magnetic field.

The general solution $m_x = a\cos(\omega t + \varphi_1)$, $m_y = b\cos(\omega t + \varphi_2)$ of (7.51) contains four constants that can easily be found from the initial conditions: $m_x(0) = m_0$, $m_y(0) = 0$,

$$\left.\frac{dm_x}{dt}\right|_{t=0} = 0, \quad \left.\frac{dm_y}{dt}\right|_{t=0} = -\omega m_0,$$

where m_0 is the magnetic moment of an electron at the left boundary of a normal metal (for simplicity, we assume that the electrons are injected with one spin direction coinciding with the direction of the x-axis). As a result, after finding the constants, one can make sure that the spin of the electron will make a rotational movement in the xy-plane:

$$m_x = m_0 \cos(\omega t); \quad m_y = m_0 \sin(\omega t). \tag{7.52}$$

For the relaxation time T_2 of the transverse components of the spin magnetization, the direction of the magnetic moment of the injected electron turns by the angle of ωT_2 radians. The experiment of [50] employed at low temperatures the following parameters: $T_2 \simeq 7 \cdot 10^{-9}$ c the cyclic frequency was $\omega \simeq 4 \cdot 10^8$ c^{-1} in the magnetic field of $2.5 \cdot 10^{-3}$ T (25 G). Thus, the vector of the electron magnetic moment rotates by an angle of the order of one radian during the lifetime of the non-equilibrium spin magnetization of the electron. Since the volume of the sample $\simeq l_s^3$ with a characteristic diffusion length l_s contains electrons injected at different times, the orientation of their magnetic moments is also different. This fact gives rise to a decrease in the stationary non-equilibrium spin magnetization of the injected electrons. Therefore, according to the expression (7.46), the spin-transport-induced spin accumulation lowers.

The picture of the phenomenon examined here is qualitative and helps understand only the essence of the method for registering spin accumulation using the Hanle effect. To achieve valid quantitative outcomes, the relaxation and diffusion terms of (7.50) need to be accounted for. In this case, the problem solving turns out to be too involved [50], and we do address it.

A smooth increase in the induction of the magnetic field applied to the normal metal raises the precession frequency ω and the angle through which the electron magnetic moment rotates in a time t. Consequently, the degree of depolarization of the injected electrons will increase. The highest polarization of the electrons and the largest value of δV are observed at $H = 0$. For values of the magnetic induction of $H \simeq 10^{-2}$ T, the degree of spin polarization of the electrons almost vanishes. The theoretical $\delta V/I$-dependence resembles a resonance curve shown in Figure 7.6. The effective resistance R of the detector circuit is plotted along the ordinate scale (potential difference is $\delta V = RI$). The abscissa scale displays values of the magnetic field induction in units of 10^{-2} T. The curve shown in Figure 7.6 also allows us to estimate the relaxation time of the transverse spin components T_2. Indeed, if we assume that the spin polarization in a normal metal is destroyed in magnetic fields satisfying the condition $(g\mu_B H/\hbar)T_2 \simeq 1$, then we can get a numerical estimate of T_2 from this expression. When measuring the relaxation time using paramagnetic resonance, the value of T_2 is usually determined based on the half-width of the resonance line of absorption of microwave radiation at the half-height. If we proceed from the same principle here, estimating the line width in Figure 7.6 at half the height of $\Delta H \simeq 4 \cdot 10^{-2}$ T, we get the value $T_2 \simeq 3 \cdot 10^{-9}$ c, which is consistent with the results found by other methods.

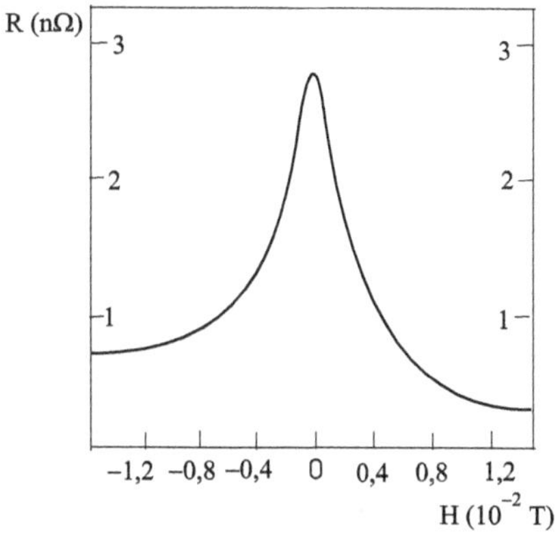

Figure 7.6: Detecting spin accumulation using the Hanle effect.

7.2.5 Coordinate-dependence of electrochemical potentials in an *F*/*N* structure

The process passing of electrons from a ferromagnetic metal to a non-magnetic metal, which takes place in the study of the giant magnetoresistance in the CPP geometry, can be described qualitatively in the framework of the two-band Mott model, that we have discussed above. This model offers the conductivity of electrons in a ferromagnetic metal as the conductivity along two spin channels characterized by their electrochemical potentials. As spin-polarized carriers approach the interface (for short, we will refer to the F/N region of the structure near the ferromagnetic metal-nonmagnetic metal boundary as to the "interface"), the difference between the spin-dependent electrochemical potentials increases, leading to spin accumulation at the interface. The passage of a spin-polarized current through the interface leads to the spin polarization (SP) of the initially unnonpolarized electron system in a nonmagnetic metal in the vicinity of the interface. This SP is preserved at a distance of about the length of the spin diffusion of electrons.

Consider the spin injection process in the ferromagnetic metal-non-magnetic metal structure. For this purpose, we resort to basic equations for describing charge and spin transport.

The ferromagnetic layer F possesses spin-dependent electrical conductivity $\sigma_\uparrow^F + \sigma_\downarrow^F = \sigma^F$ while the non-magnetic layer N has $\sigma_\uparrow^N = \sigma_\downarrow^N = \sigma^N/2$.

The starting point of the analysis is the equations (7.38) and (7.40), which allow us to express the electrochemical potentials $\bar{\zeta}_\uparrow, \bar{\zeta}_\downarrow$ through the introduced constants A, B, C, D. We solve this system of equations with respect to $\bar{\zeta}_\uparrow, \bar{\zeta}_\downarrow$ and redefine the constants, assuming that the conductivity in spin channels does not depend on z:

$$\bar{\zeta}_y^i = a^i + b^i z \pm \frac{c^i}{\sigma_y^i} \exp\left\{\frac{z}{l_{sf}^i}\right\} \pm \frac{d^i}{\sigma_y^i} \exp\left\{-\frac{z}{l_{sf}^i}\right\}. \tag{7.53}$$

The behavior of the electrochemical potentials $\bar{\zeta}_\uparrow, \bar{\zeta}_\downarrow$ both the in ferromagnet F and the normal metal N is of our interest. The upper index i marks the type of material, acquiring the two values of F or N; the index y takes two values of $y = \uparrow$ and $y = \downarrow$. The coefficients a^i, b^i, c^i, d^i in each of the regions are determined by the corresponding boundary conditions. The plus and minus signs in the formula (7.53) should be chosen for the quantity $\bar{\zeta}_\uparrow$ and $\bar{\zeta}_\downarrow$, respectively.

Let the origin of the coordinate system be located at the interface between two media, and the z-axis points deep into the normal metal. The thickness of the ferromagnetic and normal metals is assumed to be much greater than the diffusion length. Therefore, the F- and N-layers can be physically regarded as semi-infinite.

At the outer boundaries of the ferromagnetic and normal metals, the electrochemical potentials $\bar{\zeta}_\uparrow$ and $\bar{\zeta}_\downarrow$ coincide. Hence,

$$\bar{\zeta}_\uparrow^F(z = -\infty) = \bar{\zeta}_\downarrow^F(z = -\infty);$$
$$\bar{\zeta}_\uparrow^N(z = \infty) = \bar{\zeta}_\downarrow^N(z = \infty).$$

The foregoing conditions immediately enable one to calculate the values of the constants $d^F = 0$, $c^N = 0$ and simplify expressions for the electrochemical potentials of each of the regions (F, N)

$$\bar{\zeta}_y^F(z) = a^F + b^F z \pm c^F/\sigma_y^F e^{z/l_{sf}^F}, \quad z \leq 0; \tag{7.54}$$

$$\overline{\zeta}_{\gamma}^{N}(z) = a^N + b^N z \pm c^N / \sigma_{\gamma}^N e^{-z/l_{sf}^N}, \quad z \geq 0, \tag{7.55}$$

where l_{sf}^F, l_{sf}^N are spin-diffusion lengths in each of layers. Furthermore, since a^F and a^N define only the origin of the electrochemical potentials, one of these constants, without loss generality, can be treated as equal to zero. We put that $a^F = 0$

Taking into account the determination of the electric current density

$$eJ_{\gamma}^i = -\sigma_{\gamma}^i \frac{\partial \overline{\zeta}_{\gamma}^i}{\partial z}; \quad \gamma = \uparrow, \downarrow, \tag{7.56}$$

let us find the currents $J_{\uparrow}^F + J_{\downarrow}^F$ and $J_{\uparrow}^N + J_{\downarrow}^N$.

The law of conservation of charge (the current continuity condition) for any cross-section of the F/N structure says that $J_{\uparrow}^F + J_{\downarrow}^F = J$, $J_{\uparrow}^N + J_{\downarrow}^N = J$, where J is the density of the current supplied to the structure. These two conditions of constant current density allow us to find of the b^F and b^N constants

$$\begin{aligned} b^F &= -eJ/(\sigma_{\uparrow}^F + \sigma_{\downarrow}^F) = -eJ/\sigma^F; \\ b^N &= -eJ/(\sigma_{\uparrow}^N + \sigma_{\downarrow}^N) = -eJ/\sigma^N. \end{aligned} \tag{7.57}$$

Taking into account the continuity conditions for the electrochemical potentials at the interface between two media in the plane $z = 0$

$$\overline{\zeta}_{\gamma}^F(z = 0) = \overline{\zeta}_{\gamma}^N(z = 0); \quad \gamma = \uparrow, \downarrow,$$

we get two equations

$$\frac{c^F}{\sigma_{\uparrow}^F} = \frac{c^N}{\sigma_{\uparrow}^N} + a^N; \quad -\frac{c^F}{\sigma_{\downarrow}^F} = -\frac{c^N}{\sigma_{\downarrow}^N} + a^N.$$

Using the resulting set of equations, we express c^F through the constant c^N

$$c^F = c^N \frac{\sigma^N \sigma_{\uparrow}^F \sigma_{\downarrow}^F}{\sigma^F \sigma_{\uparrow}^N \sigma_{\downarrow}^N}.$$

A further transformation of the expressions for c^F and a^N requires one to introduce the quantity $\mathcal{P}$ to characterize the degree of spin polarization of the electric current in the ferromagnet. Next, we exploit identities to be satisfied by $\mathcal{P}$

$$\mathcal{P} = \frac{\sigma_{\uparrow}^F - \sigma_{\downarrow}^F}{\sigma_{\uparrow}^F + \sigma_{\downarrow}^F}; \quad (1 - \mathcal{P}) = \frac{2\sigma_{\downarrow}^F}{\sigma^F}; \quad (1 - \mathcal{P})^2 = \frac{4\sigma_{\uparrow}^F \sigma_{\downarrow}^F}{(\sigma^F)^2} = 2(1 - \mathcal{P})\frac{\sigma_{\uparrow}^F}{\sigma^F}.$$

Recalling that $\sigma_{\uparrow}^N = \sigma_{\downarrow}^N = \sigma^N/2$, we arrive at ratios for the coefficients

$$a^N = -2c^N \mathcal{P}/\sigma^N, \quad c^F = 2c^N \sigma_{\uparrow}^F (1 - \mathcal{P})/\sigma^N = \frac{c^N \sigma^F}{\sigma^N}(1 - \mathcal{P}^2). \tag{7.58}$$

The last unknown coefficient c^N is found from the condition of continuity of currents at the boundary $z = 0$. In the case of negligible spin-flip scattering processes at the interface region, the condition of equality of currents in the spin channels must be fulfilled. Consider the channel $y = \uparrow$. For this channel, the current density continuity condition

$$J_\uparrow^F(z=0) = J_\uparrow^N(z=0)$$

has the form

$$\sigma_\uparrow^F\left(-\frac{eJ}{\sigma^F} + \frac{c^F}{\sigma_\uparrow^F l_{sf}^F}\right) = \sigma_\uparrow^N\left(-\frac{eJ}{\sigma^N} - \frac{c^N}{\sigma_\uparrow^N l_{sf}^N}\right). \tag{7.59}$$

Now, in the equation (7.59), the terms proportional to the current density J can be written as

$$-eJ\left(\frac{\sigma_\uparrow^F}{\sigma^F} - \frac{\sigma_\uparrow^N}{\sigma^N}\right) = -eJ\left(\frac{\sigma_\uparrow^F}{\sigma^F} - \frac{1}{2}\right) = -eJ\mathcal{P}/2. \tag{7.60}$$

To transform the remaining terms in the equation (7.59), let us substitute the value of the coefficient c^F found above into, and transform the terms proportional to c^N:

$$-\frac{c^N}{l_{sf}^N}\left[1 + \frac{2\sigma_\uparrow^F l_{sf}^N}{\sigma^N l_{sf}^F}(1-\mathcal{P})\right] = -\frac{c^N}{l_{sf}^N}\left[1 + \frac{\sigma^F}{\sigma^N}\frac{l_{sf}^N}{l_{sf}^F}(1-\mathcal{P})^2\right]. \tag{7.61}$$

In deriving (7.61), we have used the identity

$$1/(1+\mathcal{P}) = \sigma^F/(2\sigma\uparrow^F).$$

Accounting for the results obtained, we find

$$c^N = \frac{eJl_{sf}^N\mathcal{P}}{2[1 + (l_{sf}^N\sigma^F/l_{sf}^F\sigma^N)(1-\mathcal{P}^2)]}. \tag{7.62}$$

Now, we enter the notation

$$\Sigma = \left[1 + \frac{\sigma^F}{\sigma^N}\frac{l_{sf}^N}{l_{sf}^F}(1-\mathcal{P}^2)\right]$$

to shorten the formulas for the electrochemical potentials. After substituting the deduced constants into the equations (7.54) and (7.55), we have

$$\overline{\zeta}_\uparrow^F(z) = -\frac{eJ}{\sigma^F}z + \frac{eJl_{sf}^N}{\sigma^N}\mathcal{P}(1-\mathcal{P})\Sigma^{-1}e^{z/l_{sf}^F}, \quad z \le 0; \tag{7.63}$$

$$\overline{\zeta}_\downarrow^F(z) = -\frac{eJ}{\sigma^F}z - \frac{eJl_{sf}^N}{\sigma^N}\mathcal{P}(1+\mathcal{P})\Sigma^{-1}e^{z/l_{sf}^F}, \quad z \le 0; \tag{7.64}$$

$$\overline{\zeta}_{\uparrow}^{N}(z) = -\frac{eJ}{\sigma^N}z - \frac{eJl_{sf}^N}{\sigma^N}\Sigma^{-1}\mathcal{P}(\mathcal{P} - e^{-z/l_{sf}^N}), \quad z \geq 0; \tag{7.65}$$

$$\overline{\zeta}_{\downarrow}^{N}(z) = -\frac{eJ}{\sigma^N}z - \frac{eJl_{sf}^N}{\sigma^N}\Sigma^{-1}\mathcal{P}(\mathcal{P} + e^{-z/l_{sf}^N}), \quad z \geq 0. \tag{7.66}$$

The equations (7.63)–(7.66) are responsible for the behavior of the electrochemical potentials as a function of the z-coordinate at various values of the parameters J, σ^F, σ^N, l_{sf}^F, l_{sf}^N.

Let us focus only on the simplest case when $\sigma^F = \sigma^N = \sigma$, $l_{sf}^F = l_{sf}^N = L$. Figure 7.7 shows the $\zeta_\uparrow, \zeta_\downarrow$-dependence of the electrochemical potentials $\zeta_\uparrow, \zeta_\downarrow$, measured in units of $Z = eJL/\sigma$ at a value of $\mathcal{P} = 0.6$.

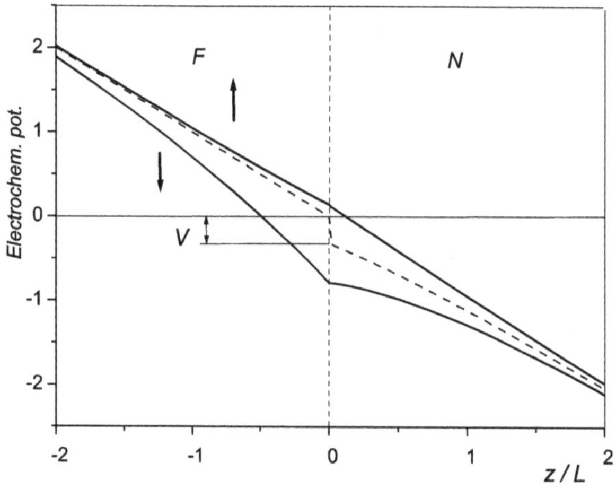

Figure 7.7: Electrochemical potentials $\overline{\zeta}_\uparrow, \overline{\zeta}_\downarrow$ in a ferromagnetic metal/ non-magnetic metal-structure: $\mathcal{P} = 0.6$; $V = eR_iJ$ is the jump of the average electrochemical potential at the interface; arrows indicate the directions of spins in channels.

Far away from the interface, the electrochemical potentials for both spin orientations coincide. Nearby the interface, the electrochemical potentials split to imbalance the spin current (spin accumulation) at the interface. The spin accumulation differs from zero on either side of the interface at a distance of the order of spin diffusion length. It should be underscored that despite the continuity of $\overline{\zeta}_\uparrow, \overline{\zeta}_\downarrow$ at the interface, the averaged values of the electrochemical potentials undergo a jump in the plane $z = 0$. We first address the F-region and define the average electrochemical potential $\overline{\zeta}_0^F$ by the relation

$$-\frac{\sigma^F}{e}\frac{d\overline{\zeta}_0^F}{dz} = J.$$

On the other hand, the definition (7.56) leads to an expression for the current in the F-region:

$$J = -\frac{\sigma_\uparrow^F}{e}\frac{d\bar{\zeta}_\uparrow^F}{dz} - \frac{\sigma_\downarrow^F}{e}\frac{d\bar{\zeta}_\downarrow^F}{dz}.$$

Setting the electrical currents equal in the last two expressions and performing integration, we can write down

$$\bar{\zeta}_0^F = \frac{\sigma_\uparrow^F}{\sigma^F}\bar{\zeta}_\uparrow^F + \frac{\sigma_\downarrow^F}{\sigma^F}\bar{\zeta}_\downarrow^F. \tag{7.67}$$

The integration constant equal to zero is determined from the boundary conditions at $z = -\infty$ (conductivities and electrochemical potentials in the spin channels are the same at $z = -\infty$). To compress writing, we introduce the coefficient $\alpha = \sigma_\uparrow^F/\sigma^F = (\mathcal{P}+1)/2$, $1-\alpha = \sigma_\downarrow^F/\sigma^F$. Then

$$\bar{\zeta}_0^F = \alpha\bar{\zeta}_\uparrow^F + (1-\alpha)\bar{\zeta}_\downarrow^F. \tag{7.68}$$

Similarly, we calculate the averaged electrochemical potential in the region N. Since $\alpha = 1/2$ in this region, the averaged electrochemical potential $\bar{\zeta}_0^N$ is just the half-sum of $\bar{\zeta}_\uparrow^N$ and $\bar{\zeta}_\downarrow^N$:

$$\bar{\zeta}_0^N = \frac{1}{2}(\bar{\zeta}_\uparrow^N + \bar{\zeta}_\downarrow^N). \tag{7.69}$$

The values of the electrochemical potentials $\bar{\zeta}_0^F$ and $\bar{\zeta}_0^N$ calculated by the formulas (7.68) and (7.69) are designated by a dotted line in the figure. It is well seen that the averaged values of the electrochemical potential experience a jump at the interface, $\delta\bar{\zeta}_0 = \bar{\zeta}_0^F(0) - \bar{\zeta}_0^N(0)$.

The emergence of the electrochemical potential jump at the interface can be interpreted as a jump in potential caused by additional interface resistance

$$R_i J = \frac{1}{e}(\bar{\zeta}_0^F(0) - \bar{\zeta}_0^N(0)). \tag{7.70}$$

After having performed fairly simple calculations, we deduce an expression for the interface resistance

$$R_i = \frac{2\mathcal{P}^2 l_{sf}^N}{\sigma^N[1 + \frac{\sigma^F}{\sigma^N}\frac{l_{sf}^N}{l_{sf}^F}(1-\mathcal{P}^2)]}. \tag{7.71}$$

Let us evaluate the spin polarization of the current passing through the F/N interface. It can be represented as the difference between the current densities in the spin channels, divided by the total current:

$$\beta = [J_\uparrow(0) - J_\downarrow(0)]/J.$$

Using the formulas (7.56), (7.63), (7.64), and the above definition of β, we get

$$\beta = \frac{\mathcal{P}}{1 + (1 - \mathcal{P}^2)(\sigma^F l_{sf}^N)/(\sigma^N l_{sf}^F)}. \tag{7.72}$$

One of the important tasks of spintronics is the creation of devices that allow performing the injection of spin-polarized electrons. The analysis of the current flow in the F/N structure and the resulting expressions (7.71), (7.72) allow us to consciously approach the problem of selecting materials that allow us to obtain a noticeable polarization of the injected electrons. As follows from the above formulas (7.71), (7.72), the spin polarization quantity and the resistance R_i have the same parameters: $\mathcal{P}$, $\sigma^F l_{sf}^N/(\sigma^N l_{sf}^F)$. In the vast majority of cases, the spin diffusion length in a ferromagnet is much less than in a non-magnetic material: $l_{sf}^F \ll l_{sf}^N$. The smallness of l_{sf}^F, in essence, is a factor that limits the possibility of generating a large degree of polarization. The problem becomes especially significant when it comes to the injection of a spin-polarized current from a ferromagnet into a semiconductor. This is because in this case $\sigma^N \ll \sigma^F$ and it makes itself dramatically felt in the degree of spin polarization of the injected electrons. This issue is known as the *conductivity mismatch problem*. For providing the largest value of the $\mathcal{P}$ parameter, the selection of a ferromagnet also plays an important role. For this purpose, a ferromagnet with the largest conductivity difference in spin channels should be chosen.

7.3 The spin Hall effect

7.3.1 Phenomenological consideration

In the first, fourth, and fifth chapters of this book, we have already discussed the standard or ordinary Hall effect for cases of classical (4.132) and quantizing (5.138) magnetic fields. The reason for the effect is that the Lorentz force depends on the value of the electron velocity, but the Coulomb force, that arises due to the redistribution of electrons, does not depend. Therefore, faster and slower electrons are deflected in opposite directions. This interpretation of the phenomenon is confirmed in conditions of adiabatic isolation of the sample, when not only the Hall voltage but also the transverse temperature gradient arises (transverse Ettinghausen effect).

As to ferromagnetic conductors, a completely different picture is observed. Here, the magnitude of the Hall effect is significantly greater, strongly depends on the temperature, and the dependence on the external magnetic field is nonlinear. The Hall effect in ferromagnets has been called spontaneous or called the anomalous Hall effect (AHE) because it also exists with no external magnetic field present. It had taken 70 years before the anomalous Hall effect was for the first time explained after its discovery. For this purpose, the idea of asymmetric electron–impurity scattering was proposed to make an allowance for spin–orbit interaction.

In 1971 Dyakonov and Perel predicted that, with no external magnetic field, an electric current in a semiconductor can induce a spin orientation in a thin layer near the samples lateral surface due to the spin–orbit effects when electron scattering takes place. However, a weak magnetic field applied parallel to the electrical current destroys this orientation. Moreover, they phenomenologically established a relationship between the electric and spin current, which is due to the spin–orbit interaction. The electric current generates a spin current, while the inhomogeneity of the spin density gives rise to an electrical current. The term "spin Hall effect" was introduced by Hirsch in his paper published in 1999.

As a small digression, it should be noted that the Hall effect turned out to be very capacious in content. Over the past 20-odd years, the Nobel Prize in Physics has been awarded twice for investigation of the quantum Hall effect in two-dimensional structures. In the first time, in 1985, Klaus von Klitzing received his Nobel Prize for the discovery of the quantum Hall effect. In the second time, in 1998, the Nobel Prize in physics was divided; one half went to Laughlin for his interpretation of the fractional quantum Hall effect, and the other half was shared between Stormer and Chee Tsui for the discovery and fundamental work on the fractional quantum Hall effect.

Let us compose phenomenological equations for the electron flux density $\vec{J}_n^0$ and spin flux density $\vec{J}_s^0$ without taking spin–orbit interactions into account:

$$\vec{J}_n^0 = -\mu n\vec{E} - D\vec{\nabla}n; \tag{7.73}$$

$$J_{s\,ij}^0 = -\mu E_i S_j - D\frac{dS_j}{dx_i}. \tag{7.74}$$

In this formula, $\mu = e\tau_p/m$ is the electron mobility, D is the diffusion coefficient, and $\vec{S}$ is the spin vector density

$$S_i = \sum_{\vec{k}} \mathrm{Sp}\{\hat{\sigma}_i \rho(\vec{r}, \vec{p}, t)\};$$

$$J_{s\,ij}^0 = \sum_{\vec{k}} \mathrm{Sp}\left\{\frac{p_i}{m}\hat{\sigma}_j \rho(\vec{r}, \vec{p}, t)\right\},$$

where $\hat{\sigma}_i$ are Pauli matrices, $\rho(\vec{r}, \vec{p}, t)$ is a statistical operator.

Let us look at how the relations (7.73), (7.74) change when accounting for the spin–orbit interaction. A proper description of the relationship between the particle flux density and the spin flux density requires bearing in mind that the flux of the number of electrons is a vector, the spin density vector is a pseudo-vector, and the spin flux density is a second-rank pseudo-tensor. If, due to the spin–orbit interaction, there are mechanisms connecting the electron flux density vector and the spin flux density vector, then in crystals that have symmetry with respect to the inversion operation, the spin flux density tensor should enter the equation (7.73) in the combination $\gamma\epsilon_{ijk}J_{s\,jk}^0$, and the electron flux density vector into the equation (7.74) in the form $\gamma\epsilon_{ijk}J^0_{nk}$:

$$J_{ni} = -\mu nE_i - D\frac{dn}{dx_i} + \gamma\epsilon_{ijk}J^0_{s\,jk}; \tag{7.75}$$

$$J_{s\,ij} = -\mu E_i S_j - D\frac{dS_j}{dx_i} - \gamma\epsilon_{ijk}J^0_{nk}. \tag{7.76}$$

Here γ is a constant proportional to the value of spin–orbit interaction (dimensions of densities of the electron flux and spin flux coincide upon the definitions chosen), ϵ_{ijk} is the unit third-rank antisymmetric tensor (the Levi-Civita tensor), i, j, k are tensor indices running through the values of x, y, z.

Substituting the definition of the spin-electron flux density (7.74) into the equation (7.75), we get

$$\vec{J}/e = \mu n\vec{E} + D\vec{\nabla}n + \gamma\mu[\vec{E} \times \vec{S}] + \gamma D[\vec{\nabla} \times \vec{S}]. \tag{7.77}$$

In this expression, we have passed from the particle flux density to the density of the electric current $\vec{J} = -e\vec{J}_n$.

Analogously, we modify the equation (7.76) using the definition of the particle number flux (7.73):

$$J_{sij} = -\mu E_i S_j - D\frac{dS_j}{dx_i} + \epsilon_{ijk}\left(\gamma\mu nE_k + \gamma D\frac{dn}{dx_k}\right). \tag{7.78}$$

The equations for the flux density should be supplemented by the continuity equations of the form (1.7) arising from the laws of conservation of the particle number density and spin density:

$$\frac{dn(\vec{r},t)}{dt} + \operatorname{div}\vec{J}_n = 0; \tag{7.79}$$

$$\frac{dS_i(\vec{r},t)}{dt} = \frac{J_{sij}}{dx_j} + [\vec{\Omega} \times \vec{S}]_i + \frac{S_i}{\tau_s} = 0, \tag{7.80}$$

where $\vec{\Omega} = g\mu_B\vec{H}/\hbar$ is the vector numerically equal to the frequency of spin precession in a magnetic field H and coincides in direction with the direction of the magnetic field, τ_s is the relaxation time of the spin components.

Thus, accounting for the spin–orbit interaction evidences the relationship between the electric and spin currents through the equations (7.77) and (7.78). The first two terms in the equation (7.77) describe the current caused by the electric field and concentration gradient. The third term $\gamma\mu[\vec{E} \times \vec{S}]$ controls the electric current in the direction perpendicular to the electric field and magnetization of the sample. In other words, it is responsible for the occurrence of the anomalous Hall effect. The last term $\gamma D[\vec{\nabla} \times \vec{S}]$ delineates the appearance of the electric current due to the spin density gradient. This stands for an inhomogeneous spin density as a result of spin accumulation, which may provoke an electric current or potential difference in the direction perpendicular to the spin density gradient. Consequently, the fourth term in the expression for the current density (7.77) describes the inverse spin Hall effect discovered experimentally in 2004.

In the expression (7.78), the first two terms indicate the generation of spin current due to the applied electric field and spin density gradient, if the spin polarization exists. The last term proportional to a spin–orbit coupling constant describes the direct spin Hall effect. Otherwise speaking, a spin flow arises in the direction perpendicular to the electric field strength or the concentration gradient even in the case when there is no initial spin polarization of electrons and no external magnetic field. Experimentally, this phenomenon was recorded by optical methods in 2004. In 2005, the phenomenon of spin accumulation produced by an electric current in semiconductor samples was investigated with direct electrical measurements.

Since in the formula (7.78) the round brackets of the last term include not only a contribution proportional to the electric field but also a term proportional to the concentration gradient, the spin accumulation in the transverse direction can be induced by an applied temperature gradient (spin analog to the Nernst–Ettinshausen effect).

In concluding the section concerning phenomenological transport equations with spin variables, brief mention should be made of experimental findings of the magnitude of the spin Hall effect. The magnitude of the normal Hall effect is usually characterized by the angle Hall θ that is given by the relation

$$\operatorname{tg}\theta_H = \frac{E_y}{E_x} = \frac{\sigma_H}{\sigma_\perp}; \quad \theta_H \simeq \frac{\sigma_H}{\sigma_\perp}, \tag{7.81}$$

where E_y is the amplitude Hall electric field, and E_x the amplitude applied electric field: $\sigma_H \equiv \sigma_{xy}$, $\sigma_\perp \equiv \sigma_{xx}$ is defined equations (4.118) and (4.123). For a degenerate electron gas, it follows from these formulas that $\theta_H \simeq \omega_0 \tau_{\bar{p}} \ll 1$.

For the spin Hall effect, we can also define the Hall angle θ_{SH} as the ratio of the spin flux density to the electron flux density conditional upon no initial spin polarization and concentration gradient. Combining formulas (7.77) and (7.78) leads to $\theta_{SH} \sim \gamma$. Thus, the magnitude of the spin Hall effect is directly proportional to the spin–orbit coupling constant.

Numerical values of θ_{SH} for various metals and alloys can be found in the paper [51]. The magnitudes of the Hall angle for the spin Hall effect are fairly wide ranged usually not exceeding fractions of a percent. A maximum value of θ_{SH}, as known by now [51], amounts to $\simeq 0,1$ and meets gold samples. However, most likely, it is not a limit value.

7.3.2 Spin–orbit interaction mechanisms

The transverse spin current, which occurs as a response of the system to an applied external electric field, is possible only in the presence of spin–orbit interaction (SOI), and therefore, to understand the nature of the spin Hall effect, it is necessary to consider the main types of corrections to the electron energy caused by SOI.

The structure of the SOI Hamiltonian for crystals with different symmetries, two-dimensional materials, and heterostructures can be obtained up to constant numerical coefficients based on symmetry considerations. The electron Hamiltonian is invariant both under the time-reversal operation ($t \to -t$; $\vec{p} \to -\vec{p}$; $\vec{s} \to -\vec{s}$, where $\vec{s}$ is the spin operator of an electron) and under symmetry operations of the point group of a crystal. For example, if the symmetry group of a crystal includes the inversion $\vec{r} \to -\vec{r}$; $\vec{p} \to -\vec{p}$; $\vec{s} \to \vec{s}$, the Hamiltonian must remain unchanged for such a transformation.

The Elliott–Yafet SOI Hamiltonian

In the simplest case, the operator of spin–orbit interaction must involve three vectors, namely, a vector of the gradient of a potential force field $\vec{\nabla} V$ (V includes the potential of the crystal field, the potential of scattering centers, and the potential of an external field), an operator of momentum $\vec{p}$ and a spin vector $\vec{\sigma}$ with the Pauli matrices σ_x, σ_y, σ_z as Cartesian components. In crystals with a symmetry center, the simplest combination of these three vectors is the scalar quantity

$$H_{\mathrm{SO}} = \lambda[\vec{\nabla} V \times \vec{p}]\vec{\sigma}. \tag{7.82}$$

This quantity remains unchanged under both the time-reversal operation and the inversion operation.

Now we derive an expression for the constant λ proceeding from semi-classical considerations. Let the charged center with the potential V be placed at the beginning of a fixed coordinate system and an electron move in it with velocity $\vec{v}$. This center creates an electric field at a point with a radius-vector $\vec{r}$:

$$\vec{E} = -\frac{\vec{r}}{r}\frac{dV}{dr}.$$

In the coordinate system associated with a moving electron, the observer will see a charged center moving at a speed of $-\vec{v}$, that will create both electric and magnetic fields with a strength of $\vec{E}'$, $\vec{H}'$. To find these fields, you need to use the Lorentz transformations for the components of the electromagnetic field.

In the Gaussian unit system, the field components appear as follows:

$$E'_x = E_x; \quad E'_y = \frac{E_y - v/cH_z}{\sqrt{1 - v^2/c^2}}; \quad E'_z = \frac{E_z + v/cH_y}{\sqrt{1 - v^2/c^2}}; \tag{7.83}$$

$$H'_x = H_x; \quad H'_y = \frac{H_y + v/cE_z}{\sqrt{1 - v^2/c^2}}; \quad H'_z = \frac{H_z - v/cE_y}{\sqrt{1 - v^2/c^2}}. \tag{7.84}$$

The dashes mark the field components in the moving coordinate system. Neglecting the terms of expansion of the radical expression in the denominator, of the order of v^2/c^2, brings to the following simple form of the magnetic field acting on the moving electron:

$$\vec{H}' = -\frac{1}{c}[\vec{v} \times \vec{E}] = -\frac{1}{mc}[\vec{p} \times \vec{E}]. \tag{7.85}$$

It is the Zeeman correction to the electron energy in the field $\vec{H}'$ that is the contribution of the spin–orbit interaction to the electron energy,

$$H'_{SO} = -g\mu_B \vec{s}\vec{H}' = \frac{e\hbar}{2m^2c^2}[\vec{\nabla}V \times \vec{p}]\vec{\sigma}. \tag{7.86}$$

Here $\vec{s}$ is the electron spin operator measured in units of $\hbar$, $\vec{\sigma}$, are the Pauli matrices,

$$\sigma_x = \begin{pmatrix} 0 & 1 \\ 1 & 0 \end{pmatrix}, \quad \sigma_y = \begin{pmatrix} 0 & -i \\ i & 0 \end{pmatrix}, \quad \sigma_z = \begin{pmatrix} 1 & 0 \\ 0 & -1 \end{pmatrix}.$$

Deduced in the quasi-classical approximation, the formula (7.86), for the contribution of spin–orbit interaction gives only a double over-estimated value. A strict quantum mechanical approach yields the result that is twice as small:

$$H_{SO}^{EY} = \frac{e\hbar}{4m^2c^2}[\vec{\nabla}V \times \vec{p}]\vec{\sigma}. \tag{7.87}$$

However, the quasi-classical consideration makes it possible to better comprehend the nature of this interaction. It is worthwhile to stress that spin–orbit interaction also occurs in a perfect impurity-free crystal lattice with no external fields present. Impurities and external fields just result in additional contributions to this interaction, which can be regarded as extra terms for the potential V. In the literature, the expression (7.87) is known as the Elliott–Yafet Hamiltonian. Elliott and Yafet were first to use it back in 1954 to calculate the spin-lattice relaxation time upon scattering of electrons by phonons and impurity centers.

The Dresselhaus SOI Hamiltonian

To write down other possible forms of the spin–orbit interaction Hamiltonian, we again use the symmetry approach. In doing so, we expand the electron Hamiltonian $H(\vec{p}, \vec{\sigma})$ in a power series, limiting ourselves to the first-order terms in spin components σ_i and the third-order terms of the components in a momentum p_i (the even powers of the momentum operator cannot enter this expansion due to the invariance of the Hamiltonian under the time-reversal operation):

$$H(\vec{p}, \vec{\sigma}) = \frac{p^2}{2m} + \beta_{ij}\sigma_i p_j + \Gamma_{ijkl}\sigma_i p_j p_k p_l, \tag{7.88}$$

where β_{ij}, Γ_{ijkl} are some constants.

Next, we can significantly simplify the form of the Hamiltonian (7.88) by introducing a pseudo-vector $\vec{\Omega}$ composed of the components of the vector $\vec{p}$ and having the dimension of an angular frequency vector,

$$H(\vec{p},\vec{\sigma}) = \frac{p^2}{2m} + \frac{\hbar}{2}\vec{\sigma}\vec{\Omega}; \tag{7.89}$$

$$\vec{\Omega} = \frac{2}{\hbar}(\beta_{ij}\sigma_i p_j + \Gamma_{ijkl}\sigma_i p_j p_k p_l), \tag{7.90}$$

while the spin–orbit interaction operator can be now written in the form

$$H_{\mathrm{SO}} = \frac{\hbar}{2}\vec{\sigma}\vec{\Omega}(\vec{p}). \tag{7.91}$$

In cubic crystals with a center of inversion $\vec{\Omega} = 0$. A_3B_5 semiconductors with the point symmetry group T_d (a symmetry group of a regular tetrahedron) have no inversion center. However, under symmetry transformations of the group T_d, the terms linear in momentum are not invariants and, therefore, in these materials $\beta_{ij} = 0$. The terms cubic in the momentum components, appearing in the formula (7.90), can constitute an invariant construction known as the Dresselhaus spin–orbit interaction Hamiltonian.

$$H_{\mathrm{SO}}^{D} = \Gamma[\sigma_x p_x(p_y^2 - p_z^2) + \sigma_y p_y(p_z^2 - p_x^2) + \sigma_z p_z(p_x^2 - p_y^2)]. \tag{7.92}$$

The Dyakonov–Perel SOI Hamiltonian

The Hamiltonian of a quantum mechanical system is invariant under the time-reversal operation, which leads to degeneration of levels of the electron energy $\varepsilon_{\vec{p}\uparrow} = \varepsilon_{-\vec{p}\downarrow}$ (the Kramers degeneracy). Crystals with no inversion center must be satisfied by the condition $\varepsilon_{\vec{p}} \neq \varepsilon_{-\vec{p}}$. Hence, it follows that $\varepsilon_{\vec{p}\uparrow} \neq \varepsilon_{\vec{p}\downarrow}$. In other words, the electron energy levels are split in spin due to the spin–orbit interaction by the value of $\Delta = \varepsilon_{\vec{p}\downarrow} - \varepsilon_{\vec{p}\uparrow}$. This splitting can be associated with a magnetic field $\vec{H}(\vec{p})$, the amplitude of which is momentum-dependent. Similarly to a usual magnetic field, $\vec{H}(\vec{p})$ causes the precession of electrons with an angular frequency $\vec{\Omega}(\vec{p}) = \delta\vec{H}(\vec{p})$, where δ is some constant. Given the above, the Dyakonov–Perel Hamiltonian can be written as an expression that coincides in form with (7.91):

$$H_{\mathrm{SO}}^{\mathrm{DP}} = \frac{\hbar}{2}\vec{\sigma}\vec{\Omega}(\vec{p}). \tag{7.93}$$

As for A_3B_5 semiconductors, the Cartesian components of the cyclic frequency $\vec{\Omega}(\vec{p})$, as the calculation shows, have the form

$$\Omega_x = \alpha p_x(p_y^2 - p_z^2); \quad \Omega_y = \alpha p_y(p_z^2 - p_x^2); \quad \Omega_z = \alpha p_z(p_x^2 - p_y^2), \tag{7.94}$$

where α is some constant. In this case, the SOI Hamiltonians (7.92) and (7.93) coincide in structure.

The Rashba SOI Hamiltonian

If the inhomogeneity of the potential energy is due to sharp changes in the parameters of the sample near its boundaries, for example, a heterojunction in a semiconductor structure, then such an asymmetry is called structural and is denoted by SIA (Structure Inversion Symmetry). In this case, the SOI is dominated by the contribution described by the Rashba Hamiltonian.

Consider a two-dimensional crystal structure. Let its plane coincide with the coordinate plane xy. Then it is obvious that there is a potential gradient along the z-axis and the spin–orbit interaction operator (7.87) can be represented in the form

$$H_{\mathrm{SO}}^{R} = \frac{e\hbar}{4m^2c^2}[\vec{\sigma} \times \vec{p}]_z \frac{dV}{dz} = \alpha_1(\sigma_x p_y - \sigma_y p_x), \tag{7.95}$$

where α_1 is some constant. The expression (7.95) is nothing but the Rashba spin–orbit interaction Hamiltonian (1960).

All of the above Hamiltonians have been constructed based on symmetry considerations. Naturally, they can be derived by applying microscopic mechanisms of accounting for spin–orbit interaction within the quantum theory of a solid body. Such a calculation enables one to express the constants in the Hamiltonians (7.87), (7.92) (7.94) and (7.95) through parameters of the band structure of materials. However, since the band structure parameters are either experimentally found or are computed with large errors, numerical values of the SOI constants in these Hamiltonians are simpler determined from experiments by measuring, for example, the spin-magnetization relaxation time.

And now to finish the section, it should be said that it is very difficult to distinguish experimentally the contribution of any SOI mechanism from the spin Hall effect. Most often, apparently, a cumulative effect of several mechanisms makes itself felt in the experiment.

7.3.3 The microscopic nature of spin transport

Let us discuss the microscopic nature of the appearance of terms that connect a flow of charged particles and spin flux in the equations (7.77) and (7.78). Early research undertaken to explain the anomalous and spin Hall effects have been carried out under the assumption that both of these phenomena are associated with the asymmetry of electron scattering by impurity centers. As a result, after colliding, the direction of the electron momentum becomes dependent on the electron spin state. In the literature, the Hall effect caused by the mechanisms of asymmetric scattering is called the *external anomalous (spin) Hall effect.* Later it was found out that there is another, more interesting, mechanism of occurrence of the discussed effects, which is not related to the presence of impurity centers, which is called the *intrinsic anomalous (spin) Hall effect.*

Quasi-classical dynamics of Bloch electrons under perturbation caused by an electric field is one of the oldest problems in physics solid. Most modern textbooks provide dynamics equations

$$\dot{\vec{p}} = -\frac{d\varepsilon}{d\vec{r}} + \frac{e}{c}[\dot{\vec{r}} \times \vec{H}];$$
$$\dot{\vec{r}} = \frac{d\varepsilon}{d\vec{p}}, \tag{7.96}$$

that have not been in doubt until recently. Back in 1954, Karplus and Luttinger were the first to succeed in interpreting the internal anomalous Hall effect by making an allowance for the expression for the velocity of Bloch electrons (7.96). This correction, later called "anomalous velocity," turned out to be perpendicular to the applied electric field and allowed them to explain the intrinsic anomalous Hall effect.

Further studies have shown that the anomalous velocity phenomenon but exactly a change in the semi-classical dynamic laws of Bloch electrons is a very widespread one and happens in systems with broken symmetry relative to either a time-reversal operation or an inversion operation, for example, as a result of switching on an external electric field. In this case, the semi-classical equations of motion of Bloch electrons are significantly modified:

$$\dot{\vec{p}} = -\frac{d\varepsilon}{d\vec{r}} + \frac{e}{c}[\dot{\vec{r}} \times \vec{H}]; \tag{7.97}$$
$$\dot{\vec{r}} = \frac{d\varepsilon}{d\vec{p}} + \frac{1}{\hbar}[\dot{\vec{p}} \times \vec{\Omega}], \tag{7.98}$$

where the quantity $\vec{\Omega}$ is called the Berry curvature. Let a sample be exposed to an external electric field $\vec{E}$ with no magnetic field $\vec{H}$ present and the system examined be an electron in a periodic potential field upon spin–orbit interaction. Then formula (7.98) implies the expression [54] for the electron velocity $\vec{v}$:

$$\vec{v} = \frac{d\varepsilon}{d\vec{p}} + \frac{e}{\hbar}[\vec{E} \times \vec{\Omega}(\sigma, \vec{k})], \tag{7.99}$$

where $\vec{k}$ is the wave vector of the electron, σ is a spin index, e is the electron charge modulus.

If it is possible to calculate the value $\vec{\Omega}(\sigma, \vec{k})$ for a given model system, then the expression (7.99) almost immediately allows us to explain the integer quantum, anomalous, and spin Hall effects. The above expression (7.99) for the velocity of Bloch electrons is only a partial, though important, result of the application of the Berry phase concept in modern physics.

The discovery of the role of the Berry phase in the analysis of the evolution of quantum mechanical systems was not related specifically to Bloch electrons, but rather to the general idea that the quantum adiabatic transfer of particles in slowly varying fields (electric, magnetic, or deformation) can, in principle, change the wave function. The Berry phase, and more generally the problem of adiabatic transport, plays a crucial role in Bloch translation-invariant systems where the parameters (Bloch pulses) change in closed manifolds (zones or Fermi surfaces) by applying electric or magnetic fields.

It should be underscored that the main content of this section is devoted to a brief acquaintance of the reader with the methodology of operating the Berry phase concept in exploring the dynamics of Bloch electrons under adiabatic perturbations. However, we first qualitatively focus on the nature of the arising the intrinsic spin Hall effect in systems with the Rashba SOI Hamiltonian. In doing so, we will utilize the standard semi-classical approach.

7.3.4 A qualitative explanation of the intrinsic spin Hall effect for systems with the Rashba Hamiltonian

Consider a system that is a two-dimensional electron gas. Let the spin–orbit interaction in this system be given by the Rashba Hamiltonian (7.95)

$$H_{2D} = \frac{\hbar^2 \hat{k}^2}{2m} + \alpha(\sigma_x \hat{k}_y - \sigma_y \hat{k}_x), \tag{7.100}$$

where $\hat{k}_i = -id/dx_i$; $i = x, y$; α is the Rashba spin–orbit interaction constant, σ_i are the matrices Pauli.

Now, we find the eigenvalues of this Hamiltonian. The Hamiltonian (7.100) containing the Pauli matrices, the solution of the stationary problem $H_{2D}\psi = E\psi$ needs to be sought in the form of two-component spinors. The coordinate part of the wave function should be determined in terms of plane waves. As a result, we find that the trial solution must look like

$$\psi(\vec{k}) = e^{i\vec{k}\vec{r}} \begin{vmatrix} c_1 \\ c_2 \end{vmatrix}, \tag{7.101}$$

where c_1, c_2 are some constants, $\vec{k}_i = k_x, k_y$; $\vec{r}_i = x, y$.

Substituting the trial solution (7.101) into the stationary Schrödinger equation, we obtain a set of two linear homogeneous equations for the coefficients c_1, c_2,

$$\begin{gathered} \left(\frac{\hbar^2 k^2}{2m} - E\right)c_1 + zc_2 = 0; \\ z^* c_1 + \left(\frac{\hbar^2 k^2}{2m} - E\right)c_2 = 0, \end{gathered} \tag{7.102}$$

where $z = \alpha(k_y + ik_x)$, $z^* = \alpha(k_y - ik_x)$.

The set of equations (7.102) has a nontrivial solution if the determinant of this set is zero. Setting equal the determinant to zero, we arrive at the dispersion relation

$$E(\lambda, k) = \frac{\hbar k^2}{2m} + \lambda \alpha k; \quad \lambda = \pm 1. \tag{7.103}$$

A plot of electron energy (7.103) as a function of the wave vector k_x at a given value of k_y is shown in Figure 7.8 (we find the eigenfunctions of the equation (7.100) later in calculating the Berry curvature). Let us pinpoint some features of the resulting spectrum of the current carriers. To start with, we hold fixed the modulus value of the wave vector and compute the energy difference

$$\Delta = E(+, k) - E(-, k) = 2\alpha k.$$

Then we can claim that the spectrum of current carriers exhibits a slit (except for the $k = 0$ point). This slit can be treated as spin splitting in a magnetic field induced by SOI

$$\vec{H}' = \frac{\hbar}{mc}[\vec{k} \times \vec{\nabla} V]; \quad \vec{\nabla} V || z. \tag{7.104}$$

The last formula also implies another, no less important result, which is that the vector $\vec{H}'$, like the vector $\vec{k}$, lies in the plane of a two-dimensional layer, and $\vec{H}'$ always remains perpendicular to the wave vector $\vec{k}$, if the change in the states of the electron occurs slowly enough.

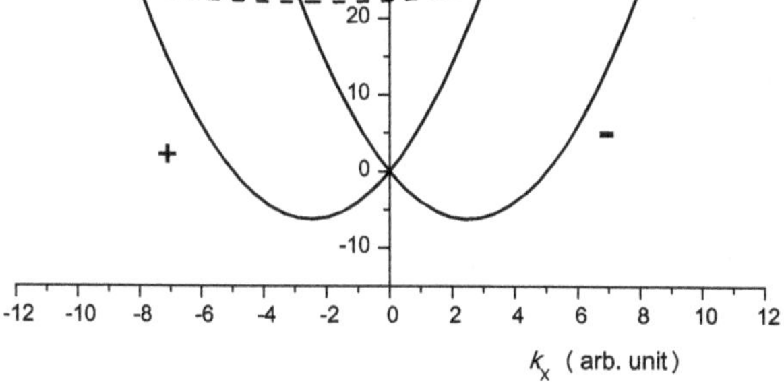

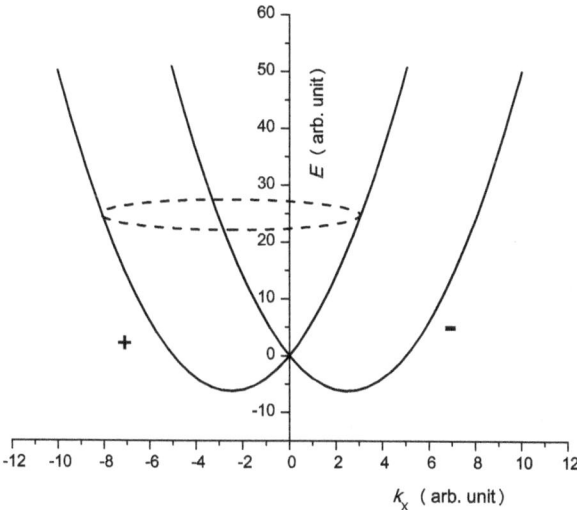

Figure 7.8: Splitting of the free-electron spectrum by the Rashba spin–orbit interaction: the signs (+) and (−) stand for bands with $\lambda = 1$ and $\lambda = -1$, respectively; the dotted line indicates the Fermi level of one of the subbands.

Since the electron's spin is aligned either along the field $\vec{H}'$, or against this field, a very important conclusion follows: when an electron moves, its spin is always perpendicular to the vector $\vec{k}$.

Thus, there is a stable relationship between the direction of the electron spin and its wave vector. The electrons from the subbands $E(\lambda = +1, \vec{k})$ and $E(\lambda = -1, \vec{k})$ that have the same wave vector will have the opposite spin orientation. We can say that as a result of the SOI, the electron acquired a new property, that is called *chirality* (a chiral object is an object that does not coincide with its representation in a flat mirror).

Let us now look at the behavior of electrons in a two-dimensional surface Fermi of one of the spin subbands exposed to an external electric the field directed along the x [52].

Figure 7.9(a) schematically depicts electrons in the Fermi surface of one of the spin subbands. The spin directions are shown by arrows. When an external electric field is turned on along the direction of the x-axis, an additional force $\dot{p}_x = -eE_x$ will act on the electrons, that causes a change in the wave vector k_x. During the time $t < T_2$, the component of the wave vector k_x in the positive direction of the axis k_x will decreases, and, for the negative direction of this axis, it will increases by the value $|eEt/\hbar|$.

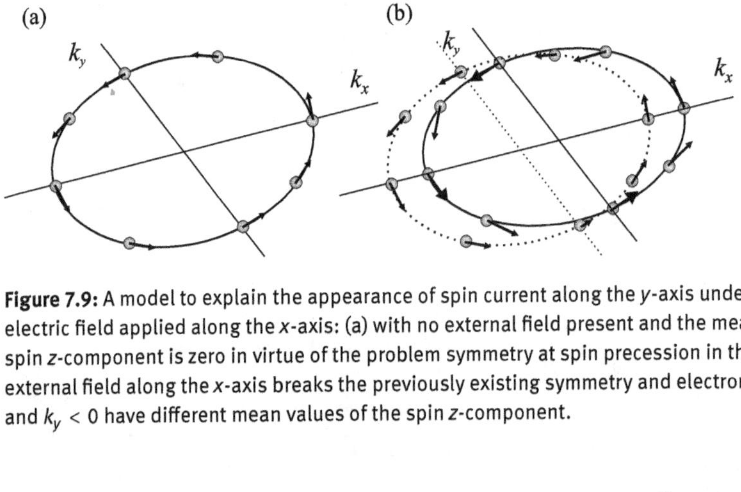

Figure 7.9: A model to explain the appearance of spin current along the y-axis under an external electric field applied along the x-axis: (a) with no external field present and the mean value of the spin z-component is zero in virtue of the problem symmetry at spin precession in the field $\vec{H}'$; (b) an external field along the x-axis breaks the previously existing symmetry and electrons with $k_y > 0$ and $k_y < 0$ have different mean values of the spin z-component.

For this reason, the values of the y-components of the field $\vec{H}'$ (see the formula $\vec{H}'$ (7.104)) for $k_y > 0$ and $k_y < 0$ are different, that leads to a nonzero mean z-component of the electron's spin. This situation is presented in Figure 7.9(b). Obviously, the contribution of another spin subband turns out to be opposite. Nevertheless, compensation for spin polarization will be not complete, if there are, for example, differences in the degree of subband population.

Summarizing a subtotal, it can be said that the spin Hall effect arises due to the time-dependent effective magnetic field $\vec{H}'$ acting on the spin as it moves in momentum space.

Problem 7.3. Using a quantum mechanical approach, find the mean value of the spin z-component in a system with the Hamiltonian

$$\hat{H}_S = \frac{1}{2} g\mu_B \vec{\sigma}\vec{H},$$

if the magnetic field is applied along the x-axis.

Solution. The magnetic field points along the x-axis, and we will keep track of the spin z-component. By $|\uparrow\rangle$ and $|\downarrow\rangle$ denote a basis for eigenfunctions of a matrix σ_z.

We write down the Zeeman Hamiltonian $\hat{H}_S$ in the matrix representation,

$$\hat{H}_S = \frac{\hbar\Omega}{2}\begin{vmatrix} 0 & 1 \\ 1 & 0 \end{vmatrix}, \quad \Omega = \frac{g\mu_B H}{\hbar}, \quad H||x.$$

The Hamiltonian's eigenfunctions that meet the eigenvalues $E_{x1} = \hbar\Omega/2$ and $E_{x2} = -\hbar\Omega/2$ are the functions

$$|x1\rangle = \frac{1}{\sqrt{2}}(|\downarrow\rangle + |\uparrow\rangle);$$
$$|x2\rangle = \frac{1}{\sqrt{2}}(|\downarrow\rangle - |\uparrow\rangle).$$

The set of the eigenfunctions $|x1\rangle, |x2\rangle$ can be used as a basis for solving the nonstationary Schrödinger problem

$$i\hbar\frac{d\psi(t)}{dt} = \hat{H}_S\psi(t).$$

Its general solution can be written as a superposition of particular solutions

$$\psi(t) = c_1 e^{i\Omega t/2}|x1\rangle + c_2 e^{-i\Omega t/2}|x2\rangle; \quad |c_1|^2 + |c_2|^2 = 1.$$

A quantum-mechanical mean value of the z-component of the spin operator is defined by the expression

$$\overline{S}_z(t) = \frac{\hbar}{2}\langle\psi(t)|\sigma_z|\psi(t)\rangle.$$

Plugging the values found above for the eigenfunctions $|x1\rangle$ and $|x2\rangle$ in matrix form into this expression and performing rather simple but cumbersome matrix calculations lead to the well-known outcome

$$\overline{S}_z(t) = S_z(0)\cos(\Omega t + \varphi),$$

where φ is the phase, which is determined by the initial value of spin projection onto the z-axis.

Thus, we obtained a well-known result: if the magnetic field is applied along the x-axis, then the spin moment will precess around this axis and the z-component of the spin will change cyclically over time.

Consider now another approach based on the application of the Bloch equations. For the Rashba spin–orbit interaction Hamiltonian, the appearance of the spin Hall effect can be obtained by studying the dynamics of spin magnetization using the Bloch equations (5.97), in which the relaxation terms can be ignored.

Following the paper [52], we assume that the Rashba spin–orbit interaction Hamiltonian (7.100) can be regarded as the Zeeman spin energy in an induced magnetic field $\vec{H}'$ depending on the electron momentum,

$$H_{SO}^R = -\frac{\alpha}{\hbar}\vec{\sigma}[\vec{n}_z \times \vec{p}] = -\frac{1}{\hbar}\vec{S}\vec{\Delta}; \quad \vec{S} = \frac{\hbar}{2}\vec{\sigma}; \quad \vec{\Delta} = \frac{2\alpha}{\hbar}[\vec{n}_z \times \vec{p}], \tag{7.105}$$

where $\vec{n}_z$ is the unit vector along the z-axis. Since the vector $\vec{\Delta}$ introduced by the relation (7.105) plays the role of the dynamic magnetic field to be followed by the electron spin vector, we write down the Bloch equations

$$\frac{d\hbar\vec{S}(t)}{dt} = [\vec{S}(t) \times \vec{\Delta}]$$

for the spin components. Indeed, this equation describes an extremely complex dynamics. As a result of the regular dynamics, the components of the vector $\vec{k}$ rotate in the xy-plane. Consequently, the vector $\vec{\Delta}$ rotates as well. However, this dynamic is of no interest since it does not lead to the appearance of spin transport. Therefore, further consideration requires restricting oneself to only the irregular dynamics of the vector $\vec{\Delta}$ due to the applied external field:

$$\dot{\Delta}_x = -\frac{2\alpha}{\hbar}\dot{p}_y = 0; \quad \dot{\Delta}_y = \frac{2\alpha}{\hbar}\dot{p}_x = -\frac{2e\alpha}{\hbar}E_x. \tag{7.106}$$

For simplifying the representation, we replace the vector $\vec{S}$ by the spin vector $\vec{s}$ measured in units of $\hbar/2$ and, for definiteness, take the initial value $s_x(0) = 1$.

Let us trace the dynamics of the spin components upon applying the electric field. For the y- and x-components, we have

$$\frac{\hbar ds_y}{dt} = s_z\Delta_x; \tag{7.107}$$

$$\frac{\hbar ds_z}{dt} = \Delta_y - s_y\Delta_x. \tag{7.108}$$

Next, we differentiate (7.108) over time and substitute the equation (7.107) in the resulting expression. As a result, the second-order inhomogeneous equation can be written relative to the quantity s_z as

$$\frac{d^2 s_z}{dt^2} + s_z \frac{\Delta_x^2}{\hbar^2} = \frac{\dot{\Delta}_y}{\hbar}. \tag{7.109}$$

The general solution to this equation is a term that oscillates with the frequency $\Delta_x/\hbar$ and is independent on the applied electric field and a term proportional to E_x:

$$s_z = \frac{\hbar \dot{\Delta}_y}{\Delta_x^2}. \tag{7.110}$$

The spin Hall effect follows from the equation (7.110). When a Bloch electron moves in the momentum space, its spin orientation changes to follow the momentum-dependent effective field; in addition, it acquires the momentum-dependent static z-component of spin magnetization. The expression (7.110) allows us to write the component of the spin current $J_{s\,zy}$ in the direction of the y-axis caused by the electric field E_x.

$$J_{s\,zy} = \int \frac{d^2\vec{p}}{(2\pi\hbar)^2} f_0(\vec{p}) \frac{\hbar s_z p_y}{2m}. \tag{7.111}$$

Since the quantity $s_z \sim \dot{\Delta}_y \sim E_x$, the formula (7.111) can really explain the origin of the intrinsic spin Hall effect caused by the Rashba spin–orbit interaction [52]. The integration is here performed over the Fermi surface of a two-dimensional system.

Nature of the external spin Hall effect

In addition to the mechanism of generation of the spin Hall effect considered above, that is associated with the splitting of the spectrum of current carriers by the spin–orbit interaction, an external spin Hall effect is also possible, as noted earlier, due to the manifestation of the spin–orbit interaction when scattering on charged impurity centers in conductors.

The mechanism of the occurrence of the spin Hall effect was proposed by Dyakonov and Perel in 1971 and rediscovered by Hirsch in 1999.

Earlier, back in 1929, Mott discovered the dependence on the spin polarization of the scattering cross-section of free relativistic electrons on heavy atoms. This effect is still used in high-energy physics to measure the degree of polarization of electron beams (Mott detectors) and is known as the *Mott skew scattering mechanism*.

The origin of the spin-polarization dependence of the electron scattering angle by charged impurity can be understood from Figure 7.10 [53].

Let us examine a two-dimensional conductive medium having negatively charged centers, one of which is depicted in the figure. When an electron moves in the field of the impurity center in the reference frame associated with the electron, along with

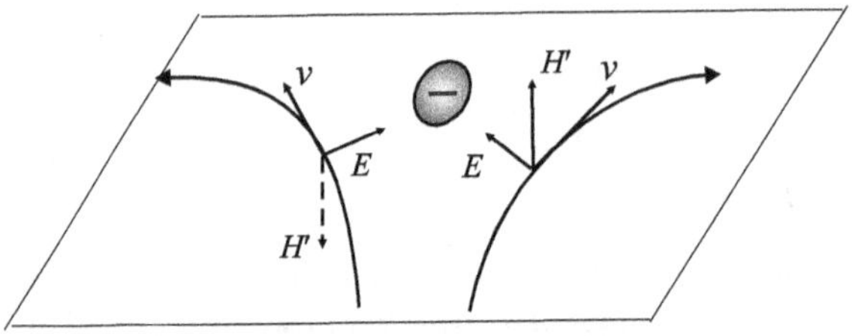

Figure 7.10: Scheme of electron scattering by a negatively charged center: electron's spin interacts with the magnetic field $\vec{H}' \sim [\vec{E} \times \vec{v}]$ perpendicular to the trajectory plane of the electron. The magnetic field has the opposite directions for electrons bouncing apart left and right.

the electric field, a magnetic field $\vec{H}'$ (7.85) arises. The latter's direction depends on at which side of the center the electron's trajectory is located. From the formula (7.85), it follows that the vector $\vec{H}'$ is perpendicular to the plane of the conductor. The magnetic field arisen determines a number of interrelated phenomena.

Firstly, as has been repeatedly discussed, the Zeeman spin energy of an electron in a dynamically changing field $\vec{H}'$ represents the contribution of the spin–orbit interaction to the electron energy, and this contribution depends on the spin orientation. The resulting magnetic field being inhomogeneous (due to the dependence of the electric field $\vec{E}$ on both the electron-center distance and varying electron velocity $\vec{v}$), a force acting on an electron's spin magnetic moment in this field depends on the orientation of the electron magnetic moment. This fact ultimately explains the dependence of the angle of the electron scattering by the charged impurity center on the electron spin orientation.

If a flow of electrons hitting the impurity center is polarized due to an intrinsic magnetic field of a ferromagnet and the electron spins are aligned along the direction of magnetization, there will be an asymmetry in the scattering and the electrons mostly deviate to one of the faces of the sample, which explains the anomalous Hall effect.

Secondly, the electron magnetic moment that had before colliding with an impurity some orientation not coinciding with the direction $\vec{H}'$ of begins precessing around $\vec{H}'$ during the collision process. The spin angle rotation upon scattering depends on the parameters of the collision. Such a mechanism of changing the orientation of the electron's spin upon scattering by a charged impurity underlies the Elliott–Yafet spin relaxation.

Finally, spin-up and spin-down unpolarized electrons incident on the impurity center deviate differently due to the scattering asymmetry. Indeed, the electron spin adjusts to the direction of the field $\vec{H}'$ for the scattering time. Consequently, under the Lorentz force, the negatively charged particles moving along a trajectory rightward from the force center scatter mainly to the right, while the particles moving leftward from the center deflect to the left. However, the electrons scattered rightward and leftward relative to the incoming flow have, respectively, a predominant spin-up and spin-down orientation. This semi-classical approach elucidates the nature of the spin Hall effect.

In the literature, a further mechanism for the external spin-Hall effect is often mentioned, the so-called *side jump mechanism* proposed by Berger in 1970. This mechanism takes into account the displacement of the center of a wave packet upon scattering by impurities but significantly changes neither the amplitude of skew scattering cross-sections nor the scattering angle measured at large distances.

7.3.5 The Berry phase. Basic definitions

Consider a quantum mechanical system described by the Hamiltonian $\hat{H}(R)$ dependent on the set of parameters $R = R_1, R_2, \ldots, R_k$. In the general case, these parameters $R_1, R_2, \ldots, R_k$ may be time-dependent, but this dependence must be adiabatically slow, so that, for each moment in time, the eigenvalues $E_n(R)$ and the eigenfunctions $|\psi_n(R)\rangle$ of the Hamiltonian $\hat{H}(R)$ can be defined

$$\hat{H}(R)\big|\psi_n(R)\big\rangle = E_n(R)\big|\psi_n(R)\big\rangle. \tag{7.112}$$

The adiabaticity criterion can be formulated as follows: if a perturbation is periodic with a frequency ω, this frequency should not cause transitions between any the nearest levels of the system. Let us analyze now the non-stationary Schrödinger equation

$$i\hbar\frac{d\Psi(t)}{dt} = \hat{H}(t)\Psi(t). \tag{7.113}$$

To start with, we expand the function $\Psi(t)$ in a series using the Hamilton operator's eigenfunctions $|\psi_n(t)\rangle$ as basic functions:

$$\Psi(t) = \sum_n A_n(t)\big|\psi_n(t)\big\rangle.$$

Substituting this expansion into the non-stationary Schrödinger equation, we arrive at

$$i\hbar\sum_n\left\{\frac{dA_n}{dt}\big|\psi_n(t)\big\rangle + A_n\frac{d}{dt}\big|\psi_n(t)\big\rangle\right\} = \sum_n E_n(t)A_n(t)\big|\psi_n(t)\big\rangle. \tag{7.114}$$

If there were no dependence of the Hamiltonian $\hat{H}(t)$ on the time t, then the coefficients $A_n(t)$ would have a rapidly oscillating dependence $A_n(t) \sim e^{i/\hbar E_n t}$.

To eliminate the fast time dependence, we need to perform the substitution

$$A_n(t) = c_n(t)\exp\left\{-i/\hbar\int_0^t E_n(\tau)\,d\tau\right\}$$

to the equation (7.114). As a result, we get the equation for the coefficients $c_n(t)$,

$$\frac{dc_n}{dt} = -\sum_m c_m(t) \exp\left\{-i/\hbar \int_0^t [E_m(\tau) - E_n(\tau)]d\tau\right\} \left\langle \psi_n(t) \middle| \frac{d\psi_m(t)}{dt} \right\rangle. \tag{7.115}$$

In deriving this expression, we have used the orthonormality of the instant basis $\langle \psi_n(t)|\psi_m(t)\rangle = \delta_{nm}$. The right side of the expression (7.115) is small because it contains derivatives of slowly (adiabatically) varying functions of the instant basis.

Let us now apply the adiabatic theorem, the essence of which is that if we put the system in a certain stationary state $\psi_m(0)$, then in the adiabatic limit $t \to \infty$, the wave function of the non-stationary problem will follow the instantaneous state of the given Hamiltonian and at time t may differ from the instantaneous eigenfunction

$$\hat{H}(t)|\psi_n(t)\rangle = E_n(t)|\psi_n(t)\rangle$$

is only a certain phase factor. In other words, with an adiabatic perturbation transitions between stationary states are unlikely. Therefore, in the right part (7.115), the summation sign should be omitted, writing

$$\frac{dc_n}{dt} = ic_n(t)\left\langle \psi_n(t) \middle| i\frac{d}{dt} \middle| \psi_n(t) \right\rangle. \tag{7.116}$$

Equation (7.116) is easy to integrate:

$$c_n(t) = c_n(0) \exp\left(i\gamma_n(t)\right); \quad \gamma_n(t) = i\int_0^t dt' \left\langle \psi_n(t') \middle| \frac{d}{dt'} \middle| \psi_n(t') \right\rangle. \tag{7.117}$$

The quantity $\gamma_n(t)$ is called the *Berry phase.*

Next, we explore the properties of this quantity. Firstly, it should be emphasized that the Berry phase is a real quantity. You can verify this by simply differentiating the normalization condition of the function $\psi_n(t)$ in time.

Secondly, if the Hamiltonian and the wave functions in (7.117) depend on time by means of a parameter R, the expression for the Berry phase can be written in the form of a curvilinear integral in the space of these parameters

$$\gamma_n = i\int dt \langle \psi_n(R)|\nabla|\psi_n(R)\rangle \frac{dR}{dt} = \int_C A_n(R)\, dR;$$

$$A_n(R) = i\langle \psi_n(R)|\nabla|\psi_n(R)\rangle. \tag{7.118}$$

The quantity $A_n(R)$ is called *the Berry connection*, and the Berry phase is a curvilinear integral in space parameters. From the above representation of the Berry phase, it is seen that it is utterly irrelevant how the parameters depend on time. However, it matters that they must change adiabatically. The magnitude of the Berry phase depends on the shape of trajectories in the parameter space. For this reason, the Berry phase is sometimes referred to as *a geometric phase*.

Finally, the Berry connection and Berry phase are not gauge-invariant quantities. The set of the functions $|\psi_n(R)\rangle$ introduced above is ambiguous. Indeed, we can always go over to an alternative set by performing the unitary transformation $|\tilde{\psi}_n(R)\rangle \to e^{i\varphi(R)}|\psi_n(R)\rangle$. In doing so, the Berry connection and Berry phase are modified as follows:

$$\tilde{A}_n(R) = A_n(R) - \nabla\varphi; \quad \tilde{\gamma}_n(R) = \gamma_n(R) + \varphi(R_f) - \varphi(R_i), \tag{7.119}$$

where $\varphi(R_f)$ and $\varphi(R_i)$ are the values of phase at the finishing and starting points of the trajectory in the parameter space. If the start and end points coincide ($R_f = R_i$), for example, when the parameters change cyclically, then the Berry connection and Berry phase are gauge-invariant and physically observable quantities. Therefore, it is for cyclic trajectories $R(t)$ that the Berry phase concept should be utilized.

The Berry connection in its properties resembles the vector potential in electrodynamics and, therefore, is often called *the Berry vector potential*. In particular, we can use Stokes' theorem and write an integral along a contour in the formula (7.118) as an integral over a surface spanned on this contour

$$\gamma_n = \int d\vec{S}\vec{\Omega}_n(R); \quad \vec{\Omega}_n(R) = \vec{\nabla}_R \times \vec{A}_n(R). \tag{7.120}$$

The quantity $\vec{\Omega}_n(R)$ introduced in this way bears the name of *the Berry curvature*.

Problem 7.4. Find the value of the connection and Berry curvature for a two-level system that is the spin of an electron in an adiabatically varying magnetic field.

Solution. The Hamiltonian of the system at hand has the form

$$\hat{H} = -\vec{h}(t)\vec{\sigma}, \tag{7.121}$$

where $\vec{h}(t)$ is a quantity proportional to the strength of the external magnetic field, with its amplitude depending on the adiabatic variable parameter t, $\vec{\sigma}$ are the Pauli matrices. The spin following the slowly cyclically changing magnetic field, the original state is reached at the end of the cycle.

To solve the problem, we will switch to a spherical coordinate system and find the components of the vector $\vec{h}$: $h\sin\theta\cos\varphi$, $h\sin\theta\sin\varphi$, $h\cos\theta$. Next, we represent the problem's Hamiltonian in explicit form, substituting the components of the Pauli matrices and the vector $\vec{h}$

$$\hat{H} = -h\begin{vmatrix} \cos\theta & \sin\theta e^{-i\varphi} \\ \sin\theta e^{i\varphi} & -\cos\theta \end{vmatrix}. \tag{7.122}$$

It is straightforward to check by simple substitution that the operator's own functions that meet the eigenvalues $E_{\pm} = \pm h$ are the functions

$$|u_+\rangle = \begin{vmatrix} e^{-i\varphi}\cos\theta/2 \\ \sin\theta/2 \end{vmatrix}; \quad |u_-\rangle = \begin{vmatrix} e^{-i\varphi}\sin\theta/2 \\ -\cos\theta/2 \end{vmatrix}. \tag{7.123}$$

There are two parameters such as the θ and φ angles, on which the wave function depends. The wave functions do not depend on the amplitude of the magnetic field. We shall find the components of the Berry connection (vector potential). Exploiting the definition (7.118), for the lower level energy, we get

$$\begin{aligned} A_\theta &= i\left\langle u_-\left|\frac{d}{d\theta}\right|u_-\right\rangle = 0, \\ A_\varphi &= i\left\langle u_-\left|\frac{d}{d\varphi}\right|u_-\right\rangle = \sin^2\theta/2 = \frac{1}{2}(1-\cos\theta). \end{aligned} \tag{7.124}$$

Determine the Berry curvature Ω and the Berry phase γ. The components of the vector $\vec{\Omega}$ can be computed from the definition (7.120):

$$\Omega_h = \frac{\partial A_\varphi}{\partial\theta} - \frac{\partial A_\theta}{\partial\varphi} = \frac{1}{2}\sin\theta; \quad \Omega_\theta = 0; \quad \Omega_\varphi = 0. \tag{7.125}$$

It should be underscored that the Berry curvature is an invariant quantity and does not change under gauge transformations similarly to the vector of induction of the magnetic field in electrodynamics, determined by the rotor of the vector potential.

The Berry phase in this problem appears as an integral over a closed contour

$$\gamma = \frac{1}{2}\oint(1-\cos\theta)d\varphi \tag{7.126}$$

and is equal to the solid angle, at which the trajectory from the unit sphere center in the parameter space is visible. To be convinced of this, we suppose that the trajectory of cyclically varying the parameters is such that $\theta = \text{const}$. In this case, $\gamma = \pi(1-\cos\theta)$ and is exactly equal to the solid angle, at which the trajectory from the unit sphere center is seen. This remark confirms the geometric meaning of the Berry phase.

Let us deduce another expression for the Berry curvature in the original Cartesian coordinate system, where the vector $\vec{h}$ has the components h_x, h_y, h_z. A Jacobian of transformation from a Cartesian coordinate system to a spherical one, as is known, is determined by the expression $D = h^2\sin\theta$. Therefore, dividing the result of $\vec{\Omega}$ (7.125) by the Jacobian of transformations D yields the Berry curvature in a Cartesian coordinate system:

$$\vec{\Omega}(h) = \frac{1}{2}\frac{\vec{h}}{h^3}. \tag{7.127}$$

Thus, the Berry curvature vector in this problem coincides in direction with an instantaneous value of the external field and inversely proportional to the square of the amplitude of this field.

7.3.6 The Berry phase for Bloch electrons

In this section, we will outline the basics of applying the Berry phase concept to the consideration of the dynamics of band electrons in a periodic crystal field. The issues raised here are described in more detail in [54].

The previous section discussed some of the propositions of the concept of the Berry phase for a system described by a Hamiltonian depending on adiabatically varying parameters. Consider how the ideas for studying the dynamics of electrons in crystalline solids can be implemented. As it turns out, the periodic structure of crystals provides a natural platform for effects associated with the Berry phase and Berry curvature to arise. We write down the Hamiltonian $\hat{H}$ of a quasi-free electron in a periodic crystal field $V(\vec{r})$

$$\hat{H} = \frac{\vec{p}^2}{2m} + V(\vec{r}); \quad V(\vec{r} + \vec{a}) = V(\vec{r}), \tag{7.128}$$

where $\vec{a}$ is a vector of the Bravais lattice. The eigenfunctions of this Hamiltonian are well-known Bloch's functions $\psi_{n\vec{q}}(\vec{r})$ meeting the conditions

$$\psi_{n\vec{q}}(\vec{r} + \vec{a}) = e^{i\vec{q}\vec{a}}\psi_{n\vec{q}}(\vec{r}),$$

where n is the band index, $\vec{q}$ is the wave vector of an electron in the Brillouin zone. Thus, the system is described by the $\vec{q}$-independent Hamiltonian with a $\vec{q}$-dependent boundary condition. Using the general Berry phase formalism, we perform a unitary transformation to obtain the $\vec{q}$-dependent Hamiltonian:

$$H(\vec{q}) = e^{-i\vec{q}\vec{r}}\hat{H}e^{i\vec{q}\vec{r}} = \frac{(\vec{p} + \hbar\vec{q})^2}{2m} + V(\vec{r}). \tag{7.129}$$

The eigenfunctions of the Hamiltonian (7.128) must be subjected to the same unitary transformation, as a result of which the Bloch wave function must be multiplied by $e^{-i\vec{q}\vec{r}}$.

As a result, we find that the eigenfunction of the Hamiltonian (7.129) is the periodic part of the Bloch function $u_{nq}(\vec{r})$. This result shows that the wave vector $\vec{q}$ plays the role of the adiabatic parameter for band electrons.

If the wave vector changes adiabatically so that the electron's motion in q-space should occur along a closed trajectory, the Berry phase is nonzero and, according to (7.118), can be given by the integral

$$\gamma = i\oint_C \langle u_{n\vec{q}}|\vec{\nabla}_{n\vec{q}}|u_{n\vec{q}}\rangle d\vec{q}. \tag{7.130}$$

This integral is taken along a closed curve C located inside the Brillouin zone. Such a closed trajectory of the motion in the q-space can be created, for example, by including an external magnetic field, as a result of which the electron will make a cyclic movement in a cyclotron orbit. Similarly, using the formulas (7.118) and (7.120), we can write down an expression for the Berry curvature of Bloch electrons:

$$\vec{\Omega} = [\vec{\nabla}_{\vec{q}} \times i\langle u_{n\vec{q}}|\vec{\nabla}_{\vec{q}}|u_{n\vec{q}}\rangle]. \tag{7.131}$$

Recall that formulas (7.130) and (7.131) contain the angular brackets $\langle| \dots |\rangle$ that stand for quantum mechanical averages involving integration in the coordinate space.

Anomalous velocity of Bloch electrons

Earlier, an adiabatic theorem was formulated, according to which, with an adiabatic change in the parameters of the Hamiltonian, the system being in some stationary state ψ_n will remain in this state indefinitely, and its wave function will only change by a certain phase factor.

Consider again the equation for the expansion coefficients of the wave function that satisfies the non-stationary Schrödinger equation (7.115). Let the system be in the state $c_n(0) = 1$, and $c_m(0) = 0$ at the initial moment of time for all $m \neq n$. In virtue of the adiabatic theorem $\dot{c}_n = 0$ but $\dot{c}_m \neq 0$, and from (7.115) it follows that

$$\frac{dc_m}{dt} = -\exp\left\{-i/\hbar\int_0^t [E_n(\tau) - E_m(\tau)]d\tau\right\}\left\langle \psi_m(t)\middle|\frac{d\psi_n(t)}{dt}\right\rangle. \tag{7.132}$$

This equation can be integrated because the expression in angle brackets changes slowly compared to the rapidly oscillating exponential term. As a result, we get

$$c_m(t) = -i\hbar\frac{\langle\psi_m|\frac{d\psi_n}{dt}\rangle}{E_n - E_m}\exp\left\{-i/\hbar\int_0^t [E_n(\tau) - E_m(\tau)]d\tau\right\}. \tag{7.133}$$

The validity of the formula (7.133) is easiest to verify by differentiating over time the left and right sides of (7.133). In doing so, we bear in mind that the expression in angular brackets changes slowly, and the term with this derivative can be neglected. If we substitute the resulting expression in the expansion of the wave function, which is the solution of the non-stationary Schrödinger equation, (7.113),

$$\Psi(t) = \sum_l c_l(t)\exp\left\{-i/\hbar\int_0^t E_l(\tau)\,d\tau\right\}|\psi_l(t)\rangle$$

and take into account that for $l = n$ $c_l = 1$, and for $l \neq n$ the coefficients are determined by the formula (7.133), we get the wave function $\Psi(t)$ taking into account the first-order correction (the second term in the curly bracket), which determines the effect of the adiabatic perturbation:

$$\Psi(t) = \exp\left\{-i/\hbar\int_0^t E_n(\tau)d\tau\right\}\left\{|\psi_n\rangle - i\hbar\sum_{m\neq n}|\psi_m\rangle\frac{\langle\psi_m|\frac{d\psi_n}{dt}\rangle}{E_n - E_m}\right\}. \tag{7.134}$$

Note that when calculating the quantum averages, the exponential factor in the formula (7.134) is canceled.

Let us determine the influence of the correction derived to the wave function under an adiabatic perturbation on dynamic characteristics of the band electrons. Consider a system described by the Hamiltonian (7.129) and being impacted by an additional time-dependent adiabatic perturbation without violating translational invariance of the Hamiltonian. As a result, the system acquires two equivalent parameters $\vec{q}$, and t on which the resulting Hamiltonian depends.

The electron velocity operator in the coordinate representation, as known, is defined by the expression $\hat{\vec{v}} = 1/i\hbar[\vec{r}, \hat{H}]$. To write the expression for the velocity $\vec{v}(q)$ in the representation in which the Hamiltonian (7.129) is written, you need to perform a unitary transformation of the commutator. Taking advantage of the properties of a commutator under the unitary transformation, for the velocity operator $\hat{\vec{v}}(q)$, we have

$$\hat{\vec{v}}(q) = \frac{1}{i\hbar} e^{-i\vec{q}\vec{r}} [\vec{r}, \hat{H}] e^{i\vec{q}\vec{r}} = \frac{\partial \hat{H}(\vec{q}, t)}{\hbar \partial \vec{q}}. \tag{7.135}$$

Let us calculate the average velocity for the states of electrons in the n band, subject to the first-order correction (7.134)

$$\vec{v}_n(q) = \frac{\partial \varepsilon_n}{\hbar \partial \vec{q}} - i \sum_{m \neq n} \left\{ \frac{\langle u_n | \frac{\partial \hat{H}}{\partial \vec{q}} | u_m \rangle \langle u_m | \frac{\partial}{\partial t} | u_n \rangle}{\varepsilon_n - \varepsilon_m} - \text{c.c.} \right\}. \tag{7.136}$$

This equation contains the term c.c., which means a complex conjugate summand. In deducing (7.136), we have taken into account that the eigenfunctions of the Hamiltonian (7.136) in the q-representation are periodic parts of the Bloch function u_n.

Using the Hermitian property of the Hamilton operator, we transform the factor in the curly brackets of (7.136), which includes the derivative of the Hamiltonian

$$\begin{aligned} \left\langle u_n \left| \frac{\partial \hat{H}}{\partial \vec{q}} \right| u_m \right\rangle &= \frac{1}{i} \{ \langle u_n | e^{-i\vec{q}\vec{r}} \vec{r} e^{i\vec{q}\vec{r}} \hat{H} | u_m \rangle \\ &\quad - \langle u_n | \hat{H} e^{-i\vec{q}\vec{r}} \vec{r} e^{i\vec{q}\vec{r}} | u_m \rangle \} = (\varepsilon_m - \varepsilon_n) \left\langle u_n \left| \frac{\partial}{d\vec{q}} \right| u_m \right\rangle \end{aligned} \tag{7.137}$$

and simplify the expression for the velocity of band electrons under an adiabatic perturbation

$$\vec{v}_n(q) = \frac{\partial \varepsilon_n}{\partial \hbar \vec{q}} - i \left\{ \left\langle \frac{\partial u_n}{\partial \vec{q}} \middle| \frac{\partial u_n}{\partial t} \right\rangle - \left\langle \frac{\partial u_n}{\partial t} \middle| \frac{\partial u_n}{\partial \vec{q}} \right\rangle \right\}. \tag{7.138}$$

For obtaining the above result, it has been assumed that, for arbitrary operators A and B, the matrix relation

$$\sum_{m \neq n} A_{nm} B_{mn} = (AB)_{nn}$$

is fulfilled.

According to the definitions (7.118) and (7.120), the expression in curly brackets in (7.138) represents the Berry curvature in space of the $\vec{q}$, t parameters, whereas the formula for the velocity of Bloch electrons upon adiabatic variation of the Hamiltonian parameters takes the simple form

$$\vec{v}_n(\vec{q}) = \frac{\partial \varepsilon_n}{\partial \hbar \vec{q}} - \vec{\Omega}^n_{qt}; \quad \vec{\Omega}^n_{qt} = i\left\{\left\langle \frac{\partial u_n}{\partial \vec{q}} \middle| \frac{\partial u_n}{\partial t} \right\rangle - \left\langle \frac{\partial u_n}{\partial t} \middle| \frac{\partial u_n}{\partial \vec{q}} \right\rangle\right\}. \tag{7.139}$$

7.3.7 Dynamics of Bloch electrons in an electric field

The formula (7.139) can be applied to the case when the adiabatic perturbation is an external homogeneous electric field $\vec{E}$. Since the setting of a uniform electric field using a coordinate-dependent potential $\varphi(\vec{r})$ violates the translational invariance of the Hamiltonian, we define a homogeneous electric field using the vector potential of the electromagnetic field $\vec{A}(t)$, which is homogeneous in space and does not violate the translational invariance of the Hamiltonian ($\vec{E} = -1/c\dot{\vec{A}}$). Consequently, with the aid of formulas (5.116)–(5.118), the Hamiltonian of an electron in the periodic potential field $V(\vec{r})$ of a crystal and an external uniform electric field may be written as

$$\hat{H} = \frac{[\vec{p} - e/c\vec{A}(t)]^2}{2m} + V(\vec{r}). \tag{7.140}$$

Passing in this Hamiltonian to the q-representation using the unitary transformation (7.129), we get a Hamiltonian that depends on the parameters $\vec{q}$, t:

$$\hat{H}(\vec{k}), \quad \vec{k} = \vec{q} - \frac{e}{c\hbar}\vec{A}(t),$$

whose properties were discussed earlier.

Using the general formula (7.139), we find expressions for the velocity of Bloch electrons and the Berry curvature for a particular case, when the adiabatic perturbation is as an external electrical field. Let us perform the following substitutions of derivatives in the formula (7.139):

$$\frac{\partial u_n(\vec{k})}{\partial q_\alpha} = \frac{\partial u_n(\vec{k})}{\partial k_\alpha}; \quad \frac{\partial u_n(\vec{k})}{\partial t} = \frac{e}{\hbar}E_\alpha \frac{\partial u_n(\vec{k})}{\partial k_\alpha}.$$

The summation is carried out over the repeated indices. When writing the last formula, it is taken into account that the electric field does not violate the translational symmetry and the wave vector $\vec{q}$ is an integral of motion. Then $\dot{\vec{k}} = -e/c\hbar\dot{\vec{A}}(t) = e/\hbar\vec{E}$. After all of the changes made, the expression for the electron velocity in the band labeled with n upon adiabatically switching-on the external field boils down to

$$\vec{v}_n(\vec{k}) = \frac{\partial \varepsilon_n}{\hbar \partial \vec{k}} + \frac{e}{\hbar}[\vec{E} \times \vec{\Omega}_n(\vec{k})]; \tag{7.141}$$

$$\Omega_n^{\alpha}(\vec{k}) = \varepsilon_{\alpha\beta\gamma}\frac{\partial A_\gamma(\vec{k})}{\partial k_\beta}; \quad A_\gamma = i\left\langle u_n(\vec{k})\left|\frac{\partial}{\partial k_\gamma}\right|u_n(\vec{k})\right\rangle, \tag{7.142}$$

where $\varepsilon_{\alpha\beta\gamma}$ is the third-rank unit antisymmetric tensor. To test the possibility of such a representation, we invite the reader, as an exercise, to find the electron velocity x-, y-, and z-components, relying on the formulas (7.139) and (7.141) and compare the results.

7.3.8 The Berry curvature and the spin Hall effect

The semi-phenomenological formula (7.141) obtained above for the velocity of band electrons allows one to immediately find both an expression for the J_y-components of an electric current upon switching on an electric field pointing along the x-axis and the σ_{xy}-component of electrical conductivity. Let a $2D$-sample be under no magnetic field. Then the simultaneous usage of the standard definition σ_{xy} of an electric current and the expression (7.141) for the velocity leads immediately to the expressions

$$\begin{aligned} J_y/E_x &= -\frac{e^2}{\hbar}\int\frac{dk_x dk_y}{(2\pi)^2} f_0(\varepsilon_{\vec{k}})\Omega_z(k_x k_y); \\ \Omega_z(k_x, k_y) &= \frac{\partial A_y(\vec{k})}{\partial k_x} - \frac{\partial A_x(\vec{k})}{\partial k_y}. \end{aligned} \tag{7.143}$$

The first summand in the formula (7.141) makes no contribution due to the symmetry considerations.

Thus, computing the Berry curvature for some model of a conductive crystal can explain the intrinsic anomalous Hall effect.

Let us address again the Rashba Hamiltonian (7.100) and determine the normalized eigenfunctions of this Hamiltonian. As is well known, these functions are the product of the coordinate part of the wave function and the spin part as a two-component spinor. We look into only the spin part $\chi(\vec{k})$ of the wave function of the Hamiltonian (7.100) and write it in a more convenient form than (7.101). For this purpose, we distinguish explicitly a normalization factor and set one of the constants c_1, c_2 equal to the unit:

$$\chi(\vec{k}) = \frac{1}{\sqrt{2}}\begin{vmatrix} 1 \\ c(\vec{k}) \end{vmatrix}, \tag{7.144}$$

where $c(\vec{k})$ is some constant to be determined, $\vec{k}$ has k_x- and k_y-components.

We write down explicitly the Schrödinger equation for the spin part of the wave function (7.100):

$$\alpha\left\{k_y\hat{\sigma}_x\begin{vmatrix} 1 \\ c \end{vmatrix} - k_x\hat{\sigma}_y\begin{vmatrix} 1 \\ c \end{vmatrix}\right\} = \alpha\begin{vmatrix} & k_y + ik_x \\ k_y - ik_x & \end{vmatrix}\begin{vmatrix} c \\ 1 \end{vmatrix} = E_R\begin{vmatrix} 1 \\ c \end{vmatrix}. \tag{7.145}$$

Here E_R is the spin part of the Rashba Hamiltonian eigenvalues. The matrix equation obtained implies that

$$\alpha(k_y + ik_x)c = E_R; \quad \alpha(k_y - ik_x) = \frac{E_R}{c};$$
$$E_R^2 = \alpha^2(k_x^2 + k_y^2); E_R = \lambda\alpha k; \quad c = \lambda\frac{k_y - ik_x}{k}; \quad \lambda = \pm 1. \tag{7.146}$$

Thus, we have arrived at the result (7.103) previously found for the dispersion law, and the eigenwave functions of the Hamiltonian (7.100) $\psi(\vec{k}, \lambda)$ can be represented as

$$\psi(\vec{k}, +) = \frac{e^{i\vec{k}\vec{r}}}{\sqrt{2}}\begin{vmatrix} 1 \\ \frac{k_y - ik_x}{k} \end{vmatrix}; \quad \psi(\vec{k}, -) = \frac{e^{i\vec{k}\vec{r}}}{\sqrt{2}}\begin{vmatrix} 1 \\ -\frac{k_y - ik_x}{k} \end{vmatrix}. \tag{7.147}$$

Furthermore, we calculate the Berry connection and the Berry curvature appearing in the formula (7.143). Given that, in calculating the Berry connection, the formula (7.142) includes only translation-invariant parts of the Bloch function, for the A_y-components of the Berry connection and z-components of the Berry curvature, we obtain

$$A_x = \frac{\lambda^2 k_y}{2k^2}; \quad A_y = -\frac{\lambda^2 k_x}{2k^2}; \quad \Omega_z = \frac{\partial A_y(\vec{k})}{\partial k_x} - \frac{\partial A_x(\vec{k})}{\partial k_y} = \frac{\lambda^2}{k^2}. \tag{7.148}$$

It should be noted that the Berry curvature is the same for both Rashba spin subbands. Next, we analyze the contribution of a band completely filled with electrons below the Fermi level and exploit the formula (7.143),

$$J_y/E_x = -\frac{e^2}{4\pi^2\hbar}\int dk_x\, dk_y f_0(\varepsilon_{\vec{k}})\left(\frac{\partial A_y}{\partial k_x} - \frac{\partial A_x}{\partial k_y}\right). \tag{7.149}$$

We perform the integration by parts, taking into account that the non-integral term vanishes,

$$J_y/E_x = -\frac{e^2}{4\pi^2\hbar}\int dk_x\, dk_y\left(\frac{\partial f_0}{\partial k_y}A_x - \frac{\partial f_0}{\partial k_x}A_y\right). \tag{7.150}$$

Now, for the above expression, the following stepwise procedure should be adopted to carry out. First, we go over to a polar coordinate system and perform integration over energy after having made the replacement $d\varepsilon = \hbar^2 k dk/m$. Next, we substitute the previously calculated values of A_x, A_y (7.148), accounting for the fact that the energy-derivative of the equilibrium distribution function can be replaced by the δ-function. In the long run, we get

$$J_y/E_x = \sigma_{xy} = -\frac{e^2}{8\pi^2\hbar}\int_0^\infty d\varepsilon\int_0^{2\pi} d\varphi\frac{df_0}{d\varepsilon} = \frac{e^2}{4\pi\hbar}. \tag{7.151}$$

The found value of J_y/E_x for the Rashba spin–orbit interaction model only confirms the well-known result of [54] that the conductivity of σ_{xy} for the anomalous Hall effect in a two-dimensional paramagnet does not depend on the spin–orbit interaction constant and is a certain invariant.

Similarly, an expression for the intrinsic spin Hall effect can be inferred. Let us define the spin Hall conductivity σ_H as the ratio between the flux J_{szy} of the spin z-component in the y-axis direction and the magnitude of the applied electric field E_x. Let us determine the contribution of the Rashba mechanism to the intrinsic spin Hall effect. This contribution differs from the above result σ_{xy}; only the obvious multiplier $\hbar/2e$ (the spin moment $\hbar/2$) is transferred rather than a charge. Finally, we present the following:

$$\sigma_H = J_{szy}/E_x = \frac{e}{8\pi}. \tag{7.152}$$

It is mandatory to emphasize that this section restricts itself to exploring an intrinsic contribution to the anomalous and spin Hall effects for a perfect impurity- and defect-free conductive medium model. As noted earlier, impurity-scattering asymmetry caused by spin–orbit interaction can also give rise to the anomalous and spin Hall effects. The relative magnitude of contributions of intrinsic and external mechanisms of the AXE and SHE is at present far from being clear. However, we have settled on the nature of the intrinsic AHE and SHE for the reason that the nature of the external AHE and SHE is quite well understood and can be explained by traditional methods of accounting for the scattering of band electrons on a charged impurity. These calculations are too involved since the spin–orbit interaction leads to the mixing of states in the conduction and valence bands.

Over the last years, the Berry phase concept has gained widespread use in various research activities of contemporary physics, including topological effects in crystals and low-dimensional structures.

The present chapter includes a small section that discusses the dynamics of band electrons in terms of the Berry phase. This section is solely necessary for the reader to draw attention to booming investigations on transfer processes in solids and accounting for the dynamical laws (7.97) and (7.108) to classically describe the behavior of electrons (particular, in particular, when using the kinetic equations) in systems with no symmetry relative to time-reversal and inversion operations. More detailed information on this issue is in the paper [54].

7.3.9 Physical principles of operation of spintronics devices

Concluding the chapter on getting to know the physical fundamentals spintronics, it is useful, at least briefly, to dwell on practical implementation of microelectronic devices to be dominated by spin transport. First and foremost, among them should be mentioned magnetic field sensors. The fundamental design of a magnetic field sensor essentially coincides with the GMR observation scheme in the CPP-geometry, as outlined in Figure 7.3. The only difference is that one of the ferromagnets is fabricated of a magnetically hard substance, and the second is made of a magnetically soft one that is capable of changing its magnetization orientation in a weak magnetic field.

When an electric current is passed through such a ferromagnetic layer system, due to the GMR effect, the resistance of the structure is highly dependent on the mutual orientation of the magnetization that can change under the influence of a weak external magnetic field. Sensors thus constructed enable one to detect fields with an amplitude of about $\simeq 10^{-8}$–10^{-2} T in times of the order of a nanosecond. GMR sensors are widely used for reading information recorded on magnetic disks. Thanks to this, over the past 20 years, the information storage capacity of the disks has increased by several orders of magnitude.

The other already implemented trend of creating microelectronic spintronics-based units is to produce cells of Magnetoresistive Random-Access Memory (MRAM). The MRAM memory cell is a magnetic tunnel junction (MTJ) as a storage element. The MTJ consists of a thin (about 1 nm) dielectric layer placed between two magnetic layers. One of them (the so-called free layer) can change the orientation of the magnetization under the action of an external magnetic field. The second ferromagnetic exhibits a constant magnetization direction. When a voltage is applied to MTJ, an electric current arises in the structure because the electrons overcome a dielectric barrier due to the tunneling process. The structure can have either low or high resistance. In the first case, the direction of magnetization of the layers coincides, and low structure resistance is interpreted as a value of the bit, equal to one. In the second case, the layer magnetizations are antiparallel (the GMR effect in a tunnel junction is higher than that in an $F/N/F$-structure), and a large resistance is interpreted as a value of the bit equal to zero.

Changing the orientation of the layer of soft magnetic material is achieved by simultaneously passing current through two mutually perpendicular conductors that position the memory cell. As the current flows (the magnetization of the free layer varies), the memory cell records information. The advantages of such a memory are non-volatility and radiation immunity. The disadvantage of MRAM includes rather large energy costs of information recording.

Nowadays, almost all the world's leading electronic companies are working at creating a new memory generation to be less energy-consumed for information recording. This new generation of non-volatile memory is called Spin-Transfer Torque Magnetic Random-Access Memory (STT-MRAM). In the coming years, such a memory can replace both semiconductor memory and regular MRAM memory.

An STT-MRAM memory cell, as much as an MRAM memory cell, contains a tunnel junction located between two magnetic layers with fixed and variable magnetization. Information reading from the cell is produced by passing an electric current through the structure: low resistance is interpreted as a logical "unit" and is realized under parallel layer magnetization. A large resistance of the structure is regarded as a logical "zero" and arises due to antiparallel magnetization of the layers.

The way of recording information (a way to change the magnetization orientation in the free layer) in the STT-MRAM memory cell is completely different. The recording is brought about by passing a spin-polarized current through a tunnel junction to change the orientation magnetization of the free magnetic layer since the spin-polarized current also carries a magnetic moment. STT-MRAM memory modules have already been manufactured by the industry since 2016.

An important direction in the development of spintronics is the creation of various designs of spin transistors. Here we will dwell on only several options for such designs.

The simplest spin transistor is the Johnson transistor. Its construction consists of two ferromagnetic layers separated by a paramagnetic insulation and is the $[F/N/F]$ structure considered earlier, in which the effect of giant magnetoresistance can be realized. As in magnetic field sensors, one of the ferromagnetic layers (an emitter) has a rigidly fixed orientation of magnetization, and the other (a collector) can change the orientation of the magnetization under sufficient weak magnetic fields. The paramagnetic (N) acts as a transistor base.

If a potential difference is created in the emitter-base circuit (the base must have a positive potential relative to the emitter), then as a result of injection, non-equilibrium electrons with a spin orientation parallel to the magnetization of the emitter will arise in the base. The current in the base-collector circuit will depend on the orientation of the magnetization of the collector film, which can be changed by an external magnetic field.

A more functional design is a spin-field transistor, the circuit of which is shown in Figure 7.11. This unit, as well as a conventional field transistor, contains a source, drain, and channel, to which the gate voltage is applied. Ferromagnetics having one direction of magnetization and a flat semiconductor with electron charge carriers are used as source/drain and channel regions, respectively. The current injected into the channel from the source is spin-polarized. The gate voltage initiates an induced magnetic field $\vec{H}'$ in the two-dimensional conductive channel (7.85).

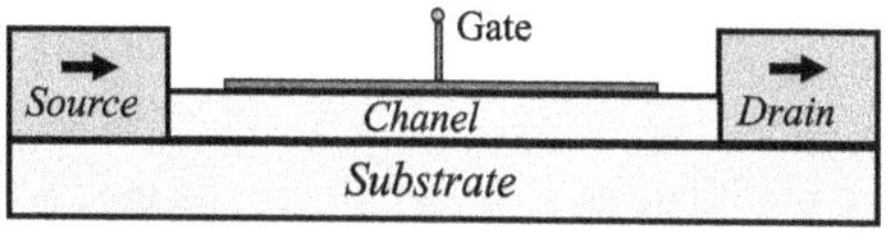

Figure 7.11: Schematic illustration of a spin-field-effect transistor.

When moving in the channel, the electron spins precess around the direction of the field $\vec{H}'$. As a result, during the time of flight t of the electrons along the channel, the mean magnetization vector of injected electrons rotates through the angle $\simeq \omega t$, where ω is the spin precession frequency in the field $\vec{H}'$. By changing the gate voltage, the electron spin orientation at the end of the channel can become opposite to the drain magnetization. In doing so, the electrons do not cross the channel–drain boundary, which leads to high resistance of the transistor. Finally, the electric field of the gate controls the electric resistance of the transistor. As a result, the spin-field-effect transistor behaves like a conventional field-effect transistor with the peculiarity that the magnetization of the source and drain (and hence the transistors' electrical characteristics) is sensitive to an external magnetic field.

More complete information about existing and future spintronics devices can be found in the review [55].

7.4 Problems to the Chapter 7

7.1. Derive an expression for the Peltier coefficient (1.36) within the two-channel Mott model, expressing it in terms of the integrals K_l (4.33) for the case of a highly degenerate electron gas.

7.2. To detect spin polarization, the Hanle effect can be used. Based on the assumptions that the rate of spin pumping into a normal metal is unchanged and the injected spins precess with angular velocity $\omega = g\mu_B H/\hbar$, prove that the stationary average magnetization of the electrons injected into the normal metal is determined by the expression

$$\langle M\rangle = T_2 \frac{dM/dt}{1 + (\omega T_2)^2}$$

and hence the magnetization as a function of the magnetic field H has the shape of a Lorentz curve (see Figure 7.6). Here dM/dt is the spin magnetization pumping rate; T_2 is the relaxation time for transverse components of the spin.
Hint: the average magnetization can be estimated by the formula

$$\langle M\rangle = \lim_{T\to\infty}\left(T\frac{dM}{dt}\right)\frac{1}{T}\int_0^T \cos(\omega t)e^{-t/T_2}\,dt.$$

7.3. Using equations (7.38) and (7.40), obtain the expression (7.53) for the z-coordinate dependence of electrochemical potentials and find the relationship between the coefficients A, B, C, D and a, b, c, d.

7.4. At the separation boundary between two media of an F/N structure, the electrochemical potentials $\overline{\zeta}^i_\gamma$, $\gamma = \uparrow, \downarrow$; $i = F, N$ remain continuous, and the averaged values

$$\overline{\zeta}^i = \alpha^i \overline{\zeta}^i_\uparrow + (1 - \alpha^i)\overline{\zeta}^i_\downarrow$$

suffer a jump

$$\delta\overline{\zeta}_0 = \overline{\zeta}^F_0(0) - \overline{\zeta}^N_0(0).$$

Using the previously obtained formulas (7.63)–(7.66) and the definition for the current polarization factor

$$\beta = [J_{\uparrow}^F(0) - J_{\downarrow}^F(0)]/J,$$

derive the formula (7.72) and estimate the value for the jump $\delta\bar{\zeta}_0$ at the following parameter values: $l_{sf}^N = 5l_{sf}^F$, $\sigma^F = \sigma^N$, $\mathcal{P} = 0,6$. For determining the values of the parameters σ^N, l_{sf}^N, use literature resources.

7.5. Selection of materials to provide a high degree of polarization of injected electrons is an important problem in spintronics. To solve this problem, a resistor model of the F/N structure can be utilized, assuming that the current flows through two connected parallel spin channels with resistances: $R_{\uparrow}^F + R_{\uparrow}^i + R^N/2$ for a channel with spin ↑ and $R_{\downarrow}^F + R_{\downarrow}^i + R^N/2$ for a channel with spin ↓. Here $R_{\uparrow}^i$, $R_{\downarrow}^i$ are spin-dependent resistances of the interface region. Prove that in the resistor model

$$\beta = \frac{R^F}{R^F + R^N + R^i}\frac{\sigma_{\uparrow} - \sigma_{\downarrow}}{\sigma_{\uparrow} + \sigma_{\downarrow}} + \frac{R^i}{R^F + R^N + R^i}\frac{\Sigma_{\uparrow} - \Sigma_{\downarrow}}{\Sigma_{\uparrow} + \Sigma_{\downarrow}},$$

where $R^F = R_{\uparrow}^F + R_{\downarrow}^F$, $\sigma_{\uparrow}$, $\sigma_{\downarrow}$ are conductivities of spin channels of a ferromagnet, $\Sigma_{\uparrow}$, $\Sigma_{\downarrow}$ are conductivities of spin channels in the interface region.
Explore under which conditions a high degree of the polarization can be achieved, if usually $R^F \ll R^F + R^N + R^i$.

7.6. Prove that the velocity of Bloch electrons (7.139) under an external perturbation associated with an external constant homogeneous electric field $\vec{E}$ can be presented in the form

$$\frac{\partial \varepsilon_{n\vec{k}}}{\partial \hbar \vec{k}} + \frac{e}{\hbar}[\vec{E} \times \vec{\Omega}_n(\vec{k})].$$

For the proof, calculate the velocity of the x, y, z-components for expressions (7.139) and (7.141) and compare them.

7.7. Find the eigenvalues and eigenfunctions of a two-dimensional electron gas upon the Dresselhaus spin–orbit interaction. The Hamiltonian of the system has the form

$$\frac{\hbar \vec{k}^2}{2m} - \beta(k_x\sigma_x - k_y\sigma_y),$$

where $\hbar k_i$ are the components of the momentum operator, σ_i are Pauli matrices.

7.8. Using the solutions to the previous problem, find the Berry connection and the Berry curvature for the case of a two-dimensional electron gas with the Dresselhaus spin–orbit interaction.

8 Response of a highly non-equilibrium system to a weak measuring field

8.1 NSO for highly non-equilibrium systems

8.1.1 Set up of the problem. A boundary condition for the NSO

By now the theory of linear response of an equilibrium system to an external mechanical perturbation is well elaborated (see Chapter 5). This theory is used successfully for solving problems of physical kinetics in the systems whose state is weakly perturbed by external influences. Such an approach allows kinetic coefficients to be expressed in terms of equilibrium correlation functions. The correlation functions in turn can be estimated by means of modern methods of statistical mechanics (see Chapters 5 and 6).

The situation is quite different in finding the response of the system being already non-equilibrium to an additional weak measuring field. Still, such problems are solved exclusively by the kinetic equation method [27] (see also Chapter 4) rather than by methods of non-equilibrium statistical mechanics.

The present chapter formulates the theory of a linear response of the non-equilibrium system to a weak measuring field. The theory contains the correct limiting transition to the case of weak non-equilibrium systems and allows one to express the kinetic coefficients via the correlation functions, which are calculated by using non-equilibrium distribution. The calculation of the coefficient of electrical conductivity of a highly non-equilibrium system of electrons is given as an example. Also, it is shown that the results obtained based on the kinetic equation in Chapter 4 for the transport coefficients of the non-equilibrium systems coincide with the data, which the linear response theory yields.

Assume that the system is already in a non-equilibrium state before switching on the measuring field and is described by the non-equilibrium statistical operator (NSO) $\rho^0(t,0)$. As opposed to the NSO-method described in the previous chapter, in this section we will introduce the reader to an alternative form of writing the NSO method proposed by D. N. Zubarev [36]:

$$
\begin{aligned}
\rho^0(t,0) &= \exp\{-\overline{\Phi(t)} - \overline{P^+F(t)}\} \equiv \exp\{\overline{-S_0(t,0)}\},\\
\overline{\Phi(t)} &= \ln \mathrm{Sp}\{\exp\{-\overline{P^+F(t)}\}\}, \quad \overline{S_0(t,0)} = \overline{\Phi(t)} + \overline{P^+F(t)},\\
\overline{P^+F(t)} &= \epsilon \int_{-\infty}^{0} dt_1 \exp\{\epsilon t_1\} \exp\{iLt_1\} P^+F(t+t_1).
\end{aligned}
\tag{8.1}
$$

It is worth paying attention that the notations being used in the formulas (8.1) coincide with the notations adopted in Chapter 6. As before, the operator P^+ denotes the

https://doi.org/10.1515/9783110727197-008

column vector of basic operators as $F(t)$ stands for the row vector of their conjugate thermodynamic forces.

In the previous chapter, the quasi-equilibrium statistical operator $\rho_q(t,0) = \exp\{-S(t,0)\}$ was subjected to the temporal smoothing operation when constructing the non-equilibrium statistical operator. However, now according to the alternative approach (8.1), the NSO is built as a canonical distribution of quasi-integrals of motion. Consequently, the temporal smoothing operation is to be performed for an entropy operator. Earlier, Kalashnikov and Zubarev showed in their work that these two methods of constructing the non-equilibrium statistical distribution were completely equivalent [43].

It is easy to agree that the form (8.1) for the NSO holds true if we change the scheme of constructing the NSO outlined in the previous chapter. Actually, as has been shown, a boundary condition that must be satisfied by the statistical operator when switching on the external field (this point is referred to $-\infty$) plays an important role in constructing the NSO. In this case, the boundary condition helps to choose a certain type of solution of the Liouville equation. In this equation, the time-dependence of physical quantities is to be a functional of quasi-integrals of motion involving the initial quasi-equilibrium distribution ρ_q. Such an idea is effective enough and has been used in the physical kinetics for a long time. Suffice it to remember the method of momenta for solving the kinetic equation. To construct the NSO in the form of (8.1) the similar conditions for $\ln \rho^0(t,0)$ should be written instead of the boundary condition (6.46):

$$\lim_{t_1\to-\infty} \exp\{it_1L\} \ln \rho_q(t+t_1,0) = \lim_{t_1\to-\infty} \exp\{it_1L\} \ln \rho^0(t+t_1,0). \tag{8.2}$$

Repeating the calculations that led us from (6.46) to (6.52), one can get the NSO-form (8.1). It is worth drawing attention to the fact that here another designation $\rho_q(t,0)$ was introduced for the quasi-equilibrium distribution. The overline is used to denote the temporal smoothing operation (see formula (8.1)).

Let us find the Liouville equation for the non-equilibrium statistical operator (NSO) (8.1). Applying the Abel theorem (6.47), the boundary condition (8.2) can be written in the integral form:

$$\lim_{\epsilon\to 0} \epsilon \int_{-\infty}^{0} \exp(\epsilon t_1) e^{iLt_1} \ln \rho_q(t+t_1,0)\, dt_1 = \lim_{\epsilon\to 0} \epsilon \int_{-\infty}^{0} \exp(\epsilon t_1) e^{iLt_1} \ln \rho^0(t+t_1,0)\, dt_1.$$

This result can be can be written more compactly using the notations for the temporal smoothing operation (8.1):

$$\overline{\ln \rho_q(t,0)} = \overline{\ln \rho^0(t,0)}. \tag{8.3}$$

After integrating the right-hand side of the equation (8.3) by parts,

$$\lim_{\epsilon\to 0}\epsilon\int_{-\infty}^{0}\exp(\epsilon t_1)e^{iLt_1}\ln\rho^0(t+t_1,0)\,dt_1$$

$$=\lim_{\epsilon\to 0}\int_{-\infty}^{0}e^{iLt_1}\ln\rho^0(t+t_1,0)\frac{d}{dt_1}\exp(\epsilon t_1)\,dt_1$$

$$=\ln\rho^0(t,0)-\lim_{\epsilon\to 0}\int_{-\infty}^{0}\exp(\epsilon t_1)e^{iLt_1}\left\{\frac{\partial}{dt_1}+iL\right\}\ln\rho^0(t+t_1,0)\,dt_1,$$

we require that $\ln\rho^0(t,0)$ is to be satisfy the Liouville equation in the sense that

$$\lim_{\epsilon\to 0}\int_{-\infty}^{0}dt_1\exp\{(\epsilon+iL)t_1\}\left\{\frac{\partial}{\partial t_1}+iL\right\}\ln\rho^0(t+t_1)=0. \tag{8.4}$$

Then we get

$$\overline{\ln\rho^0(t,0)}=\ln\rho^0(t,0). \tag{8.5}$$

It should be noted that equation (8.4) as well as the vanishing integral (6.51) is a postulate of the theory. This postulate leads to the fact that the NSO (8.1) satisfies an equation with the infinitesimal source on the right side rather than the Liouville equation. Moreover, the source in an idealized form takes into account the contact between the system and a heat bath after switching on the interaction. It also selects retarded solutions of the Liouville equation.

One may stress that the temporal smoothing operation (8.1) being applied to quantity $\ln\rho^0(t,0)$ leaves it unchanged, i. e. this operation has the property of a projection operator (in the sense that the iterated projection does not change results).

Comparing (8.3) and (8.5), we find an explicit expression for the NSO:

$$\overline{\ln\rho_q(t,0)}=\ln\rho^0(t,0). \tag{8.6}$$

Thus, it has been shown how the NSO can be constructed in the alternative form (8.1). It remains only to obtain an equation of motion satisfying the NSO.

For this purpose, it is necessary to differentiate both sides of equation (8.6) over time t. As a result, one gets

$$\lim_{\epsilon\to 0}\epsilon\int_{-\infty}^{0}\exp(\epsilon t_1)e^{iLt_1}\frac{\partial}{\partial t}\ln\rho_q(t+t_1,0)\,dt_1=\frac{\partial}{\partial t}\ln\rho^0(t,0).$$

On the left-hand side of this equation, the derivative over t may be replaced by the derivative over t_1. Then, after integrating the equation by parts and considering the result (8.6) one arrives at the desired equation of motion for $\ln \rho^0(t, 0)$:

$$\left(\frac{\partial}{\partial t} + iL\right) \ln \rho^0(t, 0) = -\epsilon(\ln \rho^0(t, 0) - \ln \rho_q(t, 0)). \tag{8.7}$$

Note that the logarithm $\ln \rho^0(t, 0)$ satisfies the equation (8.7) with sources on the right-hand side rather than the NSO $\rho^0(t, 0)$. The result obtained (8.7) is consistent with the original preposition (8.4). Consequently, the condition (8.4) holds automatically due to the equation of motion (8.7) and the boundary conditions (8.3). Thus, the equations (8.7) and (8.4) are essentially identical.

In the present chapter, the non-equilibrium distribution $\rho^0(t, 0)$ is supposed to be already known. So here there is no necessity to pose the problem of finding the quantities $F(t)$ and average values of the basic operators P^+. In Section 6.2.6, we discussed a way of doing that.

Let the system whose the non-equilibrium state is given by the distribution (8.1) be perturbed by an additional mechanical perturbation $H_{\mathcal{F}}(t) = -A^+\mathcal{F}(t)$. Here A^+ is some operator and $\mathcal{F}(t)$ is the field intensity of the external forces whose response needs to be determined. Then we assume that the perturbation emerges at the moment of time $t = -\infty$ (of course, the infinity in this context is understood as the magnitude is much larger than characteristic relaxation time scales of the problem).

Under the conditions of the external perturbation in the system there arises the new non-equilibrium state, which cannot be generally described in terms of the old basis set of the operators P^+. So this set is to be extended by adding new operators $\mathcal{M}^+$ and new thermodynamic parameters $\varphi(t)$.

We formulate the boundary condition which is satisfied the statistical operator $\rho(t, 0)$, what describes a new non-equilibrium state of the system at $t \to -\infty$. It is clear that the boundary condition (8.2) cannot simply be transferred to this case (see also expression (6.46) in the previous chapter) because the non-equilibrium distribution $\rho(t, 0)$ has to pass into the non-equilibrium distribution $\rho^0(t, 0)$ rather than into the quasi-equilibrium one as it was before.

In order to formulate an appropriate boundary condition a free relaxation $\rho(t, 0)$ should be analyzed when switching off the external action $\mathcal{F}(t)$ at some point $t \to -\infty$. In this situation thermal perturbations, which are described by the functions $\varphi(t)$, do not immediately vanish. They vary slowly with some characteristic relaxation time τ. Once the perturbation $H_{\mathcal{F}}(t) = -A^+\mathcal{F}(t)$ is switched off, some internal field $\mathcal{M}^+\varphi(t)$ acts on the system. Then the equation which is satisfied by $\ln \bar{\rho}(t, 0)$ as $t \to -\infty$ is given by

$$\frac{\partial}{\partial t} \ln \bar{\rho}(t, 0) + \frac{1}{i\hbar}[\ln \bar{\rho}, H + \mathcal{M}^+\varphi(t)] = -\epsilon(\ln \bar{\rho}(t, 0) - \ln \rho^0(t, 0)). \tag{8.8}$$

We have written the equation (8.8) by analogy with equation (8.7). It is possible to do so provided that the internal field acts as a correction to the Hamiltonian. In fact, the equation (8.8) is a postulate of the theory, and we will come back to its substantiation soon after.

Resorting to the approach described in Chapter 6 (see the derivation of the formula (6.115) in Section 6.2.3), the equation (8.8) for logarithm of the non-equilibrium statistical operator can be converted into an integral equation, which after iteration over the small parameter $\mathcal{M}^+\varphi(t)$ in the linear approximation has the form

$$\overline{S(t,0)} = \overline{S_0(t,0)} - \frac{1}{i\hbar}\int_{-\infty}^{0} dt_1 \exp\{(\epsilon + iL)t_1\}[\mathcal{M}^+, \overline{S_0(t+t_1,0)}]\varphi(t), \tag{8.9}$$

where

$$\overline{S_0(t,0)} = -\ln\rho^0(t,0); \quad \overline{S(t,0)} = -\ln\bar{\rho}(t,0).$$

Given the assumption that the functions $\varphi(t)$ are slowly varying functions of time t, in writing the equation (8.9) we have neglected their dependence on t_1. Now we can arrive at the desired boundary condition, presuming that the true distribution in the limit at $t \to -\infty$ must coincide with the result obtained by solving the equation (8.8)

$$\lim_{t_1\to-\infty} \exp\{iLt_1\}\rho(t+t_1,0) = \lim_{t_1\to-\infty} \exp\{iLt_1\}\exp\{-\overline{S(t+t_1,0)}\}, \tag{8.10}$$

where the quantity

$$\bar{\rho}(t,0) \equiv \exp\{-\overline{S(t,0)}\}$$

plays the same role as the quasi-equilibrium distribution performed in the previous chapter.

The expression for the distribution $\bar{\rho}(t,0)$ can be expanded into a series in the parameter $\mathcal{M}^+\varphi(t)$. We restrict ourselves to a linear approximation as far as only linear terms in this parameter are concerned,

$$\bar{\rho}(t,0) = \rho^0(t,0) - \frac{1}{i\hbar}\int_{-\infty}^{0} dt_1 \exp\{(\epsilon + iL)t_1\}[\mathcal{M}^+, \rho^0(t+t_1,0)]\varphi(t). \tag{8.11}$$

It should be reinforced that the distribution (8.11) is not quasi-equilibrium and a similar notation like $\bar{\rho}(t,0)$ is not to be misleading.

However, we have chosen such a long way to obtain the formula (8.11) not by chance, but to acquaint the reader with the alternative method to the NSO. It is much easier to come up with the result (8.11) within the NSO-method, developed in the previous chapter. In fact, such a treatment requires some repetitions, but they will be useful for the reader.

Consider a non-equilibrium system with the distribution

$$\rho^0(t,0) = \epsilon\int_{-\infty}^{0} dt_1 \exp\{(\epsilon + iL)t_1\}\rho_q(t+t_1),$$

where

$$\rho_q(t,0) = \exp\{-S_0(t,0)\}, \quad S_0(t,0) = \Phi(t) + P^+F(t).$$

If the correction $H_{\mathcal{F}}(t) = -A^+\mathcal{F}(t)$ defines the external field acting on the system, the new non-equilibrium state being described by a set of basic operators is produced there. Let the number of basic operators be added to the operators $\mathcal{M}^+$ and thermodynamic forces $F(t)$ be extended to new forces $\varphi(t)$. We assume that the new distribution is given by the operator $\rho(t,0)$.

The question arises of how to find the form of NSO $\rho(t,0)$. Chapter 6 deals with a method that cannot be directly applicable to this case. This is because the non-equilibrium distribution still remains (albeit somewhat modified) when switching off the external measuring field due to the presence of other disturbances which define the initial non-equilibrium state.

We may resort only to common methodology for the NSO-method to derive the distribution $\rho(t,0)$. For that, it is merely required that the analog of the expression (6.46) be properly recorded to define a boundary condition for the NSO.

To obtain such a boundary condition the system's evolution is to be regarded after switching off the external field at the moment of time $t = -\infty$. Denote the statistical distribution of the system arising after switching off the external field by the quantity $\overline{\rho}(t,0)$. Let the equation satisfied by the distribution $\overline{\rho}(t,0)$ have the form

$$\frac{\partial}{\partial t}\overline{\rho}(t,0) + \frac{1}{i\hbar}[\overline{\rho}(t,0), H + \mathcal{M}^+\varphi(t)] = -\epsilon(\overline{\rho}(t,0) - \rho^0(t,0)). \tag{8.12}$$

If one makes a linearization of the equation (8.12) in the small parameter $\mathcal{M}^+\varphi(t)$ and writes down a formal solution (see more detailed information in Section 5.1.2), one is led to

$$\overline{\rho}(t,0) = \rho^0(t,0) - \frac{1}{i\hbar}\int_{-\infty}^{0} dt_1 \exp\{(\epsilon + iL)t_1\}[\mathcal{M}^+, \rho^0(t+t_1,0)]\varphi(t),$$

which completely coincides with (8.11). As before, the function $\varphi(t)$ is slowly varying compared with the operator kernel $[\mathcal{M}^+, \rho^0(t+t_1,0)]$ and dependence of the quantity $\varphi(t+t_1)$ on t_1 has therefore been neglected.

What are the grounds for writing the equations (8.8), (8.12). It is necessary to look for that the distribution $\overline{\rho}(t,0)$, from which as a result of the evolution of the system with the total Hamiltonian $H + H_{\mathcal{F}}(t) = H - A^+\mathcal{F}(t)$ there arises the non-equilibrium distribution $\rho(t,0)$ containing new parameters of $\mathcal{M}^+\varphi(t)$. For this reason $\overline{\rho}(t,0)$ satisfies the equation with the added internal field $\mathcal{M}^+\varphi(t)$. Thus, the resulting solution for $\overline{\rho}(t,0)$ is a functional of a complete set of non-equilibrium parameters.

The final physical results should not be sensitive to the form of the particular functional dependence of $\overline{\rho}(t,0)$ on the P^+ and $\mathcal{M}^+$, parameters, $\overline{\rho}(t,0)$ was chosen so that the transition to results of the linear response theory for the equilibrium system would be carried out naturally, on the one hand, and the distribution $\overline{\rho}(t,0)$ would possess appropriate properties to construct a new NSO, on the other hand.

The approach developed below for linear response of the system as being non-equilibrium at an initial moment in time to the weak measuring field can be constructed in another way, more formally, even without touching the problem of constructing the NSO for $\rho(t,0)$ (this approach will be demonstrated later).

After finding the boundary condition (8.10) for NSO, the Liouville equation, satisfied by the distribution can be written:

$$\left(\frac{\partial}{\partial t} + iL(t)\right)\rho(t,0) = -\epsilon\big(\rho(t,0) - \overline{\rho}(t,0)\big). \tag{8.13}$$

The derivation of (8.13) bears an absolute similarity to that of equation (6.54). Given that

$$iL(t)B = (iL + iL_{\mathcal{F}})B = \frac{1}{i\hbar}[B, H + H_{\mathcal{F}}(t)],$$

we use the integral equation (6.115) for NSO and restrict ourselves to the linear terms in the small correction $H_{\mathcal{F}}(t)$, describing the system's interaction with the external field, to solve it. As a result, we arrive at the simple expression

$$\rho(t,0) = \rho^1(t,0) - \int_{-\infty}^{0} dt_1 \exp\{(\epsilon + iL)t_1\} iL_{\mathcal{F}}\rho^1(t + t_1, 0), \tag{8.14}$$

$$\text{where } \rho^1(t,0) = \epsilon \int_{-\infty}^{0} dt_1 \exp\{(\epsilon + iL)t_1\}\overline{\rho}(t + t_1, 0).$$

In the next section we will use the expression (8.14) to construct an expression for the linear response of the non-equilibrium system and to express generalized susceptibility of the system via non-equilibrium correlation functions. We will also estimate the latter by means of the statistical operator, which describes an initial non-equilibrium distribution.

8.1.2 Generalized susceptibility of a non-equilibrium system

We define the response of a non-equilibrium system as a change in the mean value of the basic operator $\mathcal{M}$,

$$\Delta\langle\mathcal{M}\rangle_t = \mathrm{Sp}\{\mathcal{M}\overline{\rho}(t,0)\} - \mathrm{Sp}\{\mathcal{M}\rho^0(t,0)\}, \tag{8.15}$$

where $\rho^0(t,0)$ is the statistical distribution describing an initial non-equilibrium process.

Once the expression (8.11) for $\bar{\rho}(t,0)$ is plugged into (8.15), the response can be written as

$$\Delta\langle\mathcal{M}\rangle_t = -\langle\mathcal{M},\mathcal{M}^+\rangle_t\varphi(t). \tag{8.16}$$

In writing this expression we have introduced a new "scalar" product of operators over a non-equilibrium state of the system. The scalar product is a generalization of Mori's scalar product and evolves into the latter for the case of an equilibrium distribution,

$$\langle\mathcal{B},\mathcal{M}^+\rangle_t = \frac{1}{i\hbar}\int_{-\infty}^{0} dt_1 e^{\epsilon t_1}\mathrm{Sp}\{\mathcal{B}e^{iLt_1}[\mathcal{M}^+,\rho^0(t+t_1,0)]\}, \tag{8.17}$$

where $\mathcal{B}$ and $\mathcal{M}$ are some operators.

Equation (8.16) provides a mean for determining the system's response to the internal field $\varphi(t)$. However, we are interested in the response to the applied external field $\mathcal{F}(t)$. For this, $\varphi(t)$ needs to be expressed through $\mathcal{F}(t)$. The connection of these functions can be easily obtained from the condition

$$\mathrm{Sp}\{\mathcal{M}[\rho(t,0)-\bar{\rho}(t,0)]\} = 0,$$

which, in accordance with general ideas of the NSO-method, is satisfied by the set of the basic operators $\mathcal{M}$.

Let us find an expression for the difference $\Delta\rho(t,0) = \rho(t,0)-\bar{\rho}(t,0)$. Integration of the expression for $\rho^1(t,0)$ in (8.14) by parts yields the intermediate result:

$$\begin{aligned}\rho^1(t,0) &= \int_{-\infty}^{0} \exp\{iLt_1\}\bar{\rho}(t+t_1,0)\frac{d}{dt_1}\exp\{\epsilon t_1\}\,dt_1\\ &= \bar{\rho}(t,0) - \int_{-\infty}^{0} \exp\{(\epsilon+iL)t_1\}\left\{\frac{\partial}{\partial t}+iL\right\}\bar{\rho}(t+t_1,0)\,dt_1.\end{aligned}$$

Now, we substitute the expression for (8.11) into the last integral and note that the condition analogous to (8.4) for $\ln\rho^0(t+t_1,0)$ is fulfilled for the non-equilibrium statistical operator $\rho^0(t+t_1,0)$. Then, performing elementary computations, we obtain

$$\begin{aligned}\rho^1(t,0)-\bar{\rho}(t,0) = \frac{1}{i\hbar}\int_{-\infty}^{0} dt_1 e^{\epsilon t_1}\int_{-\infty}^{0} dt_2 e^{\epsilon t_2}e^{iL(t_1+t_2)}\\ \times\{[\dot{\mathcal{M}}^+,\rho^0(t+t_1+t_2,0)]\varphi(t+t_1)\\ +[\mathcal{M}^+,\rho^0(t+t_1+t_2,0)]\dot{\varphi}(t+t_1)\}.\end{aligned}$$

Consequently, for the quantity $\Delta\rho(t,0) = \rho(t,0)-\bar{\rho}(t,0)$ we have

$$\begin{aligned}\Delta\rho(t,0) = \frac{1}{i\hbar}\int_{-\infty}^{0} dt_1 e^{\epsilon t_1}\int_{-\infty}^{0} dt_2 e^{\epsilon t_2}e^{iL(t_1+t_2)}\\ \times\{[\dot{\mathcal{M}}^+,\rho^0(t+t_1+t_2,0)]\varphi(t+t_1)\end{aligned}$$

$$+ [\mathcal{M}^+, \rho^0(t + t_1 + t_2, 0)]\dot{\varphi}(t + t_1)\}$$
$$- \frac{1}{i\hbar}\int_{-\infty}^{0} dt_1 e^{(\epsilon+iL)t_1}[A^+, \rho^0(t + t_1, 0)]\mathcal{F}(t + t_1). \qquad (8.18)$$

To continue the investigation, it is convenient to go over from temporary to frequency representation. Let $\mathcal{F}(t)$ be a harmonic function. As far as a linear approximation is concerned, it makes sense to adopt that the functions $\varphi(t)$ will also vary in the same manner. Introducing the notation

$$\mathcal{F}(t) = \mathcal{F}(\omega)e^{-i\omega t}, \quad \varphi(t) = \varphi(\omega)e^{-i\omega t},$$
$$\langle \mathcal{M}\rangle_t^\omega = \mathrm{Sp}\{\mathcal{M}\rho^0(t)\}e^{-i\omega t},$$

the relation between $\varphi(\omega)$ and $\mathcal{F}(\omega)$ can be obtained by considering (8.18) and the condition $\mathrm{Sp}\{\mathcal{M}\Delta\rho(t)\} = 0$:

$$[\langle \mathcal{M}, \dot{\mathcal{M}}^+\rangle_t^\omega - i\omega\langle \mathcal{M}, \mathcal{M}^+\rangle_t^\omega]\varphi(\omega) = [\langle \mathcal{M}, \dot{A}^+\rangle_t^\omega + \epsilon\langle \mathcal{M}, A^+\rangle_t^\omega]\mathcal{F}(\omega). \qquad (8.19)$$

The frequency-dependent correlation function, appearing in the expression (8.19) is given by

$$\langle \mathcal{M}, \mathcal{M}^+\rangle_t^\omega = \frac{1}{i\hbar}\int_{-\infty}^{0} dt_1 e^{(\epsilon-i\omega)t_1}\int_{-\infty}^{0} dt_2 e^{\epsilon' t_2}\mathrm{Sp}\{\mathcal{M}e^{iL(t_1+t_2)}[\mathcal{M}^+, \rho^0(t + t_1 + t_2, 0)]\}. \quad (8.20)$$

In the formula (8.20) ϵ and ϵ' tend to zero after performing the thermodynamic transition. In deriving the relation (8.19), we have converted the last term of equation (8.18) by using the identity

$$\frac{1}{i\hbar}\int_{-\infty}^{0} dt_1 e^{(\epsilon+iL)t_1}[\dot{A}, \rho^0(t + t_1, 0)]$$
$$= \frac{1}{i\hbar}[A, \rho^0(t, 0)] - \epsilon\frac{1}{i\hbar}\int_{-\infty}^{0} dt_1 e^{(\epsilon+iL)t_1}[A, \rho^0(t + t_1, 0)], \qquad (8.21)$$

which can be easily proved, provided that $\rho^0(t, 0)$ is an exact integral of the Liouville equation. Indeed, let us transform the last integral in the formula (8.21):

$$\frac{1}{i\hbar}\int_{-\infty}^{0} e^{iLt_1}[A, \rho^0(t + t_1, 0)]\frac{d}{dt_1}e^{\epsilon t_1}\, dt_1$$
$$= \frac{1}{i\hbar}[A, \rho^0(t, 0)] - \frac{1}{i\hbar}\int_{-\infty}^{0} dt_1 e^{(\epsilon+iL)t_1}\left\{\frac{\partial}{\partial t_1} + iL\right\}[A, \rho^0(t + t_1, 0)].$$

With the statistical operator $\rho^0(t + t_1, 0)$ satisfying the equation similar to (8.4), we can regard the identity (8.21) as a generalization of Kubo's identity for the case of strongly non-equilibrium systems and obtain it immediately from the last relation. The results (8.16) and (8.19) provide a mean for constructing the expression for a change in the mean value of the basic operator $\mathcal{M}$ after turning on the external field $\mathcal{F}(\omega)$

$$\Delta\langle\mathcal{M}\rangle_t^\omega = \chi_{\mathcal{M}A}(t,\omega)\mathcal{F}(\omega) \tag{8.22}$$

and to determine the components of the generalized susceptibility,

$$\chi_{\mathcal{M}A}(t,\omega) = \chi_{\mathcal{M}\mathcal{M}}(t,0)\frac{\langle\mathcal{M},\dot{A}^+\rangle_t^\omega + \epsilon\langle\mathcal{M},A^+\rangle_t^\omega}{\langle\mathcal{M},\dot{\mathcal{M}}^+\rangle_t^\omega - i\omega\langle\mathcal{M},\mathcal{M}^+\rangle_t^\omega}. \tag{8.23}$$

$\chi_{\mathcal{M}\mathcal{M}}(t,0)$ is the static admittance and is expressed through the non-equilibrium correlation function

$$\chi_{\mathcal{M}\mathcal{M}}(t,0) = -\langle\mathcal{M},\mathcal{M}^+\rangle_t. \tag{8.24}$$

In perfect analogy, one can write an expression for the change in the mean value of some other operator B, which does not belong to the set of basic operators:

$$\Delta\langle B\rangle_t = \mathrm{Sp}\{B[\rho(t,0) - \rho^0(t,0)]\}.$$

Obviously, this quantity can be written as follows:

$$\mathrm{Sp}\{B[\rho(t,0) - \rho^0(t,0)]\} = \mathrm{Sp}\{B\Delta\rho(t,0)\} + \mathrm{Sp}\{B[\overline{\rho}(t,0) - \rho^0(t,0)]\}.$$

Given that the value of $\Delta\rho(t,0)$ defined by the relation (8.18), while the quantity $\overline{\rho}(t,0) - \rho^0(t,0)$ is defined by the relation (8.11). Using the definition of the correlation function (8.20), we obtain

$$\begin{aligned}\Delta\langle B\rangle_t^\omega = &- [\langle B,\mathcal{M}^+\rangle_t - \langle B,\dot{\mathcal{M}}^+\rangle_t^\omega + i\omega\langle B,\mathcal{M}^+\rangle_t^\omega]\varphi(\omega)\\ &- [\langle B,\dot{A}^+\rangle_t^\omega + \epsilon\langle B,A^+\rangle_t^\omega]\mathcal{F}(\omega).\end{aligned}$$

Substituting the value of $\varphi(\omega)$ obtained from (8.19) in the last formula, we get

$$\Delta\langle B\rangle_t^\omega = \chi_{BA}(t,\omega)\mathcal{F}(\omega), \tag{8.25}$$

and the generalized susceptibility $\chi_{BA}(t,\omega)$ in this case is

$$\begin{aligned}\chi_{BA}(t,\omega) = &- \epsilon\langle B,\mathcal{M}^+\rangle_t^\omega \cdot [\langle\mathcal{M},\dot{\mathcal{M}}^+\rangle_t^\omega - i\omega\langle\mathcal{M},\mathcal{M}^+\rangle_t^\omega]^{-1}\\ &\times [\langle\mathcal{M},\dot{A}^+\rangle_t^\omega + \epsilon\langle\mathcal{M},A^+\rangle_t^\omega] - [\langle B,\dot{A}^+\rangle_t^\omega + \epsilon\langle B,A^+\rangle_t^\omega].\end{aligned} \tag{8.26}$$

When writing the formula (8.26) we have resorted to the following relation:

$$\langle B,\dot{\mathcal{M}}^+\rangle_t^\omega = \langle B,\mathcal{M}^+\rangle_t - (\epsilon - i\omega)\langle B,\mathcal{M}^+\rangle_t^\omega,$$

which can be easily verified by integrating by parts in the left-hand side.

It is interesting to note that despite a special role of the operators $\mathcal{M}$ in the theory considered, the expression for the dynamic susceptibility $\chi_{\mathcal{M}A}(t,\omega)$ is easily obtained from the formula (8.26) by replacing the operator B by the operator $\mathcal{M}$. In fact, if one puts $B = \mathcal{M}$, then the last formula implies that

$$\epsilon\left(\mathcal{M},\mathcal{M}^+\right)_t^\omega = \left(\mathcal{M},\mathcal{M}^+\right)_t - \left(\mathcal{M},\dot{\mathcal{M}}^+\right)_t^\omega + i\omega\left(\mathcal{M},\mathcal{M}^+\right)_t^\omega.$$

Having substituted the result obtained for $\epsilon\langle\mathcal{M},\mathcal{M}^+\rangle_t^\omega$ into (8.26), we can see that if the operator B coincides with the operator $\mathcal{M}$, the generalized susceptibility $\chi_{BA}(t,\omega)$ becomes the generalized susceptibility $\chi_{\mathcal{M}A}(t,\omega)$.

To conclude this section, it is necessary to compare the results (8.23), (8.26) with the known findings in terms of the response of equilibrium systems.

We show that the scalar product of the operators, defined by the relation (8.17) turns into the usual Mori scalar product (6.89) in the case of the equilibrium distribution. In order to make sure that this is so, it is sufficient to transform the expression (8.17), using the Kubo identity, and then to integrate by parts:

$$\left\langle B,\mathcal{M}^+\right\rangle = \beta\int_{-\infty}^{0} dt_1 e^{\epsilon t_1}\int_0^1 d\tau \mathrm{Sp}\{Be^{iLt_1}\rho_0^\tau\dot{\mathcal{M}}^+\rho_0^{1-\tau}\} = \beta(B,\mathcal{M}^+),$$

where β is reciprocal temperature. The last relation holds if the B and $\mathcal{M}$ operators satisfy the principle of correlation weakening:

$$\lim_{\epsilon\to\infty}\epsilon\int_{-\infty}^{0} dt_1 e^{\epsilon t_1}\mathrm{Sp}\{Be^{iLt_1}\mathcal{M}^+\} = \langle B\rangle_0\left\langle\mathcal{M}^+\right\rangle_0.$$

This requirement appears not to be too severe restriction to nature of the B and $\mathcal{M}$ operators for systems with mixing, where only relaxation phenomena are possible. In addition, we believe that $\langle B\rangle_0 = \langle\mathcal{M}^+\rangle_0 = 0$.

Thus, we have shown that the scalar product (8.17) becomes the usual Mori scalar product of operators, if non-equilibrium distribution $\rho^0(t,0)$ is replaced by equilibrium ρ_0.

In order to prove that the expressions (8.23), (8.26) have the correct limiting transition to the case of a linear response of equilibrium systems, it is necessary to deduce afresh formulas for the linear response using the standard technique for NSO discussed in Chapter 6. Since there are no problems to do so, we suggest the reader to solve this excellent exercise by him- of herself. However, it is worth pointing out that the result to be obtained for an isothermal response of the equilibrium system has the same structure as the formula (8.23), differing only in the scalar product (8.17) and (6.89).

8.2 Projection operator for non-equilibrium systems

8.2.1 Magnetic susceptibility

Consider the application of the general formulas (8.23), (8.26) of linear response theory for non-equilibrium systems to calculate magnetic susceptibility.

Let an alternating magnetic field $\vec{B}(t)$ act on a system of magnetic moments $\vec{M}$. To navigate to this case a replacement should be made in the general formulas of the previous section:

$$A^{+}\mathcal{F}(t) \rightarrow M^{+}B(t); \quad \mathcal{M}\varphi(t) \rightarrow M^{+}b(t),$$

where M is an operator column-vector with components of the total electron magnetic moment, B being the row-vector consisting of components of the magnetic induction of the external electromagnetic field. For the sake of simplicity, spatial inhomogeneity of the electromagnetic field should be neglected. The product $M^{+}b(t)$ is defined in a similar way. The internal non-equilibrium field $b(t)$ is the thermal perturbation induced by the external field. The thermal perturbation is associated with magnetization of the system $m(t) = \Delta\langle M\rangle_t$ by the relation (8.16):

$$m(t) = -\langle M, M^{+}\rangle_t b(t) \quad \text{or} \quad m(t,\omega) = -\langle M, M^{+}\rangle_t b(\omega). \tag{8.27}$$

We define the dynamic magnetic susceptibility by the relation

$$m(t,\omega) = \chi(t,\omega)B(\omega).$$

For this purpose, we use the formulas (8.19) and (8.27). Substituting $\mathcal{M} = M$, $\varphi(\omega) = b(\omega)$ into the formula (8.19) and expressing $b(\omega)$ with the aid of the formula (8.27) as $b(\omega) = -\langle M, M^{+}\rangle_t^{-1} m(t,\omega)$, we obtain

$$-[\langle M, \dot{M}^{+}\rangle_t^{\omega} - i\omega\langle M, M^{+}\rangle_t^{\omega}]\langle M, M^{+}\rangle_t^{-1} m(t,\omega) = [\langle M, \dot{M}^{+}\rangle_t^{\omega} + \epsilon\langle M, M^{+}\rangle_t^{\omega}]B(\omega).$$

The last formula allows one to easily write down the expression for the components of the magnetic susceptibility tensor

$$\chi(t,\omega) = \chi(t,0)\frac{1}{\overline{T}(t,\omega) - i\omega}[\overline{T}(t,\omega) + \epsilon], \tag{8.28}$$

where, as before in the case of the response of the equilibrium systems (see the formula (6.72)), we have introduced both the transport matrix

$$\overline{T}(t,\omega) = \frac{1}{\langle M, M^{+}\rangle_t^{\omega}}\langle M, \dot{M}^{+}\rangle_t^{\omega} \tag{8.29}$$

and the system's static magnetic susceptibility

$$\chi(t,0) = -\langle M, M^{+}\rangle_t. \tag{8.30}$$

The time-dependent magnetic susceptibility is related to the time-dependence of the initial non-equilibrium distribution. The role of the transport matrix introduced by the formula (8.29) is the same as that in the case of the response of equilibrium systems. In particular, in the relaxation-free regime when the amplitude of the external magnetic field is zero, the transport matrix determines the spectrum of normal modes in the system (6.75). By perfect analogy with the equilibrium case (6.76)–(6.79), a non-equilibrium Green function can be introduced by the relation

$$\overline{G}(t,\omega) = \frac{1}{\overline{T}(t,\omega) - i\omega + \epsilon} = \frac{1}{\langle M, M^+\rangle_t}\langle M, M^+\rangle_t^{\omega}. \tag{8.31}$$

Thus, a further problem of computing the magnetic susceptibility reduces to finding the transport matrix $\overline{T}(t,\omega)$ or the Green function $\overline{G}(t,\omega)$, which in turn requires the application of the projection operator technique suitable for use for the non-equilibrium systems. Such a projection operator can be constructed by analogy with the Mori projection operator (6.88), (6.91) by simply replacing the Mori scalar product by the product defined by (8.17). As a result, one arrives at

$$\begin{aligned}
\mathcal{P}_t A &= \langle A, M^+\rangle_t \frac{1}{\langle M, M^+\rangle_t} M,\\
\mathcal{P}_t A^+ &= M^+ \frac{1}{\langle M, M^+\rangle_t}\langle M, A^+\rangle_t,\\
\mathcal{P}_t(1-\mathcal{P}_t)A &= 0, \quad \mathcal{P}_t M = M, \quad \mathcal{P}_t^2 M = M.
\end{aligned} \tag{8.32}$$

In the definition (8.32), the index t of the projection operator indicates the dependence of the latter on time because the initial non-equilibrium distribution is time-dependent too. In the future, we assume that the initial non-equilibrium distribution is stationary, and the subscript t of both the projection operator and the correlation functions can be omitted.

Since the dot product (8.17) turns into Mori's scalar product in passing to equilibrium, the projection operators (8.32) become the projection operators (6.88). Therefore, the problem at hand is greatly simplified and reduced virtually to a repetition of the foregoing computations that we made in Section 6.2.4. These computations suggest replacing the operator P^+ by M^+ and the equilibrium scalar product of the operators by non-equilibrium analogue. This again yields

$$\begin{aligned}
\overline{T}(\omega) &= i\overline{\Omega} + \overline{\Sigma}(\omega),\\
i\overline{\Omega} &= \frac{1}{\langle M, M^+\rangle}\langle M, \mathcal{P}\dot{M}^+\rangle,\\
\overline{\Sigma}(\omega) &= \frac{1}{\langle M, M^+\rangle}\left\langle f, \frac{1}{-i\omega + \epsilon + (1-\mathcal{P})iL} f^+\right\rangle,
\end{aligned} \tag{8.33}$$

where $f = (1-\mathcal{P})\dot{M}$.

Thus, we can conclude that the computation of the magnetic susceptibility of the non-equilibrium system is over and draw attention to the following.

The structure of the magnetic susceptibility tensor components in the non-equilibrium case remains effectively unchanged, but the definition of the scalar product of two operators became different. For example, as to the transverse magnetic susceptibility of an electron gas, the expressions obtained from the above formulas have the same structure as the equilibrium susceptibility defined by the relations (6.147), (6.148), differing from them only in the form of the scalar product of operators and of some symbols.

The calculation of non-equilibrium correlation functions is of a certain interest. Therefore, the calculation of non-equilibrium electrical conductivity may be exemplified by analysis of these functions. There are well-known results obtained by means of the kinetic equations discussed in Section 4.2.2, which allows for comparing the two different methods in more detail.

8.2.2 Electrical conductivity of highly non-equilibrium systems

Let us give an example of the application of the developed technique of handling the linear response problem of the non-equilibrium system for a particular case to calculate the electrical conductivity. Then, in the general formulas in Section 8.1.2, we need to make the following replacement:

$$A^+ \to eX^\alpha, \qquad \mathcal{F}(t) \to E^\alpha(t),$$
$$\mathcal{M} \to eP^\beta/m, \quad \varphi(t) \to \beta V^\beta(t).$$

It is worth noting that the above notations coincide with analogous ones in Section 6.2.3.

For the sake of simplicity, the indices of the tensor quantities should not be labeled wherever it is possible because we cover only the case of the isotropic dispersion law and isotropic electron scattering. It does matter since only diagonal components of the conductivity tensor differ from zero.

Using the equation (8.23), we find

$$\sigma(\omega) = -\frac{e^2}{m}\langle P, P^+\rangle \frac{\langle P, \dot{X}^+\rangle^\omega + \epsilon\langle P, X^+\rangle^\omega}{\langle P, \dot{P}^+\rangle^\omega - i\omega\langle P, P^+\rangle^\omega}.$$

Taking into account both the Abel theorem and the principle of correlation weakening, and also that $\dot{X}^+ = P^+/m$, we have

$$\begin{aligned}\epsilon \int_{-\infty}^{0} & dt_2 e^{\epsilon t_2} \mathrm{Sp}\{[P, X^+(t_1+t_2)]\rho^0(t+t_1+t_2, 0)\} \\ &= \lim_{t_2\to-\infty} \mathrm{Sp}\{[P, X^+(t_1+t_2)]\rho^0(t+t_1+t_2, 0)\} = 0.\end{aligned}$$

Then, if the operator P is defined so that its non-equilibrium average is zero, we obtain

$$\lim_{\epsilon\to 0}\epsilon\left\langle P, X^+\right\rangle^\omega = 0.$$

Substituting these results into the formula for the electrical conductivity, we get

$$\sigma(\omega) = -\frac{e^2}{m^2}\frac{\langle P, P^+\rangle}{\overline{T}(\omega) - i\omega},$$
$$\overline{T}(\omega) = \frac{1}{\langle P, P^+\rangle^\omega}\left\langle P, \dot{P}^+\right\rangle^\omega. \tag{8.34}$$

In writing the expression (8.34) the quantities P, P^+ are supposed to be a row vector and column vector, respectively, consisting of Cartesian components of the operator of the total electron momentum.

Resorting to the projection operator method (see Section 6.2.4 and the formula (6.137)), the transport matrix $\overline{T}(\omega)$ can be represented as the sum of the frequency matrix and the memory function: $\overline{T}(\omega) = i\overline{\Omega} + \overline{\Sigma}(\omega)$ provided that

$$i\overline{\Omega} = \frac{1}{\langle P, P^+\rangle}\langle P, \mathcal{P}\dot{P}^+\rangle,$$
$$\overline{\Sigma}(\omega) = \frac{1}{\langle P, P^+\rangle}\left\langle f, \frac{1}{-i\omega + \epsilon + (1-\mathcal{P})iL}f^+\right\rangle, \tag{8.35}$$
$$\mathcal{P}A^+ = P^+\frac{1}{\langle P, P^+\rangle}\langle P, A^+\rangle, \quad f = (1-\mathcal{P})\dot{P}.$$

The formulas given above are general enough and hold true for any stationary non-equilibrium distribution.

A further treatment requires concretizing the choice of the initial non-equilibrium distribution. Let β_k, β_s, β_p be reciprocal temperatures of crystal's subsystems k, S, P to characterize the distribution, being defined by the expressions (8.36). P means a subsystem of long-wavelength phonons, the k, S stand for subsystems of kinetic and spin degrees of freedom, respectively. We have

$$\rho_q = \exp\{-\Phi - \beta_k H_k - \beta_s H_s - \beta_p H_p + \beta\zeta N\}, \tag{8.36}$$
$$\Phi = \ln \mathrm{Sp}\{\exp(-\beta_k H_k - \beta_s H_s - \beta_p H_p + \beta\zeta N)\},$$
$$H_k = \sum_{\vec{p},\sigma}\varepsilon_{\vec{p}}a^+_{\vec{p}\sigma}a_{\vec{p}\sigma}, \quad H_s = -\sum_{\vec{p},\sigma}\hbar\omega_s\sigma a^+_{\vec{p}\sigma}a_{\vec{p}\sigma},$$
$$H_p = \sum_{\vec{q},\lambda}\hbar\Omega_{\vec{q},\lambda}(b^+_{\vec{q},\lambda}b_{\vec{q},\lambda} + 1/2), \quad \varepsilon_p = \frac{p^2}{2m}, \quad \omega_s = \frac{g\mu_B H}{\hbar}.$$

Here, $a^+_{\vec{p}\sigma}$, $a_{\vec{p}\sigma}$ are creation (annihilation) operators of electrons with a momentum $\vec{p}$ and spin projection $\sigma = \pm\frac{1}{2}$ on the axis Z, respectively. $b^+_{\vec{q},\lambda}$, $b_{\vec{q},\lambda}$ are creation (annihilation) operators of phonons with a wave vector $\vec{q}$ and polarization λ, $\hbar\Omega_{\vec{q},\lambda}$ is the phonon energy, and g is the factor of spectroscopic splitting; μ_B is the Bohr magneton, H is the classical magnetic field strength.

Let us analyze the frequency matrix and the memory function defined by (8.35). It is easy to show that the numerator of the frequency matrix is equal to $\langle P, \mathcal{P}\dot{P}^{+}\rangle = 0$. Since in the given case the components of the total electron momentum operator P^{α} are basis operators, we find

$$i\overline{\Omega} = \frac{1}{\langle P, P^{+}\rangle}\langle P, \mathcal{P}\dot{P}^{+}\rangle = \frac{1}{\langle P^{\alpha}, P^{\beta}\rangle}\langle P^{\alpha}, \mathcal{P}\dot{P}^{\beta}\rangle = \frac{1}{\langle P^{\alpha}, P^{\beta}\rangle}\langle P^{\alpha}, \dot{P}^{\beta}\rangle.$$

In spite of there being the α and β tensor indices in this formula, one should bear in mind that, by virtue of isotropy of space, only the diagonal components of the conductivity tensor can be different from zero. Now it is easy enough to prove that the numerator in the last formula is zero. Suffice it to recall the definition of the correlation function $\langle P^{\alpha}, \dot{P}^{\beta}\rangle$ and the identity (8.21),

$$\begin{aligned}\langle P^{\alpha}, \dot{P}^{\beta}\rangle &= \int_{-\infty}^{0} dt_1 e^{\epsilon t_1} \mathrm{Sp}\left\{P^{\alpha} e^{iLt_1} \frac{1}{i\hbar}[\dot{P}^{\beta}, \rho^{0}(t+t_1, 0)]\right\} \\ &= \frac{1}{i\hbar}\mathrm{Sp}\{P^{\alpha}[P^{\beta}, \rho^{0}(t, 0)]\} - \epsilon \int_{-\infty}^{0} dt_1 e^{\epsilon t_1} \\ &\quad \times \mathrm{Sp}\left\{P^{\alpha} e^{iLt_1} \frac{1}{i\hbar}[P^{\beta}, \rho^{0}(t+t_1, 0)]\right\} = 0.\end{aligned}$$

The first term in the last expression is zero as the components of the total momentum operator commute among themselves. When commuting any two operators taken at different moments of time, the second term is equal to zero too, provided that the difference in the times tends to infinity (it is assumed that the operators satisfy the principle of correlation weakening).

Consider the correlation function

$$\begin{aligned}\langle P^{\alpha}, P^{\beta}\rangle &= \int_{-\infty}^{0} dt_1 e^{\epsilon t_1} \mathrm{Sp}\left\{P^{\alpha} e^{iLt_1} \frac{1}{i\hbar}[P^{\beta}, \rho^{0}(t+t_1, 0)]\right\} \\ &= m \int_{-\infty}^{0} dt_1 e^{\epsilon t_1} \mathrm{Sp}\left\{P^{\alpha} e^{iLt_1} \frac{1}{i\hbar}[\dot{X}^{\beta}, \rho^{0}(t+t_1, 0)]\right\} \\ &= m\mathrm{Sp}\left\{P^{\alpha} \frac{1}{i\hbar}[X^{\beta}, \rho^{0}(t, 0)]\right\} \\ &\quad - \epsilon \int_{-\infty}^{0} dt_1 e^{\epsilon t_1} \mathrm{Sp}\left\{P^{\alpha} e^{iLt_1} \frac{m}{i\hbar}[\dot{X}^{\beta}, \rho^{0}(t+t_1, 0)]\right\} = -nm.\end{aligned}$$

Let us analyze the electrical conductivity in the Born approximation of the scattering theory. This means that we have to restrict ourselves only to terms of the second order in the interaction with scatterers (phonons, for example) when calculating the inverse relaxation time of the total momentum of the electron system (or the memory function $\overline{\Sigma}$). Consequently, for interaction terms of the fourth order and higher not to be retained, the projection operators in the memory function can be omitted.

We make one further simplification and restrict ourselves to computation of the static conductivity by putting $\omega = 0$. Such an approximation works well indeed provided that the frequency of an external electric field is $\omega \ll 1/\tau$. This requirement is valid for ordinary materials up to frequencies of the optical range. Furthermore, as noted above, the initial non-equilibrium distribution is assumed to be stationary and time-independent through the macro parameters $\beta_k, \beta_s, \beta_p$, which accounts for the statement $\rho^0(t + t_1 + t_2, 0) = \rho^0$. Substituting the results obtained into (8.34) for the conductivity tensor components, we obtain

$$\sigma = \frac{e^2 n\tau}{m}; \quad \frac{1}{\tau} = \bar{\Sigma} = -\frac{1}{nm}\int_{-\infty}^{0} dt_1 \int_{-\infty}^{0} dt_2 e^{\epsilon t_1 + \epsilon' t_2} \mathrm{Sp}\left\{\dot{P}^{\alpha} e^{iL(t_1+t_2)} \frac{1}{i\hbar}[\dot{P}^{\beta}, \rho^0]\right\},$$

$$\dot{P}^{\alpha} = \frac{1}{i\hbar}[P^{\alpha}, H], \quad iLA = \frac{1}{i\hbar}[A, H_k + H_s + H_p + H_{el}], \tag{8.37}$$

where H_{el} is an interaction Hamiltonian with the scatterers.

The expression (8.37) for the inverse relaxation time of the non-equilibrium system already incorporates the second-order interaction terms of H_{el} since the commutator of the total momentum operator commutes both with the Hamiltonians of the subsystems k, S and with the phonon Hamiltonian:

$$[P^{\alpha}, H_k + H_s + H_p + H_{el}] = [P^{\alpha}, H_{el}].$$

For this reason, the interaction Hamiltonian H_{el} can be ignored, and the statistical operator $\rho^0(t, 0)$ may be replaced by the quasi-equilibrium distribution (8.36). When recording the latter the electron–phonon interaction Hamiltonian has been prudently omitted. Therefore, the formula (8.37) can be directly used to calculate the inverse relaxation time of the non-equilibrium system.

Looking ahead to, it should be noted that the next chapter contains a different expression derived for the inverse relaxation time of the non-equilibrium system. The form of the new formula differs from the record (8.37). Therefore, it would be advisable to immediately transform the expression (8.37) in accordance with the master equation method (see Chapter 9).

First, using the identity (8.21), we can represent the integral in the expression (8.37) in the following form:

$$\int_{-\infty}^{0} dt_1 \int_{-\infty}^{0} dt_2 e^{\epsilon t_1 + \epsilon' t_2} \mathrm{Sp}\left\{\dot{P}^{\alpha} e^{iL(t_1+t_2)} \frac{1}{i\hbar}[\dot{P}^{\beta}, \rho^0]\right\}$$

$$= \frac{1}{i\hbar}\int_{-\infty}^{0} dt_1 \int_{-\infty}^{0} dt_2 e^{\epsilon t_1 + \epsilon' t_2} \mathrm{Sp}\left\{\dot{P}^{\alpha} \frac{d}{dt_2} e^{iL(t_1+t_2)} [P^{\beta}, \rho^0]\right\}$$

$$= \frac{1}{i\hbar}\int_{-\infty}^{0} dt_1 e^{\epsilon t_1} \mathrm{Sp}\{\dot{P}^{\alpha} e^{iLt_1} [P^{\beta}, \rho^0]\}.$$

In deriving this equation we have taken the principle of correlation weakening into account. Next, given that the stationary non-equilibrium distribution satisfies the Liouville equation (8.13),

$$iL\rho_0 = -\epsilon(\rho^0 - \rho_q),$$

and $P^\beta/m = iLX^\beta$, we write $1/\tau$ in such a way that

$$\overline{\Sigma} = -\frac{1}{n}\frac{1}{i\hbar}\Bigg\{\int_{-\infty}^{0} dt_1 e^{\epsilon t_1}\mathrm{Sp}\{\dot{P}^\alpha e^{iLt_1} iL[X^\beta,\rho^0]\}$$
$$+\,\epsilon\int_{-\infty}^{0} dt_1 e^{\epsilon t_1}\mathrm{Sp}\{\dot{P}^\alpha e^{iLt_1}[X^\beta,(\rho^0-\rho_q)]\}\Bigg\}.$$

Using Abel's theorem and the principle of correlation weakening, one can prove that the second summand in the last expression vanishes, since

$$\lim_{t_1\to-\infty}\mathrm{Sp}\{[\dot{P}^\alpha(-t_1),X^\beta]\rho^0\} = \lim_{t_1\to-\infty}\mathrm{Sp}\{[\dot{P}^\alpha(-t_1),X^\beta]\rho_q\} = 0.$$

Thus the expression for $1/\tau$ can be written as follows:

$$\overline{\Sigma} = \overline{\Sigma}_1 + \overline{\Sigma}_2,$$
$$\overline{\Sigma}_1 = -\frac{1}{n}\int_{-\infty}^{0} dt_1 e^{\epsilon t_1}\mathrm{Sp}\left\{\dot{P}^\alpha e^{iLt_1} iL_v\frac{1}{i\hbar}[X^\beta,\rho^0]\right\}; \tag{8.38}$$
$$\overline{\Sigma}_2 = -\frac{1}{n}\int_{-\infty}^{0} dt_1 e^{\epsilon t_1}\mathrm{Sp}\left\{\dot{P}^\alpha e^{iLt_1} iL_0\frac{1}{i\hbar}[X^\beta,\rho^0]\right\}. \tag{8.39}$$

The first summand under the spur sign on the right-hand side of the last expression contains a second order in interaction. Consequently, the interaction both in the evolution operators and in the non-equilibrium statistical operator ρ^0 needs to be omitted with respect to the Born approximation of the scattering theory.

In the second summand the second-order interaction terms can be collected by keeping interaction with the scatterers either within the time evolution operator $\exp(iLt_1)$ or within the statistical operator ρ^0. If one omits the interaction in the operator ρ^0, this distribution becomes ρ_q, and then

$$iL_0[X^\beta,\rho^0] = iL_0[X^\beta,\rho_q] \sim iL_0\frac{\beta_k P^\beta}{m}\rho_q = 0,$$

as the P^β and ρ_q operators commute with the Hamiltonian H_0, and in this case, $\overline{\Sigma}_2$ is zero.

If one keeps the interaction in the statistical operator but omits it in the evolution operator, the expression for $\overline{\Sigma}_2$ can be represented in the form

$$\overline{\Sigma}_2 = -\frac{1}{n}\int\limits_{-\infty}^{0} dt_1 e^{\epsilon t_1} \mathrm{Sp}\left\{\dot{P}^\alpha \frac{d}{dt_1} e^{iL_0 t_1} \frac{1}{i\hbar}[X^\beta, \rho^0]\right\}.$$

Integrating this expression by parts again, we obtain

$$\overline{\Sigma}_2 = -\frac{1}{n}\mathrm{Sp}\left\{\dot{P}^\alpha \frac{1}{i\hbar}[X^\beta, \rho^0]\right\} + \epsilon\frac{1}{n}\int\limits_{-\infty}^{0} dt_1 e^{\epsilon t_1} \mathrm{Sp}\left\{\dot{P}^\alpha e^{iL_0 t_1} \frac{1}{i\hbar}[X^\beta, \rho^0]\right\} = 0. \tag{8.40}$$

The vanishing of the first summand becomes obvious if one reconstructs the commutator $\dot{P}^\alpha = iLP^\alpha$,

$$\mathrm{Sp}\left\{\dot{P}^\alpha \frac{1}{i\hbar}[X^\beta, \rho^0]\right\} = -\mathrm{Sp}\left\{P^\alpha \frac{1}{i\hbar}[P^\beta, \rho^0]\right\} + \epsilon \mathrm{Sp}\left\{P^\alpha \frac{1}{i\hbar}[X^\beta, (\rho^0 - \rho_q)]\right\},$$

and takes into account that $[P^\alpha, P^\beta] = 0$,

$$\frac{1}{i\hbar}\mathrm{Sp}\{[P^\alpha, X^\beta]\rho^0\} = \frac{1}{i\hbar}\mathrm{Sp}\{[P^\alpha, X^\beta]\rho_q\},$$

by virtue of the normalization condition.

The integral term in the formula (8.40) is equal to zero because after applying Abel's theorem (omitting inessential constants) it can be written as

$$\lim_{t_1\to-\infty} \mathrm{Sp}\{[\dot{P}^\alpha(-t_1), X^\beta]\rho^0\} = 0.$$

Although evolution of the operator $\dot{P}^\alpha$ in this expression is determined by the Hamiltonian H_0, which does not contain the interaction, the operator $\dot{P}^\alpha$ is not invariant under such an evolution. Therefore, the principle of correlation weakening is justifiably applicable to this situation.

Thus, we have proved that the inverse relaxation time for the non-equilibrium system can be represented in the form (8.38).

Before proceeding to direct calculations of the inverse relaxation time of the non-equilibrium system by the formula (8.38), it should be again paid attention to the condition of its applicability. Undoubtedly, the expression (8.35) is correct for the memory function, containing the projection operators. However, if we reject them, the correct expression for the inverse relaxation time can be obtained only in the Born approximation of the scattering theory. Moreover, we can show that, as in the case of linear response of the equilibrium system, the memory function is exactly zero by discarding the projection operators in (8.35) and taking into account evolution of the system with the total Hamiltonian. Pursuant to the material set forth in Chapters 5, and 6 of the present book, this result should not come as a surprise since exact Hamiltonian dynamics cannot lead to irreversible behavior.

As it follows from the expression (8.38), the calculation of the non-equilibrium conductivity reduced to the computation of the inverse relaxation time. If one assumes that the resistance is determined by the scattering of electrons by phonons, the electron–phonon interaction Hamiltonian H_{ep} in the second-quantization representation with respect to electron variables can be written in the form (4.76):

$$H_{ep} = \sum_{\vec{q},\lambda,\vec{p}',\vec{p},\sigma} (U^{\vec{q}\lambda}_{\vec{p}'\vec{p}} b_{\vec{q}\lambda} + U^{-\vec{q}\lambda}_{\vec{p}'\vec{p}} b^{+}_{\vec{q}\lambda}) a^{+}_{\vec{p}'\sigma} a_{\vec{p}\sigma},$$

where the matrix elements $U^{\vec{q}\lambda}_{\vec{p}'\vec{p}}$ are defined by

$$U^{\vec{q}\lambda}_{\vec{p}'\vec{p}} = C_{\vec{q}\lambda} \langle \vec{p}' | e^{i\vec{q}\vec{r}} | \vec{p} \rangle,$$

$|\vec{p}\rangle$ being a normalized system of eigenfunctions of the conduction electrons, $C_{\vec{q}\lambda}$ being the electron–phonon coupling constant.

Let us show that the inverse relaxation time (8.38) for the phonon scattering mechanism provides exactly the same result (4.157) as we obtained from the kinetic equation.

Using the Kubo identity (5.81), one can record an expression for the inverse relaxation time in the form to be used for further calculations:

$$\bar{\Sigma}_1 = \frac{1}{n} \int_{-\infty}^{0} dt_1 e^{\varepsilon t_1} \int_{0}^{1} d\tau \mathrm{Sp}\left\{ \dot{P}^{\alpha} e^{iLt_1} iL_v \rho_q^{\tau} \frac{1}{i\hbar} [X^{\beta}, S^0] \rho_q^{1-\tau} \right\}.$$

In this formula $\rho_q = \exp\{-S^0\}$

$$S^0 = \Phi + \beta_k H_k + \beta_s H_s + \beta_p H_p - \beta\zeta N$$

is the entropy operator of the non-equilibrium system. Since, for the case under consideration

$$\frac{1}{i\hbar}[X^{\beta}, S^0] = \frac{\beta_k}{i\hbar}[X^{\beta}, H_k] = \frac{\beta_k}{m} P^{\beta}$$

and the operator P^{β} commutes with the operator ρ_q, the integration over the variable τ can be easily performed. Then this gives

$$\bar{\Sigma}_1 = \frac{\beta_k}{nm} \int_{-\infty}^{0} dt_1 e^{\varepsilon t_1} \mathrm{Sp}\left\{ \dot{P}^{\alpha} e^{iLt_1} \frac{1}{i\hbar} (P^{\beta} \rho_q H_{ep} \rho_q^{-1} - H_{ep} P^{\beta}) \rho_q \right\}.$$

In deriving this formula we have used the fact that in our case

$$iL_v A = \frac{1}{i\hbar}[A, H_{ep}].$$

Going over to the second-quantization representation over electron variables, we rewrite the expression for the spur:

$$\begin{aligned}
&\mathrm{Sp}\left\{\dot{P}^{\alpha}e^{iLt_1}\frac{1}{i\hbar}(P^{\beta}\rho_q H_{ep}\rho_q^{-1} - H_{ep}P^{\beta})\rho_q\right\} \\
&= \frac{1}{i\hbar}\sum_{\nu'\nu\mu'\mu\kappa}\{\langle \dot{P}^{\alpha}_{(ep)\nu'\nu}(-t_1)P^{\beta}_{\kappa}H_{ep\mu'\mu}(i\hbar\beta)\rangle_{\mathrm{scat}} \\
&\times \langle a^{+}_{\nu'}(-t_1)a_{\nu}(-t_1)a^{+}_{\kappa}a_{\kappa}a^{+}_{\mu'}(i\hbar\beta)a_{\mu}(i\hbar\beta)\rangle \\
&- \langle \dot{P}^{\alpha}_{(ep)\nu'\nu}(-t_1)H_{ep\mu'\mu}P^{\beta}_{\kappa}\rangle_{\mathrm{scat}} \\
&\times \langle a^{+}_{\nu'}(-t_1)a_{\nu}(-t_1)a^{+}_{\mu'}a_{\mu}a^{+}_{\kappa}a_{\kappa}\rangle\}.
\end{aligned} \tag{8.41}$$

Here, the indices $\nu', \nu, \mu', \mu, \kappa$ have been applied for denoting the quantum numbers $(\vec{p}, \sigma)$, characterizing the electron states. The large angular brackets stand for the averaging over the states of the scatterers (phonons), $A(i\hbar\beta) = \rho_q A \rho_q^{-1}$.

Using the statistical Wick–Bloch–de Dominicis theorem [38] (see also (5.75)), we express the average values of six electron creation (annihilation) operators in the last formula via the Fermi–Dirac distribution function. Considering only the non-zero pairing, we have

$$\begin{aligned}
\langle a^{+}_{\nu'}a_{\nu}a^{+}_{\kappa}a_{\kappa}a^{+}_{\mu'}a_{\mu}\rangle &= f_{\nu'}(1-f_{\nu})^2\delta_{\nu'\mu}\delta_{\nu\kappa}\delta_{\kappa\mu'} - f^2_{\nu'}(1-f_{\nu})\delta_{\nu'\kappa}\delta_{\mu\kappa}\delta_{\nu\mu'} \\
&= f_{\nu'}(1-f_{\nu})(1-f_{\nu}-f_{\nu'})\delta_{\nu'\mu}\delta_{\nu\mu'}.
\end{aligned} \tag{8.42}$$

Similarly,

$$\langle a^{+}_{\nu'}a_{\nu}a^{+}_{\mu'}a_{\mu}a^{+}_{\kappa}a_{\kappa}\rangle = f_{\nu'}(1-f_{\nu})(1-f_{\nu}-f_{\nu'})\delta_{\nu'\mu}\delta_{\nu\mu'}. \tag{8.43}$$

If the electron creation (annihilation) operators are time-dependent, then this relationship must be clearly distinguished by means of the commutation relations

$$a^{+}_{\nu'}(t_1+i\hbar\beta)a_{\nu}(t_1+i\hbar\beta) = a^{+}_{\nu'}a_{\nu}e^{i/\hbar(\varepsilon_{\nu'}-\varepsilon_{\nu})t_1-\beta_k(\varepsilon_{\nu'}-\varepsilon_{\nu})}.$$

Given that in the formula (8.41)

$$\dot{P}^{\alpha}_{(ep)\nu'\nu}(t_1) = -i\sum_{\vec{q},\lambda}q^{\alpha}\{U^{\vec{q}\lambda}_{\nu'\nu}b_{\vec{q}\lambda}e^{-i\Omega_{\vec{q}\lambda}t_1} - U^{-\vec{q}\lambda}_{\nu'\nu}b^{+}_{\vec{q}\lambda}e^{i\Omega_{\vec{q}\lambda}t_1}\},$$

we obtain the following result for the memory function:

$$\begin{aligned}
\bar{\Sigma}_1 &= \frac{\beta_k}{nm}\frac{1}{i\hbar}\int_{-\infty}^{0}dt_1 e^{\varepsilon t_1}\sum_{\nu'\nu\vec{q}\lambda} -iq^{\alpha}\langle\{U^{\vec{q}\lambda}_{\nu'\nu}b_{\vec{q}\lambda}e^{i\Omega_{\vec{q}\lambda}t_1-\beta\hbar\Omega_{\vec{q}\lambda}} \\
&- U^{-\vec{q}\lambda}_{\nu'\nu}b^{+}_{\vec{q}\lambda}e^{-i\Omega_{\vec{q}\lambda}t_1+\beta\hbar\Omega_{\vec{q}\lambda}}\}\{U^{\vec{q}\lambda}_{\nu\nu'}b_{\vec{q}\lambda} + U^{-\vec{q}\lambda}_{\nu\nu'}b^{+}_{\vec{q}\lambda}\}\rangle_{\mathrm{scat}} \\
&\times \{[P^{\beta}_{\nu}(1-f_{\nu}) - P^{\beta}_{\nu'}f_{\nu'}]\cdot e^{\beta_k(\varepsilon_{\nu'}-\varepsilon_{\nu})} - [P^{\beta}_{\nu'}(1-f_{\nu'}) - P^{\beta}_{\nu}f_{\nu}]\} \\
&\times f_{\nu'}(1-f_{\nu})\cdot e^{i/\hbar(\varepsilon_{\nu}-\varepsilon_{\nu'})t_1}.
\end{aligned}$$

It should be noted that for the averaging over the states of scatterers, to be reduced to the calculation of quantum-statistical averages of the phonon creation (annihilation) operators, the new selection rules arise:

$$\langle b^{+}_{\vec{q}\lambda} b_{\vec{q}'\lambda'} \rangle = N_{\vec{q}\lambda}\delta_{\vec{q}\vec{q}'}\delta_{\lambda\lambda'}, \quad N_{\vec{q}\lambda} = \frac{1}{e^{\beta_p \hbar\Omega_{\vec{q}\lambda}} - 1};$$
$$\langle b_{\vec{q}\lambda} b^{+}_{\vec{q}'\lambda'} \rangle = (N_{\vec{q}\lambda} + 1)\delta_{\vec{q}\vec{q}'}\delta_{\lambda\lambda'}.$$

Next, putting that $\vec{q} = \vec{q}'$, $\lambda = \lambda'$ and averaging over the states of scatterers and not forgetting that

$$f_\nu(1 - f_\nu) = -\beta_k^{-1}\frac{\partial}{\partial\varepsilon_\nu} f_\nu = -\beta_k^{-1} f'_\nu,$$
$$e^{\beta_k(\varepsilon'_\nu - \varepsilon_\nu)} f_{\nu'}(1 - f_\nu) = f_\nu(1 - f_{\nu'}),$$

we can get the expression for the inverse relaxation time:

$$\begin{aligned}
\bar{\Sigma}_1 = &-\frac{1}{nm}\frac{1}{i\hbar}\int_{-\infty}^{0} dt_1 e^{\epsilon t_1} \sum_{\nu'\nu\vec{q}\lambda} -iq^\alpha\{[|U^{\vec{q}\lambda}_{\nu'\nu}|^2 (N_{\vec{q}\lambda} + 1) e^{i\Omega_{\vec{q}\lambda} t_1} \\
&\times e^{-\beta_p\hbar\Omega_{\vec{q}\lambda}} - |U^{-\vec{q}\lambda}_{\nu'\nu}|^2 N_{\vec{q}\lambda} e^{-i\Omega_{\vec{q}\lambda} t_1} \cdot e^{\beta_p\hbar\Omega_{\vec{q}\lambda}}][P^\beta_\nu f'_\nu (1 - f_{\nu'}) \\
&- P_{\nu'} f'_{\nu'} f_\nu] - [|U^{\vec{q}\lambda}_{\nu'\nu}|^2 (N_{\vec{q}\lambda} + 1) e^{i\Omega_{\vec{q}\lambda} t_1} - |U^{-\vec{q}\lambda}_{\nu'\nu}|^2 N_{\vec{q}\lambda} e^{-i\Omega_{\vec{q}\lambda} t_1}] \\
&\times [P^\beta_{\nu'} f'_{\nu'}(1 - f_\nu) - P^\beta_\nu f'_\nu f_{\nu'}]\} \cdot e^{i/\hbar(\varepsilon_{\nu'} - \varepsilon_\nu) t_1}.
\end{aligned}$$

We evaluate the integral over t_1 in the last expression by considering that

$$(N_{\vec{q}\lambda} + 1) e^{-\beta_p\hbar\Omega_{\vec{q}\lambda}} = \left(\frac{1}{e^{\beta_p\hbar\Omega_{\vec{q}\lambda}} - 1} + 1\right) e^{-\beta_p\hbar\Omega_{\vec{q}\lambda}} = N_{\vec{q}\lambda}.$$

The result of these simple calculations is

$$\begin{aligned}
\bar{\Sigma}_1 = &\frac{1}{nm\hbar}\sum_{\nu'\nu\vec{q}\lambda} q^\alpha \Big\{ |U^{\vec{q}\lambda}_{\nu'\nu}|^2 [N_{\vec{q}\lambda}(P^\beta_\nu f'_\nu (1 - f_{\nu'}) - P^\beta_{\nu'} f'_{\nu'} f_\nu) \\
&- (N_{\vec{q}\lambda} + 1)(P^\beta_{\nu'} f'_{\nu'}(1 - f_\nu) - P^\beta_\nu f'_\nu f_{\nu'})] \frac{\hbar}{i(\varepsilon_{\nu'} - \varepsilon_\nu + \hbar\Omega_{\vec{q}\lambda} - i\epsilon)} \\
&- |U^{-\vec{q}\lambda}_{\nu'\nu}|^2 [(N_{\vec{q}\lambda} + 1)(P^\beta_\nu f'_\nu (1 - f_{\nu'}) - P^\beta_{\nu'} f'_{\nu'} f_\nu) \\
&- N_{\vec{q}\lambda}(P^\beta_{\nu'} f'_{\nu'}(1 - f_\nu) - P^\beta_\nu f'_\nu f_{\nu'})] \frac{\hbar}{i(\varepsilon_{\nu'} - \varepsilon_\nu - \hbar\Omega_{\vec{q}\lambda} - i\epsilon)} \Big\}.
\end{aligned}$$

We make the change of variables $\nu' \leftrightarrows \nu$ in the second term and take into account that

$$|U^{\vec{q}\lambda}_{\nu'\nu}|^2 = |U^{-\vec{q}\lambda}_{\nu\nu'}|^2.$$

Then, using the definition of the delta function

$$\delta(x) = \lim_{\epsilon \to 0} \frac{1}{2\pi i} \left\{ \frac{1}{x - i\epsilon} - \frac{1}{x + i\epsilon} \right\},$$

we get

$$\bar{\Sigma}_1 = \frac{1}{nm} \frac{2\pi}{\hbar} \sum_{v' v \vec{q} \lambda} \hbar q^\alpha |U_{v'v}^{\vec{q}\lambda}|^2 \{ (N_{\vec{q}\lambda} + 1)(P_{v'}^\beta f'_{v'}(1 - f_v)$$

$$- P_v^\beta f'_v f_{v'}) - N_{\vec{q}\lambda}(P_v^\beta f'_v (1 - f_{v'}) - P_{v'}^\beta f'_{v'} f_v) \} \delta(\varepsilon_{v'} - \varepsilon_v + \hbar\Omega_{\vec{q}\lambda}).$$

Now we recall that

$$|U_{v'v}^{\vec{q}\lambda}|^2 = |U_{\vec{p}'\vec{p}}^{\vec{q}\lambda}|^2 = |C_{\vec{q}\lambda}|^2 |\langle \vec{p}'|e^{i\vec{q}\vec{r}}|\vec{p}\rangle|^2 = |C_{\vec{q}\lambda}|^2 \delta(\vec{p}' - \vec{p} - \hbar\vec{q}).$$

After applying the law of conservation of momentum, the final expression for the relaxation frequency of the momentum of a non-equilibrium system appears as

$$\frac{1}{\tau} = -\frac{1}{3nm} \frac{2\pi}{\hbar} \sum_{\vec{q}\lambda\vec{p}'\vec{p}\sigma} |C_{\vec{q}\lambda}|^2 (\hbar q)^2 \{ (N_{\vec{q}} + 1) f'_{\vec{p}'\sigma'}(1 - f_{\vec{p}\sigma})$$

$$+ N_{\vec{q}} f'_{\vec{p}'\sigma'} f_{\vec{p}\sigma} \} \delta(\varepsilon_{\vec{p}'\sigma'} - \varepsilon_{\vec{p}\sigma} + \hbar\Omega_{\vec{q}\lambda}). \tag{8.44}$$

It should be emphasized that the odd contributions with respect to the total momentum components vanish when summing over the momentum within the Brillouin zone, and $q^\alpha q^\beta = 1/3q^2 \delta_{\alpha\beta}$. Moreover, as far as the non-equilibrium electrons are concerned, the expression (8.44) for the inverse relaxation time and the result (4.157) obtained by means the method of kinetic equation for the momentum relaxation frequency correspond each other completely.

Finally, it is necessary to point out that the results obtained here relating to the electrical conductivity can be found as already noted in another way without mentioning the reason of the emergence of a new non-equilibrium distribution when switching on the additional measuring field. In essence, this is just the generalization of Kubo's formal linear response theory for a non-equilibrium system.

Let $H_F(t) = -AF(t)$ be an additional weak external field, acting on the non-equilibrium system, which is described by the Hamiltonian H. Then we write the Liouville equation satisfied by the new non-equilibrium distribution $\rho(t, 0)$:

$$\frac{\partial \rho(t, 0)}{\partial t} + [iL + iL_F(t)]\rho(t, 0) = -\epsilon(\rho(t, 0) - \rho^0(t, 0)).$$

Here $\rho^0(t, 0)$ is the initial non-equilibrium distribution of the system, iL, $iL_F(t)$ are the Liouville operators corresponding to the H and $H_F(t)$ Hamiltonians, respectively. Next, it is natural to think that the initial condition for the distribution $\rho(t)$ coincides with the original non-equilibrium distribution $\rho^0(t, 0)$ at the moment in time $t = -\infty$

after switching on the external field. In this case, the formula for the non-equilibrium admittance is expressed in terms of a commutator Green's function in full accordance with the Kubo theory. For example, in the case of the electrical conductivity, by analogy with the linear case, we obtain

$$\sigma(t,\omega) = -\frac{e^2}{m}\int_{-\infty}^{0} dt_1 e^{(\epsilon-i\omega)t_1}\mathrm{Sp}\left\{Pe^{iLt_1}\frac{1}{i\hbar}[X^+,\rho^0(t+t_1,0)]\right\}.$$

This formula can be easily transformed to the result (8.34) obtained earlier. For that, the principle of correlation weakening and the operator identity (8.21) can be applied. The latter is the generalization of the Kubo identity as to the non-equilibrium distribution. The simple calculations lead to

$$\sigma(t,\omega) = -\frac{e^2}{m^2}\langle P,P^+\rangle_t\overline{G}(t,\omega), \quad \overline{G}(t,\omega) = \frac{1}{\langle P,P^+\rangle_t}\langle P,P^+\rangle_t^{\omega}.$$

Given the relationship between the Green functions $\overline{G}(t,\omega)$ and the transport matrix $\overline{T}(t,\omega)$ defined by (8.31), it becomes clear that the above expression for non-equilibrium conductivity coincides with the result (8.34) previously obtained.

Problem 8.1. Obtain an expression for the inverse momentum relaxation time of non-equilibrium electrons using the Hamiltonian (4.81), describing the interaction between charge carriers and charged impurity centers, and also the formulas (8.35), (8.37) for a memory function.

Solution. Using the definition (8.35) and the fact that $\langle P^\alpha, P^\beta\rangle = -nm$ we obtain for the inverse relaxation time

$$\frac{1}{\tau} = -\frac{1}{nm}\int_{-\infty}^{0} dt_1 e^{(\epsilon-i\omega)t_1}\int_{-\infty}^{0} dt_2 e^{\epsilon' t_2}\mathrm{Sp}\left\{\dot{P}^\alpha e^{iL(t_1+t_2)}\frac{1}{i\hbar}[\dot{P}^\beta,\rho^0]\right\}. \tag{8.45}$$

As far as this expression already incorporates the second-order coupling constants of the scatterers, the non-equilibrium distribution ρ^0 can be replaced by a quasi-equilibrium distribution ρ_q. The quasi-equilibrium distribution can be represented as

$$\rho_q = e^{-S^0}, \quad S^0 = \Phi + \beta_k H_k + \beta_s H_s - \beta\zeta N.$$

Thus, the quasi-equilibrium distribution ρ_q describes the non-equilibrium distribution of electrons with reciprocal temperatures of the kinetic and spin degrees of freedom β_k and β_s, respectively. In addition, we can ignore the interaction in the evolution operator iL. So,

$$[\dot{P}^\beta,\rho^0] = -\int_0^1 d\tau \rho_q^\tau[\dot{P}^\beta_{(ei)}, S^0]\rho_q^{1-\tau}, \quad \dot{P}^\beta_{(ei)} = \frac{1}{i\hbar}[P^\beta, H_{ei}].$$

After writing both $\dot{P}^\alpha = \dot{P}^\alpha_{(ei)}$ and the commutator $[\dot{P}^\beta_{(ei)}, S^0]$ in the second-quantization representation, we find

$$\frac{1}{\tau} = \frac{1}{nm}\int_{-\infty}^{0} dt_1 e^{(\epsilon - i\omega)t_1}\int_{-\infty}^{0} dt_2 e^{\epsilon' t_2}\int_0^1 d\tau \sum_{\nu'\nu}\Big\langle \dot{P}^\alpha_{(ei)\nu'\nu} \times \frac{1}{i\hbar}[\dot{P}^\beta_{(ei)}, S^0]_{\mu'\mu}\Big\rangle_{imp}\langle a^+_{\nu'}a_\nu a^+_{\mu'}(z)a_\mu(z)\rangle, \tag{8.46}$$

$z = t_1 + t_2 + i\hbar\beta$. The angle brackets in this expression mean the averaging over the states of the impurities. Given the explicit form of the electron–impurity interaction Hamiltonian (4.81), the matrix elements of the operator $\dot{P}^\alpha_{(ei)\nu'\nu}$ have the form

$$\dot{P}^\alpha_{(ei)\nu'\nu} = -i\sum_{\vec{q}} q^\alpha G_{\vec{q}}\rho_{-\vec{q}}\langle\nu'|e^{i\vec{q}\vec{r}}|\nu\rangle, \quad \rho_{\vec{q}} = \sum_{j=1}^{N_i} e^{i\vec{q}\vec{R}_j},$$
$$[\dot{P}^\beta_{(ei)}, S^0]_{\mu'\mu} = \dot{P}^\beta_{(ei)\mu'\mu}(S^0_\mu - S^0_{\mu'}), \tag{8.47}$$

where $\vec{R}_j$ is the coordinate of the jth impurity center. In this case, the averaging over states of the scatterers reduces to the averaging of the quantities $\rho_{\vec{q}}$:

$$\langle\rho_{\vec{q}'}\rho_{\vec{q}}\rangle = N_i, \delta_{-\vec{q}\vec{q}'},$$

where N_i is the number of the scattering centers.

Substituting the result (8.47) into the expression for relaxation frequency, we obtain

$$\frac{1}{\tau} = \frac{N_i}{3nm}\int_{-\infty}^{0} dt_1 e^{(\epsilon - i\omega)t_1}\int_{-\infty}^{0} dt_2 e^{\epsilon' t_2}\sum_{\vec{q}\nu'\nu} q^2|G_{\vec{q}}|^2|\langle\nu'|e^{i\vec{q}\vec{r}}|\nu\rangle|^2 \times \frac{1}{i\hbar}\int_0^1 d\tau(S^0_{\nu'} - S^0_\nu)e^{i/\hbar(\varepsilon_\nu - \varepsilon_{\nu'})(t_1+t_2)}e^{(S^0_{\nu'} - S^0_\nu)\tau} f_{\nu'}(1 - f_\nu).$$

Integrating over τ, we arrive at

$$\int_0^1 d\tau(S^0_{\nu'} - S^0_\nu)e^{(S^0_{\nu'} - S^0_\nu)\tau} f_{\nu'}(1 - f_\nu) = (e^{(S^0_{\nu'} - S^0_\nu)} - 1)f_{\nu'}(1 - f_\nu)$$
$$= f_\nu(1 - f_{\nu'}) - f_{\nu'}(1 - f_\nu) = f_\nu - f_{\nu'}.$$

Next, we assume that the frequency ω of the external field is zero, and the integration over t_1 and t_2 gives

$$I = \mathrm{Re}\frac{1}{i\hbar}\int_{-\infty}^{0} dt_1 e^{\epsilon t_1}\int_{-\infty}^{0} dt_2 e^{\epsilon' t_2}e^{i/\hbar(\varepsilon_\nu - \varepsilon_{\nu'})(t_1+t_2)}$$
$$= -\mathrm{Re}\frac{\hbar}{i}\lim_{\epsilon\to 0}\lim_{\epsilon'\to 0}\left\{\frac{1}{\varepsilon_\nu - \varepsilon_{\nu'} - i\epsilon}\cdot\frac{1}{\varepsilon_\nu - \varepsilon_{\nu'} - i\epsilon'}\right\}.$$

We introduce the notation $\varepsilon_\nu - \varepsilon_{\nu'} = x$. Then, given that the following equality holds true in the limit $\epsilon \to 0$:

$$\lim_{\epsilon\to} \frac{1}{x - i\epsilon} = \frac{1}{x} + i\pi\delta(x),$$

we get the representation for the integral I

$$I = -\operatorname{Re}\frac{\hbar}{i}\left[\frac{1}{x} + i\pi\delta(x)\right]\left[\frac{1}{x} + i\pi\delta(x)\right] = -2\pi\hbar\frac{1}{x}\delta(x) = 2\pi\hbar\delta'(x).$$

The quantity $-1/x \cdot \delta(x)$ is usually determined as the derivative of the delta function $\delta'(x)$. To make sure that the representation is valid, it is necessary to consider an integral containing the product of the normal function $F(x)$ and the generalized function $\delta'(x)$. Computation of such integrals is produced by integrating by parts, assuming that $\delta(x) = 0$ if $x \neq 0$. Thus it is usually accepted that

$$\int F(x)\delta'(x)\,dx = -\int F'(x)\delta(x)\,dx.$$

Let us put $F(x) = xf(x)$. Then we have

$$\int xf(x)\delta'(x)\,dx = -\int f(x)\delta(x)\,dx - \int xf'(x)\delta(x)\,dx.$$

Since the last integral is always zero, hence the definition of derivative for the delta function $x\delta'(x) = -\delta(x)$ follows.

Once we plug the above results into the last expression for the inverse relaxation time, we have

$$\frac{1}{\tau} = -\frac{N_i}{3nm}\frac{2\pi}{\hbar}\sum_{\vec{q}\nu'\nu}(\hbar q)^2|G_{\vec{q}}|^2\left|\langle\nu'|e^{i\vec{q}\vec{r}}|\nu\rangle\right|^2 f'_\nu\delta(\varepsilon_\nu - \varepsilon_{\nu'}). \tag{8.48}$$

This expression coincides up to notations with the result previously obtained by the kinetic equation (4.204) for the inverse relaxation time of hot electrons.

Without doubt, we could have used the memory function presentation (8.38) to obtain the formula (8.48). It would be appropriate to dwell concisely on this method to deduce an expression for the inverse relaxation time of non-equilibrium electrons. Using the expression (8.38) as an original definition, one is led to

$$\begin{aligned}
\bar{\Sigma}_1 &= -\frac{1}{n}\int_{-\infty}^{0} dt_1 e^{\epsilon t_1}\operatorname{Sp}\left\{\dot{P}^\alpha e^{iLt_1} iL_\nu \frac{1}{i\hbar}[X^\beta, \rho^0]\right\} \\
&= \frac{1}{n}\int_{-\infty}^{0} dt_1 e^{\epsilon t_1}\int_0^1 d\tau \operatorname{Sp}\left\{\dot{P}^\alpha e^{iLt_1} iL_\nu \rho_q^\tau \frac{1}{i\hbar}[X^\beta, S^0]\rho_q^{1-\tau}\right\} \\
&= \frac{\beta_k}{nm}\int_{-\infty}^{0} dt_1 e^{\epsilon t_1}\operatorname{Sp}\left\{\dot{P}^\alpha_{(ei)} e^{iLt_1}\frac{1}{i\hbar}(P^\beta\rho_q H_{ei}\rho_q^{-1} - H_{ei}P^\beta)\rho_q\right\}.
\end{aligned}$$

Passing on to the second-quantization representation, we come with up the following result instead of (8.46):

$$\frac{1}{\tau} = \frac{\beta_k}{nm} \frac{1}{i\hbar} \int_{-\infty}^{0} dt_1 e^{\epsilon t_1} \sum_{v'v\vec{q}} (-iq^{\alpha})|G_q|^2 N_i |\langle v'|e^{i\vec{q}\vec{r}}|v\rangle|^2$$

$$\times \{P_v^{\beta} f_{v'}(1-f_v) e^{S_{v'}^0 - S_v^0} - P_{v'} f_{v'}(1-f_v)\} e^{i/\hbar(\varepsilon_v - \varepsilon_{v'})t_1}.$$

Furthermore, after making a change in the summation indices of the first term $v \leftrightarrows v'$, $\vec{q} \to -\vec{q}$, and integrating over time t_1, we have

$$\frac{1}{\tau} = -\frac{\beta_k}{nm} N_i \sum_{v'v\vec{q}} iq^{\alpha}|G_q|^2 |\langle v'|e^{i\vec{q}\vec{r}}|v\rangle|^2$$

$$\times P_{v'}^{\beta} f_{v'}(1-f_v) \left\{ \frac{1}{\varepsilon_v - \varepsilon_{v'} + i\epsilon} - \frac{1}{\varepsilon_v - \varepsilon_{v'} - i\epsilon'} \right\},$$

since

$$f_{v'}(1-f_v) e^{S_{v'}^0 - S_v^0} = f_v(1-f_{v'}).$$

We use the fact that by virtue of the momentum conservation law $P_{v'}^{\beta} = P_v^{\beta} + \hbar q^{\beta}$, and taking the definition of the delta function into account, we get

$$\frac{1}{\tau} = \frac{2\pi}{\hbar} \frac{\beta_k}{3nm} N_i \sum_{v'v\vec{q}} (\hbar q)^2 |G_q|^2 |\langle v'|e^{i\vec{q}\vec{r}}|v\rangle|^2 f_{v'}(1-f_v)\delta(\varepsilon_v - \varepsilon_{v'}). \qquad (8.49)$$

The expression (8.49) is diagonal in spin indices. Therefore, in spite of being non-equilibrium functions, the functions $f_v, f_{v'}$ in fact differ only in the kinetic energy of electrons.

Therefore, the distribution functions f_v and $f_{v'}$ are equal to each other provided that the kinetic energies are equal too. Consequently,

$$f_{v'}(1-f_v) = -\beta_k^{-1} f'_{v'},$$

and we again obtain the result (8.48).

Problem 8.2. Obtain an expression for transverse components of paramagnetic spin susceptibility of non-equilibrium electrons in conductive crystals.

Solution. Let a non-equilibrium state of the electronic system be stationary and be described by the initial non-equilibrium distribution

$$\rho^0 = \epsilon \int_{-\infty}^{0} dt_1 e^{\epsilon t_1} e^{iLt_1} \rho_q, \quad \rho_q = e^{-S^0},$$

$$S^0 = \Phi + \beta_k H_k + \beta_s H_s + \beta_l H_l + \beta_d H_d - \beta\zeta N,$$
$$\Phi = \ln \operatorname{Sp} \exp\{-\beta_k H_k - \beta_s H_s - \beta_l H_l - \beta_d H_d + \beta\zeta N\}, \tag{8.50}$$

where H_l and H_d are the Hamiltonians of the phonon subsystem and the subsystem of d-electrons, respectively (the presence of the d-subsystem is important for magnetic semiconductors). β_l and β_d are the corresponding inverse temperatures;

$$H_k = \sum_{\vec{k}\sigma} \varepsilon_{\vec{k}} a^+_{\vec{k}\sigma} a_{\vec{k}\sigma}, \varepsilon_{\vec{k}} = \frac{\hbar^2 k^2}{2m}, \quad H_s = -\hbar\omega_s S^z = -\hbar\omega_s \sum_{\vec{k}\sigma} \sigma a^+_{\vec{k}\sigma} a_{\vec{k}\sigma},$$

ω_s being the Zeeman spin precession frequency in an external constant magnetic field $H \parallel Z$. Allowance for the d-subsystem of the local magnetic moments without analyzing both the shape of the specimen and problems of critical dynamics leads only to some renormalization of the external magnetic field. So we may exclude the d-electrons from further consideration. H_l is the Hamiltonian of the phonon subsystem. As far as in the future we will not analyze processes of the energy transfer from the electron system to the phonon subsystem and then into a heat bath, then H_l in the operator entropy S^0 can be omitted without any loss of generality.

As is well known [61], for finding the system response to the external static magnetic field it is sufficient to find the response to a single Fourier component of this field. Therefore, the Hamiltonian describing the interaction of electrons with a weak external polarized perpendicular to the Z-axis magnetic field can be written as

$$H_{eF} = -\frac{g\mu_B}{2} \sum_i [S_i^+ h^-(\vec{q}) + S_i^- h^+(\vec{q})] e^{i\vec{q}\vec{r}_i}, \tag{8.51}$$

$S_i^\pm = S_i^x \pm i S_i^y$, $h^\pm(\vec{q}) = h^x(\vec{q}) \pm i h^y(\vec{q})$, $h^x(\vec{q})$, $h^y(\vec{q})$ are the Fourier components of the inhomogeneous magnetic field in a Cartesian coordinate system. Introducing the notation

$$S_q^\pm = \sum_i S_i^\pm e^{i\vec{q}\vec{r}_i},$$

we study the system response to only one of two circular components in the plane-polarized external field. Then the Hamiltonian of interaction with the external field (8.51) can be simplified by leaving only one component of the circularly polarized external field

$$H_{eF} = -\frac{g\mu_B}{2} S_q^- h^+(\vec{q}). \tag{8.52}$$

In accordance with the general response theory (see Section 8.1.2), the non-equilibrium system response to the perturbation (8.51) is determined by the transverse components of the static susceptibility tensor (8.24). Therefore, the static paramagnetic susceptibility χ_q^{+-} per lattice node is computed as follows:

$$\chi_q^{+-} = -\frac{(g\mu_B)^2}{2N} \langle S_q^+, S_{-q}^- \rangle. \tag{8.53}$$

The static magnetic susceptibility differs from zero in the zeroth-order interaction with the scatterers. The small corrections due to the interaction being ignored, the following changes in determining the equilibrium correlation function (8.17) can be made:

$$iL \to iL_0, \quad \rho^0(t,0) \to \rho_q.$$

As a result, we arrive at an expression for the transverse component of the static spin susceptibility:

$$\begin{aligned} \chi_q^{+-} &= -\frac{(g\mu_B)^2}{2N} \int_{-\infty}^{0} dt_1 e^{\epsilon t_1} \mathrm{Sp}\left\{ S_q^+ e^{iL_0 t_1} \frac{1}{i\hbar} [S_{-q}^-, \rho_q] \right\} \\ &= -\frac{(g\mu_B)^2}{2N} \frac{1}{i\hbar} \int_{-\infty}^{0} dt_1 e^{\epsilon t_1} [\mathrm{Sp}\{S_q^+ S_{-q}^-(t_1)\rho_q\} - \mathrm{Sp}\{S_{-q}^-(t_1) S_q^+ \rho_q\}]. \end{aligned} \tag{8.54}$$

In deriving this expression we have taken into account the fact that $[\rho_q, H_0] = 0$. We write the operators $S_q^{\pm}$ in the second-quantization representation:

$$S_q^{\pm} = \sum_i S_i^{\pm} e^{i\vec{q}\vec{r}_i} = \sum_{\vec{k}'\vec{k}\sigma'\sigma} S_{\sigma'\sigma}^{\pm} \langle \vec{k}' | e^{i\vec{q}\vec{r}} | \vec{k} \rangle a_{\vec{k}'\sigma'}^{+} a_{\vec{k}\sigma}.$$

Then χ_q^{+-} can be rewritten as

$$\begin{aligned} \chi_q^{+-} &= \frac{(g\mu_B)^2}{2N} \sum_{\vec{k}'\vec{k}\sigma'\sigma} \left[|S_{\sigma'\sigma}^{+}|^2 |\langle \vec{k}' | e^{i\vec{q}\vec{r}} | \vec{k} \rangle|^2 \frac{1}{\varepsilon_{\vec{k}\sigma} - \varepsilon_{\vec{k}'\sigma'} - i\epsilon} \right. \\ &\quad \left. - |S_{\sigma'\sigma}^{-}|^2 |\langle \vec{k}' | e^{-i\vec{q}\vec{r}} | \vec{k} \rangle|^2 \frac{1}{\varepsilon_{\vec{k}'\sigma'} - \varepsilon_{\vec{k}\sigma} - i\epsilon} \right] f_{\vec{k}'\sigma'}(1 - f_{\vec{k}\sigma}). \end{aligned} \tag{8.55}$$

Replacing the summation indices $\vec{k}'\sigma' \leftrightarrows \vec{k}\sigma$ of the second term, and noting that

$$S_{\sigma'\sigma}^{+}|^2 |\langle \vec{k}' | e^{i\vec{q}\vec{r}} | \vec{k} \rangle|^2 = |S_{\sigma\sigma'}^{-}|^2 |\langle \vec{k} | e^{-i\vec{q}\vec{r}} | \vec{k}' \rangle|^2,$$

we get

$$\begin{aligned} \chi_q^{+-} &= \frac{(g\mu_B)^2}{2N} \sum_{\vec{k}'\vec{k}\sigma'\sigma} |S_{\sigma'\sigma}^{+}|^2 |\langle \vec{k}' | e^{i\vec{q}\vec{r}} | \vec{k} \rangle|^2 \\ &\quad \times \left[\frac{f_{\vec{k}'\sigma'}(1 - f_{\vec{k}\sigma})}{\varepsilon_{\vec{k}'\sigma'} - \varepsilon_{\vec{k}\sigma} - i\epsilon} - \frac{f_{\vec{k}\sigma}(1 - f_{\vec{k}'\sigma'})}{\varepsilon_{\vec{k}'\sigma'} - \varepsilon_{\vec{k}\sigma} - i\epsilon} \right]. \end{aligned} \tag{8.56}$$

When separating them out from the resolvent in the first and second terms of (8.56), the two contributions (in the principal value and singular) then allow one to divide the non-equilibrium static susceptibility into the real and imaginary parts:

$$\mathrm{Re}\chi_q^{+-} = \frac{(g\mu_B)^2}{2N} \sum_{\vec{k}} \frac{f_{\vec{k}+\vec{q}\uparrow} - f_{\vec{k}\downarrow}}{\varepsilon_{\vec{k}+\vec{q}\uparrow} - \varepsilon_{\vec{k}\downarrow}}, \tag{8.57}$$

$$\operatorname{Im}\chi_q^{+-} = \frac{\pi(g\mu_B)^2}{2N}\sum_{\vec{k}}[f_{\vec{k}+\vec{q}\uparrow} - f_{\vec{k}\downarrow}]\delta(\varepsilon_{\vec{k}+\vec{q}\uparrow} - \varepsilon_{\vec{k}\downarrow}). \tag{8.58}$$

While obtaining these results we have used the fact that the non-zero matrix element

$$|S^+_{\sigma'\sigma}|^2 = |S^+_{\uparrow\downarrow}|^2 = 1.$$

Here the arrows $\uparrow\downarrow$ stand for states with a projection of the z-component of spin momentum. The states are oriented along and opposite the direction to the external magnetic field H. Resorting to the formulas (8.56)–(8.58) we can easily prove that

$$\operatorname{Re}\chi_{-q}^{-+} = \operatorname{Re}\chi_q^{+-}, \quad \operatorname{Im}\chi_q^{+-} = -\operatorname{Im}\chi_{-q}^{-+}.$$

As it follows from (8.57), the real component of the transverse static susceptibility of non-equilibrium electrons has the same form as the analogous quantity in the equilibrium case (see, for example, the monograph [61]). The difference consists only in replacing the equilibrium distribution functions by non-equilibrium, as in our case

$$f_{\vec{k}+\vec{q}\uparrow} = \left[\exp\{\beta_k\varepsilon_{\vec{k}+\vec{q}} - \beta_s\hbar\omega_s/2 - \beta\zeta\} + 1\right]^{-1}.$$

Since the further procedure for computing the real part $\operatorname{Re}\chi_q^{+-}$ is known well enough, we should not dwell on it.

The imaginary component χ_q^{+-} (8.58) is of much greater interest. In the non-equilibrium case, the imaginary component of the static susceptibility becomes non-zero, since equality of the energies $\varepsilon_{\vec{k}+\vec{q}\uparrow} = \varepsilon_{\vec{k}\downarrow}$ due to the presence of a delta function in (8.57) does not mean equality of the distribution functions $f_{\vec{k}+\vec{q}\uparrow}$ and $f_{\vec{k}\downarrow}$. Therefore, there occurs energy dissipation in such a system (energy exchange between the k- and s-subsystems of a crystal is caused by a static inhomogeneous magnetic field). It is easy to see that

$$\lim_{q\to 0}\operatorname{Im}\chi_q^{+-} = 0,$$

because the kinetic energies of initial and final states are equal at $q = 0$, which means nonexistence of spin-flip transitions.

To perform further calculations in the formula (8.58) we pass from summation over the wave vector $\vec{k}$ to integration over the momentum $\vec{p} = \hbar\vec{k}$. As a result, we obtain

$$\operatorname{Im}\chi_q^{+-} = \frac{\pi(g\mu_B)^2}{2N}\frac{V}{(2\pi\hbar)^3}\int d\vec{p}\int d\varepsilon\left[f\left(\beta_k\frac{(\vec{p}+\hbar\vec{q})^2}{2m} - \beta_s\frac{\hbar\omega_s}{2} - \beta\zeta\right)\right.$$
$$\left. - f\left(\beta_k\frac{\vec{p}^2}{2m} + \beta_s\frac{\hbar\omega_s}{2} - \beta\zeta\right)\right]\delta\left(\varepsilon - \frac{(\vec{p}+\hbar\vec{q})^2}{2m} + \frac{\hbar\omega_s}{2}\right)\delta\left(\varepsilon - \frac{\vec{p}^2}{2m} - \frac{\hbar\omega_s}{2}\right).$$

Using delta functions, we replace the arguments of the distribution functions by avoiding the dependence on the momentum components. Then integration over $\vec{p}$ covers only the delta-functions and we shall deal with the integral

$$I = \int d\vec{p}\,\delta\left(\varepsilon - \frac{(\vec{p}+\hbar\vec{q})^2}{2m} + \frac{\hbar\omega_s}{2}\right)\delta\left(\varepsilon - \frac{\vec{p}^2}{2m} - \frac{\hbar\omega_s}{2}\right).$$

The procedure for estimating the integral was discussed in Chapter 4 (see (4.162)–(4.166)). So it makes sense to give at once the result:

$$I = \frac{2\pi m^2}{\hbar q}, \quad \hbar q_- \le \hbar q \le \hbar q_+, \quad \hbar q_\mp = \sqrt{2m(\varepsilon - \hbar\omega_s/2)}\left[\sqrt{1 + \frac{\hbar\omega_s}{\varepsilon - \hbar\omega_s/2}} \mp 1\right].$$

If the quantity q does not satisfy these inequalities, then $I = 0$. Substituting the value of the integral I into the definition (8.58) we get for $\operatorname{Im}\chi_q^{+-}$

$$\operatorname{Im}\chi_q^{+-} = \frac{V}{N}\frac{(g\mu_B)^2 m^2}{8\pi\hbar^4 q}\int_0^\infty d\varepsilon\left[f\left(\beta_k\varepsilon + (\beta_k - \beta_s)\frac{\hbar\omega_s}{2} - \beta\zeta\right)\right.$$
$$\left. - f\left(\beta_k\varepsilon - (\beta_k - \beta_s)\frac{\hbar\omega_s}{2} - \beta\zeta\right)\right]. \tag{8.59}$$

Assuming that the parameter $(\beta_k - \beta_s)\hbar\omega_s/2$ is small, we expand the distribution functions of the expression (8.59) into a series over this parameter, restricting ourselves to linear terms,

$$f(\beta_k\varepsilon + (\beta_k - \beta_s)\hbar\omega_s/2 - \beta\zeta) \simeq f(\beta_k\varepsilon - \beta\zeta) + \frac{(\beta_k - \beta_s)}{\beta_k}\frac{\hbar\omega_s}{2}\frac{df}{d\varepsilon}.$$

Then, if one assumes that an electron gas is under conditions of degeneration and

$$\frac{df}{d\varepsilon} \simeq -\delta(\varepsilon - \zeta),$$

we can obtain the simple estimate

$$\operatorname{Im}\chi_q^{+-} \simeq \frac{V}{N}\frac{(g\mu_B)^2 m^2\omega_s}{8\pi\hbar^3 q}\frac{(\beta_s - \beta_k)}{\beta_k}. \tag{8.60}$$

It should be noted that the emergence of the imaginary part of the static magnetic susceptibility of non-equilibrium electrons only evidences the possible influence of an external inhomogeneous magnetic field on processes of the energy transfer between non-equilibrium kinetic and spin degrees of freedom of the crystal. The energy change of the kinetic degrees of freedom of the conduction electrons occurs due to the action of force fields, which determine the original non-equilibrium state of the system.

8.3 Problems to Chapter 8

8.1. If a basic set of operators initially includes all necessary dynamical variables for constructing the NSO, the analogue of the expressions (8.23), (8.26) for the isothermal response of a non-equilibrium system to an external mechanical perturbation, defined by the Hamiltonian $H_f = -Ah(\omega)\exp\{-i\omega t\}$, can be obtained by using the modified expressions (6.62), (6.71). Plugging the summand (see the expression (6.115))

$$-\int_{-\infty}^{0} dt_1 \exp(\epsilon t_1)\exp(iLt_1)iL_f\bar{\rho}(t+t_1)$$

into the right-hand side of the expression (6.62), the formula (6.71) can be rewritten as

$$i\omega(P,P^+)^{\omega}\delta F(\omega) - (P,\dot{P}^+)^{\omega}\delta F(\omega) = (P\dot{A})^{\omega}h(\omega).$$

The formula derived previously allows the internal fields $F(\omega)$ as a result of the system's reaction to be expressed through the external field $h(\omega)$.
For this case, deduce an expression for isothermal susceptibility of the non-equilibrium system, using the definition

$$\chi_{BA}(\omega) = \delta\langle B\rangle^{\omega}h(\omega).$$

8.2. Argue that the identities

$$\mathcal{P}_t(1-\mathcal{P}_t)A = 0, \quad \mathcal{P}_t M = M, \quad \mathcal{P}_t^2 M = M,$$

are valid for the projection operators given by the relation (8.32).

8.3. Using Wick–Bloch–Dominicis's statistical theorem, prove that all nonvanishing-contribution pairings obtained by averaging in the formula (8.41) can be determined by the expression

$$\begin{aligned}\langle a_{\nu'}^{+}a_{\nu}a_{\kappa}^{+}a_{\kappa}a_{\mu'}^{+}a_{\mu}\rangle &= f_{\nu'}(1-f_{\nu})^2\delta_{\nu'\mu}\delta_{\nu\kappa}\delta_{\kappa\mu'} - f_{\nu'}^2(1-f_{\nu})\delta_{\nu'\kappa}\delta_{\mu\kappa}\delta_{\nu\mu'}\\ &= f_{\nu'}(1-f_{\nu})(1-f_{\nu}-f_{\nu'})\delta_{\nu'\mu}\delta_{\nu\mu'}.\end{aligned}$$

8.4. Using the definitions (8.36), (8.37) one needs to derive an expression for the reverse relaxation time of the average momentum of non-equilibrium electrons, interacting with long-wavelength phonons, and determine its temperature dependence.

8.5. Using the expression for the memory function

$$\bar{\Sigma}_1 = \frac{1}{n}\int_{-\infty}^{0} dt_1 e^{\epsilon t_1}\int_{0}^{1} d\tau \mathrm{Sp}\left\{\dot{P}^{\alpha}e^{iLt_1}iL_{\nu}\rho_q^{\tau}\frac{1}{i\hbar}[X^{\beta},S^0]\rho_q^{1-\tau}\right\},$$

derive an expression for the reverse relaxation time of the average momentum of non-equilibrium electrons in scattering of the latter by charged impurity centers.

9 Master equation approach

9.1 The basic idea of the method

9.1.1 Problem statement

The purpose of the present chapter is to give an overview of the master equation method and details of its application for solving problems of physical kinetics. The equation of motion for some part of a statistical operator is referred to as *a master kinetic equation*. A finding of this part of the statistical operator is not arbitrary and must satisfy the principles formulated in the previous chapters. Some issues are likely to be repeated, but it makes sense to re-discuss a program to construct a theory of irreversible processes [62].

Firstly, the Boltzmann equation and Fokker–Planck equations are closed Markovian equations (i. e. they do not make an allowance for any lagging), describing the establishment of thermal equilibrium in a system. As shown in Chapter 6, it is impossible to construct an equation of motion to describe irreversible evolution for the total statistical operator. Indeed, the material set forth in Chapters 5, 6, 8 requires addressing the projection operator method to define appropriate expressions for transport coefficients even in spite of using the non-equilibrium statistical operator, which satisfies a time-irreversible equation of motion. For this reason, an attempt to design immediately the statistical operator to describe the system's irreversible evolution is natural. In addition, it makes sense to restrict oneself to a simple assumption, under which the statistical operator is to be represented as the sum of two terms:

$$\rho(t) = \mathcal{P}\rho(t) + (1 - \mathcal{P})\rho(t). \tag{9.1}$$

These two summands are formed so that the quantity $\mathcal{P}\rho(t)$ should permit to formulate a closed equation. All existing theories are based on the fact that the remainder of the statistical operator $(1 - \mathcal{P})\rho(t)$ does not contribute to the observed dynamics at all.

It should be reinforced that the division of the statistical operator into two parts is trivial by itself, and gives nothing new at all, because a quantity A can always be represented as $B + (A - B)$. The representation (9.1) can serve as a basis for constructing the theory provided that this division is natural and consistent with parts of slow kinetics and fast oscillating dynamics. Further, in order for the theory to be self-consistent, the operators $\mathcal{P}$ and $(1 - \mathcal{P})$ must have the properties of the projection operators

$$\mathcal{P}^2 = \mathcal{P}, \quad (1 - \mathcal{P})^2 = (1 - \mathcal{P}), \quad \mathcal{P}(1 - \mathcal{P}) = 0. \tag{9.2}$$

The relations (9.2) provide orthogonality of the operators in some sense and create prerequisites to separate the quantities $\mathcal{P}\rho(t)$ and $(1 - \mathcal{P})\rho(t)$ in the dynamics.

Secondly, the most important property of splitting the statistical operator is to construct a closed equation for the kinetic part $\mathcal{P}\rho(t)$. In other words, there should

https://doi.org/10.1515/9783110727197-009

arise sub-dynamics of the quantity $\mathcal{P}\rho(t)$. For that, the projection operator has to possess some additional properties. Indeed, let $U(t)$ be an evolution operator which describes a change in the statistical operator $\rho(t) = U(t)\rho(0)$ in time, and $W(t)$ be an evolution operator describing the Markov kinetic part of the statistical operator $\overline{\rho}(t) = \mathcal{P}\rho(t)$. The quantity introduced by the last relation $\overline{\rho}(t)$ plays the role of the "relevant" part of the statistical operator. If $W(t)$ is the evolution operator for $\overline{\rho}(t)$, the equation $\overline{\rho}(t) = W(t)\overline{\rho}(0)$, must hold true. Consequently, recollecting the definition of the $\overline{\rho}(t)$, we get

$$\mathcal{P}\rho(t) = W(t)\mathcal{P}\rho(0).$$

This relation can be written differently by considering the equation of motion of the statistical operator $\rho(t) = U(t)\rho(0)$. Indeed, there is equality $\mathcal{P}U(t)\rho(0) = W(t)\mathcal{P}\rho(0)$. Hence there follows the "intertwining" relation

$$\mathcal{P}U(t) = W(t)\mathcal{P}, \tag{9.3}$$

which allows one to control the correctness of the theory developed.

The algorithm stated above can lead to completely different equations. The reason for this is enough obvious since there exists only a single state of thermodynamic equilibrium in every system. However, there are countless non-equilibrium states. As far as a choice of the projection operator is responsible for the "class" of the non-equilibrium states, it is clear that there are innumerable projection operators. The previous chapters discussed the projection operators, projecting dynamical variables on a certain basic set of operators. The following sections make one familiar with other definitions of the projection operator for the statistical distribution. Moreover, we will see how to calculate kinetic coefficients by means of this approach.

9.1.2 The Zwanzig kinetic equation

To familiarize the reader with the method of master equations, it would be proper to start with the equation obtained by Zwanzig [57]. To illustrate the method, Zwanzig has used the projection operator, which defines system's dynamics in momentum space by averaging completely the motion in coordinate space. However, this fact makes it impossible for the Zwanzig equation to be applied directly to calculate kinetic coefficients, but nevertheless, Zwanzig's paper [57] reviews basic ideas of the projection operator method.

We begin by looking at the Liouville equation (5.19)

$$\frac{\partial}{\partial t}\rho(t) + iL\rho(t) = 0 \tag{9.4}$$

for a statistical operator. Although the equation holds true both in the classical and in the quantum case, the latter is worth considering for the sake of definiteness.

We introduce a time-independent projection operator $\mathcal{P}$ and divide the statistical operator $\rho(t)$ into two summands:

$$\rho(t) = \bar{\rho}(t) + \tilde{\rho}(t), \quad \bar{\rho}(t) = \mathcal{P}\rho(t), \quad \tilde{\rho}(t) = (1 - \mathcal{P})\rho(t). \tag{9.5}$$

We apply the operators $\mathcal{P}$ and $(1 - \mathcal{P})$ both sides of the Liouville equation (9.4). As a result, we obtain

$$\frac{\partial \bar{\rho}(t)}{\partial t} = -\mathcal{P}iL[\bar{\rho}(t) + \tilde{\rho}(t)], \tag{9.6}$$

$$\frac{\partial \tilde{\rho}(t)}{\partial t} = -(1 - \mathcal{P})iL(\bar{\rho}(t) + \tilde{\rho}(t)). \tag{9.7}$$

In order for the set of equations (9.6), (9.7) to have a single-valued solution, it is necessary to define the statistical operator at some point of time. This at first glance formal mathematical procedure indeed has a deep physical meaning to be discussed later.

To obtain a closed equation for $\bar{\rho}(t)$, $\tilde{\rho}(t)$ needs to be eliminated from the right-hand side of the expression (9.6). We perform a formal integration of the equation (9.7). It is easier to do this in the following way: multiply both sides of the equation (9.7) by the operator $\exp\{i(1 - \mathcal{P})Lt\}$ from the left and write it in the form

$$\frac{d}{dt} \exp\{i(1 - \mathcal{P})Lt\}\tilde{\rho}(t) = -i \exp\{i(1 - \mathcal{P})Lt\}(1 - \mathcal{P})L\bar{\rho}(t). \tag{9.8}$$

The formal integration of the equation (9.8) in the range from some initial time t_0 to time t, which is of interest to us, yields

$$\begin{aligned} &\exp\{i(1 - \mathcal{P})Lt\}\tilde{\rho}(t) - \exp\{i(1 - \mathcal{P})Lt_0\}\tilde{\rho}(t_0) \\ &\quad = -i \int_{t_0}^{t} \exp\{i(1 - \mathcal{P})Lt'\}(1 - \mathcal{P})L\bar{\rho}(t')\, dt'. \end{aligned} \tag{9.9}$$

Re-multiply both sides of the equation (9.9) by the operator $\exp\{-i(1 - \mathcal{P})Lt\}$ from the left. Performing the necessary calculations, we obtain

$$\begin{aligned} \tilde{\rho}(t) = &-i \int_{t_0}^{t} \exp\{i(1 - \mathcal{P})L(t' - t)\}(1 - \mathcal{P})L\bar{\rho}(t')\, dt' \\ &+ \exp\{i(1 - \mathcal{P})L(t_0 - t)\}\tilde{\rho}(t_0). \end{aligned} \tag{9.10}$$

After substituting the expression (9.10) into the right side of (9.6), we can arrive at the equation for the part of the statistical operator $\bar{\rho}(t)$, describing the system's irreversible evolution:

$$\frac{\partial \bar{\rho}(t)}{\partial t} + i\mathcal{P}L\bar{\rho}(t) = \int_{t_0}^{t} \Sigma(t' - t)\bar{\rho}(t')dt' - i\mathcal{P}L \exp\{i(1 - \mathcal{P})L(t_0 - t)\}\tilde{\rho}(t_0). \tag{9.11}$$

$$\Sigma(t'-t) = i\mathcal{P}L\exp\{i(1-\mathcal{P})L(t'-t)\}i(1-\mathcal{P})L. \tag{9.12}$$

In expressions (9.8), (9.12) the exponential functions of the operator quantities iL and $\mathcal{P}$ mean the corresponding power series. Equation (9.11) is still not a closed equation since it contains the quantity $\tilde{\rho}(t_0)$ at some initial point of time t_0.

Let us return to the problem of defining an initial condition for the Liouville equation (9.4). The setting of the statistical operator at some initial moment of time is of paramount importance because it is equivalent to defining an ensemble of identical systems whose evolution the Liouville equation describes. Moreover, the choice of initial conditions may determine the class of solutions of the Liouville equation.

It is clear that there is no correct mathematical procedure to write down this initial distribution for any complex system. Certainly, coordinates and velocities of all particles making up a system in the classical case, or a wave function of a particle system in the quantum case may serve as initial conditions, but it will be a formal and useless assignment.

As mentioned earlier in the previous chapters, the initial distribution for internally stochastic systems should not determine anything already over a short time interval of the order of a mixture time in the system. Therefore, the initial distribution can be chosen quite arbitrarily, for example, to define its dependence on dynamical quantities via slowly changing variables such as integrals or quasi-integrals of motion. The reason for this is that the particular type of the projection operator and the initial distribution are to be consistent, so that the projection operator does not change the initial distribution.

Following these recommendations, we choose an initial distribution for the equation (9.11):

$$\rho(t_0) = \overline{\rho}(t_0), \quad \tilde{\rho}(t_0) = 0. \tag{9.13}$$

Then the Zwanzig master equation can be written in the form

$$\frac{\partial\overline{\rho}(t)}{\partial t} + i\mathcal{P}L\overline{\rho}(t) = \int_{t_0}^{t} \Sigma(t'-t)\overline{\rho}(t')dt'. \tag{9.14}$$

The kernel of the integral equation (9.14)

$$\Sigma(t'-t) = i\mathcal{P}L\exp\{i(1-\mathcal{P})L(t'-t)\}i(1-\mathcal{P})L \tag{9.15}$$

is responsible for the “memory” about all previous states of the system (similar to lagging in electrodynamics).

Thus, we have derived a closed equation, describing both non-Markovian and irreversible evolution of the statistical operator $\overline{\rho}(t)$. If we define a particular form of the projection operator and an expression for the averages values of operators of physical quantities, the equations (9.14), (9.15) can be used to compute kinetic coefficients. The next sections concern more interesting, from a practical standpoint, applications of the projection operator technique. In particular, we will obtain the master equation for a quasi-equilibrium distribution and show how to deduce an expression for the kinetic coefficients for a high-non-equilibrium system by using it.

9.2 Master equation for the quasi-equilibrium distribution

9.2.1 Robertson projection operator

Chapters 6 and 8 dealt with the quasi-equilibrium distribution $\rho_q(t,0)$ that is of some part of the NSO, on the one hand, and, on the other hand, allows one to calculate averages of basic operators. This is evidenced by the fact that the NSO-method provides equality of the averages, computed by using the true non-equilibrium distribution and quasi-equilibrium distribution. (See equation (6.6)).

Thus, if we succeed in constructing a closed equation to determine the quasi-equilibrium distribution and in finding a practical way to solve this equation for recovering the form of $\rho_q(t,0)$, it enables one immediately to express the kinetic coefficients in terms of correlation functions of operators of dynamic variables, computed with the use of the quasi-equilibrium distribution.

It is appropriate to recall once again the differences in the programs of constructing the kinetic theory based on techniques of the kinetic equation, statistical operator and master equation.

As to the kinetic equation, the main difficulty is to find the non-equilibrium distribution function, in other words to construct a solution of the Boltzmann equation. Such a function being found, computation of the kinetic coefficients reduces to quadratures.

Within the Kubo method in the quantum-statistical approach, it is relatively easy to obtain a formal solution, for example, of the Liouville equation. So, to calculate the kinetic coefficients, it is necessary to compute appropriate correlation functions correctly. This problem can be correctly solved if and only if one replaces equations of motion for the operators of dynamic variables by Langevin-like equations of motion. The latter to be derived require applying the projection operator technique.

It should be emphasized that the projection operators are used here to construct the correct dynamic equations of the equilibrium system.

When using the NSO-method we have an intermediate situation at some sense. On the one hand, the NSO is built only from quasi-integrals of motion, i. e. slowly changing dynamical variables obtained as a result of the temporal averaging operation (6.52). The replacement itself of the exact statistical operator (6.52) by NSO is a projection operation to separate out some part of the statistical operator. This approach allows for deriving closed equations to find the non-equilibrium thermodynamic parameters $F_n(t)$ of the system (see Section 6.2.6).

That some roughening in description, caused by the temporal smoothing is being used here, is the fact that the number of the non-equilibrium parameters turned out to be finite. To describe the system's dynamics exactly, it is necessary that the number of the non-equilibrium parameters should be of the order of the number of particles in the system.

On the other hand, the dynamic equations satisfied by the basic operators are standard equations of Newtonian dynamics, or the Schrödinger equations. That is why the conception of projection operators needs to be brought about for constructing the correct dynamical equations in systems with a stirring.

Finally, there is an approach underlying the constructing the equation of motion for the quasi-equilibrium distribution using the technique of projection operators from scratch.

Let us consider the derivation of this equation. It is convenient to start with the Liouville equation for NSO (6.54):

$$\frac{\partial \rho(t,0)}{\partial t} + iL\rho(t,0) = -\epsilon\big(\rho(t,0) - \rho_q(t,0)\big); \quad \epsilon \to +0. \tag{9.16}$$

Subtract from both sides of this equation the operator

$$\left[\frac{\partial}{\partial t} + iL(t)\right]\rho_q(t).$$

As a result, we obtain

$$\left[\frac{\partial}{\partial t} + iL\right]\big(\rho(t,0) - \rho_q(t,0)\big) = -\epsilon\big(\rho(t,0) - \rho_q(t,0)\big) - \left[\frac{\partial}{\partial t} + iL\right]\rho_q(t,0). \tag{9.17}$$

Consider the time derivative of the operator $\rho_q(t,0)$. As noted in Chapter 6, the quasi-equilibrium distribution is a functional of the observed averages $\langle P_n\rangle^t$, taken at one and the same moment of time t. Therefore, we have

$$\frac{\partial \rho_q(t,0)}{\partial t} = \sum_n \frac{\partial \rho_q(t,0)}{\partial \langle P_n\rangle^t}\frac{\partial}{\partial t}\langle P_n\rangle^t. \tag{9.18}$$

It should be emphasized that the expression (9.18) differs from (6.7) only by other designation of the quasi-equilibrium distribution, but for convenience we have written this formula again. Recall that P_n is a set of basic operators, which are of quasi-integrals of motion relevant to problem under consideration.

Since

$$\langle P_n\rangle^t = \mathrm{Sp}\{P_n\rho(t,0)\}, \quad \frac{\partial}{\partial t}\langle P_n\rangle^t = \mathrm{Sp}\left\{P_n\frac{\partial}{\partial t}\rho(t,0)\right\},$$

then, using the equation of motion for NSO (9.16), we can write down the last equality in the form

$$\frac{\partial}{\partial t}\langle P_n\rangle^t = -\mathrm{Sp}\{P_n iL(t)\rho(t,0)\}.$$

For simplicity, it is convenient to introduce the Robertson projection operator $\mathcal{P}_q$ being defined by the relation

$$\mathcal{P}_q(t)A = \sum_n \frac{\delta\rho_q(t)}{\delta\langle P_n\rangle^t}\mathrm{Sp}\{P_nA\}. \tag{9.19}$$

Using the foregoing outcomes and the definition of the Robertson projection operator, one is led to

$$\frac{\partial\rho_q(t)}{\partial t} = -\sum_n \frac{\partial\rho_q(t)}{\partial\langle P_n\rangle^t}\mathrm{Sp}\{P_n iL(t)\rho(t)\} = -\mathcal{P}_q(t)iL(t)\rho(t).$$

We add and subtract simultaneously the term $\mathcal{P}_q(t)iL\rho_q(t)$ on the right-hand side of the last formula, which allows one to write it as follows:

$$\frac{\partial\rho_q(t)}{\partial t} = -\mathcal{P}_q(t)iL(t)\rho_q(t) - \mathcal{P}_q(t)iL(t)(\rho(t) - \rho_q(t)). \tag{9.20}$$

Substituting this result into the last term on the right-hand side of the equation (9.17), we arrive at an equation that is still not closed for $\rho_q(t)$ because it contains the NSO $\rho(t)$:

$$\begin{aligned}&\left[\frac{\partial}{\partial t} + iL(t)\right][\rho(t,0) - \rho_q(t,0)]\\ &\quad= -\epsilon[\rho(t,0) - \rho_q(t,0)] - [1 - \mathcal{P}_q(t)]iL(t)\rho_q(t) + \mathcal{P}_q(t)iL(t)[\rho(t) - \rho_q(t)].\end{aligned} \tag{9.21}$$

Let us transform the equation (9.21) to a form that admits integration,

$$\left[\frac{\partial}{\partial t} + \epsilon + [1 - \mathcal{P}_q(t)]iL(t)\right][\rho(t,0) - \rho_q(t,0)] = -[1 - \mathcal{P}_q(t)]iL(t)\rho_q(t). \tag{9.22}$$

The Liouville operator here is time-dependent; therefore one is required to introduce a generalized evolution operator $U(t)$ to integrate the equation (9.22). It should describe the evolution of an arbitrary dynamical quantity from the point of time $t_0 = -\infty$ to the moment t when a Hamiltonian of the system depends on the time.

We define the generalized evolution operator, describing non-Hamiltonian dynamics of the system by the equation [36]

$$\frac{\partial}{\partial t}U(t) = U(t)[1 - \mathcal{P}_q(t)]iL(t).$$

This expression is a natural generalization of the equation of motion for the evolution operator $\exp\{iLt\}$ when the Liouville operator depends on time, and evolution is determined by only some projection of the full Hamiltonian. Equality of the evolution operator to unity is a natural initial condition for this equation provided that the temporal arguments coincide. In this case, we can get the simple result for the evolution operator:

$$U(t) = T\exp\left\{\int_{-\infty}^{t} dt_1[1 - \mathcal{P}_q(t_1)]iL(t_1)\right\}. \tag{9.23}$$

In (9.23), the integral is the sum of operators, taken at different moments in time, with the exponent in terms of the corresponding power series. Since it is assumed that the operators taken at the different points in time cannot commute with each other, it is important to define their sequence order. For this purpose, we use the symbol T, indicating the temporal ordering of the operators. It is assumed that the time-argument of the operators increases from right to left.

Using the generalized evolution operator (9.23), we write the solution of the equation (9.22):

$$\rho(t) - \rho_q(t) = -\int_{-\infty}^{0} dt_1 e^{\varepsilon t_1} U(t_1)[1 - \mathcal{P}_q(t + t_1)] iL(t + t_1)\rho_q(t + t_1).$$

Finally, the substitution of the above result into (9.20) gives the desired closed equation. Thus, we have derived the master equation (9.24), containing only the quasi-equilibrium statistical operator $\rho_q(t)$:

$$\frac{\partial \rho_q(t)}{\partial t} = -\mathcal{P}_q(t) iL(t)\rho_q(t) + \mathcal{P}_q(t) iL(t) \times \int_{-\infty}^{0} dt_1 e^{\varepsilon t_1} U(t_1)[1 - \mathcal{P}_q(t + t_1)] iL(t + t_1)\rho_q(t + t_1). \tag{9.24}$$

Before concluding the section, devoted to the derivation of the master equation for the quasi-equilibrium distribution, however, it would be advisable to think of a way to practical application of the result (9.24).

Of course, we can try to integrate this equation, but it is evident that these attempts are doomed to fail except for the simplest cases. It is much easier to write a set of generalized kinetic equations for basic dynamic variables and then solve this system. At least, as to the stationary case, this program appears not to be too complicated. The next section demonstrates the application of the master equation approach for finding the electrical conductivity of a non-equilibrium system.

9.2.2 Use of the master equation to calculate kinetic coefficients

To describe a system of non-equilibrium electrons, consider the derivation of the momentum balance equation which is based on the use of the integro-differential equation (9.24) for $\rho_q(t)$. Let the system of the non-equilibrium conduction electrons be described by reciprocal temperature β_k of kinetic degrees of freedom of the electrons, a non-equilibrium chemical potential ζ and drift velocity $\vec{V}$.

To simplify the problem, we assume that the non-equilibrium temperature and non-equilibrium chemical potential of the electronic system are already known. It remains to determine only the drift velocity. Such a situation may arise when the system reaches a non-equilibrium state due to the action of an external field. However, we have to find a response to another weak measuring field. Besides, the last condition is not crucial, for we can also consider a full statement of the problem when, for example, a strong electric field applied to the system leads to heating of the electronic system, and to the emergence of non-zero drift momentum components. In this case, one would have to write three balance equations: energy, momentum, and the number of particles.

To arrive at the balance equation for momentum of the electronic system, we multiply both sides of the equation (9.24) by the component P^{α} of the momentum operator and calculate the spur of the sides. Carrying out the acts mentioned above, we get

$$\frac{\partial}{\partial t}\langle P^{\alpha}\rho_q(t)\rangle = -\operatorname{Sp}\{P^{\alpha}\mathcal{P}_q iL\rho_q\} + \int_{-\infty}^{0} dt_1 e^{\varepsilon t_1}\operatorname{Sp}\{P^{\alpha}\mathcal{P}_q iLe^{(1-\mathcal{P}_q)iLt_1}[1-\mathcal{P}_q]iL\rho_q\}. \quad (9.25)$$

In deriving this equation, it is assumed that the system's Hamiltonian is time-independent, and the non-equilibrium state of the system is stationary. Then the quasi-equilibrium distribution does not depend on time as well. It has been taken into account on the right side of the equation. Furthermore, if the system's Hamiltonian is independent on time (the applied electric field that causes the electron drift is constant), the evolution operator is greatly simplified:

$$U(t) = T\exp\left\{\int_{-\infty}^{t} dt_1(1-\mathcal{P}_q(t_1))iL(t_1)\right\} = e^{(1-\mathcal{P}_q)iLt}.$$

Equation (9.25) is the desired balance equation for momentum of the non-equilibrium system of the electrons. However, we have written down this equation in general form, and it requires some clarification for particular applications.

Firstly, suppose that the Hamiltonian of the system H can be written as

$$H = H_e + H_p + H_{ep} + H_F; \quad H_0 = H_e + H_p,$$

where H_e, H_p are the Hamiltonians of the electron and phonon subsystems of a crystal, respectively; $H_{\rm ep}$ is the electron–phonon interaction Hamiltonian; H_F is the interaction Hamiltonian of electrons with the constant uniform electric field. The explicit form of these Hamiltonians was already discussed in Chapters 4–6 and 8, so we will not come back to this issue.

Secondly, the entropy operator of the system can be written as

$$S = \phi + \beta_k H_e + \beta H_p - \beta_k P^{\alpha}V^{\alpha} - \beta\zeta N,$$

where N is the particle number operator, ϕ is the Massieu–Planck functional determined from the condition:

$$\mathrm{Sp}\{\rho_q\} = \mathrm{Sp}\{e^{-S}\} = 1,$$
$$\phi = \ln \mathrm{Sp}\{e^{-(\beta_k H_e + \beta H_p - \beta_k P^\alpha V^\alpha - \beta\zeta N)}\}.$$

We return to the momentum balance equation (9.25) and simplify its individual terms.

The expression on the left side of the equation is just equal to zero because the statistical operator ρ_q is time-independent.

Consider the term outside the integral on the right-hand side of the equation (9.25). Using the definition of the Robertson projection operator (9.19), we obtain

$$-\mathrm{Sp}\{P^\alpha \mathcal{P}_q iL\rho_q\} = -\sum_n \mathrm{Sp}\left\{P^\alpha \frac{\delta\rho_q}{\delta\langle P_n\rangle}\right\}\mathrm{Sp}\{P_n iL\rho_q\}. \tag{9.26}$$

Here, the summation is being performed over the entire set of the basis operators involving in the definition of the entropy operator (besides P^α in our case, the operators H_e and N). By virtue of the symmetry properties of correlation functions, only the term whose the operator P^α, thermodynamically conjugated to the drift velocity V^α, makes a non-zero contribution to the sum in the expression (9.26).

Indeed, in accordance with the results obtained in Chapter 6 [see Formula (6.45), (6.60)],

$$\frac{\delta\rho_q}{\delta\langle P_n\rangle} = \frac{\delta\rho_q}{\delta\langle F_m\rangle}\frac{\delta F_m}{\delta\langle P_n\rangle} = \sum_m \int_0^1 d\tau \rho_q^\tau \Delta P_m \rho_q^{1-\tau}\frac{1}{(P_m, P_n)}, \tag{9.27}$$

where δ, as before, means the functional derivative, and $\Delta P_m = P_m - \mathrm{Sp}\{P_m\rho_q\}$.

Substituting this result into (9.26) and reducing the same terms both in the numerator and in the denominator, we get

$$-\mathrm{Sp}\{P^\alpha \mathcal{P}_q iL\rho_q\} = -\mathrm{Sp}\{P^\alpha iL\rho_q\} = \mathrm{Sp}\{\dot{P}^\alpha \rho_q\}, \tag{9.28}$$

where

$$\dot{P}^\alpha = \frac{1}{i\hbar}[P^\alpha, H_0 + H_{\mathrm{ep}} + H_F].$$

The operator P^α commutes with the Hamiltonian H_0. Further, as far as the statistical operator ρ_q, in structure, does not contain interaction, and the Hamiltonian H_{ep} has no diagonal matrix elements, then $\mathrm{Sp}\{[P^\alpha, H_{\mathrm{ep}}]\rho_q\} = 0$. Thus, the only non-zero contribution comes from the commutator of the operators P^α and H_F. Given the explicit form of the operator $H_F = -e\sum_i X_i^\beta E^\beta$, where X_i^β is the coordinate of the ith electron, but the summation is produced over all electrons, finally we have

$$-\mathrm{Sp}\{P^\alpha \mathcal{P}_q iL\rho_q\} = enE^\alpha. \tag{9.29}$$

Consider now the integral term on the right-hand side of the equation (9.25). The integral term describes collisions between the electrons and scatterers; consequently, it corresponds in terms of the kinetic equation to a collision integral. When considered within the kinetic theory, the usual approximation permits the influence of an electric field to the collision processes to be neglected. From the above, it follows that the summand H_F in the Hamiltonian of the system can be omitted.

We start by choosing the expression under the integral sign in the formula (9.25). Carrying out the projection by means of the operator $\mathcal{P}_q$, standing first in the curly bracket, we obtain

$$\begin{aligned} &\mathrm{Sp}\{P^\alpha \mathcal{P}_q iLe^{(1-\mathcal{P}_q)iLt_1}[1-\mathcal{P}_q]iL\rho_q\} \\ &\quad = \mathrm{Sp}\left\{P^\alpha \frac{\delta\rho_q}{\delta\langle P^\beta\rangle}\right\}\mathrm{Sp}\{P^\beta iLe^{(1-\mathcal{P}_q)iLt_1}[1-\mathcal{P}_q]iL\rho_q\}. \end{aligned}$$

Let I denote the integral term on the right-hand side in (9.25). Then, given

$$\mathrm{Sp}\left\{P^\alpha \frac{\delta\rho_q}{\delta\langle P^\beta\rangle}\right\} = \delta_{\alpha\beta},$$

we get

$$I = \int_{-\infty}^{0} dt_1 e^{\epsilon t_1}\mathrm{Sp}\{P^\alpha iLe^{(1-\mathcal{P}_q)iLt_1}[1-\mathcal{P}_q]iL\rho_q\}. \tag{9.30}$$

It is important to note that the momentum balance equation has the simple meaning: the force acting on the conduction electrons from the external electric field is equal to the rate of change of electron momentum due to their collisions with the scatterers. That is why the collision integral in (9.25) should be linearized with respect to the drift velocity V^α.

To perform the linearization it is necessary to expand the quasi-equilibrium statistical operator. Using the expansion (6.60), in our case, we have

$$\rho_q = \rho_q^0 + \int_0^1 d\tau \rho_q^{0\ \tau}\beta_k V^\beta P^\beta \rho_q^{0\ 1-\tau}.$$

Substituting this result into the collision integral (9.30), one gets

$$I = \beta_k \int_{-\infty}^{0} dt_1 e^{\epsilon t_1}\int_0^1 d\tau \times \mathrm{Sp}\{P^\alpha iLe^{(1-\mathcal{P}_q)iLt_1}[1-\mathcal{P}_q]iLP^\beta(\tau)\rho_q^0\}V^\beta. \tag{9.31}$$

Here $P^\beta(\tau) = \rho_q^{0\ \tau}P^\beta\rho_q^{0\ 1-\tau}$.

For further transformation of the expression (9.31), it is convenient to progress to another representation by replacing the Robertson projection operator by a more convenient projection operator, which in turn is the generalization of the Mori projection operator for the case of non-equilibrium systems.

Consider the correlation function

$$\int_0^1 d\tau \mathrm{Sp}\{B\mathcal{P}_q[CA(\tau)\rho_q]\}.$$

Here, A, B and C are some arbitrary operators, the meaning of the designation $A(\tau)$ is the same as that defined above.

After carrying out the projection by the Robertson operator (9.19) and using (9.27), the correlation function in hand can be written as

$$\begin{aligned}\int_0^1 d\tau \mathrm{Sp}\{B\mathcal{P}_q(CA(\tau)\rho_q)\} &= \sum_{nm}\int_0^1 d\tau \mathrm{Sp}\{B\rho_q^\tau \Delta P_n \rho_q^{1-\tau}\}\frac{1}{(P_n, P_m)}\int_0^1 d\tau \mathrm{Sp}\{P_m CA(\tau)\rho_q\} \\ &= \int_0^1 d\tau \mathrm{Sp}\{B\mathcal{P}_\tau CA(\tau)\rho_q\}.\end{aligned} \tag{9.32}$$

In (9.32) we have introduced the new projection operator $\mathcal{P}_\tau$, being defined by the relation

$$\mathcal{P}_\tau CA(\tau) = \sum_{nm} P_n(\tau)\frac{1}{(P_n, P_m)}(P_m C, A). \tag{9.33}$$

It follows from formula (9.32) that the correlation functions allow one to replace the Robertson projection operator $\mathcal{P}_q$ by the projection operator $\mathcal{P}_\tau$, which is the generalization of the Mori projection operator for the case of non-equilibrium systems. This enables to transform the collision integral further.

We perform integration in the expression (9.31) over time t_1. As a result, we have a representation of the integral collision in the form of the correlation function of the resolvent

$$I = -\beta_k \int_0^1 d\tau \mathrm{Sp}\left\{\dot{P}^\alpha \frac{1}{\epsilon + (1-\mathcal{P}_q)iL}(1-\mathcal{P}_q)iLP^\beta(\tau)\rho_q\right\}V^\beta. \tag{9.34}$$

Consider the operator

$$M(\tau)\rho_q = \frac{1}{\epsilon + (1-\mathcal{P}_q)iL}(1-\mathcal{P}_q)iLP^\beta(\tau)\rho_q,$$

involved in the expression under the spur sign in the formula (9.34). It is easy to check that the identity for this operator holds:

$$(\epsilon + iL)M(\tau)\rho_q = \mathcal{P}_q iLM(\tau)\rho_q + (1-\mathcal{P}_q)iLP^\beta(\tau)\rho_q. \tag{9.35}$$

If we resort to the relation (9.32), the following two equations can be proved:

$$\int_0^1 \mathrm{Sp}\{B(\mathcal{P}_q iLM(\tau)\rho_q)\}d\tau = \int_0^1 \mathrm{Sp}\{B(\mathcal{P}_\tau iLM(\tau)\rho_q)\}d\tau;$$

$$\int_0^1 \mathrm{Sp}\{B(\mathcal{P}_q iLP^\beta(\tau)\rho_q)\}d\tau = \int_0^1 \mathrm{Sp}\{B(\mathcal{P}_\tau iLP^\beta(\tau)\rho_q)\}d\tau.$$

Hence, by virtue of arbitrariness of the operator B, the identity (9.35) can be rewritten by replacing the Robertson projection operator $\mathcal{P}_q$ by the new projection operator $\mathcal{P}_\tau$:

$$(\epsilon + iL)M(\tau)\rho_q = \mathcal{P}_\tau iLM(\tau)\rho_q + (1 - \mathcal{P}_\tau)iLP^\beta(\tau)\rho_q.$$

The last expression can have another form, if one moves the first term on the right-hand side to the left side and solves the resulting equation with respect to the operator $M(\tau)\rho_q$. In the end, one gets

$$M(\tau)\rho_q = \frac{1}{\epsilon + (1 - \mathcal{P}_\tau)iL}(1 - \mathcal{P}_\tau)iLP^\beta(\tau)\rho_q. \tag{9.36}$$

The last equality is valid only if the operator $M(\tau)\rho_q$ is under the spur sign of the correlation functions (see formula (9.32)). Given the original definition of the operator $M(\tau)\rho_q$, we plug the result obtained (9.36) into the collision integral (9.34). This gives an expression reminiscent in structure the expression for the memory function (6.137), (6.137):

$$I = -\beta_k\left(\dot{P}^\alpha \frac{1}{\epsilon + (1 - \mathcal{P}_\tau)iL}(1 - \mathcal{P}_\tau)iL, P^\beta\right)V^\beta. \tag{9.37}$$

In this expression, as before, the definition of the scalar product of the operators A and B has been used:

$$(A, B) = \int_0^1 d\tau \mathrm{Sp}\{A, \rho_q^\tau B \rho_q^{1-\tau}\}.$$

In our definition of the entropy operator, operator $P^\alpha(\tau)$ commutes with the Hamiltonian H_0. If one divides the Liouville operator into two parts, $iL = iL_0 + iL_{\mathrm{ep}}$, where iL_0 is the Liouville operator corresponding to the Hamiltonian H_0, and iL_{ep} corresponds to the Hamiltonian H_{ep}, then the equality $iL_0P^\beta(\tau)\rho_q = 0$ holds. Consequently, the collision integral (9.37) in the Born approximation in the scattering theory can be written as

$$I = -\beta_k \int_{-\infty}^0 dt_1 e^{\epsilon t_1} \int_0^1 d\tau \mathrm{Sp}\{\dot{P}^\alpha_{(\mathrm{ep})} e^{iL_0 t_1} iL_{\mathrm{ep}} P^\beta(\tau)\rho_q\}. \tag{9.38}$$

It is worth noting that as far as the second-order explicit incoming interaction H_{ep} in the expression (9.38) has already been taken, the projection operators are omitted. Otherwise, there would be a necessity to retain terms of the fourth and even higher powers in the electron–phonon interaction Hamiltonian, when account is taken of these projection operators.

Next, we return again to the momentum balance equation (9.25) to establish a relationship between the quantity I in the expression (9.38) and phenomenological characteristics. Based on the phenomenological relations, the momentum balance equation for a stationary case can be represented as

$$enE^{\alpha} = \frac{\langle P^{\alpha}\rangle}{\tau}, \quad \langle P^{\alpha}\rangle = nmV^{\alpha},$$

where τ is the momentum relaxation time of non-equilibrium electrons. Using the relations (9.25), (9.38) and (9.29), and the above definition of the relaxation time τ, one is led to

$$\frac{1}{\tau} = -\frac{\beta_k}{nm}\int_{-\infty}^{0} dt_1 e^{\varepsilon t_1}\int_{0}^{1} \text{Sp}\{\dot{P}^{\alpha}_{(\text{ep})} e^{iL_0 t_1} iL_{\text{ep}} P^{\beta}(\tau)\rho_q\}\, d\tau. \tag{9.39}$$

The expression (9.39) determines the momentum relaxation time of the non-equilibrium electrons. At the end of Chapter 8, we dwelled in detail on the method of calculating non-equilibrium correlation functions and showed that the result obtained above gives the same expression for the inverse relaxation time as the kinetic equation.

Thus, we have demonstrated that the use of the master equation for the quasi-equilibrium distribution allows one to effectively solve problems to compute kinetic coefficients of strong non-equilibrium systems by applying the quantum-statistical approach.

9.3 Problems to Chapter 9

9.1. Show that the master equation (9.24) and the average momentum balance equation built with the help of the former (9.25) are not invariant under the time reversal operation.

9.2. If a system's Hamiltonian is time-independent, the temporal evolution of an operator A is given by $A(t) = \exp\{iLt\}A$.
Write down a similar expression for a time-dependent Hamiltonian of the system.

9.3. Verify the identities of the Robertson projection operator

$$\mathcal{P}_q(t)A = \sum_n \frac{\delta\rho_q(t)}{\delta\langle P_n\rangle^t}\text{Sp}\{P_n A\}:$$

$$\mathcal{P}_q^2(t)A = \mathcal{P}_q(t)A; \quad \mathcal{P}_q(t)(1 - \mathcal{P}_q(t))A = 0.$$

9.4. The entropy operator for a system of phonons and conduction electrons having the average drift momentum takes the form

$$S = \phi + \beta_k H_e + \beta H_p - \beta_k P^\alpha V^\alpha - \beta \zeta N.$$

What is the projection $\mathcal{P}_q(t)P^\beta$, where P^β is the total momentum of the electronic system?

9.5. Demonstrate that there is another way to reduce the collision integral (9.31),

$$I = \beta_k \int_{-\infty}^{0} dt_1 e^{\epsilon t_1} \int_0^1 d\tau \times \mathrm{Sp}\{P^\alpha iLe^{(1-\mathcal{P}_q)iLt_1}[1 - \mathcal{P}_q]iLP^\beta(\tau)\rho_q^0\}V^\beta,$$

to the more convenient form for further calculations. Indeed, since $iL = iL_0 + iL_{\mathrm{ep}}$ and the entropy operator does not contain any interactions, then $iLP^\beta(\tau)\rho_q^0 = iL_{\mathrm{ep}}P^\beta(\tau)\rho_q^0$. Then the collision integral (9.31) can be written in the form

$$I = -\beta_k \int_{-\infty}^{0} dt_1 e^{\epsilon t_1} \int_0^1 d\tau \mathrm{Sp}\{\dot{P}^\alpha_{(\mathrm{ep})} e^{(1-\mathcal{P}_q)iLt_1}[1 - \mathcal{P}_q]iL_{\mathrm{ep}}P^\beta(\tau)\rho_q^0\}V^\beta.$$

Argue that if no higher than second-order summands are kept in the electron–phonon interaction, the summand $\mathcal{P}_q iL_{\mathrm{ep}}P^\beta(\tau)\rho_q^0$ in the last expression can be neglected in the Born approximation in scattering theory.

Bibliography

[1] I.P. Bazarov, *Thermodynamics*, MacMillan, 1964.
[2] I. Prigogine and F. Kondepudi, *Modern Thermodynamics*, John Wiley and Sons, New York, 1999.
[3] R. Kubo, *Thermodynamics*, North-Holland Company, Amsterdam, 1968.
[4] L.D. Landau and E.M. Lifshitz, *Statistical physics*, 3rd edn, Pergamon, 1980.
[5] K.P. Gurov, *Phenomenological Thermodynamics of Irreversible Processes*, Nauka, Moscow, 1978.
[6] I.P. Bazarov, E.V. Gevorkyan and P.N. Nikolaev, *Non-equilibrium Thermodynamics and Physical Kinetics*, Moscow State University Press, Moscow, 1989.
[7] I. Gyarmati, *Non-Equilibrium Thermodynamics. Field Theory and variational principles*, Springer-Verlag, 1970.
[8] B.M. Askerov, *Kinetic Effects in Semiconductors*, Nauka Publishers, 1970.
[9] W. Ebeling, *Structurbildung bei irreversiblen prozessen*, BSB B. G. Teubner Verlagsgesellschaft, 1976.
[10] G. Nicolis and I. Prigogine, *Self-organization in non-equilibrium systems*, J. Wiley and Sons, New York, 1972.
[11] H. Haken, *Synergetics: Instability Hierarchies of Self-Organizing Systems and Devices*, Mir, Moscow, 1985.
[12] A.Y. Loskutov and A.S. Mikhailov, *Introduction into Synergetics*, Nauka, Moscow, 1990.
[13] J. Keizer, *Statistical thermodynamics of non-equilibrium processes*, Springer-Verlag, New York Inc., 1987.
[14] S.P. Kuznetsov, *Dynamic Chaos: Lecture course: Study Guide for Universities*, 2nd edn, Fizmatlit, Moscow, 2006.
[15] H.J. Stockmann, *Quantum Chaos*, Cambridge University Press, 2000.
[16] H.G. Shuster, *Deterministe Chaos*, Physik-Verlag, Weinheim, 1984.
[17] I.A. Kvasnikov, *Thermodynamics and Statistical Mechanics. Theory of Non-equilibrium Systems*, Moscow State University Press, Moscow, 1987.
[18] I.I. Olhovsky, *Course of Theoretical Mechanics for Physicists*, Nauka, Moscow, 1970.
[19] V.I. Arnold, *Mathematical Methods of Classical Mechanics: Study Guide for Universities*, Nauka, Moscow, 1989.
[20] N.N. Bogolyubov, *Problems of a Dynamical Theory in Statistical Physics*, Selected Works in Three Volumes, Vol. 2, Naukova Dumka, Kiev, 1970.
[21] E.M. Lifshitz and L.P. Pitaevskii, *Physical Kinetics*, Nauka, Moscow, 1979.
[22] V.P. Silin, *Introduction to the Kinetic Theory of Gases*, Nauka, Moscow, 1971.
[23] C. Cercignani, *Theory and application of the Boltzman equation*, Scottish Academic Press, Edinburgh & London, 1975.
[24] J. Ferziger and G. Kaper, *Mathematical Theory of Transport Processes in Gases*, Mir, Moscow, 1976.
[25] N. Kogan, *Rarefied Gas Dynamics*, Nauka, Moscow, 1967.
[26] F.J. Blatt, *Theory of mobility of electrons in solids*, Academic Press Inc., Publishers, New York, 1957.
[27] E.M. Conwell, *High field transport in semiconductors*, Academic Press, New York and London, 1967.
[28] A.A. Abrikosov, *Fundamentals of the Theory of Metals*, Nauka, Moscow, 1987.
[29] X. Aymerich-Humet, F. Serra-Mestres and J. Millan, *A generalized approximation of the Fermi–Dirac integrals*, J. Appl. Phys. **54** (1983), 2850.
[30] F.J. Blatt, P.A. Schroeder, C.L. Foiles and D. Greig, *Thermoelectric power of metals*, Plenum Press, New York, 1976.
[31] A.I. Anselm, *Introduction to the Theory of Semiconductors*, Fizmatgiz, 1962.

https://doi.org/10.1515/9783110727197-010

[32] I.M. Lifshitz, M.J. Azbel and M.I. Kaganov, *Electron Theory of Metals*, Nauka, Moscow, 1971.
[33] A.F. Volkov and S.M. Kogan, *Physical Phenomena in Semiconductors with Negative Differential Conductivity*, Sov. Phys., Usp. **96** (1968), 633.
[34] A.M. Zlobin and P.S. Zyryanov, *Hot electrons in semiconductors subjected to quantizing magnetic fields*, Sov. Phys. Usp. **14** (1972), 379.
[35] I. Prigogine, *From being to becoming: time and complexity in the physical sciences*, W. H. Freeman and Company, San Francisco, 1980.
[36] D.N. Zubarev, *Non-equilibrium statistical thermodynamics*, Nauka, Moscow, 1978.
[37] D. Zubarev, V. Morozov and G. Ropke, *Statistical mechanics of non-equilibrium processes*, Vols. 1, 2, Academie Verlag, 2002.
[38] S.V. Tyablikov, *Methods in the Quantum Theory of Magnetism*, Nauka, Moscow, 1965.
[39] P.S. Zyryanov and G.I. Guseva, *Quantum Theory of thermomagnetic phenomena in metals and semiconductors*, Sov. Phys. Usp. **11** (1969), 538.
[40] P.S. Zyryanov, *Quantum Theory of Electron Transport Phenomena in Crystalline Semiconductors*, Nauka, Moscow, 1976.
[41] T. Ando, A.B. Fowler and F. Stern, *Electronic proreties of two-dimensional sistems*, Rev. Mod. Phys. **54** (1982), 2.
[42] G. Ropke, *Statishe mechanik fur das Nichtgleichgewicht*, VEB Deutscher Verlag der Wissenschaften, Berlin, 1987.
[43] I.I. Lyapilin and V.P. Kalashnikov, *Non-equilibrium statistical operator*, Yekaterinburg, 2008.
[44] D. Forster, *Hydrodynamic fluctuations, Broken symmetry, and correlation functions*, W. A. Benjamin, Inc, 1975.
[45] A. Fert and I.A. Campbell, *Two-current conduction in nickel*, Phys. Rev. Lett. **21** (1968), 16.
[46] A. Fert, *Nobel Lecture: Origin, development, and future of spintronics*, Rev. Mod. Phys. **80** (2007), 4.
[47] P.A. Grünberg, *Nobel Lecture: From spin waves to giant magnetoresistance and beyond*, Rev. Mod. Phys. **80** (2007), 4.
[48] X. Zhang and W. Butler, *Theory of Giant Magnetoresistance and Tunneling Magnetoresistance*, Handbook of Spintronics, Springer Science+Business Media, Dordrecht, 2015.
[49] T. Valet and A. Fert, *Theory of the perpendicular magnetoresistance in magnetic multilayers*, Phys. Rev. B **48** (1993), 10.
[50] M. Johnson and R.H. Silsbee, *Coupling of electronic charge and spin at a ferromagnetic-paramagnetic metal interface*, Phys. Rev. B **37** (1988), 10.
[51] A. Hoffman, *Spin Hall Effects in Metals*, IEEE Trans. Magn. **49** (2013), 10.
[52] J. Sinova, D. Culcer, Q. Niu, N.A. Sinitsyn, T. Jungwirth and A.H. MacDonald, *Universal Intrinsic Spin Hall Effect*, Phys. Rev. Lett. **92** (2004), 12.
[53] M.I. Dyakonov, *Spin Hall Effect*, Proc. SPIE 7036, Spintronics, Vol. 12 (2008).
[54] D. Xiao, M.C. Chang and Q. Niu, *Berry phase effects on electronic properties*, Rev. Mod. Phys. **82** (2010), 3.
[55] A. Hirohata and K. Takanashi, *Future perspectives for spintronic devices*, J. Phys. D, Appl. Phys. **47** (2014), 193001.
[56] H. Mori, *Transport, collective motion, and Browinian motion*, Prog. Theor. Phys. **33** (1965), 3.
[57] R. Zwanzig, *Ensemble Method in the theory of irrversibility*, J. Chem. Phys. **3** (1960), 3.
[58] J. Mathews and R. Walker, *Mathematical Methods of Physics*, Atomizdat, Moscow, 1972.
[59] N.N.J. Bogolyubov and B.I. Sadovnikov, *Some Problems of Statistical Mechanics*, Vysshaya Shkola, Moscow, 1975.
[60] N.M. Hugenholtz, *Quantum Theory of many-body system*, Rep. Prog. Phys. **XXVIII** (1965), 201–247.
[61] R.M. White, *Quantum Theory of magnetism*, Mc Graw-Hill Book Company, New York, 1970.
[62] R. Balescu, *Equilibrium and non-equilibrium statistical mechanics*, A Wiley Interscience Publication, J. Wiley and Sons, New York, 1975.

Index